TYPICAL RANGE OF ^{13}C CHEMICAL SHIFTS

δ_{13C}

216   198   180   162   144   126   108   90   72   54   36   18   0

-C-
\>C=C\<
\>C=C=C\<
-C≡C-
(benzene ring)
R-C(=O)H
C=C-C(=O)H
R_2C=O
C=C-C(=O)-R
R-CO$_2$H
RCO$_2$CH$_3$
R-CO$_2$R
C=C-COCH$_3$
-C≡N

Experiments and Techniques in Organic Chemistry

DANIEL J. PASTO
University of Notre Dame

CARL R. JOHNSON
Wayne State University

MARVIN J. MILLER
University of Notre Dame

Prentice Hall, Upper Saddle River, New Jersey 07458

Editorial/production supervision: Kathleen M. Lafferty
Interior design: Lisa Domínguez
Cover design: Ed Butler/Butler-Udell Design
Pre-press buyer: Paula Massenaro
Manufacturing buyer: Lori Bulwin
Acquistions editor: Dan Joraanstad

© 1992 by Prentice-Hall, Inc.
Simon & Schuster Company / A Viacom Company
Upper Saddle River, New Jersey 07458

Printed in the United States of America

10 9 8 7

ISBN 0-13-298860-7

Prentice-Hall International (UK) Limited, *London*
Prentice-Hall of Australia Pty. Limited, *Sydney*
Prentice-Hall Canada Inc., *Toronto*
Prentice-Hall Hispanoamericana, S.A., *Mexico*
Prentice-Hall of India Private Limited, *New Delhi*
Prentice-Hall of Japan, Inc., *Tokyo*
Prentice-Hall of Southeast Asia Pte. Ltd., *Singapore*
Editora Prentice-Hall do Brasil, Ltda., *Rio de Janeiro*

Contents

**PART III   CHEMICAL AND PHYSICAL TECHNIQUES
FOR IDENTIFICATION OF ORGANIC COMPOUNDS**

PART IV   LABORATORY EXPERIMENTS

10   Laboratory Experiments   387

Table of Experiments

Preface

In *Experiments and Techniques in Organic Chemistry* we have attempted to provide a text of exceptional versatility—one that will guide students through laboratory experiments and provide a wealth of basic information on the various techniques of separation, purification, and structure determination using modern spectroscopic and chemical techniques that will be of value throughout their studies and career.

Part I provides detailed discussions of the various techniques that are used today to carry out chemical reactions, including the isolation and purification of products and their physical characterization. Part II provides a thorough introduction to the major spectroscopic methods (UV, IR, NMR, and MS) available for the identification of the structures of organic compounds and the analysis of complex reaction mixtures. Part III integrates the spectroscopic and chemical methods used in the identification of the structures of specific classes of organic compounds. Tables of physical properties of many common organic compounds and their derivatives are included for the convenience of the students in comparing the physical properties of their isolated compound with the properties reported in the literature.

Part IV contains a wide variety of tested laboratory experiments that should be adaptable to almost all laboratory environments. Although carrying out experiments on the microscale has received considerable attention recently, the authors' experience suggests that this is not as pedagogically sound as a comprehensive introduction to experimental organic chemistry. The microscale experience appears to place the student more in the role of an observer than a doer. Exclusive exposure to microscale organic chemistry at the introductory level does not, in our opinion, build a foundation of skills and knowledge of laboratory techniques to carry forward to further practice of chemistry. The experiments contained in Chapter 10 are, in general, on the miniscale level, using gram or fractional gram amounts of starting reagents. In many cases alternative microscale experiments are described. Microscale techniques are included for some reactions, in particular for the formation of derivatives and the determination of physical properties, such as boiling and melting points.

The experiments are arranged according to the type of functional group or type of reaction being studied, for example, the chemistry of aldehydes or ketones, aromaticity, concerted reactions, rearrangement reactions, and so on. The sequence of the experiments is not necessarily designed to be employed in the order given. Instead, the design of Chapter 10 is meant to provide the student with a description of the various important reactions of a particular class of compounds. The individual experiments are to be selected to coincide with the material being covered in the lecture course, the facilities available, and the interests of the instructor.

This text has evolved from two previous texts, *Organic Structure Determination,* published in 1969, and *Laboratory Text for Organic Chemistry,* published in 1979. The former was written at the senior/first-year graduate level and was designed to be used as a core text for a course in the determination of structures of organic compounds using chemical and spectroscopic techniques. The latter version was substantially modified to make it more suitable as a sophomore-level, organic laboratory core text which contained material at such a level that it would hold continuing value as a reference text. In the 1979 text explicit laboratory experiments were not included, although experimental procedures for functional group interconversion and the preparation of derivatives were included. The use of that text relied on the individual instructors to provide handouts for the specific experiments each instructor wanted to include in his or her laboratory course. In developing *Experiments and Techniques in Organic Chemistry,* we have further refined the procedures discussed in our earlier texts and have provided a range of carefully tested experiments. Having kept in mind the needs of the sophomore student in the organic chemistry laboratory, we are happy with this newly integrated presentation of techniques and experiments. Any comments about the book that you wish to make would be sincerely appreciated.

The authors wish to acknowledge the contributions of Xavier Creary, Jeremiah P. Freeman, Conrad J. Kowalski, and James Duncan of the University of Notre Dame and Norman A. LeBel and Kim Albizati of Wayne State University, as well as the many graduate teaching assistants who have been instructional in developing the laboratory experiments included in Chapter 10. We are also grateful to the following individuals who reviewed the manscript during various stages of development: Winfield Baldwin, University of Georgia; Myron Bender, Northwestern University; Joseph Casanova, California State University, Los Angeles; Wilmer Fife, Purdue University; John Gilbert, University of Texas; Christina Noring Hammond, Vassar College; Leland Harris, University of Arizona; Ulrich Hollstein, University of New Mexico; James Hoobler, Chemeketa Community College; Michael Jung, University of California Los Angeles; Floyd Kelly, Casper College; Harold Pinnick, Bucknell University; George Wahl, North Carolina State University; Ron Wilde, Transylvania University; James Worman, Dartmouth College; and Henry Zimmerman, New York City Technical College. The late Bret Watkins assisted in the preparation of the tables in Chapter 9. The authors also wish to thank their families for their patience during the preparation of this text.

DANIEL J. PASTO
University of Notre Dame

CARL R. JOHNSON
Wayne State University

MARVIN J. MILLER
University of Notre Dame

1

Introduction to the Organic Laboratory

1.1 SUCCESS IN THE ORGANIC LABORATORY

The first step to success in an organic laboratory course is to be prepared *before* you enter the laboratory! Carefully study each experiment before you come to the lab. Outline in a schematic way in your laboratory notebook (Section 1.4) the reactions and/or operations you are expected to carry out. Include in your outline a sketch or brief description of any special apparatus that will need to be assembled. When you enter the lab, you should be prepared to begin your experimentation both rapidly and efficiently. The organic chemistry laboratory is perhaps the most expensive space on a square-foot basis that you will occupy during your academic studies. Laboratory time is valuable and limited. The laboratory is neither the time nor the place to be wondering about what to do or to be reading an experiment for the first time!

The second step to success is to be aware of *what* you are doing and *why* at all times. There is nothing more frustrating to the laboratory supervisor nor more humiliating to the student than to have conversations such as the following occur:

SUPERVISOR: What reaction are you running now?

STUDENT: Uh—I don't remember—just a second, let me look in my laboratory instructions.

Such students are not likely to become accomplished chemists, surgeons, or cooks.

Neatness and organization are important. Keep your laboratory area and locker neat and organized. Keep your glassware clean (Section 1.6) and ready to use. Use odd moments—while a reaction mixture is refluxing or a solution filtering, and so on—to clean up your glassware. It is especially important that you allow enough

1

time at the end of each session to clean your equipment so that you may begin experimentation promptly in the next session.

Record your operations and observations in your laboratory book as you go along. Observe good safety practices at all times (Section 1.2).

1.2 LABORATORY SAFETY AND FIRST AID

As in most things we do, there is some risk associated with working in an organic laboratory. This risk may be immediate — toxic effects of chemicals, fire, explosions, burns by highly acidic or caustic materials, cuts by broken glassware — or insidious — exposure to chemicals that may cause allergic reactions or may be carcinogenic. Since there is no substitute for hands-on experience in the organic laboratory, risks must be minimized by strict adherence to safety practices and the recommendations for the handling and disposal of chemical wastes (Section 1.3).

The first thing a chemist should do when beginning work in a new laboratory is to learn the locations and methods of use of the emergency facilities:

exits	fire blankets
eye wash facilities	gas masks
fire extinguishers	first aid supplies
safety showers	emergency phone

Of equal importance is knowing how to secure help quickly when it is needed.

1.2.1 Eye Protection

1. Wear recommended safety glasses or goggles at all times in the laboratory. If you normally wear prescription lenses, consult with your instructor as to their suitability. (Contact lenses are *not* eye protection devices; in fact, in an accident they may increase the degree of injury to the eye. It is recommended that contact lenses not be worn in the laboratory or that full eye protection be used in conjunction with them at all times.)

2. Never look directly into the mouth of a flask containing a reaction mixture, and never point a test tube or reaction flask at yourself or your neighbor.

3. Avoid measuring acids, caustics, or other hazardous materials at eye level. Place a graduated cylinder on the bench and add liquids a little at a time.

1.2.2 Fire

1. Use flames only when absolutely necessary. Before lighting a flame, make sure there are no highly flammable materials in the vicinity. Promptly extinguish any flame not being used.

2. Learn the location and use of fire extinguishers. For wood, paper, or textile fires, almost any kind of extinguisher is suitable. For grease or oil fires, avoid the use of water extinguishers—they simply spread the burning material. For fires involving electrical equipment, use carbon dioxide or dry chemical extinguishers. For fires involving active metals or metal hydrides, use dry chemical extinguishers or sand.

   To put out a fire, first cool the area immediately surrounding the fire with the extinguishers to prevent the spread of the flames: then extinguish the *base* of the blaze. Remember to aim the extinguisher at the base of the fire and not up into the flames.

3. When clothing is afire, the victim should not run any distance. This merely fans the flames. Smother the fire by wrapping the victim in a fire blanket. Use a coat or roll the victim on the floor if a fire blanket is not readily available, or douse the flames under the safety shower.

1.2.3 First Aid

Although severe injuries seldom occur in the chemistry laboratory, it is wise for all chemists to be familiar with such important first-aid techniques as stoppage of severe bleeding, artificial respiration, and shock prevention. Why not consult a first-aid manual today?

Treatment of Chemical Injuries to Eyes

The most important part of the treatment of a chemical injury to the eye is that done by the victim himself or herself in the first few seconds. Get to the eye wash fountain or any source of water immediately, and wash the injured eye thoroughly with water for at least 15 minutes. Thorough and long washing is particularly important in the case of alkaline materials. A physician should be consulted at once.

Burns from Fire and Chemicals

Chemical burns of all types should be immediately and thoroughly washed with water. Ethanol may prove more effective in removing certain organic substances from the skin.

For simple thermal burns, ice cold water is a most effective first-aid measure. If cold water or a simple ice pack is applied until the pain subsides, healing is usually more rapid.

For extensive burns, place the cleanest available cloth material over the burned area to exclude air. Have the victim lie down and call a physician and/or ambulance immediately. Keep the head lower than the rest of the body, if possible, to prevent shock. Do not apply ointments to severe burns.

The most common minor laboratory accident involves cuts on the hand. Such cuts can usually be treated by applying an antiseptic and a bandage. If the cut is deep and possibly contains imbedded glass, a physician should be consulted.

For severe wounds—don't waste time! Use pressure directly over the wound to stop bleeding. Use a clean cloth over the wound and press with your hand or both hands. If you do not have a pad or bandage, close the wound with your hand or fingers. Raise the bleeding part higher than the rest of the body unless broken bones are involved. Never use a tourniquet except for a severely injured arm or leg. Keep the victim lying down to prevent shock. Secure professional help immediately.

1.3   THE HANDLING AND DISPOSAL OF CHEMICALS

Any person working in a chemical laboratory must exercise precautions in the handling of chemicals and in the disposal of waste chemicals. Of paramount importance is the recognition of certain chemicals and classes of compounds that are known to be toxic and/or carcinogenic.

The Occupational Safety and Health Administration (OSHA) has issued a list of chemicals for which there is some evidence of carcinogenicity. *Special care should be taken to minimize exposure to these substances.* The list that follows contains compounds selected from the complete OSHA tentative Category I carcinogen list (*Chemistry and Engineering News,* July 31, 1978, p. 20) on the basis of their common occurrence in organic laboratories. Of particular importance is the appearance of the solvents *benzene, carbon tetrachloride, chloroform,* and *dioxane* on the list. In many procedures toluene can be substituted for benzene and dichloromethane can be used in place of chloroform or carbon tetrachloride. Tetrahydrofuran or 1,2-dimethoxymethane can often replace dioxane. Since the publication of the original list several other chemicals have been indicated to be carcinogenic. A list of these chemicals appears in Table 1.1. *However, the laboratory worker must consider all chemicals to be potentially hazardous.*

General classes of compounds for which the carcinogenic risk is apparently high are listed in Table 1.2.

1.3.1   Handling of Chemicals

The following precautions are recommended when working in the chemical laboratory.

1. Consider all chemicals and solvents to be potentially hazardous. Be cautious at all times when handling chemicals, especially those about which you know little. Avoid direct contact with all chemicals. Wear rubber gloves, particularly when handling hazardous chemicals.
2. Work in the hood as much as possible. Keep chemical vapors, particularly of solvents, to a minimum in the laboratory. Handle all chemicals that produce corrosive, toxic, or obnoxious vapors in a hood.
3. Always wear a lab coat or apron; it will protect you and your clothing.

TABLE 1.1 Potential Carcinogens

Acetamide	Ethyl methanesulfonate
Asbestos	1,2,3,4,5,6-Hexachlorocyclohexane
Aziridine	Hexamethylphosphoric triamide (HMPA)*
Benzene	Hydrazine and its salts
Benzidine	Lead(2+) acetate
4-Biphenylamine	Methyl methanesulfonate
Bis(2-chloroethyl) sulfide	N-Methyl-N-nitrosourea
Bis(chloromethyl) ether	4-Methyl-2-oxetanone
1,3-Butadiene	1-Naphthylamine
Carbon tetrachloride	2-Naphthylamine
Chloroform	4-Nitrobiphenyl
Chromic oxide	1,2-Oxathiolane 2,2-dioxide
Coumarin	2-Oxetanone
Diazomethane	Phenylhydrazine and its salts
1,2-Dibromo-3-chloropro-	Polychlorinated biphenyls
pane	Thioacetamide
1,2-Dibromoethane	Thiourea
Dimethyl sulfate	o-Toluidine
p-Dioxane	Trichloroethylene
Ethyl carbamate	Vinyl chloride
Ethyl diazoacetate	

* 1,3-Dimethyl-3,4,5,6-tetrahydro-2(1H)-pyrimidone is recommended as a nontoxic, noncarcinogenic substitute (see Section 1.15.11).

4. Handle compressed gas cylinders with care. Always transport them on a cart and strap them in place.
5. Use a rubber or plastic bucket when carrying glass bottles of acids, bases, and other corrosive materials.
6. Never pour quantities of non–water-miscible solvents into the sink. They should be placed in suitable containers for appropriate disposal. Nontoxic water-miscible solvents, such as acetone, methanol, and so on, may be flushed down the sink with copious quantities of water.
7. Use special precautions when handling sealed vials of volatile materials. Never

TABLE 1.2 General Classes of Potentially Carcinogenic Compounds

Alkylating reagents	Nickel and its compounds
Arsenic and its compounds	Nitrogen mustards (β-halo amines)
Azo compounds	N-Nitroso compounds
Beryllium and its compounds	N-Methyl amides
Cadmium and its compounds	Polychlorinated substances
Chromium and its compounds	Polycyclic aromatic amines
Estrogenic and androgenic steroids	Polycyclic aromatic hydrocarbons
Hydrazine derivatives	Sulfur mustards (β-halo sulfides)
Lead (II) compounds	

subject a sealed glass container to severe thermal shock, such as by placing the vial directly into liquid nitrogen.

8. Never use your mouth to pipette chemicals. Always use a rubber bulb for pipetting.

9. When working with any potentially dangerous reaction, use adequate safety devices — safety shields, gloves, goggles, and so on.

10. Do not smoke, eat, or drink in the laboratory.

1.3.2 Disposal of Chemical Wastes

Considerable attention has been focused recently on the methods of disposal of chemical wastes, in particular the disposal of large quantities of industrial wastes and smaller laboratory quantities of highly toxic wastes. This attention and its effects have filtered down to the smaller laboratory operations such as academic teaching and research laboratories. It has been made abundantly clear that the disposal methods used in the past, that is, the use of landfill dumps, or simply flushing into sewer systems, cannot, and will not, be tolerated in the future.

The problem of chemical waste disposal has been addressed by the Committee on Hazardous Substances in the Laboratory, the Commission of Physical Sciences, Mathematics, and Resources, and the National Research Council, resulting in the publication of a monograph entitled *Prudent Practices for Disposal of Chemicals from Laboratories* (National Academy Press, Washington, D.C., 1983). Major important recommendations are:

1. More careful planning of experiments in order to reduce the amounts of left-over starting materials and products and by-products;

2. Reduction in the scale of the experiments in order to reduce the amounts of chemical wastes and solvents; and

3. Exchange and recovery of unused or recoverable chemicals and solvents.

Recommendations have been made for the proper disposal of chemical wastes. The most common method, flushing down the drain, is recommended only for organic chemicals that may be defined as soluble in water (see Section 8.4) and possessing a low vapor pressure and low ignitability (diethyl ether *does not* qualify under these conditions). Certain highly toxic, heavy-metal residues should not be flushed down the drain. Chemical wastes that do not fall under the above classification of being flushable should be placed in containers for ultimate disposal by acceptable methods.

Laboratory chemicals have been classified by the Environmental Protection Agency (EPA) regulations into the following categories.

reactive	corrosive, acid
toxic	corrosive, base
miscellaneous laboratory chemicals	oxidizers
ignitable	

Waste laboratory chemicals of one category should not be mixed with those of another category, and even at times with another of the same category. Caution must be exercised. The miscellaneous class is reserved for small quantities of materials that have not been well characterized in terms of their chemical and physiological properties. These wastes should be placed in small individual containers and labeled.

Solvents used for chemical reactions, extractions, and chromatographic separations should be purified and reused if possible. If this is not feasible, the solvents should be stored in acceptable containers for later disposal by recommended procedures. Chlorinated hydrocarbons should be stored in a separate container because of their toxic properties (see the section on solvent properties, Section 1.15). Hydrocarbons, ethers, and alcohols can be mixed, but the contents of the container must be indicated completely on the label.

The National Academy of Science publication also contains procedures for the chemical destruction of various classes of hazardous chemicals. *A copy of this book should be present in each chemical laboratory.*

1.4 LABORATORY NOTEBOOK AND RECORD KEEPING

A bound notebook with numbered pages should be obtained for use as a laboratory notebook. The first page of the notebook, or first few pages if the notebook is large, should be used for an index of the experiments. This index should include the title of the experiment, the experiment number, and the page number of the notebook where that experiment may be found.

Before going into the laboratory, the following items must be in your notebook for the experiment that you are about to do: The title of the experiment and the date must be at the top of the page; a balanced chemical equation representing the reaction should be present, followed by a listing of all reagents, solvents, and products with their melting and boiling points, solubility, molecular weight, and amounts to be used in the experiment; there should also be, if needed, the theoretical yield based on the limiting reagent(s); a sketch of any apparatus to be used in the experiment should be present; and, finally, there should be a brief description of the experiment and the procedures to be used.

In keeping a laboratory notebook, whether for class or research purposes, there is one cardinal rule: When one makes an observation, it should be written down immediately. Neatness and order, though important, are secondary. Chemists should never get into the habit of recording experimental observations on loose sheets of paper to be transcribed later into a bound notebook. Loose pages tend to get lost, and one's immediate impressions are often tempered with time. One poor recall can be very costly in terms of time, materials, and reputation. The laboratory notebook should be kept at your side in the laboratory at all times. Observations should be recorded when they are made, and plenty of space should be allowed in your notebook for comments, additions of later information, and computations. Adequate references for the procedures used and data cited should be included.

Students should get into the habit of coding all samples, spectra, analyses, and so on, with their initials and notebook page numbers. For example, the code RAF-2-147-D appearing on an infrared spectrum weeks, months, or years after the spectrum was originally run indicates that the spectrum is that of compound D described on page 147 of laboratory notebook 2 of chemist RAF. This system will allow the chemist or anyone else to look up the history and source of the compound immediately. Label all samples that you are to turn in or that you intend to store until later. *Never leave unlabeled chemicals in your locker.*

Your laboratory supervisor will probably have specific instructions concerning the format of your notebook: the format used, in part, will depend on the nature and purpose of the experiment. For example, in an experiment involving the synthesis of a particular compound, the following format is useful.

Before Coming to Lab

TITLE   DATE

BALANCED EQUATIONS FOR ALL REACTIONS

DATA ON REACTANTS, CATALYSTS, AND SOLVENTS
Molecular weight, grams, moles, solvent volume, useful physical constants

PRODUCT DATA
Molecular weight, theoretical yield, literature value of melting point or boiling point

APPARATUS
Sketch or brief description

BRIEF SCHEMATIC OUTLINE OF EXPERIMENT

REFERENCES

During the Lab

RUNNING ACCOUNT OF WHAT YOU DO AND WHAT YOU OBSERVE

SIGN YOUR NOTEBOOK

After the Lab

CALCULATION OF YIELD (SECTION 1.5)

DISCUSSION OF RESULTS

ANSWERS TO QUESTIONS

1.5 CALCULATION OF YIELD

For practical reasons related to costs, reaction rates, positions of equilibria, side reactions, and so on, organic reactions are most frequently run with reactants in nonstoichiometric quantities. One reactant is chosen to be the *limiting reagent* which ultimately restricts the theoretical amount of product that can be formed. For example, the oxidation of cyclohexanol to cyclohexanone with sodium dichromate in aqueous sulfuric acid proceeds according to the following *balanced* equation:

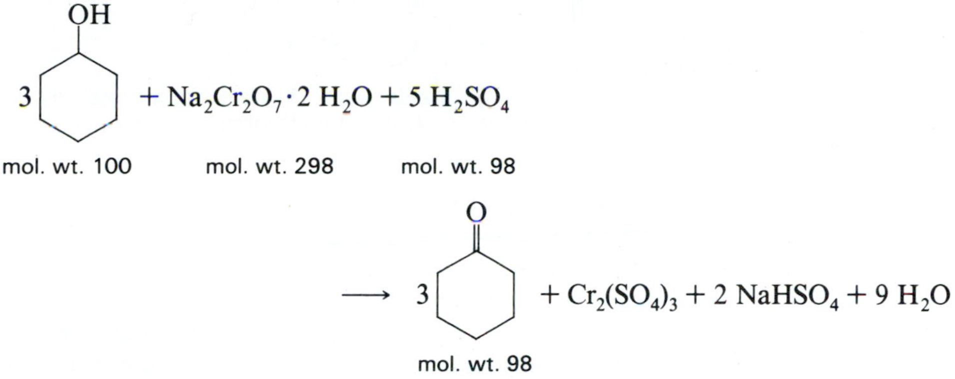

$$3\ \text{(cyclohexanol, OH)} + Na_2Cr_2O_7 \cdot 2\ H_2O + 5\ H_2SO_4$$

mol. wt. 100   mol. wt. 298   mol. wt. 98

$$\longrightarrow 3\ \text{(cyclohexanone, O)} + Cr_2(SO_4)_3 + 2\ NaHSO_4 + 9\ H_2O$$

mol. wt. 98

If a reaction mixture contains 20 g (0.20 mol) of cyclohexanol, 21 g (0.07 mol) of sodium dichromate dihydrate, and 20 mL (37 g, 0.37 mol) of sulfuric acid, it is clear that the cyclohexanol is the limiting reagent (0.07 mol of dichromate is capable of oxidizing 0.21 mol of cyclohexanol). The theoretical yield of cyclohexanone is then 0.20 mol or 19.6 g. If an experimental yield of 15 g is obtained, this corresponds to 15 g/19.6 g $\times$ 100 = 76.5% yield.

In a reaction where some reactant is recovered unchanged after termination of the reaction, it is sometimes useful to compute the percent conversion. For example, if in the work-up of the above reaction 4.0 g of cyclohexanol were recovered, the consumption of cyclohexanol would have been 20 g − 4 g = 16 g (0.16 mol). That amount of cyclohexanol could lead to 0.16 mol $\times$ 98 g/mol = 15.7 g of cyclohexanone; a yield of 15 g would correspond to 15 g/15.7 g $\times$ 100 = 96% conversion.

1.6 GLASSWARE

Most laboratories use glassware with standard-taper ground glass joints (Figure 1.1). Such joints have the advantage of rapid assembly and disassembly and of providing secure seals. (The principal disadvantage of ground glass joints is their expense; each component of a joint is worth several dollars.)

Ground glass joints are obviously made to fit together snugly. Sometimes the joints freeze and disassembly becomes difficult or impossible; this can be avoided by

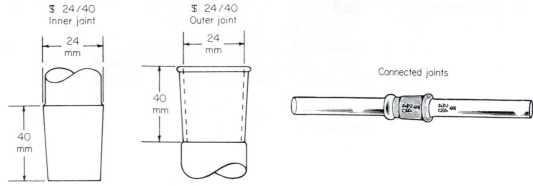

Figure 1.1   Standard-taper ($) ground glass joints. The most common sizes are 24/40, 19/39, 14/20, and 10/22 (diameter/length in mm).

following proper procedures. If joints are kept clean, they normally can be assembled without the use of a lubricant. Disassemble the apparatus immediately after use; if the apparatus is hot, disassemble it before it cools, if possible. Lubricants should always be used when the apparatus is to be evacuated with a vacuum pump or when strongly basic solutions (which tend to etch glass) are used. Lubricants should be used by sparse but even application to the top of the male (inner) joint. Grease lubricants are preferred, as the joints can be easily cleaned by washing with organic solvents after disassembly. Silicone lubricants are almost impossible to remove completely by any means; about the best that can be done is to wipe them with a towel. Should a joint become frozen, gentle tapping will sometimes loosen it; otherwise, after removing all organic material from the apparatus, the outer joint can be gently heated with a Bunsen flame to cause expansion and release.

Glassware that will not become clean simply by rinsing with water can usually be cleaned with the aid of warm water, detergents, and a brush. Organic residues can frequently be removed by dissolving in acetone. For extremely difficult-to-remove organic residues the piece of glassware can often be cleaned by allowing it to stand in a chromic acid bath (prepared by the cautious addition of 30 mL of saturated aqueous sodium dichromate to 100 mL of concentrated sulfuric acid contained in an Erlenmeyer flask cooled in an ice bath).

Wet glassware can be allowed to dry by standing or, more conveniently, by the use of a heated oven. If a piece of glassware is needed for use immediately after washing, a final rinse can be made with a little acetone; the glassware may then be air-dried more quickly, or dried by the use of a heat gun. Air from compressed air lines is generally not very useful for drying glassware, since dirt and oil are often blown from the lines.

Figure 1.2 shows the common pieces of glassware that will be encountered in the undergraduate organic laboratory. The name of each piece is indicated below the drawing of the piece of glassware. There are several sizes of glassware that are available; the size that any experiment will require depends on the quantities (vol-

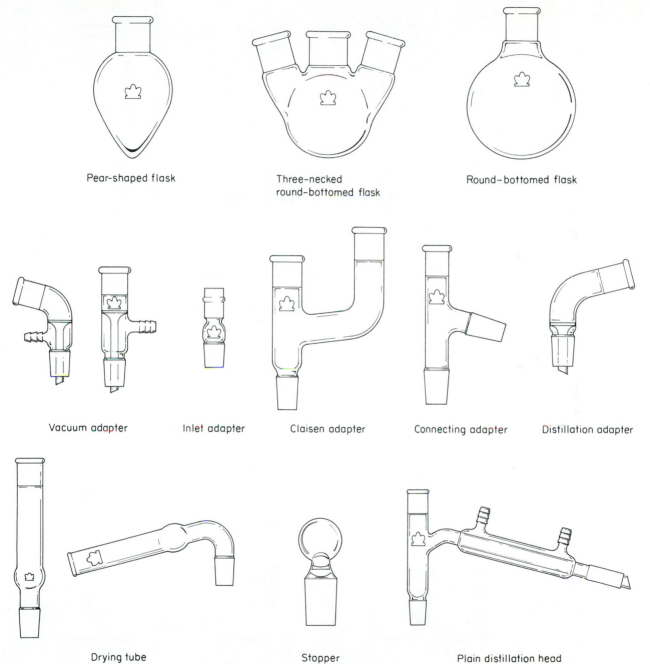

Pear-shaped flask

Three-necked round-bottomed flask

Round-bottomed flask

Vacuum adapter

Inlet adapter

Claisen adapter

Connecting adapter

Distillation adapter

Drying tube

Stopper

Plain distillation head

Figure 1.2 Examples of mini-sized glassware. (Reproduced by permission of Kontes Glass Company, Vineland, N.J.)

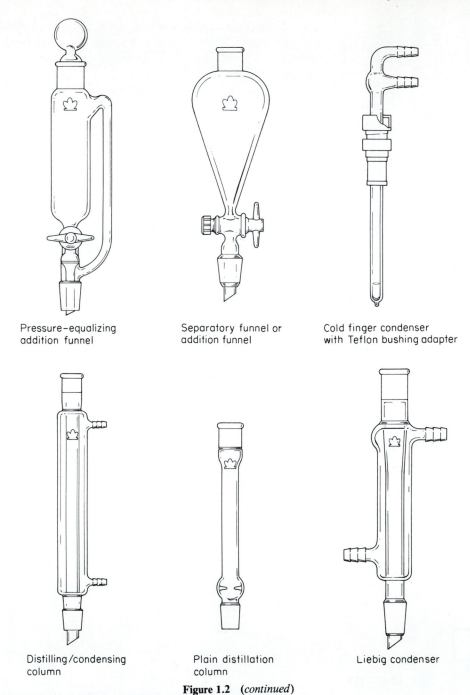

Pressure-equalizing
addition funnel

Separatory funnel or
addition funnel

Cold finger condenser
with Teflon bushing adapter

Distilling/condensing
column

Plain distillation
column

Liebig condenser

Figure 1.2   (*continued*)

umes) of materials used in the experiment. Most of the experiments that you will encounter in this laboratory are designed on a miniscale that requires the use of commonly available 14/20 glassware. Some experiments are also designed on a microscale that often uses vials instead of flasks with 14/10 threaded joints (see Figure 1.3). Actual setups of glassware will be illustrated in later sections.

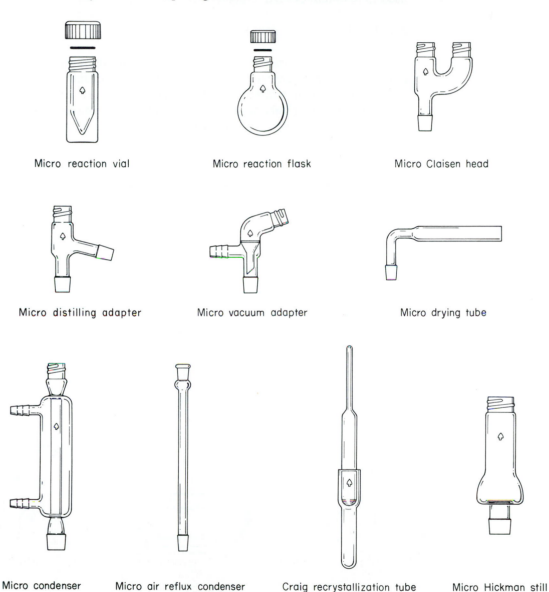

Micro reaction vial   Micro reaction flask   Micro Claisen head

Micro distilling adapter   Micro vacuum adapter   Micro drying tube

Micro condenser   Micro air reflux condenser   Craig recrystallization tube   Micro Hickman still

Figure 1.3 Examples of micro-sized glassware. (Reproduced by permission of Ace Glass Incorporated, Vineland, N.J.)

1.7   TEMPERATURE CONTROL

During the course of conducting experiments it is often necessary to heat or cool a reaction mixture or solution. In this section we will describe the various techniques used to maintain desired temperatures.

1.7.1   Methods of Heating

The use of bunsen burners, or other sources of open flames, should be avoided if at all possible, as most organic substances are volatile and flammable. Open flames should never be used when flammable solvents (see Section 1.15) are being used by *you* or *others near you* in the laboratory. The methods for heating described in the following sections do not involve the use of open flames; however, certain precautions must be exercised when using any of the techniques.

Heating Mantles

Heating mantles have a resistance wire buried either in glass fabric or ceramic (see Figure 1.4). The temperature is controlled by a variable electric transformer (Variac). The upper temperature limit is about 450°C. Mantles find common use in laboratories because of their safety feature of having no exposed heating coil and

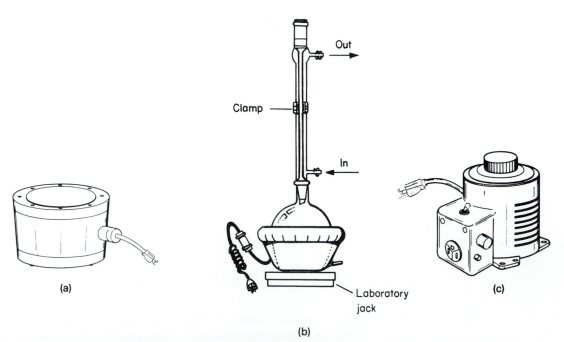

Figure 1.4   Ceramic (a) and glass fabric (b) heating mantels (supported on a laboratory jack) with a variable transformer (c).

because of their convenience; however, there are several disadvantages. The sizes and shapes of the mantles and flasks must match, so for general use a large range of sizes of mantles must be available (at a significant cost per mantle). Mantles are useful for refluxing reaction mixtures, but are not recommended for use in distillation. During the use of a heating mantle the electrical energy input remains constant, and when the amount of material in a distillation flask gets very low, overheating will occur that may result in the rapid, and sometimes catastrophic, decomposition of residues remaining in the flask. When using a heating mantle for refluxing, care must be taken to make sure that the flow of cold water through the condenser is maintained (hoses must be carefully wired to the water tap and the inlet and outlet of the condenser).

Heating Baths

A heating bath consists of a container of a substance in which is immersed an electrical heating device. The substance in the bath can be either a fluid or sand. Mineral oil is a commonly used heating fluid and can be used up to 150°C. Above that temperature mineral oil presents a fire and smoke hazard. The more expensive silicone oil (e.g., Dow 550) can be used up to about 250°C. A polyethylene glycol, Carbowax 600, which solidifies near room temperature and is water soluble, can be used up to 175°C. Oil baths have a number of advantages: rather precise temperature control, bath temperature easily monitored with a thermometer, and reduced hazard. On the other hand, oil baths can be dangerous if overheated and can be easily spilled or splashed. Contamination by water results in severe splattering when the temperature is raised over 100°C. Heating devices useful with oil baths include immersion heaters and hot plates. Stirring bars are used in the bath to maintain a constant temperature throughout the bath (see Figure 1.5).

Sand baths have the advantages that they are not easily spilled and do not result in splattering when contaminated with water. The major disadvantage in the use of sand baths is that they do not provide a uniform temperature throughout the bath as there is no reasonable way in which the heating medium can be stirred (the temperature of the sand decreases markedly with increasing distance from the heating source), and the bulb of the thermometer must carefully be placed close to the reaction vessel to record an accurate temperature. Sand baths are recommended for sealed-tube reactions in that any explosion of the sealed tube results in the blowing off of the sand which rapidly loses its temperature and energy.

Hot Plates

Electric hot plates are often used to warm solvents, solutions, or mixtures in flat-bottomed containers, particularly Erlenmeyer flasks whose sloping sides are designed to prevent vapor loss. Hot plates are particularly useful for recrystallizations. Highly flammable solvents with low flash points (ignition temperatures) should not be used with hot plates, even though many of the modern hot plates have sealed electrical switches. Hot plates are often used to heat oil and sand baths, which, in turn, are used to control reaction or distillation temperatures.

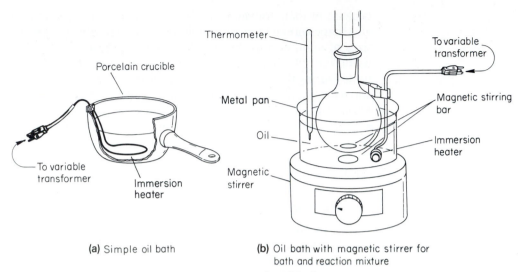

(a) Simple oil bath

(b) Oil bath with magnetic stirrer for bath and reaction mixture

Figure 1.5   Oil baths.

Steam Baths

Steam baths offer the highest safety factor at the lowest cost. Heat exchange from steam is reasonably efficient, and the maximum temperature attainable is 100°C. It is not necessary to immerse the flask or container into the steam bath. The flask or container is placed on top of a suitably sized ring so that the flask or container is adequately supported on the top of the steam bath (see Figure 1.6). The flow of steam is adjusted so that no excess steam goes out into the laboratory atmosphere. Steam that you can see is worthless steam. When using a steam bath it is necessary to protect the sample from condensing moisture, either by controlling the amount of excess steam emanating from the steam bath or by placing a loose cotton plug in the neck of the flask.

Infrared Lamps and Heat Guns

Infrared lamps and heat guns also provide very useful means of heating reaction mixtures or drying glassware (see Figure 1.7). Infrared lamps offer good safety and provide a rather low temperature limit. One disadvantage is that the intense infrared radiation makes it difficult for visual observation of the contents of the container

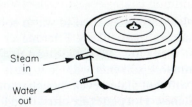

Figure 1.6   A steam bath. Remove rings to accommodate flasks or beakers of different sizes.

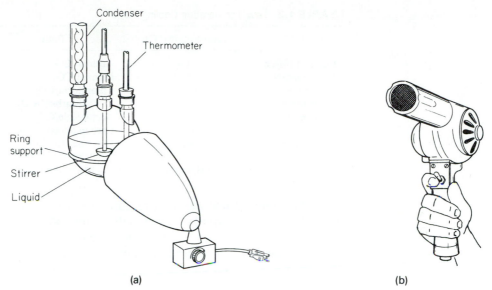

Figure 1.7 (a) An illustration showing the use of an infrared heating lamp. The infrared bulb is housed in a reflector equipped with a dimmer switch. The lamp should be pointed away from the laboratory worker. (b) A heat gun.

being heated. Heat guns, which are much like electric hair dryers, are often useful to heat objects. They must be used with caution, however, as heat guns can generate flowing air streams with temperatures of up to 600°C and have exposed heating elements.

1.7.2 Methods of Cooling

It is often necessary to cool a reaction, particularly when a reaction is rapid and highly exothermic, or when an intermediate formed in a reaction is thermally labile. Temperatures near 0°C can be maintained by the use of an ice-water bath. Temperatures down to about −20°C can be maintained by the use of ice-salt baths formed by the addition of salt (sodium chloride) to ice. Lower temperatures can be maintained by the addition of Dry Ice (solid carbon dioxide) to various solvents. Very low temperatures can be maintained by the use of liquid nitrogen (bp −196°C).

The addition of Dry Ice or liquid nitrogen to any solvent must be carried out with caution; the addition causes excessive bubbling by the evaporation of gaseous carbon dioxide or nitrogen. The Dry Ice or liquid nitrogen is slowly added to the solvent contained in a wide-mouth vacuum flask (a Dewar flask wrapped with tape) or any other suitably insulated container until a slush is formed. Additional quantities of Dry Ice or liquid nitrogen are added to maintain the slush. Solvents of low flammability are recommended for use in such cooling baths, although some of the most useful cooling baths employ highly flammable solvents because of the desired

TABLE 1.3 Low-Temperature Cooling Baths*

Bath	Temperature (°C)	Bath	Temperature (°C)
Liquid nitrogen	-196	Acetone/CO_2	-77
Isopentane/N_2	-160	Ethanol/CO_2	-72
Pentane/N_2	-131	Chloroform/CO_2	-61
Isooctane/N_2	-107	Diethyl carbitol/CO_2	-52
Methanol/N_2	-98	Acetonitrile/CO_2	-41
Toluene/N_2	-95	Carbon tetrachloride/CO_2	-23
Ethyl acetate/N_2	-84	Ethylene glycol/CO_2	-15

* For additional combinations see reference 1 at the end of this chapter.

accessible temperatures and due precautions must be exercised. Caution must also be used in the use of liquid nitrogen in that if the cooled system is not protected from the air, liquid oxygen (bp $-176°C$) may be condensed in the apparatus which on warming may result in the formation of very high internal pressures. Useful combinations of coolant and solvent are listed in Table 1.3.

1.8   REFLUXING

Refluxing refers to boiling a solution and condensing the vapors in a manner that allows return of the condensate to the reaction flask. The technique of refluxing is used in preparative organic chemistry to maintain a reaction mixture at a constant and appropriate temperature. Two reflux setups are shown in Figure 1.8. Setup (a) shows the use of miniscale glassware with a water-cooled condenser that is useful for low boiling solvents. Setup (b) shows the use of microscale glassware using a vial and an air-cooled condenser. Air-cooled condensers should be used only with higher boiling solvents. *Boiling chips or magnetic stirring should be used to promote smooth boiling and to prevent bumping!*

1.9   ADDITIONS TO REACTION MIXTURES

1.9.1   Addition of Liquids

A reaction that requires a controlled rate of addition of a liquid is usually run in a three-necked, round-bottomed flask equipped with an addition funnel (dropping funnel) (see Figure 1.9). The simplest addition funnel is similar to a separatory funnel equipped with a male standard-taper joint at the outlet; during dropwise addition of a liquid reactant or solution, it is used with the cap off. Pressure equilibrated addition funnels can be used with the stopper in place; these are convenient for

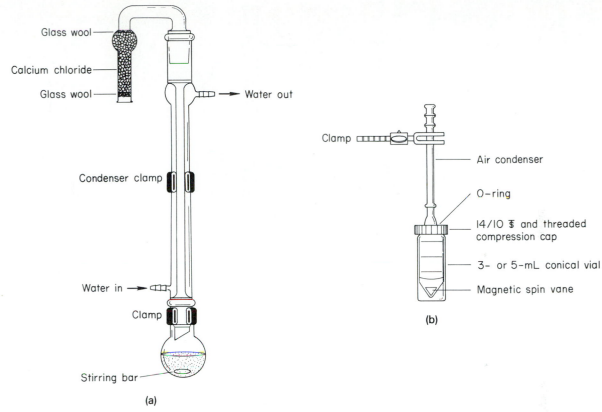

Glass wool

Calcium chloride

Glass wool

Water out

Condenser clamp

Water in

Clamp

Stirring bar

(a)

Clamp

Air condenser

O-ring

14/10 ⑃ and threaded compression cap

3- or 5-mL conical vial

Magnetic spin vane

(b)

Figure 1.8 (a) Simple reflux apparatus using a water-cooled condenser with a drying tube to protect the reaction from moisture. (b) Microscale reflux apparatus using a vial reaction vessel and air-cooled condenser. The reaction flasks may be heated by oil or sand baths.

inert atmosphere operations (Section 1.11). Teflon stopcocks are preferred on addition funnels so that use of stopcock grease, which can be leached by organic liquids, is avoided.

The transfer of extremely air-sensitive solutions, such as solutions of alkyl lithium and magnesium reagents, can be accomplished by the use of an oven-dried, nitrogen- or argon-purged syringe, or by the use of a long hollow steel tube called a cannula inserted through the top of the reagent bottle. In the latter case the transfer of solution is accomplished by the use of a slight pressure of an inert gas injected into the reagent bottle. A typical setup is shown in Figure 1.10. It is difficult to measure the volume of solution transferred using this technique. Alternatively, the solution may be transferred by the cannula to a calibrated addition funnel for more accurate measurement of the volume of the reagent solution.

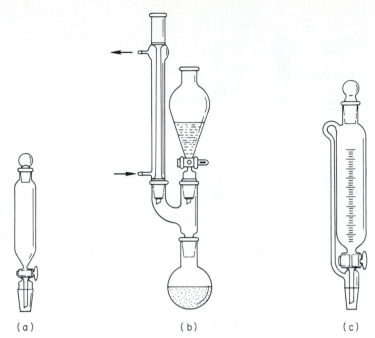

Figure 1.9   Simple addition funnels (a) and separatory funnels (b) must have stoppers removed during additions to reaction flasks. Pressure-equilibrating funnels (c) can be used with stoppers in place.

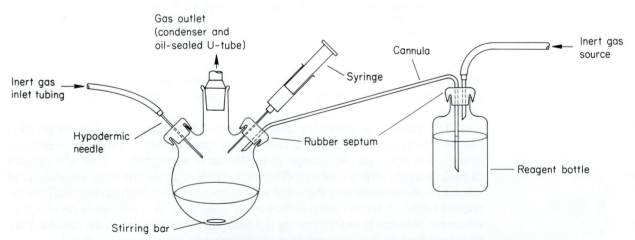

Figure 1.10   The addition of an air-sensitive solution using a syringe or cannula. Rubber septa that fit standard-taper joints and cannula are available from the Aldrich Chemical Company, Inc., Milwaukee, Wisc.

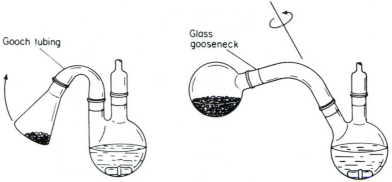

Figure 1.11   Methods for the addition of a solid to a reaction mixture.

1.9.2   Addition of Solids

Solids can often be added directly to a reaction flask by briefly removing a stopper. If the reaction mixture has been heated to near reflux, heating should be ceased and the reaction mixture allowed to cool slightly before the addition of solids is made; otherwise, boilover is likely to occur. For inert atmosphere operations, solids may be added with the aid of Gooch tubing. This is a thin flexible rubber tubing that will fit over a neck of the reaction flask and over the neck of an Erlenmeyer flask. Solids can periodically be poured through the tubing into the reaction flask; when not in use, the Erlenmeyer flask is allowed to hang in order to close the tubing (see Figure 1.11). The rate of addition of solids may be difficult to control, and caution must be exercised in making such additions in highly exothermic reactions. Addition of solutions is usually preferable to addition of solids.

1.9.3   Addition of Gases

A gas may be added to a reaction mixture through a glass tube (often fitted with a glass dispersion frit) or hypodermic needle that dips below the surface of the liquid. If the gas is quite reactive or highly soluble, it may only be necessary or desirable to lead the gas to the surface of a stirred reaction mixture. If the gas reacts to form a precipitate, special provision should be made to avoid clogging of the inlet tube. A safety trap (see Figure 2.3) should be placed between the gas source and the reaction vessel.

1.10   STIRRING AND MIXING

The most commonly used device for stirring and mixing in the organic laboratory is the magnetic stirrer. This simple device consists of a variable speed motor, which spins a magnet (see Figure 1.12). This is placed under the vessel, and a glass or inert plastic-covered magnetic bar is placed inside the vessel. This magnetic system is ideal

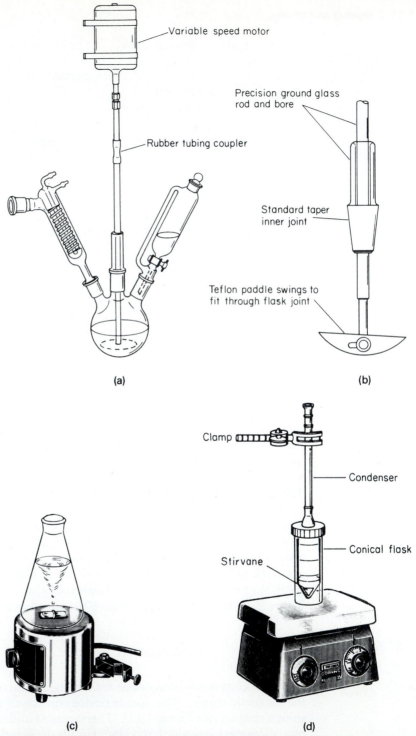

Variable speed motor

Rubber tubing coupler

Precision ground glass rod and bore

Standard taper inner joint

Teflon paddle swings to fit through flask joint

(a)

(b)

Clamp

Condenser

Conical flask

Stirvane

(c)

(d)

Figure 1.12 Stirring devices. (a) Variable speed motor and paddle stirrer. (b) Precision-bore glass stirrer with Teflon paddle. (c) Magnetic stirrer with magnetic stirring bar. (d) Combination hot plate and magnetic stirrer.

22

for stirring reaction mixtures that are not too viscous or that do not contain heavy precipitates.

For reaction mixtures that are more difficult to stir or that need very vigorous mixing, a variety of direct-drive stirrers are available. One commonly used device is the Trubore stirrer, which consists of a precision-ground glass shaft and housing with a male standard-taper joint to fit the flask and a glass or Teflon paddle, which can be tilted to fit into the flask (see Figure 1.12). This device is rather expensive. In use it should carefully be aligned and secured; the glass bearing should be lubricated with silicone or mineral oil.

There are numerous other types of stirring devices for special application, including mercury-sealed stirrers for inert gas operations and very high-speed stirrers (up to 10,000 rpm) with water-cooled bearings.

1.11 INERT ATMOSPHERES

It is frequently necessary or desirable to protect reaction mixtures from moisture or oxygen, or both. The degree of protection necessary depends on the reactivity of the reagents. During the course of reactions run under reflux, some protection of the reaction mixture is afforded by displacement of air from the system by the solvent vapor. Moderate protection can be achieved in the following manner. A clean flask and auxiliary equipment (addition funnel, stirring bar, etc.) with one neck open are heated with a blue Bunsen flame until just too hot to touch. (Moisture condensate can usually be observed as it is driven from the glass walls.) A dry condenser capped with a drying tube filled with calcium chloride or Drierite is fitted into the open neck. As the system cools, the air that enters the system passes over the drying agent. When the flask has cooled sufficiently, it is charged with the appropriate solvents and reagents.

More elaborate protection is usually achieved by running reactions under a slight positive pressure of an inert gas; dry nitrogen is usually sufficient, but occasionally argon must be used (e.g., metallic lithium will react slowly with nitrogen). A typical setup is illustrated in Figure 1.13. The entire system is assembled; however, the condenser should not contain water. The U-tube should contain just enough mineral oil to fill the bottom and to form a liquid seal. The system is flame-dried, while a small steady stream of dry inert gas is flowing. During the course of the reaction a slight positive pressure is maintained, as denoted by the slow release of gas at the U-tube. The inert gas inlet can be glass, or a septum and short hypodermic needle can be used. [Rubber laboratory tubing ($\frac{3}{16}$ in. diameter) will fit securely over the large end of a steel hypodermic needle.]

1.12 CONTROL OF EVOLVED NOXIOUS GASES

All reactions that involve volatile, toxic, or irritating substances should be run in an efficient hood. Toxic or irritating gases (e.g., HCl, HBr, SO_2, NH_3, mercaptans, sulfides, etc.) that may be used in an experiment, or generated as a product, should

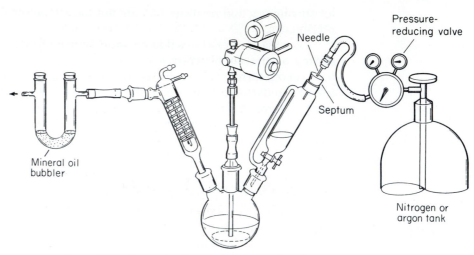

Figure 1.13   Apparatus for carrying out reactions in an inert atmosphere.

not be allowed to escape into the atmosphere. Such gases can be trapped using the setup shown in Figure 1.14.

A slow stream of air, or an inert gas such as nitrogen or argon when oxygen-sensitive reagents are being used, is passed through the T-tube inserted in the top of the condenser which carries the evolved gas through the bubbler that is kept below the surface of the solution contained in the flask. For acidic gases the solution in the flask should be an aqueous solution of sodium hydroxide, and for basic gases a solution of

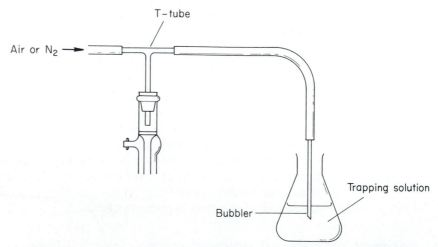

Figure 1.14   A setup for trapping noxious gases evolved from a reaction mixture. (A more convenient T-tube for fitting into standard-taper condensers can be made by attaching a standard-taper joint to the bottom of the T-tube that inserts into the top of the condenser.)

hydrochloric or sulfuric acid should be used. For mercaptans and sulfides the flask should contain a basic solution of potassium permanganate that oxidizes the sulfur-containing compounds to less noxious, water-soluble compounds. If a stream of air or inert gas is not used to sweep the evolved gases through the solution in the trap, the reaction of the gas with the solution may cause the solution to be sucked back up and into the reaction mixture.

1.13   SOLVENT REMOVAL

It is often necessary to concentrate a solution or completely remove solvents to obtain a desired product from a reaction mixture, an extract, a chromatographic fraction, or from mother liquors from a recrystallization. This may be done by simple or fractional distillation techniques at atmospheric pressure or under reduced pressure (Section 2.3). Sometimes atmospheric-pressure distillation is used to remove the bulk of the solvents, and last traces are then removed at reduced pressure. In any event, due consideration must be given to the stability and volatility of the desired product. Care must be exercised to prevent product and/or byproduct decomposition due to overheating of the residue. This can usually be avoided by use of water,

Figure 1.15   A Büchi rotary evaporator. (Reproduced by courtesy of Büchi/Brinkman Instruments, Inc., Westbury, N.Y.)

steam, or oil baths (Section 1.7.1). The residue should end up in a flask of appropriate size; when small amounts of product are present in dilute solutions, sequential
additions of portions of the solution to be concentrated to the distillation flask are
often convenient. (Note: Erlenmeyer flasks should not be used for reduced-pressure
evaporations.)

Rotary evaporators (see Figure 1.15) provide one of the most efficient methods
for the removal of volatile solvents. The flask filled to half capacity or less is immersed in a water bath, pressure in the system is reduced by an aspirator, and the flask
is rotated. The rotation not only prevents bumping, but continuously wets the
warmed walls of the flask from which evaporation is rapid.

Figure 1.16 illustrates two simple, but reasonably effective, means of removal of
solvent from a sample. In Figure 1.16a, a boiling chip should be added to the flask to
effect continuous bubbling and to avoid violent bumping. Also, the flask should be
swirled as frequently as is convenient. (It should be remembered that a water aspirator operating with water at 20°C will reduce boiling points by approximately 100°C.
Solvents should not be removed from volatile samples by this method; fractional
distillation is recommended.) Smaller quantities of solvent can be removed by blowing a stream of air, or preferably nitrogen, over the solution that is contained in a test
tube or a small flask (Figure 1.16b).

1.14 DRYING PROCEDURES

1.14.1 Drying Liquids and Solutions

Small amounts of water can be removed from liquids and solutions by allowing them
to stand in direct contact with a drying agent for a suitable period of time followed by
decantation or filtration. Some common drying agents that function by forming

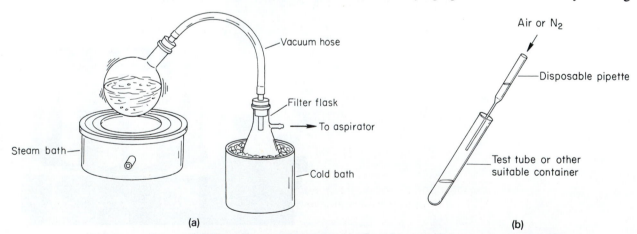

Figure 1.16 (a) Simple apparatus for removal of small amounts of volatile solvent
under reduced pressure. (b) A micro procedure for removal of very small amounts of
solvent.

hydrates, for example,

$$Na_2SO_4 + 10\ H_2O \rightleftharpoons Na_2SO_4 \cdot 10\ H_2O$$

are briefly reviewed below.

CaSO₄   Sold under the trade name Drierite. Low capacity (forms a hemi-hydrate), fast acting, efficient general utility. Colorless form is used for drying liquids. The blue indicating form (pink when hydrated) is used in drying tubes and so on.

CaCl₂   Generally fast and efficient. Eventually forms a hexahydrate. Can react or form complexes with carboxylic acids, amines, alcohols, and, occasionally, carbonyl compounds.

Na₂SO₄   Neutral drying agent. Slow, but of high capacity (forms a deca-hydrate). Best for initial drying of rather wet samples.

MgSO₄   Slightly acidic drying agent. Not as rapid or intense as CaSO₄ or CaCl₂. Capacity lower than that of Na₂SO₄.

K₂CO₃   Basic drying agent, excellent for organic bases or acid-sensitive compounds. Cannot be used with acidic compounds.

Dry liquids can be kept water-free by storage over molecular sieves (3A or 4A). Molecular sieves are synthetic zeolites containing cavities of uniform size which are capable of incorporating small molecules. Water molecules fit into the 3 Å and 4 Å pores of the type 3A or 4A sieves. Small quantities of solvents or reagents can be dried effectively by passing through a short column of activated alumina. Ultra-dry inert solvents (hydrocarbons and ethers) are often prepared by allowing the solvents to stand over CaH_2, $LiAlH_4$, or sodium (usually as a wire or dispersion). After hydrogen evolution is complete, they may be distilled. All of this latter group react vigorously with water, and there is a high explosion and fire risk if not handled properly. *Do not use active metal or metal hydrides as drying agents unless you have received proper instructions from your laboratory supervisor.* Never use active metals or metal hydrides to dry acidic or halogenated liquids. Never use these agents with any solvent before predrying with ordinary drying agents, unless assured of a low water content.

The low-boiling toluene-water azeotrope[1] (bp 85°C) (Section 2.3) can be used to advantage to drive water out of wet samples or to remove water from an equilibrium reaction. For example, consider the following equilibrium reaction:

$$\text{\includegraphics reaction}$$

This reaction can be carried out in toluene at reflux in the apparatus shown in Figure 1.17. As the toluene-water azeotrope condenses and falls into the water separator (Dean–Stark trap), water separates as a lower phase (the trap is calibrated in mL). Toluene fills the trap and spills back into the reaction flask.

---

[1] Benzene has been most frequently used for this purpose [benzene-water azeotrope (bp 69°C)]; however, due to the high toxicity of benzene, toluene is now preferred for this purpose.

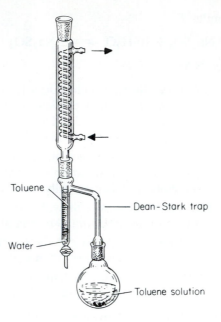

Toluene

Dean-Stark trap

Water

Toluene solution

Figure 1.17 Apparatus for removing water by azeotropic distillation.

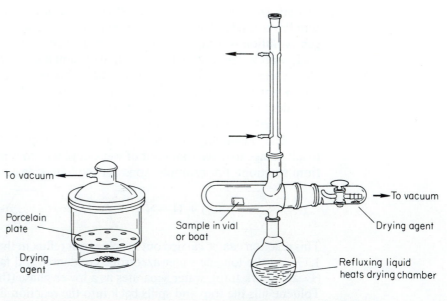

To vacuum

Porcelain plate

Drying agent

Sample in vial or boat

To vacuum

Drying agent

Refluxing liquid heats drying chamber

(a) Desiccator   (b) Abderhalden drying pistol

Figure 1.18 Devices for drying solids.

1.14.2 Drying Solids

For some purposes it is possible to dry solids simply by allowing them to stand in an open container (an evaporating dish, Petri dish, or beaker, but not an Erlenmeyer flask). More rigorous drying can be achieved by use of a desiccator (Figure 1.18). Drying agent is placed in the bottom of the desiccator, and the solid in an open container is placed on a porous plate above the desiccant. Greater efficiency is obtained by evacuation of the desiccator. Drying agents used in desiccators include phosphorous pentoxide, magnesium perchlorate, concentrated sulfuric acid, and calcium sulfate (Drierite). Large solid samples can be dried in vacuum ovens. Very small samples for microanalysis and so on are usually dried by placing the sample in a small open vial in the inner chamber of a drying pistol [called an Abderhalden (Figure 1.18)]. The inner chamber containing sample and desiccant is evacuated, and the system is heated to the desired temperature by the refluxing of an appropriate solvent.

1.14.3 Drying Gases

Water vapor can be removed from a gas by passing the gas through a drying tower filled with pellets of a desiccant (Drierite, $CaCl_2$, or KOH) or through a gas-wash bottle containing concentrated sulfuric acid (Figure 1.19). Reactions or containers are often protected from atmospheric moisture by use of drying tubes. Indicating Drierite is often used for this purpose, as is calcium chloride.

1.15 COMMON ORGANIC SOLVENTS

Organic solvents are the organic substances that are worked with in the largest quantities in the organic laboratory. There are many risks associated with their indiscriminate use. In the compilation that follows, we have listed the properties, advantages, and disadvantages of the solvents that you most frequently encounter in the practice of and reading about organic chemistry.

1.15.1 Acetone

CH_3COCH_3	bp 56°C	Highly flammable.
	fp −95°C	Miscible with water.
	density 0.78	Health hazard: low.

Technical-grade acetone is useful as an inexpensive, water-miscible, highly volatile solvent for cleaning glassware in the organic laboratory. Acetone-Dry Ice baths are used to maintain a temperature of −78°C.

Acetone is a moderately reactive solvent. Mild bases (e.g., alumina) or acids

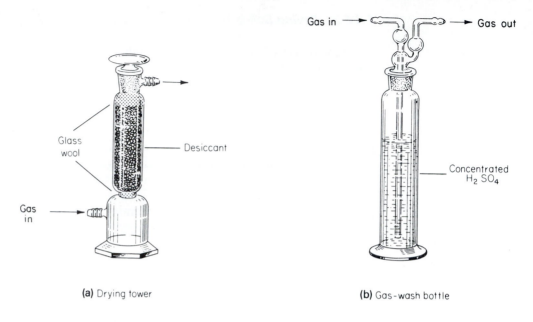

(a) Drying tower

(b) Gas-wash bottle

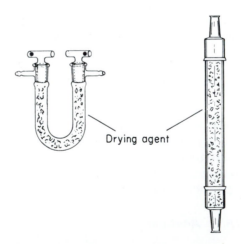

(c) Drying tubes

Figure 1.19   Devices for drying gases.

(e.g., $MgSO_4$) can cause aldol condensation. Acetone is a good medium for S_N2 reactions (provided the nucleophilic component does not react with acetone). Acetone is often used as a cosolvent with water in chromic acid oxidations.

Reagent-grade acetone is of high purity. Water content can be reduced to less than 0.005% with 4A molecular sieves or Drierite.

1.15.2   Acetonitrile

CH_3CN bp 82°C   Flammable.

                     fp -44°C   Miscible with water.

                     density 0.78   Health hazard: modestly toxic.

Acetonitrile is a relatively polar, non–hydrogen-bonding material of high solvent power. It is an excellent solvent for many ionic and inorganic substances.

Acetonitrile may be purified by distilling fairly dry samples from phosphorous pentoxide ($<$5 g/L).

1.15.3   Benzene

C_6H_6 bp 80°C   Flammable.

                     fp 5.5°C   Immiscible with water.

                     density 0.87   Health hazard: high.
Cumulative poison.
Is claimed to be carcinogenic.

Because of the leukogenic risk, benzene should be replaced by other solvents (e.g., toluene, hexane) in routine solvent applications. If its use is necessary, it should be used only in an excellent hood.

Benzene, a nonpolar solvent, is used as a basic starting material in aromatic chemistry. It has been frequently used for the azeotropic removal of water; the benzene-water azeotrope boils at 69°C and contains 9% water.

High-purity grades are readily available. Drying may be accomplished by distillation from phosphorus pentoxide or by refluxing for several days over calcium hydride, followed by slow distillation.

1.15.4   Carbon Disulfide

CS_2 bp 46°C   Highly flammable.

                     fp -112°C   Immiscible with water.

                     density 1.2   Health hazard: high.

C A U T I O N : The high vapor pressure, low minimum ignition temperature (120°C), high flammability (about 1% in air), toxicity, and skin and mucous membrane irritating properties make carbon disulfide a hazardous substance.

Carbon disulfide is occasionally used as a solvent in infrared spectroscopy and as a solvent for Friedel–Crafts reactions. It reacts readily with nucleophiles such as amines, alkoxides, and so on.

1.15.5 Carbon Tetrachloride

CCl$_4$ bp 77°C   Nonflammable.
                                 fp −23°C   Immiscible with water.
                                 density 1.58   Health hazard: acute and
                                                           chronic toxicity.
                                                        Possibly carcinogenic.

As a general rule, halogenated hydrocarbons are fairly toxic. They can enter the body by inhalation, ingestion, or absorption through the skin. From a health viewpoint, the concentration of carbon tetrachloride in the air is too high if the odor is detectable.

Carbon tetrachloride is a nonpolar solvent; it is useful as a solvent in infrared and magnetic resonance spectroscopy. It is readily available in high purity and can be dried over small quantities of phosphorus pentoxide.

1.15.6 Chloroform

CHCl$_3$ bp 61°C   Nonflammable.
                                 fp −6.4°C   Immiscible with water.
                                 density 1.48   Health hazard: toxic.
                                                        Possibly carcinogenic.

Chronic exposure may cause kidney and liver damage and heart irregularities. When exposed to air and light, chloroform can slowly oxidize to the highly toxic phosgene. Chloroform is irritating to the skin.

Chloroform is an excellent solvent for most organic compounds, including many ammonium, sulfonium, and phosphonium salts. Chloroform is widely used as a solvent in infrared spectroscopy. Deuterochloroform is often used as a solvent in magnetic resonance spectroscopy. Chloroform forms an azeotrope with water (bp 61°C).

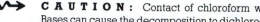 **C A U T I O N :** Contact of chloroform with basic reagents and solutions should be avoided. Bases can cause the decomposition to dichlorocarbene. The decomposition can become violent or, at the least, cause product contamination by chlorine-containing byproducts. Do not use chloroform as a solvent for amines or to extract highly basic solutions.

Chloroform contains small amounts of water and about 0.75% ethanol as an antioxidant. Small quantities *for immediate use* can be purified by passing through an activated alumina column (~ 100 g of alumina/200 mL of chloroform).

1.15.7 Dichloromethane (Methylene Chloride)

CH_2Cl_2	bp 40°C	Poorly flammable.
	fp −95°C	Immiscible with water.
	density 1.42	Health hazard: low.
		Least toxic of the chlorinated methanes.

Dichloromethane is an excellent solvent for most organic substances, including many ionic compounds, and is recommended for use instead of chloroform. It is highly volatile, and its solutions can be concentrated with ease. Dichloromethane is inert to most common reagents but will react with strong bases (e.g., butyllithium).

Dichloromethane can be purified by washing with water, then by sodium carbonate solution, followed by drying over calcium chloride and fractionally distilling.

1.15.8 Diethyl Ether (Ether)

$CH_3CH_2OCH_2CH_3$	bp 35°C	Highly flammable.
	fp −116°C	Immiscible with water.
	density 0.70	Health hazard: low except for explosibility.

Diethyl ether vapor concentrations in the range of 4% to 7% (by volume) produce anesthesia; 10% concentrations can be fatal. Repeated-dose inhalation can cause serious physical- and mental-health problems. Ether is flammable to concentrations as low as 2% and ether in air mixtures can ignite at temperatures as low as 174°C.

All ethers form peroxides with prolonged exposure to oxygen. Residues from the distillation or evaporation of ethers present high explosive hazards. Never distill an ether to near dryness unless peroxides are known to be absent.

Peroxide Test

Place 10 mL of the ether in a test tube and add 1 mL of a freshly prepared 10% aqueous potassium iodide solution. Acidify with several drops of dilute sulfuric acid. Add a drop of starch indicator. A blue color is indicative of the presence of peroxides. (Starch-iodide paper is not reliable for this test.)

Diethyl ether is a highly volatile solvent. The anhydrous grade (sometimes called absolute ether) is used as solvent for many organometallic and related reactions (e.g., Grignard and organolithium reactions, lithium aluminum hydride reductions) and as a cosolvent in column chromatography. Other grades, which may contain ethanol, are often used in extractions.

To purify diethyl ether, first test a small sample in a test tube with lithium aluminum hydride ($LiAlH_4$). *Do not use $LiAlH_4$ without proper supervision!* If a vigorous reaction occurs (hydrogen evolution), the water content should be reduced with conventional drying agents before proceeding. Diethyl ether containing small amounts of water can be purified by refluxing for 12 to 24 hours over $LiAlH_4$, followed by distillation (not to dryness because of the thermal instability of $LiAlH_4$). The $LiAlH_4$ will also destroy any peroxides present. The distilled solvent can be stored over molecular sieves and under nitrogen. The $LiAlH_4$ residues should be destroyed with great care by the slow and cautious addition of ethyl acetate or 10% sodium hydroxide. *Do not add water directly.*

1.15.9   Dimethylformamide (DMF)

$HCON(CH_3)_2$	bp 153°C	Flammable.
	fp -60°C	Miscible with water.
	density 0.94	Health hazard: possible carcinogen, irritant.

The vapors of dimethylformamide may cause nausea and headaches and has been implicated as a possible carcinogen. The liquid and vapor are absorbed through the skin.

Dimethylformamide has high solvent power for a wide variety of organic and inorganic compounds. It is a good dipolar, aprotic medium for nucleophilic reactions. DMF will react with Grignard reagents, alkyllithiums, lithium aluminum hydride, and so on.

Molecular sieves can be used to reduce the water content from 1% down to a few ppm.

For an excellent substitute solvent see 1,3-dimethyl-3,4,5,6-tetrahydro-2(1H)-pyrimidone.

1.15.10   Dimethyl Sulfoxide (DMSO)

$\overset{\displaystyle O}{\underset{\displaystyle CH_3SCH_3}{\|}}$	bp 189°C	Flammable.
	fp 64°C (4 mm)	Miscible with water.
		Health hazard: low toxicity.

CAUTION: Although dimethyl sulfoxide itself has low toxicity, it readily penetrates the skin and can carry toxic solutes with it into body fluids. When working with DMSO, many people experience a strange and persistent taste (some say garlic!) caused by the absorption of DMSO liquid and/or vapors through the skin.

Dimethyl sulfoxide has extraordinary solvent power for organic and inorganic substances. DMSO is an excellent solvent for S_N2 reactions and reactions of stabilized carbanions (e.g., Wittig reactions). *DMSO should not be used with oxidizing or reducing reagents.* (Numerous explosions have been reported with perchlorates, etc.) DMSO reacts with sodium hydride or other strong bases to produce methylsulfinyl carbanion, which is of use as a base and a reagent. Because it is difficult to remove, DMSO is usually a poor choice as a recrystallization solvent.

Reagent-grade dimethyl sulfoxide can be used as is for most purposes. Drying can be achieved by stirring overnight with calcium hydride and *distilling at reduced pressure.* Distillation should be conducted below 90°C to prevent disproportionation to the sulfide and sulfone.

1.15.11   1,3-Dimethyl-3,4,5,6-tetrahydro-2(1H)-pyrimidone

bp 146°C/44 mm
fp −20°C
density 1.060

Flammable.
Miscible with water.
Health hazard: low
          toxicity.

1,3-Dimethyl-3,4,5,6-tetrahydro-2(1H)-pyrimidone (DMP) is an excellent solvent for replacing the carcinogenic hexamethylphosphoramide and the N-methylamides. It is used as a cosolvent with tetrahydrofuran for reactions involving highly nucleophilic and basic reagents. The 1 : 1 mixture of DMP and tetrahydrofuran has a melting point <70°C.

DMP can be dried over calcium hydride.

1.15.12   Ethanol (Absolute Ethanol)

CH_3CH_2OH

bp 78.5°C
fp −114°C
density 0.78

Flammable.
Miscible with water.
Health hazard: very low.

Ethanol is often used as a solvent, a cosolvent, or reactant. It is often used in combination with sodium ethoxide (prepared *in situ* by addition of sodium) or potassium hydroxide in base-catalyzed or base-promoted reactions. Both polar and nonpolar compounds have reasonable solubility in ethanol. It is an excellent solvent for many recrystallizations. The less expensive 95% *ethanol* (ethanol-water azeotrope, bp 78.2°C) should be used for all applicants where the water is not a problem.

Traces of water can be removed from commercial absolute ethanol (water content 0.01% or less) by use of magnesium turnings. To dry 1 liter, about 5 g of Mg turnings are placed in a 2-L round-bottomed flask, and 60 mL of ethanol and several drops of carbon tetrachloride are added. The mixture is refluxed until the Mg is vigorously reacting. The remaining ethanol (940 mL) is added to the flask, and the reflux is continued for 1 hour. The ethanol is distilled into a receiver while being protected from atmospheric moisture.

1.15.13 Ethyl Acetate

$CH_3COOCH_2CH_3$ bp 77°C   Flammable.
                            fp -98°C   Immiscible with water.
                            density 0.92   Health hazard: low.
                                                           Vapors can cause
                                                           dizziness.

Ethyl acetate is a useful solvent in thin-layer and column chromatography, particularly in combination with hexane or other hydrocarbons. It is sometimes used in recrystallizations, and it is a good solvent for polymers and resins. Commercial material usually has a purity of 99.5% or better.

1.15.14 Hexamethylphosphoramide (HMPA)

$[(CH_3)_2N]_3P{=}O$ bp 233°C   Flammable.
                            fp 7.2°C   Miscible with water.
                            density 1.0   Health hazard: may be a
                                                           potent carcinogen.

C A U T I O N : *Hexamethylphosphoramide has been shown to cause cancer of the nasal passage in rats exposed to its vapors. Its use should be avoided whenever possible. It should be used only with extreme safety precautions.*

Hexamethylphosphoramide is a dipolar aprotic solvent that, even as a cosolvent in small quantities, has been found to have profound effects on certain nucleophilic and strong base reactions.

1.15.15 Hexane

$CH_3(CH_2)_4CH_3$ bp 69°C   Flammable.
                            fp -95°C   Immiscible with water.
                            density 0.65   Health hazard: low.

The volatile, saturated alkanes are all highly flammable, have anesthetic properties by inhalation, and are skin irritants.

Hexane and related hydrocarbon solvents are useful in extraction, recrystallization, and chromatography of nonpolar organic compounds. Because they are transparent to ultraviolet and visible light, saturated hydrocarbons are often used as solvents in ultraviolet spectroscopy.

The principal impurities in saturated hydrocarbons are alkenes and aromatics. Saturated hydrocarbons can be purified by stirring or shaking several times with concentrated sulfuric acid, and then with $0.1M$ potassium permanganate in 10% sulfuric acid. The hydrocarbon layer is then washed with water, dried over calcium chloride, refluxed over calcium hydride or sodium wire, and distilled.

1.15.16   Methanol

CH_3OH	bp 65°C	Flammable.
	fp −98°C	Miscible with water.
	density 0.79	Health hazard: low by inhalation, high by ingestion.

Ingestion of methanol can lead to blindness and death. Death from ingestion of less than 30 mL has been reported.

Methanol is a useful solvent for the recrystallization of a wide range of compounds. It is often used in combination with sodium methoxide or sodium hydroxide in base-promoted reactions.

Methanol can be purified by refluxing with magnesium turnings (10 g/L) for 4 hours followed by distillation. The distilled material can be stored over 3A molecular sieves.

1.15.17   Pentane

$CH_3(CH_2)_3CH_3$	bp 36°C	Highly flammable.
	fp −128°C	Immiscible with water.
	density 0.62	Health hazard: low.

See comments under hexane.

1.15.18   Petroleum Ether (Skelly Solvents)

Mixtures of alkanes	Highly flammable.
	Immiscible with water.
	Health hazard: low.

See comments under hexane. The petroleum ethers are mixtures of hydrocarbons and can be used interchangeably with the pure substances for many solvent purposes. Petroleum ethers are sold in various boiling ranges: 20 to 40°C, 30 to 60°C, 30 to 75°C, 60 to 70°C, 60 to 110°C, and so on.

1.15.19   2-Propanol (Isopropyl Alcohol)

$CH_3CHOHCH_3$ bp 82°C   Flammable.
                             fp −88°C   Miscible with water.
                             density 0.78   Health hazard: low.

Solvent properties of 2-propanol are much the same as those of ethanol. Dry Ice and 2-propanol baths are often used to maintain temperatures near −80°C.

C A U T I O N : 2-Propanol that has been exposed to the air for extended periods of time may contain peroxides. Test for the presence of peroxides using the procedure described for the peroxide test in Section 1.15.8. If present, reduce the peroxides before distillation.

1.15.20   Tetrahydrofuran (THF)

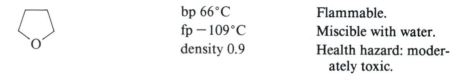

                             bp 66°C   Flammable.
                             fp −109°C   Miscible with water.
                             density 0.9   Health hazard: moder-
                                                       ately toxic.

Tetrahydrofuran is an excellent solvent for organometallic and metal hydride reactions. Grignard reagents can be prepared from vinyl halides and aryl chlorides in this solvent, whereas they do not form in diethyl ether.

Tetrahydrofuran forms peroxides very readily. (See peroxide test under diethyl ether.) Small amounts of peroxide can be removed by shaking TMF with cuprous chloride. THF containing large amounts of peroxide should be discarded by flushing down the drain with tap water. For instructions and precautions concerning the purification of THF, see *Org. Syn.,* Coll. Vol. 5, 976 (1973). Many of the explosions reported with THF are probably due to the exothermic cleavage of the THF by reaction with excess metal hydride at high temperature zones produced on dry portions of the wall of the flask. Purified THF should be stored over molecular sieves and under nitrogen.

1.15.21   Toluene

$C_6H_5CH_3$ bp 111°C   Flammable.
                             fp −95°C   Immiscible with water.
                             density 0.86   Health hazard:
                                                       moderate. Less
                                                       acutely toxic than
                                                       benzene.

Although toluene is less acutely toxic than benzene, exposure to low concentrations of its vapor probably involves less chronic risks (benzene has been claimed to be carcinogenic).

Toluene is an excellent, nonpolar solvent for low-temperature operations. Toluene and water form an azeotrope (bp 85°C) that contains 20% water.

1.16 REFERENCES

General

1. Gordon, A. J., and R. A. Ford, *The Chemist's Companion. A Handbook of Practical Data, Techniques, and References.* New York: Wiley-Interscience, 1972.

Inert Atmosphere Operations

2. Brown, H. C., *Organic Syntheses via Boranes.* New York: Wiley-Interscience, 1975.
3. Lane, C. F., and G. W. Kramer, "Handling Air-Sensitive Reagents," *Aldrichimica Acta,* **10,** 11 (1977).
4. Shriver, D. F., *The Manipulation of Air-Sensitive Compounds.* New York: McGraw-Hill, 1969.

Solvents

5. Riddick, J. A., and W. B. Bunger, *Organic Solvents. Physical Properties and Methods of Purification,* 3rd ed. New York: Wiley-Interscience, 1970.

Physical Properties of Chemicals

6. *Aldrich Catalog Handbook of Fine Chemicals,* Aldrich Chemical Company, Inc., Milwaukee, Wisc.

2

Separation
and Purification
Techniques

2.1 FILTRATION

Filtration is the method of separation of insoluble solids from a liquid by use of a porous barrier known as a filter; the liquid passes through the filter while the solid is retained by it. The liquid can pass through the filter by gravity (gravity filtration) or by the use of suction (suction filtration). The process of filtration is used to collect solid material from a heterogeneous mixture and/or to clarify a solution.

2.1.1 Gravity Filtration

The apparatus for gravity filtration consists simply of a sheet of filter paper, a conical funnel, and a collection flask. The filter paper can be folded into quarters and placed in the funnel or can be fluted (see Figure 2.1b) to increase the rate of the flow of the liquid through the filter paper. The size of the filter paper should be chosen so that the paper projects just above the top of the funnel. Filtration is accomplished by pouring, in portions if necessary, the sample to be filtered into the filter paper. For filtering a hot solution from which crystallization is likely to occur upon cooling, a wide-bore, short-stem funnel is used, and the solution in the collection flask is refluxed off from the funnel to keep the funnel warm.

Micro filtrations used to clarify a solution are easily accomplished by pouring the solution through a disposable pipette in which a small plug of cotton has been pressed into the bottom of the pipette (see Figure 2.1c). To force all the solution out

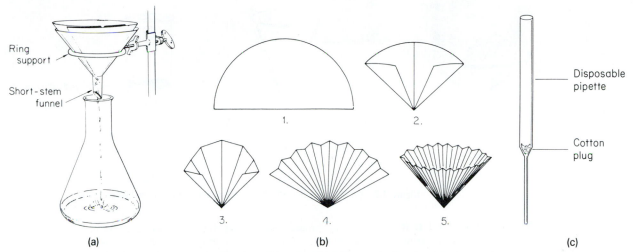

Ring support

Short-stem funnel

Disposable pipette

Cotton plug

1.

2.

3.

4.

5.

(a)   (b)   (c)

Figure 2.1 (a) Simple gravity filtration apparatus. (b) Fluted filter paper. To flute filter paper, first fold into quarters; then by alternating the direction of each fold and limiting each fold to two thicknesses of paper, continue folding the paper. (c) A disposable pipette containing a cotton plug for use as a micro filter.

of the cotton plug and the pipette, a small rubber bulb is placed over the top of the pipette and squeezed. This procedure is not useful for collecting crystals after a recrystallization.

2.1.2 Suction Filtration

In suction filtration, pressure is reduced below the filter by means of a water aspirator; the atmospheric pressure then forces the liquid through the filter. Following a re-crystallization (Section 2.2), the crystals are usually collected by suction filtration.

For suction filtration three types of funnels are available for use: Büchner, Hirsch, and sintered glass funnels (see Figure 2.2). Büchner and Hirsch funnels require the use of circular filter paper of correct diameter to cover the perforations at the bottom of the funnel. Care must be exercised in the removal of the sample from the sintered glass funnel so as not to scrape off small particles of the sintered glass. These funnels are inserted through rubber stoppers or Neoprene adapters and placed in the top of a vacuum filter flask (which is shaped like an Erlenmeyer flask but with thick walls and a vacuum attachment). Filter flasks should be clamped for stability because the assembled apparatus is top-heavy and is easily tipped over. Filtration is accomplished by placing a piece of filter paper into the funnel, starting a gentle vacuum, wetting the filter paper with a little of the solvent being used (this seals the paper in place), and pouring the solution to be filtered into the funnel. A small portion of cold solvent is then poured over the crystals to wash away any adhering solution. The crystals can be air-dried by keeping the vacuum on which draws air through the crystals.

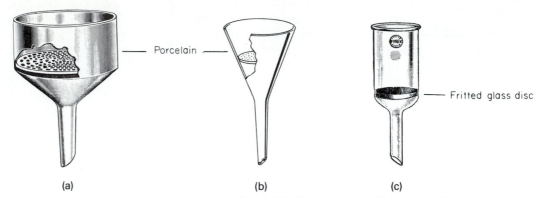

Figure 2.2   (a) Büchner funnel. (b) Hirsch funnel. (c) Sintered glass funnel.

C A U T I O N : If the water pressure decreases while a system is under aspirator vacuum, water may be sucked back into the system. Although many aspirators have one-way check valves designed to prevent this, these valves may occasionally fail. To prevent water from backing up into the system, a trap should be placed between the aspirator and the filtration flask as shown in Figure 2.3. The vacuum should always be released by opening the system with the three-way stopcock before the flow of water is shut off or by disconnecting the vacuum tubing.

2.1.3 Filter Aids

When the object is to clarify a solution by removal of finely suspended particles, the use of a more efficient filter than filter paper may be necessary. An effective filter can be created in a Büchner funnel by depositing a pad of some porous, inert filter aid on top of the paper prior to filtration of the suspension. Such a filter pad is not easily

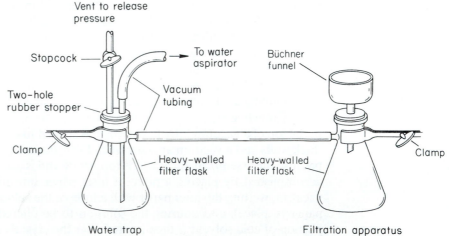

Figure 2.3   Filter flask with a Büchner funnel attached to a water trap and water aspirator.

clogged by fine particles. The filter aid most often used is a diatomaceous earth product sold under the trade name Celite.

To prepare a filter pad, place the filter paper in the Büchner, wet with solvent, add filter aid as a heavy slurry in the solvent (use enough filter aid to create a pad about 0.5 cm thick), turn on the aspirator to remove the excess solvent, disconnect the aspirator and rinse the collection flask, reconnect the aspirator, and gently pour the solution to be filtered into the Büchner funnel in order to cause minimum disturbance to the filter pad.

2.2 RECRYSTALLIZATION

Crystallization is the deposition of crystals from a solution or melt of a given material. During the process of crystal formation, a molecule will tend to become attached to a growing crystal composed of the same type of molecules because of a better fit in a crystal lattice for molecules of the same structure than for other molecules. If the crystallization process is allowed to occur under near-equilibrium conditions, the preference of molecules to deposit on surfaces composed of like molecules will lead to an increase in the purity of the crystalline material. Thus the process of recrystallization is one of the most important methods available to the chemist for the purification of solids. Additional procedures can be incorporated into the recrystallization process to remove impurities. These include filtration to remove undissolved solids and adsorption to remove highly polar impurities.

Recrystallization depends on the differential solubility of a substance in a hot and a cold solvent. It is desirable that the solubility of the substance be high in the hot solvent and low in the cold solvent to facilitate the recovery of the starting material. The solution remaining after crystals have deposited is known as the *mother liquor*. The proper choice of solvent is critical and may require trial tests with small quantities of the material in a variety of solvents or solvent pairs (combinations of two solvents).

Solvents used in recrystallizations should be relatively low-boiling so that solvent adhering to the crystals can be readily removed by evaporation. If structural characteristics of the compound to be recrystallized are known, the adage "like dissolves like" should be kept in mind; polar compounds are more soluble in polar solvents, and nonpolar compounds are more soluble in nonpolar solvents. Some common solvents used in recrystallization are listed in Table 2.1.

Recrystallization Procedure

The solvent, or solvent pair, to be used in the recrystallization of a substance is chosen in the following manner. A small amount of the substance is placed in a small test tube and a few drops of solvent are added. The test tube is gently heated to see if the sample dissolves in the heated solvent. In general, one should first use a nonpolar solvent, for example, hexane or petroleum ether. If the sample does not dissolve, try

TABLE 2.1 Common Solvents* for Recrystallization Listed in Order of Decreasing Polarity

Solvent	bp (°C)	Useful for	Common Cosolvents
Water	100	Salts, amides, some carboxylic acids	Acetone, methanol, ethanol
Methanol	65	General use	Water, diethyl ether, benzene
Ethanol	78	General use	Water, diethyl ether, benzene
Acetone	56	General use	Water, hexane
Ethyl acetate	77	General use	Hexane
Dichloromethane	40	General use, low-melting compounds	Ethanol, hexane
Diethyl ether	35	General use, low-melting compounds	Methanol, ethanol, hexane
Chloroform†	61	General use	Ethanol, hexane
Toluene	111	Aromatic compounds	
Benzene†	80	Aromatic compounds	Ethanol, diethyl ether, ethyl acetate
Hexane (or petroleum ether)	69	Hydrocarbons	All solvents on this list except water and methanol

* For additional information on organic solvents see Section 1.15.

† Not recommended for general use because of health hazard (Sections 1.15.3 and 1.15.6).

using a more polar solvent such as ethanol or acetone. Should the sample completely dissolve in any solvent, chill the solution to see whether crystals will form. (Sometimes it is necessary to chill the solution using a Dry Ice-acetone bath in order to cause crystallization.) If no crystals appear, the material is too soluble in that solvent, and that solvent should not be used for the recrystallization. If no single solvent provides suitable results, a mixture of two solvents can be employed, one of the solvents being a good solvent for the sample, and the other being a poor solvent for the sample. The sample is first dissolved in the solvent in which the sample is most soluble, and then small portions of the other solvent are added until a cloudiness is formed upon addition of the second solvent. A small amount of the better solvent is added to remove the cloudiness, and the solution is allowed to cool. The correct proportion of the two solvents must be determined by trial and error. Once the proper solvent has been chosen, the remainder of the sample is recrystallized.

For gram- or multigram-scale recrystallizations, the material to be recrystallized is placed in a suitable container such as an Erlenmeyer flask. Solvent is added slowly, maintaining a gentle reflux of the solvent in the flask, until no more of the material dissolves. Occasionally, highly insoluble materials may be present in the sample, which, regardless of the amount of solvent added, will not dissolve. When it appears that no more material dissolves, the addition of further solvent is stopped.

If the sample is colored, a small amount of activated carbon (charcoal) can be added to adsorb highly polar, generally colored contaminants. The addition of activated carbon must be carried out with great care, for if the saturated solution has become superheated, the addition of the finely divided charcoal will induce violent boiling with the possible loss of material. The solution should be below its boiling point before the addition of the activated carbon. After the addition of the charcoal, the mixture is gently boiled for about 30 seconds while rapidly swirling or stirring to avoid the rapid boiling (bumping) of the solution. The solids are allowed to settle to the bottom of the container. Any change in the color of the solution is noted. If the color of the solution has decreased, an additional amount of activated carbon is added and the process is repeated until no further change in color is noted.

CAUTION: Be very careful when recrystallizing highly polar or colored compounds in order to avoid a great excess of the activated carbon that may cause a considerable loss of material.

The hot, saturated solution is filtered through a hot, solvent-saturated filter that is kept warm, by refluxing the solvent, to prevent the premature formation of crystals (see Figure 2.1). The filter is washed with a small amount of the hot solvent, and the volume of the solution is reduced by boiling until a saturated solution is again attained. When removing the activated carbon by filtration it is often necessary to use a filter aid, for example Celite, to remove all of the very small particles of the activated carbon completely. Suction filtration techniques are generally employed when such a filter aid is used to decrease the filtering time (Section 2.1.2). The filter aid is added to the solution just prior to filtering; the solution is quickly mixed with the filter aid and poured into the filter, or the hot solution is filtered through a preformed layer of the filter aid in the funnel.

The rate of crystal growth from the saturated solution is critical. Too rapid a precipitation rate does not allow equilibrium conditions to exist, and the very slow precipitation with the formation of large crystals often leads to extensive solvent inclusion in the crystals. The hot, saturated solution should be allowed to cool at a rate between the two extremes. This can generally be accomplished by allowing the solution to cool on the bench top, followed by chilling in an ice bath. Occasionally, with low-boiling solvents such as ether or pentane, it may be desirable to chill the sample in a Dry Ice-acetone bath to induce crystal growth. In such cases the collection funnel should be chilled prior to filtration by pouring a portion of chilled solvent through the filter funnel.

Some materials tend to "oil out" as a liquid instead of forming crystals. This usually occurs when the melting point of the material is below the temperature at which a saturated solution is attained. Additional solvent should be added, maintaining a clear solution, until crystals finally begin to form. If crystals are reluctant to form, scratching the side of the container with a glass stirring rod or adding a seed crystal will generally induce crystallization. The crystals are then collected by suction filtration and washed with a small portion of cold solvent to remove any adhering mother liquor. When using filter paper to collect the sample, care should be taken in removing the sample from the paper to avoid dislodging any fibers of the paper, which may contaminate the sample. These fibers may interfere with subsequent elemental or spectral analysis.

The recrystallization of small quantities of materials can be carried out in small test tubes or centrifuge tubes. Centrifuge tubes are the most convenient in that the crystalline mass can be forced to the bottom of the tube and the mother liquors

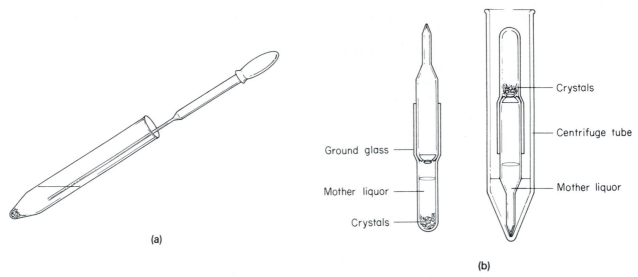

(a)

Ground glass

Mother liquor

Crystals

Crystals

Centrifuge tube

Mother liquor

(b)

Figure 2.4   Illustrations showing the use of a centrifuge tube (a) and a Craig tube (b)
for recrystallization. In (a), the crystals have been centrifuged to the bottom of the
tube and the solvent is being removed with a disposable pipette. In (b), the Craig tube
in which the crystals have formed (left) is placed in a centrifuge tube and the mother
liquor is forced from the Craig tube by centrifugation (right).

removed by a capillary pipette (see Figure 2.4). The crystalline mass can then be
washed with several small quantities of cold solvent and then removed from the tube
to be dried.

The recrystallization of milligram quantities of material can be conveniently
carried out using a Craig tube (see Figure 2.4). The Craig tube consists of an outer
part in which the material to be recrystallized is dissolved in the solvent. The inner
piece has a ground glass area that forms a seal with the outer part. The inner part is
inserted and the contents of the Craig tube are allowed to cool to induce crystallization. After crystallization is completed, the Craig tube is placed upside down in a
centrifuge tube and centrifuged. The spinning of the centrifuge forces the mother
liquors through a sintered glass frit in the inner part of the apparatus. After centrifugation the inner tube is removed and adhering crystals are scraped off into the outer
part of the apparatus. A small amount of cold solvent can be added to the crystals to
wash away any adhering mother liquor, and then removed by the use of a disposable
pipette.

The crystalline material obtained in the procedure outlined above is thoroughly
dried to remove adhering solvent. Many compounds can be dried by allowing the
samples to set in the open air (care must be exercised to protect the sample from
contamination). Hygroscopic compounds or compounds recrystallized from high-
boiling solvents must be dried in a vacuum drying apparatus (e.g., a vacuum dessicator, Section 1.14.2). The entire recrystallization procedure should be repeated until
a constant melting point of relatively narrow range is obtained (see Section 3.2 on
melting points for the details of melting-point determinations).

2.3 DISTILLATION

Distillation can be defined as the partial vaporization of a liquid with transport of these vapors and their subsequent condensation in a different portion of the apparatus. Distillation is one of the most useful methods for the separation and purification of liquids. The successful application of distillation techniques depends on several factors. These include the difference in vapor pressure (related to the difference in the boiling points) of the components present, the size of the sample and the distillation apparatus (as well as the type of apparatus employed), the occurrence of codistillation or azeotrope formation, and the care exercised by the experimentalist. Distillation relies on the fact that the vapor above a liquid mixture is richer in the more volatile component than in the liquid, the composition being controlled by Raoult's law.

Raoult's law states that the partial pressure (P_A) of component A in an ideal solution at a given temperature is equal to the vapor pressure (at the same temperature) of pure A (P_A°) multiplied by the mole fraction of A (N_A) in solution. Consider an ideal solution of A and B:

$$N_A = \frac{\text{moles of } A}{\text{moles of } A + \text{moles of } B} \qquad N_B = \frac{\text{moles of } B}{\text{moles of } A + \text{moles of } B}$$

$$N_A + N_B = 1 \tag{2.1}$$

$$P_A = P_A^\circ N_A \qquad P_B = P_B^\circ N_B$$

$$P_T \text{ (total vapor pressure)} = P_A + P_B$$

The boiling point of the solution is reached when P_T is equal to the pressure applied to the surface of the solution [see Equation (2.1)].

Phase diagrams are helpful in illustrating simple and fractional distillation. The diagram shown in Figure 2.5a plots the equilibrium composition of vapor and liquid phases against temperature. It can be seen that at 105°C, liquid with composition 90% B and 10% A is in equilibrium with vapor of composition 74% B and 26% A. If this vapor were removed and condensed, the condensate, with a composition of 74% B and 26% A, is enriched in the lower-boiling component. Now, if this condensate is allowed to reach equilibrium with its vapor (at 97°C), the vapor will contain 53% B and 47% A, as will the liquid when it is condensed. Each of these vaporization-condensation steps, indicated by the dashed lines in Figure 2.5a, is called a theoretical plate. In a fractional distillation these vaporization-condensation cycles occur on the wall of the distillation column, or on the packing that is often used in the distillation column. Theoretically, an infinite number of these steps would be required to obtain 100% pure material. In practice this is not possible. The early fractions coming out of the distillation column will be essentially pure A, intermediate fractions will be mixtures of A and B, and the later fractions will be essentially pure B. The purity of A and B that can be obtained will depend on the difference in the boiling points of A and B and the efficiency of the distillation column. In a simple distillation there may be only a very few theoretical plates and the purity of the

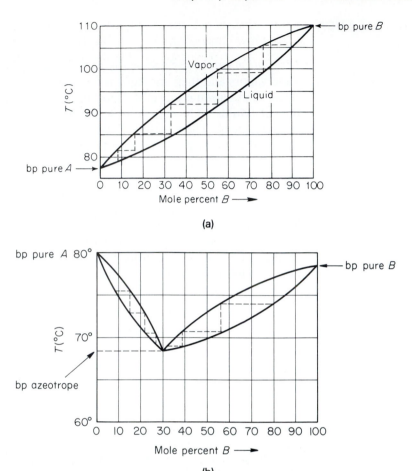

Figure 2.5 (a) Boiling point diagram for an ideal solution of two liquids: *A* (bp 76°C) and *B* (bp 110°C). (b) Boiling point diagram for a mixture of liquids that form an azeotrope: *A* (bp 80°C), *B* (bp 78.5°C) and azeotrope (bp 68°C).

fractions attainable is limited. The use of efficient fractionating columns may produce more effective separation when sufficient quantities of material are available. In considering distillations, the following points are pertinent:

1. The boiling points of mixtures are always intermediate between those of the highest- and lowest-boiling constituents.
2. The closer the boiling points of the pure constituents of a mixture, the more difficult their separation by distillation.
3. A mixture of two or more substances with the same boiling point will have the same boiling point as the constituents and cannot be separated by distillation.

There are a number of mixtures of liquids that do not obey Raoult's law, but form an *azeotrope* (also called an azeotropic mixture). An azeotrope is a constant boiling mixture of definite composition. The boiling point of an azeotrope is always outside the range of the boiling points of the pure components. A typical phase diagram for an azeotropic system is shown in Figure 2.5b. If one distills a mixture of 15% *B* and 85% *A*, the early fractions will contain the azeotropic mixture of 30% *B* and 70% *A*, while the last fractions will contain essentially pure *A*. Pure *B* cannot be obtained by the fractional distillation of this starting mixture. On the other hand, as one fractionally distills a mixture containing 55% *B* and 45% *A*, one first obtains the azeotrope and then essentially pure *B*. In the case of azeotrope formation, the separation of the azeotrope into its pure components must be achieved by other means, such as preparative gas chromatography (Section 2.6). Examples of azeotropes and their boiling points are given in Table 2.2.

2.3.1 Simple (Nonfractional) Distillation

If a sample is known to contain essentially only one volatile component, a simple, nonfractional distillation can be used to accomplish a suitable purification of the component. A suitable apparatus for such a distillation is illustrated in Figure 2.6 and employs standard-taper 14/20 precision-ground glassware (the correct handling

TABLE 2.2 Some Common Azeotropes

Components	Percent by Weight	bp Pure Component (°C)	bp Azeotrope (°C)
Acetone	88	56	
Water	12	100	56
Benzene	91	80	
Water	9	100	69
Benzene	68	80	
Ethanol	32	78.5	68
Chloroform	93	61	
Ethanol	7	78.5	59
Chloroform	97	61	
Water	3	100	56
Dichloromethane	98	40	
Water	2	100	39
Ethanol	68	78.5	
Toluene	32	111	77
Ethanol	96	78.5	
Water	4	100	78
Toluene	80	111	
Water	20	100	85
Benzene	74	80	
Ethanol	19	78.5	65
Water	7	100	
Toluene	51	111	
Ethanol	37	78.5	74
Water	12	100	

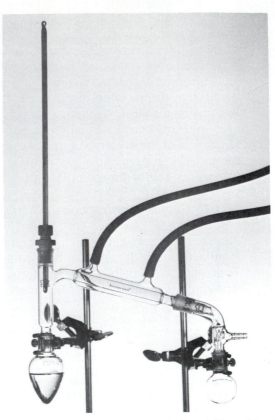

Figure 2.6   Simple distillation apparatus employing 14/20 precision-ground glass-ware. (Photograph by courtesy of the Kontes Glass Company, Vineland, N.J.; photo-graph by Bruce Harlan, Notre Dame, Ind.)

of precision-ground glassware is discussed in Section 1.6). The apparatus shown in Figure 2.6 can be used for atmospheric or reduced pressure (vacuum) distillations and is suitable for sample sizes down to 0.5 g. Vacuum distillation is recommended for most organic compounds to prevent decomposition.

A small portion of the sample should be subjected to a micro boiling point determination (Section 3.4) to determine the approximate boiling point of the sample and to determine whether the sample is stable under the conditions of the distillation. In general, if the boiling point of the sample is above 180°C, a reduced pressure distillation should be employed. This can be accomplished by the use of a water aspirator capable of producing pressures down to 12 to 15 mm-Hg, or a vacuum pump capable of producing pressures down to 0.01 mm-Hg. Figure 2.7 shows the recommended setup for a vacuum distillation, including a vacuum gauge for determining the pressure at which the distillation is being carried out, a cold trap (to be used only when a vacuum pump is employed) to prevent vapors being drawn into the vacuum pump, and an air bleed to introduce air into the system to bring the internal

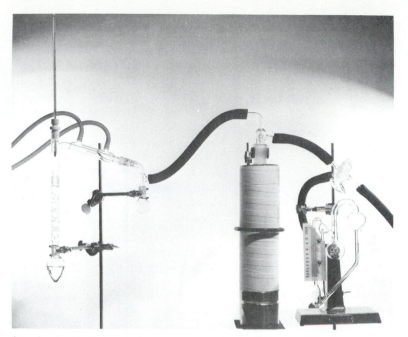

Figure 2.7 Typical setup for a vacuum distillation including a fractionating column, cold trap, and vacuum gauge. (Photograph by courtesy of the Kontes Glass Company, Vineland, N.J.; photograph by Bruce Harlan, Notre Dame, Ind.)

pressure back to atmospheric pressure before disassembling the apparatus. The vacuum should never be released by the removal of a flask or the thermometer. A safety shield should be routinely used to shield the experimentalist carrying out a vacuum distillation. The comparison of boiling points obtained at different pressures is readily accomplished by use of a boiling point nomograph (Appendix).

The sample is placed in the distilling flask; a boiling chip or a boiling stick is then added to promote a smooth and continuous boiling action. In many cases, boiling chips or sticks do not provide sufficient agitation of the sample to promote a smooth and continuous boiling action. In such cases it is necessary to provide the necessary agitation by means of a slow passage of nitrogen or argon through a capillary into the liquid. A magnetic stirrer can also be used. The distillation flask should be heated by means of an oil or wax bath, never with a heating mantle or the flame of a Bunsen burner. Adequate control of the temperature of the distillation flask generally cannot be maintained with the latter two methods of heating.

2.3.2 Fractional Distillation

Fractional distillation is employed when the separation of two or more volatile components is required. The principle of fractional distillation is based on the establishment of a large number of theoretical vaporization-condensation cycles

55847

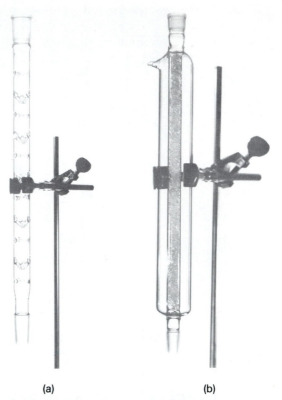

(a)   (b)

Figure 2.8 Typical fractionating columns. (a) Vigreux column. (b) Glass helices-packed, vacuum-jacketed fractionating column. (Photographs by Bruce Harlan, Notre Dame, Ind.)

(Section 2.3). A fractionating column that allows equilibrium of the descending condensed liquid with the ascending vapors is used, thus producing the effect of a multiple vaporization-condensation cycle.

The length and type of fractionating column required depends on the boiling points of the components to be separated. Suitable separations of components differing in boiling points by 15 to 20°C can be accomplished by means of a Vigreux column (see Figure 2.8a), a column containing indentations that increase the wall area of the column. For separations of components with closer boiling points, packed columns (see Figure 2.8b) or spinning band columns can be used.

The material used in a packed column should be finely divided and must be chemically inert. Small glass helices are commonly used in such columns. Steel wool may also be used providing it does not react with the components being distilled. A spinning band column contains a central spiral band that is rapidly rotated. The spinning column creates a very thin film of the condensate on the surface of the column that provides for very rapid equilibration to be attained. Spinning band columns have up to several hundred theoretical plates and are capable of separating liquids that differ in boiling points by as little as 0.5°C.

Equilibrium conditions must be maintained in the fractionating column at all times in order to achieve a successful separation. The ratio of distillate to the amount of condensed material returning to the distillation flask (referred to as the reflux ratio) should always be much larger than 1, generally in the region of 5 to 10 for relatively easily separated components. Maintaining the reflux ratio in this region requires a very careful control of the amount of heat applied to the distillation flask. Flooding of the column should be avoided at all times. Fractional distillations carried out under reduced pressures should involve the use of fraction cutting devices. These devices do not require the breaking of the vacuum, which would in turn destroy the equilibrium conditions established on the fractionating column.

2.3.3 Microdistillation

Frequently a chemist is faced with the problem of distilling small quantities of a liquid. Less-than-gram quantities of a substance may be distilled using a short-path distilling head (see Figure 2.9a). Short-path distilling heads provide little or no fractionation. Milligram quantities of a substance can be distilled using a microscale Hickman molecular still shown in Figure 2.9b. The vaporized liquid condenses on

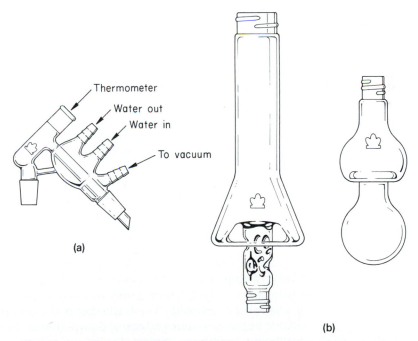

Figure 2.9 (a) A short-path distillation head. (b) Two microscale Hickman molecular stills. (Reproduced courtesy of the Kontes Glass Company, Vineland, N.J.)

the sides of the upper portion of the still, the liquid flowing down into the annular ring where it is retained. The distilled material is removed by the use of a micro pipette. The distillation head shown in Figure 2.9a can be used for larger-scale distillations of up to a few grams.

2.3.4 Steam Distillation

A *heterogeneous* mixture of two liquids (*A* and *B*) does not obey Raoult's law, but each component exerts a partial vapor pressure (P_A° or P_B°) that is the same as that for the pure substance at any given temperature. In other words, the partial pressure of each component of a heterogeneous mixture is dependent only on temperature. When $P_A^\circ + P_B^\circ$ equals the applied pressure, the mixture boils. Since P_A° and P_B° are additive, the boiling point of the mixture is always *below* the boiling point of the lowest-boiling component (cf., ideal homogeneous solutions, Section 2.3). The boiling point of the mixture and the composition of the distillate will remain constant as long as there is some of each constituent present. Since the molar concentration of each component in the vapor is proportional to its vapor pressure, the composition of the distillate is given by the following expression, where P_A° and P_B° are the partial pressures of the components at the temperature at which distillation occurs [see Equation (2.2)].

$$\frac{\text{moles of } A}{\text{moles of } B} = \frac{P_A^\circ}{P_B^\circ} \quad \text{or} \quad \frac{\text{wt. } A}{\text{wt. } B} = \frac{P_A^\circ \cdot \text{mol. wt. } A}{P_B^\circ \cdot \text{mol. wt. } B} \tag{2.2}$$

If water is employed as one phase in the distillation of a two-phase immiscible system, the method is called steam distillation. A practical application of steam distillation is provided by the example of bromobenzene. Bromobenzene (bp 156°C) and water (bp 100°C) are, for all practical purposes, insoluble in each other. At 95°C the combined vapor pressures of bromobenzene (120 mm) and water (640 mm) equal 760 mm, and the mixture boils. The composition of the distillate, as long as some bromobenzene remains in the distillation flask, is given by Equation (2.3).

$$\frac{\text{wt. C}_6\text{H}_5\text{Br}}{\text{wt. H}_2\text{O}} = \frac{P_{\text{C}_6\text{H}_5\text{Br}} \cdot \text{mol. wt. C}_6\text{H}_5\text{Br}}{P_{\text{H}_2\text{O}} \cdot \text{mol. wt. H}_2\text{O}}$$

$$= \frac{120 \cdot 157}{640 \cdot 18} = \frac{1.64 \text{ g of bromobenzene}}{\text{g of water}} \tag{2.3}$$

Steam distillations are very useful for separating volatile components from relatively nonvolatile components, particularly when the more volatile component possesses a very high boiling point and may be subject to decomposition if a direct distillation is attempted. This technique is also useful if the presence of other lesser volatile impurities causes extensive destruction of the desired fraction under normal distillation conditions. Steam distillation can often be used to separate isomeric compounds, particularly when one isomer is capable of extensive intramolecular hydrogen bonding; hence, that isomer is made more volatile, whereas the other isomers can participate only in intermolecular hydrogen bonding. A mixture of *o*-and *p*-nitrophenol can be separated by steam distillation. The ortho isomer, which exhibits intramolecular hydrogen bonding, readily steam distills, whereas the para isomer remains in the distillation flask.

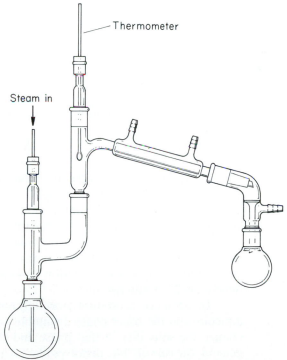

A typical steam distillation setup is illustrated in Figure 2.10. Small quantities of materials can be steam-distilled by the direct distillation of a mixture of the sample and water, deleting the external stream source and a water entrainment trap. The organic material is then recovered from the distillate by extraction with an organic solvent.

Figure 2.10 A typical steam distillation setup. (Courtesy of the Kontes Glass Company, Vineland, N.J.)

2.4 SUBLIMATION

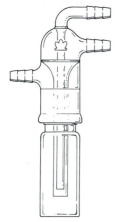

Figure 2.11 A vacuum sublimation apparatus. (Courtesy of the Kontes Glass Company, Vineland, N.J.)

Many solid substances pass directly into the vapor phase when heated and, on cooling, pass directly back to the solid phase. Such a process is termed *sublimation* and, as such, provides a useful method for the purification of materials. Unfortunately, however, relatively few organic substances undergo sublimation at atmospheric pressure. Generally, sublimations are carried out under reduced pressures. Types of compounds that readily sublime include many α-amino acids, ketones, carboxylic acids, and most acid anhydrides and quinones.

A typical vacuum sublimation apparatus is shown in Figure 2.11. The finely divided sample is placed in the bottom of the cup, and the cold-finger condenser is then inserted into the container cup. A vacuum is applied, and the temperature of the sample is slowly increased until sublimation occurs; however, the temperature of the sample *must* be kept well below the melting point of the sample. The flow of coolant to the condenser is shut off, the condenser is removed, and the sample is scraped from the cold finger. Repeated sublimations may be required to obtain suitably pure material. If the apparatus such as that shown in Figure 2.11 is not available, the sample can be placed in a side-arm test tube, with a small test tube inserted to act as the condenser. The inner tube can be filled with ice or constructed as a condenser.

2.5 EXTRACTION

The distribution of a substance between two immiscible phases is the basis of extraction. The distribution equilibrium constant K is defined in Equation (2.4),

$$K = \frac{[C_A]}{[C_B]} \tag{2.4}$$

where $[C_A]$ is the equilibrium concentration of a substance in phase A and $[C_B]$ is the equilibrium concentration in phase B. The distribution coefficient can be calculated using the limiting equilibrium solubilities of the material in the two phases. In actuality, the equilibrium constant expression may be more complex than is illustrated in Equation (2.4) in that the substance may exist in dimeric or higher polymeric forms in one or both of the phases, necessitating the introduction of coefficients and exponents in Equation (2.4) to describe adequately the concentration dependencies in the two phases.

In extraction procedures normally employed in organic chemistry, one phase is aqueous and the other phase a suitable organic solvent. A change in solvent will change the solubility in that phase and thus alter the distribution coefficient. In general, the rule of "like dissolves like" is a suitable guideline in choosing the appropriate solvent. The organic solvent should be relatively volatile for ease of removal after the extraction has been accomplished.

Procedure for Simple Extraction

A separatory funnel supported by a ring is charged with the cool aqueous solution to be extracted. Add the organic solvent. (Do not add volatile solvent to a hot solution or fill the separatory funnel to more than three-quarters full.) Place the stopper in the funnel. Remove the funnel from the supporting ring. Grasp the funnel with both hands so that the stopper and stopcock are held firmly in place. (The stopper should be against the palm of one hand, and the fingers of the other hand should be around the handle of the stopcock.) Shake the funnel once or twice so that the phases mix. With the stopcock end pointed upward and away from all persons, vent the stopcock to release excess pressure. Repeat several times, making sure that the phases have been vigorously intermixed. Replace the funnel in the support ring, remove the stopper, and allow the phases to separate. Drain the lower phase slowly through the stopcock, and just as the interface of the two phases reaches the stopcock, close it. Be sure that you can distinguish between the two phases so you do not discard your product. In aqueous extractions, organic layers heavier than water will be the lower phase, and those lighter than water will be the upper phase (Table 2.3). A simple test that will usually allow you to distinguish between an organic phase and an aqueous phase is to place a drop of each phase on the edge of a circle of filter paper; the spot that is wet with water will tear easily while the other spot will resist tearing.

In any extraction, each phase becomes saturated with respect to the other phase. When extracting an aqueous phase with an organic phase, it is necessary to remove the dissolved water before recovering the extracted material by evaporation of the solvent. Generally most of the dissolved water in the organic phase can be removed by washing the organic phase with a saturated aqueous solution of sodium chloride. Final drying of the organic phase is accomplished by allowing the organic phase to stand for a period of time over an anhydrous inorganic salt (see Section 1.14.1).

Various materials can be added to either phase, either to increase or decrease the solubility of a substance in that phase. The solubility of a material in one phase can be increased by the addition of a reagent that is capable of forming a stable, soluble complex in that phase. For example, many reactive alkenes and polyenes can be extracted into an aqueous phase in the presence of silver ion, or chelating agents

TABLE 2.3 Common Solvents for Extraction of Aqueous Solutions

Lighter than Water	Heavier than Water
Ether	Dichloromethane
Pentane, hexane, petroleum ether	Chloroform
Ethyl acetate, Toluene	Carbon tetrachloride

can be extracted either into or from an aqueous phase in the presence of a complexing metal ion. These complexes can be subsequently degraded, and the desired material recovered.

The solubility of a substance can be decreased, particularly in aqueous phases, by the addition of neutral salts that reduce the solubility of the substance in that phase. Various salts can be used, sodium chloride probably being the most widely used. Other salts that can be similarly used are sodium sulfate, potassium carbonate, and calcium chloride.

The distribution coefficient for acidic or basic materials can be greatly altered by changing the pH of one phase. For acidic materials, one can show that the distribution coefficient between two phases, one of which is the aqueous phase with a hydrogen ion concentration of $[H^+]$, can be represented by an effective distribution coefficient K_{eff} [Equation (2.5)], where K_D is the distribution coefficient between an organic phase and pure water, and K_A is the ionization constant of the acid.

$$K_{eff} = \frac{K_D K_A}{[H^+]} \tag{2.5}$$

This approach is the basis for the facile separation of strong acids from weak acids by extraction with aqueous bicarbonate. After extraction of the acid or base from an organic phase, the compound can be recovered from the aqueous phase by acidification or basification.

In many situations, however, it is not possible to obtain a favorable distribution coefficient by any of the foregoing procedures. When an unfavorable distribution coefficient is encountered, it can be shown by Equation (2.6) that several extractions with small volumes of the extracting phase will be more efficient than one extraction utilizing all of the extracting phase.

$$[C_A] = [C_A^o]\left(\frac{V_A}{KV_B + V_A}\right)^n \tag{2.6}$$

$[C_A]$ and $[C_A^o]$ are the final and original concentrations of the material in phase A, K is the distribution coefficient, V_A is the volume of phase A, V_B is the volume of the extracting phase used in each extraction step, and n is the number of extraction steps.

Continuous liquid-liquid extractions may be required when particularly unfavorable distribution coefficients are encountered. Continuous liquid-liquid extractors suitable for use with lighter- and heavier-than-water organic phases are shown in Figures 2.12 and 2.13. More elaborate continuous extractors are commercially available. (One source is Aldrich Chemical Company, Inc., 940 West Saint Paul Avenue, Milwaukee, WI 53233.)

Continuous liquid-solid extractions can be carried out with a Soxhlet extractor (see Figure 2.14). The finely ground solid is placed in the extraction shell, and fresh solvent is allowed to percolate through the sample, the extracted material accumulating in the solvent distillaton flask.

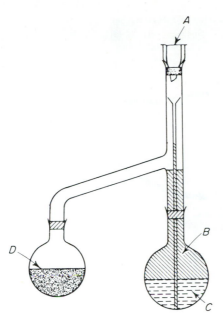

Figure 2.12 Diagram of a simple continuous extractor for use with lighter-than-water extracting phase: (*A*) condenser, (*B*) lighter-than-water extracting phase, (*C*) water phase being continuously extracted, and (*D*) distillation reservoir. The solvent is vaporized from *D*, condensed in the condenser *A*, and conducted to the bottom of the extracting flask by the small inner tube. The organic phase percolates up through the water phase and returns to the distillation flask, where the extracted material concentrates.

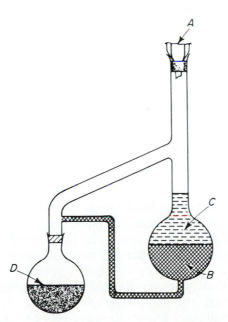

Figure 2.13 Diagram of a continuous extractor for use with heavier-than-water extracting phases: (*A*) condenser, (*B*) heavier-than-water extracting phase, (*C*) water phase being continuously extracted, and (*D*) distillation reservoir. The extracting solvent is vaporized from the distillation flask, condensed in the condenser, then drips down through the phase being extracted. The extracting phase returns to the distillation flask by the small return tube, and the material extracted concentrates in the distillation flask.

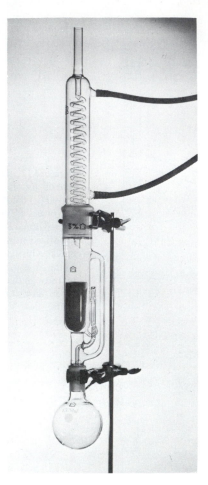

Figure 2.14 Soxhlet extractor for continuous liquid-solid extractions. (Photograph by courtesy of the Kontes Glass Company, Vineland, N.J.; photograph by Bruce Harlan, Notre Dame, Ind.)

2.6 CHROMATOGRAPHY

2.6.1 Introduction

Chromatography can be defined as the separation of a mixture into various fractions by distribution between two phases, one phase being stationary and essentially two-dimensional (a surface), and the remaining phase being mobile. The underlining principle of chromatography is that different substances have different partition coefficients between the stationary and mobile phases. A compound that interacts weakly with the stationary phase will spend most of its time in the mobile phase and move rapidly through the chromatographic system. Compounds that interact strongly with the stationary phase will move very slowly. In the ideal case, each component of a mixture will have a different partition coefficient between the mobile and stationary phases, and consequently each will move through the system at a different rate, resulting in complete separations. Various types of chromatography are possible, depending on the physical states of the phases. Employing a gas as the mobile phase is termed *gas chromatography* (*gc*) or *vapor phase chromatography* (*vpc*). Separations using gas chromatography involve vapor phase versus adsorption and/or solution equilibria. *Liquid chromatography* (*lc*) refers to any chromatographic process that employs a mobile liquid phase.

Liquid chromatography is inherently more versatile in that it is not limited by sample volatility, and separations are the result of specific interactions of sample molecules with *both* the stationary and mobile phases. (Such specific interactions are absent in the mobile phase of gc.) There are four basic types of liquid chromatography. *Liquid-liquid*, or *partition chromatography*, involves a stationary phase of an immiscible liquid coated on particles of an inert support, or, in some special cases, a stationary phase chemically bonded to the support surface. *Liquid-solid*, or *adsorption chromatography*, involves partition between solution in the mobile liquid phase and adsorption on high-surface-area particles acting as the stationary phase. *Gel*, or *exclusion chromatography*, employs stationary phases that possess cavities or pores of rather uniform size into which molecules of appropriate size can enter and be retained. Molecules of size inappropriate to the cavities move rapidly with the mobile phase. The two basic types of materials used in exclusion chromatography are molecular sieves (inorganic silicates with pore sizes of 3 Å to 10 Å) and organic polymeric gels (e.g., Bio-Gels-P and Sephadex Gels, which have an operating range of molecular weights of 200 to 400,000). Gel chromatography is an important technique in biochemistry. The stationary phase in *ion-exchange chromatography* is an organic resin substituted with sulfonic or carboxylic acid, amine, or quaternary ammonium groups that act as ion-exchange donor groups. The position of ion-pairing equilibria involving ionic samples and the ionic groups on the resin is controlled by changing the pH or some ion concentraton in the mobile phase. Ion exchange resins find common use in water purification, in water-softening operations, in amino acid analyses, and in inorganic separations.

All types of chromatography are useful for analytical purposes. Under appropriate conditions, all types of chromatography can be used for preparative scale separations. In every type of chromatography there are three elements to be considered:

Load (sample size)
Resolution (relative separation of components)
Speed

Of course, it would be ideal if all three elements could be maximized so that complete separation of samples of any desired size could be quickly achieved. In practice, generally any two of these elements can be maximized at the expense of the third. For routine analytical work, resolution and speed are maximized at the expense of load. In preparative scale separations load and speed can be maximized, but then separations are usually incomplete. Complete separations of large samples can be achieved, but the overall operation is likely to be slow and tedious, and may involve the use of large quantities of solvent that must be distilled for reuse, or discarded. Keep these factors in mind as we examine the chromatographic techniques used in organic chemistry in the sections that follow.

2.6.2 Adsorption Chromatography

The separation of the components of a mixture by adsorption chromatography depends on adsorption-desorption equilibria between compounds adsorbed on the surface of the solid stationary phase and dissolved in the moving liquid phase. The extent of adsorption of a component depends on the polarity of the molecule, the activity of the adsorbent, and the polarity of the mobile liquid phase. The actual separation of the components in a mixture is dependent on the relative values of the adsorption-desorption equilibrium constants for each of the components in the mixture.

In general, the more polar the functional groups of a compound, the more strongly it will be adsorbed on the surface of the solid phase. Table 2.4 lists typical classes of organic compounds in the order of elution (increasing polarity). Minor alterations in this series may occur if a functional group is highly sterically protected by other less-polar portions of the molecule.

The activity of the adsorbent depends on its chemical composition, the particle size, and the porosity of the particles. Table 2.5 lists the more common adsorbents in the usual order of increasing activity.

The choice of the proper adsorbent will depend on the types of compounds to be chromatographed. Cellulose and starch are used primarily with very labile, polyfunctional plant and animal products. Silica gel is the adsorbent with the widest variety of organic applications. It is available over a large range of particle sizes (10 to 150 μm) and porosities. An advantage of a silica gel column is that it is translucent in chloroform and in chloroform containing up to 5% ethanol. Colorless compounds

TABLE 2.4 Compound Elution Sequence

Saturated hydrocarbons
Alkenes
Aromatic hydrocarbons
Halogen compounds
Sulfides
Ethers
Nitro compounds
Aldehydes, ketones, esters
Alcohols, amines
Sulfoxides
Amides
Carboxylic acids

TABLE 2.5 Adsorbents for Adsorption Chromatography

Cellulose Starch Sugars Silicic acid (silica gel) Florisil (magnesium silicate) Aluminum oxide (alumina) Activated charcoal	General order of increasing activity

can often be observed as opaque bands as they move down the column. Above 5% ethanol in chloroform the column becomes opaque and this advantage is lost. Many chromatograms can be started with pure chloroform and eluted with chloroform containing increasing amounts of ethanol. Florisil has properties similar to those of silica gel, and it can be used for the separation of most functional classes, including carboxylic acid derivatives and amines.

Alumina, a widely used adsorbent, can be obtained in three forms: acidic, basic, and neutral. Acidic alumina is an acid-washed alumina giving a suspension in water with pH of approximately 4. This alumina is useful for the separation of acidic materials such as acidic amino acids and carboxylic acids. Basic alumina (pH of approximately 10) is useful for the separation of basic materials such as amines. Neutral alumina (pH of approximately 7) is useful for the separation of nonacidic and nonbasic materials.

In addition to varying the acid-base properties of the adsorbent, the activity of alumina can be varied by controlling the moisture content of the sample. For example, alumina can be prepared as activity grades I through V (Brockman scale), in which the activity decreases as the activity grade number increases. In general, the more active grades of alumina should be used when the separation of the components of a mixture is difficult. Detailed procedures are available for the preparation and standardization of the various types and grades of alumina (see reference 11 in Section 2.7). Acidic, basic, and neutral activity I aluminas are commercially available and can be converted to the other activity grades by the addition of 3% (activity II), 6% (III), 10% (IV), and 15% (V) by weight of water.

The use of the more active absorbents, particularly alumina, may lead to the destruction of certain types of compounds during chromatography. Owing to the presence of water in the lower-activity grades of alumina, many esters, lactones, and acid halides may undergo hydrolysis, the resulting carboxylic acids being very strongly adsorbed. The low-molecular-weight aldehydes and ketones undergo extensive aldol and ketol condensations on the surface of the more active adsorbents. The correct choice of adsorbent will be contingent on some knowledge of the types of compounds to be separated. Even then, a small trial chromatographic separation on a small portion of the sample is recommended to determine the proper conditions for separation of the remainder of the sample.

In addition to controlling the pH of the adsorbent, the properties of the adsorbent can be altered by coating the adsorbent with various chemicals. For example,

TABLE 2.6 Eluotropic Series

Petroleum ether	
Cyclohexane	
Carbon tetrachloride	
Benzene	
Dichloromethane	
Chloroform (alcohol free)	
Diethyl ether	
Ethyl acetate	Increasing
Pyridine	polarity
Acetone	
1-Propanol	
Ethanol	
Methanol	
Water	
Acetic acid	

silver nitrate-coated alumina, prepared by dissolving 20% by weight silver nitrate in aqueous methanol and slurrying with the alumina, followed by removal of the solvent, is particularly useful for the separation of alkenes. A stationary liquid phase, for example, water, ethylene glycol, or a low-molecular-weight carboxylic acid, can be applied to some adsorbents. Separations employing such preparations are termed liquid-liquid chromatographic separations.

The recovery of the material from a chromatogram can be accomplished in either of two ways. The individual components of the mixture can be separated and developed into distinct bands on the column by allowing a solvent, or solvent mixture, of sufficient polarity to proceed down the column affecting the separation. If the individual bands can be discerned by their color, fluorescence under the influence of ultraviolet light, or reaction with a colored indicator, the developed chromatogram can be extruded from the column and the separated components recovered by leaching cut portions of the column with a solvent of high polarity. The general procedure is, however, to flush the column continually with solvents of increasing polarity to remove each component individually. This type of procedure is generally referred to as elution chromatography.

The solvents generally used as eluents are listed in Table 2.6 in the order of increasing polarity. Often a pure solvent or single solvent system (e.g., 5% diethyl ether in hexane) can be used to elute all components. In other cases gradient solvent systems are used. Since the entire separation is dependent on the establishment and maintenance of equilibrium conditions, in gradient elution the polarity of the solvent system is gradually increased by the slow increase of the concentration of a more polar solvent. The rate of change of solvent polarity will depend upon the similarity or dissimilarity of the components in the mixture to be separated. If closely related compounds are to be separated, for example, isomeric alkenes or alcohols, the change in solvent composition should be very gradual. The separation of compounds having distinctly different functional groups can often be accomplished by a much more rapid change in solvent composition.

The choice of eluent solvents will depend on the type of adsorbent used and the nature of the compounds to be chromatographed. The solvents must be of high purity. The presence of traces of water, alcohols, or acids in the lesser polar solvents will alter the adsorption activity of the adsorbent.

In many respects, the application of chromatographic separation procedures is more an art than a science. Experience gained in the use of these techniques leads to a more efficient use of these techniques in future work. When faced with a separation problem involving a totally unknown mixture, it is usually best to carry out a crude and rapid trial chromatographic separation with a small portion of the mixture, using the information and experience gained in the trial run to carry out a more efficient separation on the remainder of the material. Such a procedure usually results in a saving of time and material. Thin-layer chromatography (Section 2.6.4) can often serve as an excellent guide to the conditions for column chromatography. For a discussion of the relationship between thin-layer and column chromatography see Section 2.6.4.

Techniques of Column Chromatography

The success of an attempted chromatographic separation depends on the care exercised in the preparation of the column and in the elution procedure. Figure 2.15 shows a typical adsorption chromatography column. The container is usually a glass tube provided with some means of regulating the flow of eluent through the column (stopcock preferably made of Teflon). If a ground-glass stopcock is used to control the rate of flow of solvent, all lubricating grease should be removed. Stopcock grease is quite soluble in many of the organic solvents used for elution purposes and may be leached from the stopcock, contaminating the material being eluted from the column.

The actual packing of the column is very important. The adsorbent must be uniformly packed, with no entrainment of air pockets. A plug of glass wool or cotton is placed in the bottom of the column, and a layer of sand is added on top of the plug to provide a square base for the adsorbent column. Failure to provide a square base for the column may well lead to elution of more than one fraction at a time if the separation is not very great, as is illustrated in Figure 2.16.

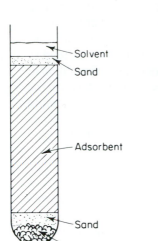

Solvent

Sand

Adsorbent

Sand

Glass wool plug

Stopcock

Figure 2.15 Construction of a chromatography column.

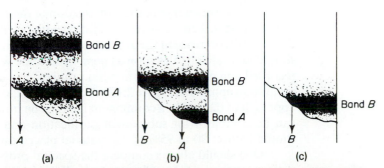

Band B

Band A

A

(a)

Band B

Band A

B

A

(b)

Band B

B

(c)

Figure 2.16 Elution of successive bands from a chromatography column that is not square at the bottom. (a) Elution of A. (b) Elution of A and B. (c) Elution of B.

The packing of the column can be accomplished in several ways. The adsorbent can be dry-packed, the solvent being added after the adsorbent. However, this method is generally not suitable because air pockets become lodged within the column and are difficult to remove; in addition, some adsorbents undergo an expansion in volume when wetted with a solvent, thus bursting the column (for example, silica gel and many ion-exchange resins). The recommended method involves filling the container with petroleum ether, or the least polar solvent to be used, and then adding the adsorbent in a continuous stream until the desired amount of adsorbent has been added. The tube should be maintained in a vertical position to promote uniform packing. Adsorbents that are not of uniform size and possess a low density tend to produce columns in which layering of large and small particles occurs when filled in this manner. Florisil and silica gel are notorious in this respect. In such cases, the adsorbent is slurried with the solvent, and the slurry is rapidly added to the empty column. This usually prevents segregation of particle sizes in the column. A layer of sand is added to the top of the column to protect the top of the column from being disturbed during the addition of solvent during the elution process.

The quantity of adsorbent and the column size required will depend on the type of separation to be carried out. Generally 20 to 30 g of adsorbent is required per gram of material to be chromatographed. Ratios as high as 50 or 100 to 1 may be required in cases when the components of a mixture are all quite similar. The height and diameter of the column are also important. Too short a column may not provide a sufficient length of column to effect the separation. In general, a height-to-diameter ratio of 8:1 to 10:1 is recommended.

After the column has been prepared, the solvent level is lowered to the top of the column by draining from the bottom of the column. The material to be separated is dissolved in the least polar solvent where solubility can be attained (10 to 15 mL/g of material) and carefully added to the top of the column. Elution is begun with a solvent system corresponding to the composition of the solvent used to dissolve the sample. Occasionally some highly polar components present in mixtures may require a solvent system of such highly polar character for complete solubility that the lesser polar components are immediately eluted from the column with no separation being achieved. In such cases, the sample may have to be added to the column as a two-phase system; however, this may lead to problems of column congestion.

The rate of change of the composition of the eluting solvent will depend on the type of mixture being separated. Elution with one solvent composition should continue until a decrease in the amount of material being eluted is noted. The fractions should be evaporated as soon as possible and the weight recovered in each fraction recorded and plotted versus fraction number. A typical plot of a chromatographic separation appears in Figure 2.17. A decrease in the weight per fraction indicates the approach of the end of a component band. The fractions appearing under one peak can be combined for subsequent purification and identification. In general, one should recover 90 to 95% of the material placed on the chromatographic column.

One should always compare the spectral properties (such as the nuclear magnetic resonance spectra, Chapter 6) of all the fractions eluted off a column with the

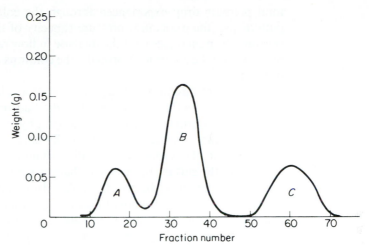

Figure 2.17 Typical plot of weight versus fraction number for a chromatographic separation.

spectral properties of the material originally placed on the column to make sure that all the components that were placed on the column were eluted unchanged. Occasionally, compounds may undergo chemical change on the highly active surface of the adsorbent.

2.6.3 Medium- and High-Pressure Liquid Chromatography

One of the limiting factors that determines the degree of resolution obtained in a chromatographic separation on a per unit weight of sample size is the total surface area of the adsorbent. As one decreases the particle size of the adsorbent, the total surface area greatly increases. However, owing to the closer packing of the smaller-sized particles of the adsorbent, the rate of flow of the mobile phase markedly decreases. There becomes a point at which the rate of flow of the mobile phase is so slow that the chromatographic process requires too much time to complete. There are two ways in which the rate of flow of the mobile phase may be increased: (1) Apply pressure to the mobile phase at the top of the column and (2) apply a vacuum at the outlet of the column. Both techniques find use.

Both medium- and high-pressure liquid chromatography involve applying pressure at the inlet (top) to the column and require special pumps and columns. Medium-pressure chromatography, operating at up to 300 pounds per square inch (psi), can be carried out in specially constructed, thick-walled glass columns. These columns vary in size from 15 to 40 mm in diameter and 24 to 44 cm in length. The particle size of the adsorbent, usually silica gel, is generally in the $40-63$ μm range.

High-pressure liquid chromatography (HPLC) may involve pressures up to 6000 pounds psi. The columns are generally made of stainless steel and range in diameter from $\frac{1}{8}$ to $\frac{1}{4}$ in. The maximum length of the column is determined by the total pressure drop experienced through the column, which cannot exceed 6000 pounds psi (the maximum pressure capacity of the usually available pump). The pressure drop increases with the increase in flow rate and the viscosity of the solvent used to elute the column. Normally, the columns range from 1 to 3 ft (up to 250 cm) in length.

Silica gel is a commonly used adsorbent available in 3-, 5-, 7-, and 10-μm-diameter particles. As the separation is based on adsorption, the elution sequence is the same as with normal column chromatography given in Table 2.4. The characteristics of the adsorbent can be dramatically changed by chemically bonding various substances to the surface of the silica. The bonding of long-chain hydrocarbon residues to the surface of the silica changes the surface properties of the silica from highly polar and adsorbing to nonpolar. In this case highly polar solvents, such as mixtures of water and methanol or acetonitrile, are used to elute the column, in which case the elution sequence in Table 2.4 is reversed; that is, polar compounds are eluted first followed by those with decreasing polarity. This type of chromatography is called "reversed phase" chromatography, and it finds extensive use in the separation of highly polar, functionalized compounds. The bonding of a chiral substrate to the silica results in a chiral environment that can be used to separate enantiomers. The student is encouraged to consult the catalogs of manufacturers of chromatographic equipment and supplies for detailed descriptions of column materials and their applications.

High-pressure liquid chromatographs are generally equipped with ultraviolet absorption and/or refractive index detectors, with the choice of the detector being determined by the properties of the components being separated and the solvent being used to elute the column. Elaborate computer-controlled systems are available for controlling the rate of flow of the mobile phase, as well as its composition (gradient elution). A schematic of a high-pressure liquid chromatograph is shown in Figure 2.18.

High-pressure liquid chromatography offers major advantages over traditional column chromatographic techniques, such as an increase in the ability to achieve difficult separations, speed, and the use of smaller volumes of solvent. The columns

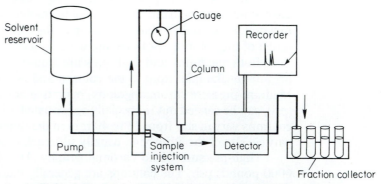

Figure 2.18   Schematic of a high-pressure liquid chromatograph.

are generally reusable providing that not too much "junk" becomes irreversibly adsorbed on the stationary phase. It is always advisable to subject the sample to a prefiltration through a short, inexpensive, disposable filter containing the adsorbent used in the column. At the end of a separation the column is flushed with a polar solvent to remove polar components that may have been present in the sample. Then flushing with a nonpolar solvent will return the column to its original state for reuse.

The size of a sample that can be preparatively separated depends on the size of the column and the degree of resolution attained on the column. Preparative separations of 0.5 to 1.0 g are possible, although generally only quantities of less than 100 mg can be successfully separated.

A technique known as "flash chromatography"[1] may use either a slight positive pressure at the top of the column or a reduced pressure at the bottom. In both methods specific volumes of solvent are added to the top of the column and the solvent is forced or drawn through the column, allowing the column to "run" dry. The eluted fraction is removed, and a second portion of solvent of the same or different composition is added and the process is repeated. The required apparatus is rather simple, consisting of a column with a sintered glass disk in the bottom and equipped with a stopcock that is carefully packed with the finely divided adsorbent, a solvent reservoir, and a pressurized-gas inlet connection (see Figure 2.19). Alternatively, the adsorbent can be placed in a funnel with a sintered glass bottom that is connected to a vacuum flask. The vacuum is applied and the solvent is sucked through the adsorbent. This technique has been called "dry-column" chromatography.[2]

2.6.4 Thin-Layer Chromatography

Thin-layer chromatography (TLC) is a special application of adsorption chromatography in which a thin layer of the adsorbent supported on a flat surface is utilized instead of a column of the adsorbent. Elution, more properly referred to as the development of the chromatogram, is accomplished by the capillary movement of the solvent up the layer of the adsorbent. Unfortunately, generally only one solvent, or solvent mixture, can be used to develop a single plate instead of the use of gradient solvent systems as in column chromatography.

The most commonly used adsorbents in TLC are silica gel and alumina. The adsorbents used in TLC are of much smaller particle size than those used in column chromatography, providing for a much greater surface area and greater resolving capability. Ready-to-use TLC plates on which the adsorbent is layered on a thin sheet of plastic or aluminum are commerically available. These plates generally come in 5 cm × 20 cm and 20 cm × 20 cm sizes and are easily cut to the desired dimensions with scissors.

[1] W. C. Still, M. Kahn, and A. Mitra, *J. Org. Chem.,* **43,** 2923 (1978).

[2] L. M. Harwood, *Aldrichimica Acta,* **18,** 25 (1985).

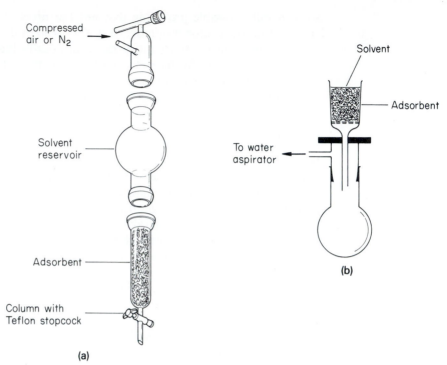

Figure 2.19 (a) Typical apparatus for positive-pressure flash chromatography. The individual pieces of the apparatus are held together with appropriate clamps. (b) Simple apparatus for vacuum flash chromatography.

The choice of eluent for TLC will depend on the adsorbent used and the types of compounds to be separated. The same general rules for column chromatography also apply to TLC. However, since only a single solvent mixture can be used for each plate, it may be necessary to try several solvent combinations in order to find the one that gives the best separation.

Visualization of the chromatogram to locate the position of each of the components of a mixture depends on the type of molecules present in the mixture. If all the components of the mixture are colored, then visual inspection of the developed plate is sufficient to locate the spots. If this is not the case, the two most commonly used methods for visualizing compounds are the use of an adsorbent that contains a fluorescent dye and the use of iodine vapor. With the former technique, the developed plate is irradiated with a "black" light (a low-intensity ultraviolet lamp), the fluoresence of the dye being quenched by the compounds on the plate producing a dark spot against a light blue background of the fluorescing dye. This method requires that the components of a mixture all are capable of quenching the fluoresence of the dye. This is not always the case, particularly with hydrocarbons and other relatively nonpolar compounds. The use of iodine vapor for visualization depends on the ability of the components of the mixture to form charge-transfer complexes with iodine. Compounds that do not contain π or Lewis basic functions do not form charge-transfer complexes with iodine.

The value of the use of TLC lies in the relatively small amount of time and material required to carry out an analysis, generally on the order of 10 minutes or less and less-than-milligram quantities of material. TLC is very useful for monitoring the progress of reactions, in the detection of reaction intermediates, and for checking the efficiency of larger-scale separations such as chromatographic separations or fractional distillations.

TLC can be used as a tentative means of identification of a compound. If plate conditions are kept constant, a compound will progress up the plate at the same rate relative to an added standard or with respect to the solvent front (the position on the plate to which the solvent has migrated). The relative displacement is referred to as

$$R_x = \frac{\text{displacement of the compound}}{\text{displacement of the standard}}$$

or

$$R_f = \frac{\text{displacement of the compound}}{\text{displacement of the solvent front}}$$

For identification purposes, a comparison of a known and unknown should be carried out under identical conditions with a number of different solvent systems and, if possible, different adsorbents. R_x and R_f values from the literature are not reliable enough for use in final identification; they should be used only as general guides. Whenever possible, a direct comparison between an unknown and a possible known compound should be made.

TLC can be used in a preparative manner by using thicker layers (1 – 3 mm in thickness) of the adsorbent on the supporting plate, generally a glass plate. The mixture to be separated is applied in a band along the bottom of the plate, and the plate is developed. After visualization, the bands across the plate are individually scraped off from the plate and are extracted with an appropriate solvent.

Finally, the use of TLC is often very useful in helping to select a proper solvent system for larger-scale separations by column chromatography.

Techniques of Thin-Layer Chromatography

Application of the sample. For analytical separations, a dilute solution of the sample is prepared and one or more small drops of the solution are applied to the plate about $\frac{1}{4}$ in. from the bottom of the plate (see Figure 2.20). The plate is then placed in a small screw-capped bottle containing enough solvent to come up about $\frac{1}{8}$ in. above the bottom of the plate. (The solvent level in the container should never come above the point at which the sample was applied to the plate.) The plate is left undisturbed in the closed container until the solvent front, as indicated by the "wet"-looking portion at the top of the plate, reaches approximately $\frac{1}{4}$ in. from the top of the plate. The plate is then removed and allowed to dry in the open atmosphere.

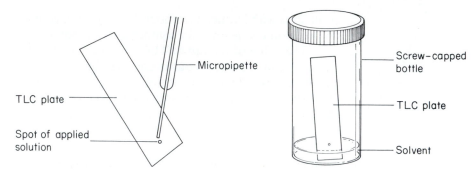

Figure 2.20 The application of a solution to a TLC plate (left) and development of the plate in a screw-capped bottle (right).

The plate is then visualized by irradiating with a "black" lamp or by placing the plate in a screw-capped bottle containing a few crystals of iodine.

Preparative-scale thin-layer chromatography.   Preparative-scale thin-layer chromatography can be carried out on commercially available plates or on plates prepared by oneself.   Preparative thin-layer chromatography is generally carried out on 5 cm × 20 cm or 20 cm × 20 cm plates.   A slurry of the adsorbent is prepared in a volatile solvent and is poured on the surface of the plate.   The slurry is spread evenly over the plate by the use of a broad spatula or by gently rolling the plate.   The plate is dried in an oven.   The sample is applied to the plate by the use of a capillary pipette, or a disposable pipette, in a band $\frac{1}{4}$ in. from the bottom of the plate.   Care must be exercised so one does not scrape and damage the surface of the plate.   The plate is developed as described above, and the bands indicating the presence of compounds are carefully cut and scraped from the plate.   The removed material is then extracted with an appropriate solvent to recover the organic compound.

2.6.5 *Thin-Layer Rotating Disk Chromatography*

A recent innovation in thin-layer chromatography is to carry out the thin-layer chromatography on a rapidly rotating disk of an adsorbent supported on a circular glass plate.   A slurry of the adsorbent containing a binder ($CaSO_4$) and an ultraviolet sensitive phosphor are poured onto a circular plate and allowed to dry.   The surface and edges of the plate are carefully trimmed to form a uniform circular, thin layer of the adsorbent 0.5 to 4.0 mm thick.   [This process can be carried out by using an old record player to support the plate while drying and by trimming the plate with edging tools provided with the instrument called the Chromatotron® (see Figure 2.21).]

The thin-layer disk is rapidly rotated and a solution of the mixture of substances to be separated is added to the center of the plate.   The eluent is slowly added to the center of the disk and migrates outward in a circular manner due to the centrifugal force created by the rapidly rotating plate.   The separation and elution of the bands is easily monitored by irradiating the disk with a long-wavelength ultraviolet light

Figure 2.21   The Chromatotron®; a thin-layer, rotating-disk chromatography apparatus. The eluting solvent is pumped onto the center of the rapidly rotating disk. The eluted sample drains from the bottom of the apparatus. (Reproduced by courtesy of Harrison Research, Inc., Palo Alto, Calif.)

source. Quantities of up to 0.5 g of material can be efficiently separated using this technique. Figure 2.21 shows such an apparatus, which is known as a Chromatotron®.

2.6.6   Paper Chromatography

Paper chromatography is somewhat similar to thin-layer chromatography except that high-grade filter paper is used as the adsorbent or solid stationary phase. In actual fact, paper chromatography is not strictly adsorption chromatography but a combination of adsorption and partition chromatography. A partitioning of the solute occurs between the water of hydration of the cellulose and the mobile organic phases. Paper chromatography is used primarily when extremely polar or polyfunctional compounds, for example, sugars and amino acids, are to be separated. Such materials cannot be chromatographed on more active adsorbents. The method is used as an analytical procedure much like analytical thin-layer chromatography; sample sizes for paper chromatography are usually in the range of 5 to 300 μg.

The selection of solvent systems for paper chromatography is very important. For a discussion of solvent systems, refer to the books listed in Section 2.7. In most cases the solvent system should contain some water. The solvent development can be accomplished by suspending the paper strip so that the end of the strip is immersed in a container of solvent (ascending paper chromatography) or by immersing the top of the strip in a small trough of solvent and allowing the solvent movement to occur by a combination of capillary action and gravity (descending paper chromatography). It is very important in all paper chromatography procedures that the development of the chromatogram occurs in an atmosphere saturated with solvent vapors. Otherwise, the solvent will evaporate from the paper faster than it is replaced by capillary action and separation of the components will not be achieved. After development is complete, the position of the solvent front is noted and the paper strip is allowed to dry. The paper is then sprayed with a visualizing reagent. R_f and R_x values are calculated.

Test-Tube Technique for Paper Chromatography

As in thin-layer chromatography, both elaborate and simple techniques and equipment can be employed in paper chromatography. By far the simplest technique involves the use of an ordinary test tube as the developing chamber; the method can be recommended for most identification work or, at least, for exploratory work in determining conditions for using larger strips or sheets.

A 6- or 8-in. test tube can be used. The paper strip cut from Whatman No. 1 filter paper can be suspended in the test tube by suspending it from a slit cut in the stopper, by putting a fold in the paper, or by cutting the paper with a broad end at the top, which will serve to suspend the paper in the tube (see Figure 2.22). Handle the paper only at the extreme top edge since fingerprints will contaminate the chromatogram.

Place a light pencil mark about 1 cm from the bottom of the strip. The sample is applied as a small spot with a micropipette (made by pulling a capillary) at the center of the line and allowed to dry. The sample size should be about 10 μg. A small amount of solvent is placed in the bottom of the tube by means of a pipette in such a

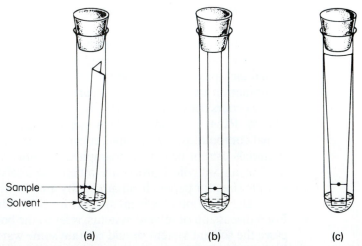

Sample
Solvent

(a)   (b)   (c)

Figure 2.22   Paper chromatography in test tubes. (a) Paper suspended in tube by folding. (b) Paper suspended from slit in rubber or cork stopper. (c) Paper beveled to suspend from sides of tube at the broad end.

manner that the sides of the tube above the surface of the solvent remain dry. The spotted and dried paper strip is carefully inserted into the tube so that the end dips into the solvent, but the surface of the solvent is below the pencil line. The tube is stoppered and allowed to stand until the solvent ascends near the top of the strip. This process may take up to 3 hours. The paper is removed with forceps and the solvent front marked. The strip is allowed to dry and the compounds are visualized with an appropriate spray.

2.6.7   Gas Chromatography

Gas chromatography (gc), sometimes called vapor phase chromatography (vpc), utilizes a moving gas phase with a solid or liquid stationary phase. With a solid stationary phase, the separation of the components of a mixture is dependent on the establishment of adsorption equilibria as in column adsorption chromatography. When a liquid stationary phase is supported on the surface of an inert solid, the separation is dependent on solubility equilibria between the components in the gas phase and liquid phase. Herein lies the great utility of gc. Proper choices of the stationary liquid phases allows the experimentalist to vary the sequence of elution of compounds in a mixture from the column greatly.

Gas chromatography has developed into one of the most powerful analytical tools available to the chemist. The technique allows separation of extremely small quantities of material, of the order of 10^{-4} to 10^{-6} g. The quantitative analysis of mixtures can be readily accomplished (as discussed in greater detail later in this section). The possibility of using long columns, producing a great number of theoretical plates, increases the efficiency of separation. The technique is applicable over a wide range of temperatures (-70 to $+500°C$), making it possible to chromatograph materials covering a wide range of volatilities. Finally, gas chromatographic analysis requires very little time compared with most other analytical techniques.

The instrumentation involved may be quite simple or may contain a high degree of sophistication and complexity. The basic components required are illustrated in Figure 2.23. A relatively high-pressure gas source, 30 to 100 psi, is required to provide the moving gas phase. The gas is introduced into a heated injector block.

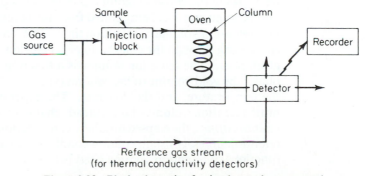

Figure 2.23   Block schematic of a simple gas chromatograph.

The sample is introduced into the gas stream in the injector block by means of a microsyringe forced through a syringe septum. Liquids can be injected directly, whereas solids are generally dissolved in a volatile solvent. The injector block is maintained at temperatures ranging from room temperature to approximately 350°C. The operating temperature is chosen to ensure complete vaporization of all components present in the mixture. The temperature of the injector, or the column, need not be higher than the boiling point of the least volatile substance. The partial vapor pressure of most substances, in the presence of the high-pressure carrier gas, is sufficient to ensure complete vaporization. It must be kept in mind that high injector block temperatures may lead to decomposition of heat-sensitive compounds.

The gas stream, after leaving the injector block, is conducted through the chromatographic column, which is housed in an oven. If all the retention times (see later discussion for definition) of the components are quite similar, the column can be maintained at a constant temperature (isothermal operation). If the retention times are greatly different, it may be desirable to increase the temperature of the column during the chromatographic process slowly. Instruments are available for constant-temperature and/or variable (programmed)-temperature operations. The length of the column required will depend on the difficulty of separation of the components in a given mixture; this is usually determined by trial runs. Packed columns are usually 0.25 in. in diameter and 2 to 50 ft in length. Normal column lengths are usually 6 to 10 ft; the longer columns are required for the more difficult separations. Larger diameter columns, up to 2 in., are used for preparative purposes (collection of the individual fractions). Capillary columns of up to 300 ft in length are also available.

The effluent gas stream containing the separated components is conducted through a detector, which measures some physical phenomenon and sends an electronic signal to a strip chart recorder. Many types of detectors are available. These include thermal-conductivity devices and ionization devices. The actual interpretation of the data recorded by the strip chart recorder depends on the type of detector used. The types of detectors and the method of handling the data will be discussed later.

The strip chart recorder provides a chromatogram (not a spectrum!) such as that illustrated in Figure 2.24. The first question to be asked is: What do the individual peaks in the chromatogram represent? First of all, the peak represents the detection of some material being eluted from the column. The peak can be characterized by a retention volume or retention time. The *retention volume* is the total volume of gas passing through the column to effect elution of the compound responsible for a given peak. The retention volume is a function of the volume of stationary liquid phase, the void volume of the column (volume of gas in the column at a given time), the temperature, and the flow rate. The experimentalist generally does not determine retention volumes but, instead, characterizes the peaks with respect to their *retention times,* the elapsed time between injection and elution. In the comparison of retention times, constant column conditions are implied, those conditions being the functions of retention volume cited before.

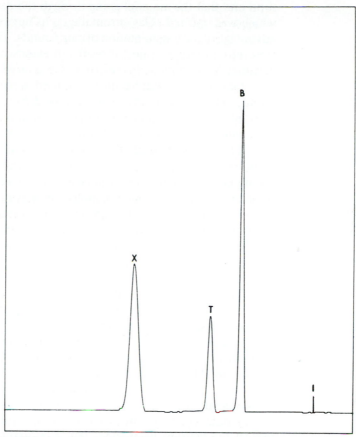

Figure 2.24   Gas-liquid chromatogram of benzene (*B*), toluene (*T*), and xylene (*X*) on a $5\frac{1}{2}$-ft 20% Carbowax 20M column at 140°C. The small peak labeled *I* indicates the point of injection of the sample.

The identification of an unknown giving rise to a peak in a chromatogram can be tentatively inferred by the comparison of retention times of the unknown with a possible known on a number of columns differing widely in polarity. Additional evidence can be obtained by recording the chromatogram of an admixture of the unknown and known samples. A single peak whose appearance (symmetry) is the same as the unknown and known samples indicates the possible identity of the unknown and known samples. A rigorous identification of an unknown cannot be made by comparison of retention times since it is very conceivable that many other materials may possess similar elution properties.

Rigorous identification of the material giving rise to a peak can be accomplished by collecting the fraction as it emerges from the exit port and then characterizing the material by its physical properties. However, it must be remembered that isomerizations and fragmentations can and do occur in the injector block and column and that the material going through the detector may not be the same as that which was injected. Gas chromatographs have been developed specifically to take advantage of the fragmentation of compounds. Injection of highly nonvolatile materials into a pyrolysis injector results in extensive cracking of the material, and the chromatogram obtained is referred to as a *cracking pattern.*

Gas chromatographs can also be used in conjunction with other types of instruments. The emerging gas stream can be directed through an infrared cell of a fast-scan infrared spectrometer, or directly into a mass spectrometer. Such procedures eliminate the necessity of collecting the individual fractions.

The area of the peak of a gas chromatogram represents a quantitative measurement of some physical property of the material going through the detector. For a given material, the response will be linear with concentration; however, the physical constant for a series of different molecules may not have the same absolute value, and the relative response of one compound with respect to another will not be directly proportional to the partial pressures of the materials in the eluent gas stream. The relative response of different compounds depends on the type of detector used, and appropriate corrections in calculations must be made if quantitative data are to be derived.

The effluent gas stream containing the separated components is conducted through a detector that measures some property of the gas stream and sends an electronic signal to a strip chart recorder. The two most commonly used types of detectors are the thermal conductivity and flame ionization detectors. The thermal conductivity detector measures changes in the thermal conductivity of the effluent gas stream. Helium is used as the carrier gas. (The use of nitrogen as the carrier gas may result in some of the peaks appearing as negative peaks.) Thermal conductivity detectors must be used when preparative gas-liquid chromatographic separations are carried out. The flame ionization detector actually combusts the compounds coming off the column producing ions, the quantity of which are measured by the detector. *Flame ionization detectors cannot be used in preparative gas-liquid chromatographic separations.* The response of either detector is not directly related to the amount of material coming off the column. The thermal conductivities of organic compounds vary greatly, and thus their responses will not be linear with the amount of material present. Flame ionization detectors measure the amounts of ions formed during the combustion of compounds, again the number of which will vary from compound to compound. For quantitative analyses, standard mixtures must be used to calibrate the responses of the compounds present, and corrections must be made to transform area ratios into weight or molar ratios. An example of this approach is given in the following sample calculation.

Example

Reaction of A (2.50 g, 134 mol. wt.) with reagent X gave 2.40 g of a product mixture containing A and two products B (148 mol. wt.) and C (162 mol. wt.). The gas chromatogram of the product mixture showed peaks for A, B, and C, with areas of 77, 229, and 276, respectively. Addition of 36.4 mg of B to a 120.2-mg aliquot of the reaction mixture produced a gas chromatogram with areas of 111, 567, and 396 for A, B, and C. A mixture of 52.7 mg of A, 47.3 mg of B, and 63.2 mg of C produced a gas chromatogram with areas of 287, 275, and 423, respectively.

Calculate the yields of B and C based on reacted starting material according to the following equation:

$$A + X \longrightarrow B + C$$

Solution

$$\text{Response ratio } \frac{A}{B} = \frac{\text{wt. A/wt. B}}{\text{area A/area B}} = \frac{52.7/47.3}{287/275} = \frac{1.113}{1.044} = 1.067 \left(\frac{A}{B} \right)$$

$$\text{Response ratio } \frac{C}{B} = \frac{\text{wt. C/wt. B}}{\text{area C/area B}} = \frac{63.2/47.3}{423/275} = \frac{1.337}{1.540} = 0.867 \left(\frac{C}{B} \right)$$

The area of B in admixture chromatogram due to added B is

$$\text{Area B}_{\text{admix}} - \text{area A}_{\text{admix}} \cdot \frac{\text{area B}_{\text{orig}}}{\text{area A}_{\text{orig}}} = 567 - 111 \cdot \frac{229}{77} = 567 - 330 = 237$$

From the proportion

$$\frac{\text{wt. B}_{\text{added}}}{\text{area B}_{\text{added}}} = \frac{\text{wt. B}_{\text{orig}}}{\text{area B}_{\text{orig}}}$$

we have

$$\frac{36.4 \text{ mg B}}{237} = \frac{\text{wt. B}_{\text{orig}}}{330}$$

$$\text{wt. B}_{\text{orig}} \text{ in aliquot} = \frac{36.4 \cdot 330}{237} = 50.7 \text{ mg}$$

$$\text{wt. of B in total sample} = \frac{50.7 \text{ mg}}{120.2 \text{ mg}} \cdot 2.40 \text{ g} = \underline{1.014 \text{ g of B}}$$

$$\text{wt. of A in sample} = \frac{\text{area A}}{\text{area B}} \cdot \text{response ratio} \left(\frac{A}{B} \right) \cdot \text{wt. B}$$

$$= \frac{77}{229} \cdot 1.067 \cdot 1.014 \text{ g A} = \underline{0.364 \text{ g of A}}$$

$$\text{wt. of C in sample} = \frac{\text{area C}}{\text{area B}} \cdot \text{response ratio} \left(\frac{C}{B} \right) \cdot \text{wt. B}$$

$$= \frac{276}{229} \cdot 0.867 \cdot 1.014 \text{ g C} = \underline{1.060 \text{ g of C}}$$

The percent yields are calculated in the usual manner.

In this particular case, the entire sample proves to be volatile. This may not be true in all cases.

A similar approach would be used if an internal standard had been added instead of employing additional B.

Column Solid Supports and Stationary Liquid Phases
for Gas-Liquid Chromatography

The great utility of gas chromatography lies in the large number of combinations of solid supports and stationary liquid phases that are available for the preparation of columns. Table 2.7 lists some of the solid supports that are available. It should be noted that several of the entries are porous polymer beads which may be used without a liquid stationary phase. These materials are available in a variety of pore sizes and separate mixtures on the basis of molecular shape and size. Table 2.8 lists a number of common stationary liquid phases and the maximum temperatures at which they can be used. (The use of a column at a temperature above the indicated maximum will result in the evaporation of the stationary liquid phase from the column, which may then condense in the detector portion of the instrument causing potential damage to the detector.) Table 2.9 lists the type of stationary liquid phase that is useful for the separation of various classes of compounds. Chiral stationary liquid phases have also been developed which have proven useful for the separation of mixtures of enantiomers. The student is encouraged to consult the catalogs of manufacturers of gc equipment and supplies for more detailed descriptions of the solid supports and stationary liquid phases and their application.

The column packings are prepared by dissolving the liquid phase in a volatile solvent, usually methylene chloride or methanol, and slurrying with the solid support. The solvent is evaporated under reduced pressure, with continuous agitation of the slurry. The dried packing is poured into the column and firmly packed by vibration. The ends of the column are plugged with cotton or fine glass wool. Before use, the column must be purged by a stream of gas while being maintained at the maximum column operating temperature until equilibrium conditions are reached as indicated by a steady base line on the recorder. After each use, the column should be briefly purged at the maximum operating temperature to remove any relatively nonvolatile materials that may have been left on the column.

TABLE 2.7 Solid Supports

Name	Description
Chromosorb P	Highly adsorptive, pink calcined diatomaceous silica, pH 6 to 7
Chromosorb W	Medium adsorptive, white diatomaceous silica, which has been flux-calcined with sodium carbonate, pH 8 to 10
Chromosorb G	Dense, low adsorptivity, flux-calcined material, pH 8.5 (maximum liquid phase loading, 5%)
Fluoropak 80 Haloport F Chromosorb T	Fluorocarbon polymer, maximum usable temperature 260°C (reduces tailing of highly polar compounds)
Glass beads	Used only with very low amounts of liquid phase (0.5% and less)
Poropak Tenax Chromosorb 100 series	Porous beads of polymer available in varying pore sizes

TABLE 2.8 Examples of Liquid Stationary Phases

Number	Abbreviation	Trade Name or Description	Maximum Temperature (°C)
1	Apiezon L	Hydrocarbon grease	300
2	Apiezon M	Hydrocarbon grease	275
3	Carbowaxes (400–6000)	Polyethylene glycols	100–200
4	Carbowax 20M	Polyethylene glycol	250
5	DC-200	Dow Corning methyl silicone fluid	225
6	DC-550	Dow Corning phenyl methyl silicone fluid	225
7	DC-710	Dow Corning phenyl methyl silicone fluid	250
8	DEGS or LAC 728	Diethyleneglycol succinate	225
9	LAC 446	Diethyleneglycol adipate	225
10	Flexol Plasticizer 10-10	Didecyl phthalate	175
11	Nujol	Paraffin oil	200
12	QF-1-6500	Dow Corning fluorinated silicone rubber	260
13	SE-30	G.E. methyl silicone rubber	300
14	SE-52	G.E. phenyl silicone rubber	300
15	SF-96	G.E. fluoro silicone fluid	300
16	—	Silicone gum rubber	375
17	—	Silver nitrate-propylene glycol	75
18	TCP	Tricresyl phosphate	125
19	TCEP	1,2,3-Tris(2-cyanoethoxy)propane	180
20	THEED	Tetrakis(2-hydroxyethyl)ethylenediamine	130
21	Ucon polar	Polyalkylene glycols and derivatives	225
22	Ucon nonpolar	Polyalkylene glycols and derivatives	225
23	—	4,4'-Dimethoxyazobenzene	120–135*

* The stationary phase exists as liquid crystals, referred to as nematic or smectic phases depending on crystal orientations. The temperature range is limited to the regions where these phases exist. Other alkoxyderivatives of azobenzene have also been used [M. J. S. Dewar and J. P. Schroeder, *J. Am. Chem. Soc.*, **86,** 5235 (1964)].

TABLE 2.9 Suggested Column Uses

Class of Compound	Column Number (from Table 2.8)
Acetates	8, 10, 12, 19, 20, 21
Acids	6 (on acid-washed support), 16
Alcohols	3, 4, 5, 6, 16, 19, 20, 21, 22
Aldehydes	3, 4, 10, 11
Amines	3, 4, 6, 20 (on fluoropak or haloport or potassium hydroxide-coated support)
Aromatics	1, 2, 10, 16, 18, 19, 20, 21
Esters	3, 4, 8, 9, 10, 12, 16, 19, 20
Ethers	6, 9, 16, 18, 19
Halides	5, 10, 11, 18, 19
Hydrocarbons	1, 2, 6, 10, 16, 19, 20
Ketones	3, 4, 6, 10, 16, 18, 20
Nitriles	9, 18, 20
Alkenes	10, 17, 19, 20
Phenols	4, 12
Isomeric aromatics	23

2.7 REFERENCES

General

1. *Advances in Chromatography.* New York: Marcel Dekker, 1965.

2. Berg, E. W., *Physical and Chemical Methods of Separation.* New York: McGraw-Hill, 1963.

3. Sixma, F. L. J., and H. Wynberg, *A Manual of Physical Methods in Organic Chemistry.* New York: John Wiley, 1964.

4. Wiberg, K. B., *Laboratory Technique in Organic Chemistry.* New York: McGraw-Hill, 1960.

5. Weissberger, A., ed., *Technique of Organic Chemistry,* Vol. III. New York: Interscience, 1956.

Liquid Chromatography

6. Bobitt, J. M., A. E. Schwarting, and R. J. Gritter, *Introduction to Chromatography.* New York: Reinhold, 1968.

7. Brown, P. R., *High Pressure Liquid Chromatography: Biochemical and Biomedical Applications.* New York: Academic Press, 1973.

8. Hamilton, R. J. and P. A. Sevill, *Introduction to High Performance Liquid Chromatography,* 2nd ed. London: Chapman and Hall, 1982.

9. Heftmann, E., ed., *Chromatography: A Laboratory Handbook of Chromatographic and Electrophoretic Methods.* New York: Van Nostrand Reinhold, 1975.

10. Kirkland, J. J., ed., *Modern Practice of Liquid Chromatography.* New York: Wiley-Interscience, 1971.

11. Lederer, E., and M. Lederer, *Chromatography,* 2nd ed. Amsterdam: Elsevier, 1957.

12. Perry, S. G., R. Amos, and P. I. Brewer, *Practical Liquid Chromatography.* New York: Plenum, 1972.

13. Scott, R. P. W., ed., *Small Bore Liquid Chromatography Columns: Their Properties and Uses. Chemical Analysis,* Vol. 72. New York: John Wiley, 1984.

14. Simpson, C. F., *Practical High Performance Liquid Chromatography.* London: Heyden and Sons, Ltd., 1976.

15. Snyder, L. R., and J. J. Kirkland, *Introduction to Modern Liquid Chromatography.* New York: Wiley-Interscience, 1974.

16. Stock, R., and C. B. F. Rice, *Chromatographic Methods,* 3rd ed. London: Chapman and Hall, 1974.

Thin-Layer Chromatography

17. Bobitt, J. M., *Thin-Layer Chromatography.* New York: Reinhold, 1963.

18. Randerath, K., *Thin-Layer Chromatography.* New York: Academic Press, 1963.

19. Stahl, E., ed., *Thin-Layer Chromatography. A Laboratory Handbook,* 2nd ed. New York: Springer, 1969.

20. Truter, E. V., *Thin-Film Chromatography.* New York: Interscience Publishers, 1966.

Paper Chromatography

21. Block, R. J., E. L. Durrum, and G. Zweig, *Paper Chromatography and Paper Electrophoresis.* New York: Academic Press, 1958.
22. Hais, I. M., and K. Macek, eds., *Paper Chromatography.* New York: Academic Press, 1963.

Gas Chromatography

23. Burchfield, H. P., and E. E. Storrs, *Biochemical Applications of Gas Chromatography.* New York: Academic Press, 1962.
24. Crippen, R. C., *Identification of Organic Compounds with the Aid of Gas Chromatography.* New York: McGraw-Hill, 1973.
25. Grant, D. W., *Gas-Liquid Chromatography.* New York: Van Nostrand Reinhold, 1971.
26. Hardy, C. J., and E. H. Pollard, "Review of Gas-Liquid Chromatography," in *Chromatographic Reviews,* Vol. 1, M. Lederer, ed. Amsterdam: Elseveir, 1960.
27. Jennings, W. G., ed., *Applications of Glass Capillary Gas Chromatography; Chromatographic Science Series,* Vol. 15. New York: Marcel Dekker, 1981.
28. Kaiser, R., *Gas Phase Chromatography.* London: Butterworths Scientific Publications, 1963.
29. Keulemans, A. I. M., *Gas Chromatography,* 2nd ed., C. G. Verver, ed. New York: Reinhold, 1959.
30. Littlewood, A. B., *Gas Chromatography, Techniques and Applications.* New York: Academic Press, 1962.
31. Nogare, S. D., and R. S. Juvet, Jr., *Gas-Liquid Chromatography.* New York: Interscience, 1962.
32. Purnell, H., *New Developments in Gas Chromatography.* New York: Wiley-Interscience, 1973.
33. Perry, J. A., *Introduction to Analytical Gas Chromatography; Chromatographic Science Series,* Vol. 14. New York: Marcel Dekker, 1981.

3

Physical Characterization

3.1 INTRODUCTION

Substances possess a number of physical properties that are very useful for the characterization of a substance. Some of these physical properties are the melting point, boiling point, density, and optical rotation, and the ulraviolet and visible, infrared and nuclear magnetic resonance spectral properties. The identification of a substance can often be made by the comparison of the experimentally observed physical properties of a substance with those previously reported in the literature. The various physical properties of many substances appear in *CRC Handbook of Chemistry and Physics, Lange's Handbook of Chemistry,* and *Aldrich Handbook of Fine Chemicals.* The *Aldrich Handbook of Fine Chemicals* also contains references to published infrared and nuclear magnetic resonance spectra of many substances. This chapter is devoted to the discussion of the measurement of the various physical properties of a substance, while the discussion of the spectral properties of substances appears in the following chapters.

3.2 MELTING POINTS

The melting point of a solid, or conversely, the freezing point of a liquid, can be defined as the temperature at which the liquid and solid phases are in equilibrium at a given pressure. Implied in this statement is the more quantitative definition of melting point or freezing point as that temperature at which the vapor pressures of the solid and liquid phases are equal.

Although the intersection of the vapor pressure curves for the liquid and solid phases gives a single point temperature for the melting point of a solid, in an actual experimental determination this is rarely observed owing to the experimental methods employed. In general, very small samples of solids are placed in a capillary tube in an oil bath, or between glass plates on a hot stage, whose temperature is continuously raised by 1 to 2°C/minute. Since a finite time is required for complete melting to occur, a finite temperature range will be traversed during this process. Therefore, it is necessary to report a temperature range for the melting point of any substance, for example, benzoic acid—121.0 to 121.5°C. Handbooks and tables of melting points list a single temperature, usually an average of several reported, indicating where a sample of the pure substance should melt. However, in reporting the melting point of a specific sample, a melting point range should be reported, that is, the temperature at which the first melting is observed and the temperature at which the last crystal disappears. Both the temperature and range of an observed melting point are important attributes. Both reflect the purity of the sample. A pure substance will melt over a very narrow temperature range.

The addition of a nonvolatile soluble substance to a liquid produces a decrease in the vapor pressure of the liquid phase (an application of Raoult's law). From the definition of melting or freezing point, a decrease in the vapor pressure of the liquid phase will result in a lowering of the melting or freezing point (see Figure 3.1). The vapor pressure of the solid in equilibrium with the liquid will, for all purposes, be unaffected, since the liquid in the solid phase will tend to be present as pure small liquid regions with crystalline solute dispersed throughout. The liquid will not, in general, be incorporated into the crystal lattice of the crystalline material. Therefore, if the solute remains soluble in the liquid phase, the solute concentration will decrease as the amount of liquid phase increases during the melting process. The net result is

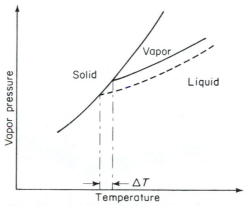

Figure 3.1 Vapor pressure versus temperature diagram. Solid lines represent the pure substance A, and the dashed line represents a solution of a nonvolatile material A.

that the melting point of the mixture will increase as melting proceeds, giving a melting range much greater (several degrees) than that observed for a pure substance.

During the purification of a solid material, the melting range should decrease, and the melting point should increase as the impurities are removed during successive stages of purification. When no further change in the melting range and point is observed on further purification, the material is probably as pure as can be obtained using that purification technique.

Several methods can be used to determine melting points. The capillary tube method is one of the most common methods employed. A portion of a finely ground sample is introduced into a fine glass capillary, 1 mm $\times$ 100 mm, sealed at one end. (Glass capillaries are commercially available, or you can prepare your own by drawing a *clean* piece of 8- to 12-mm soft glass tubing.) Enough sample is placed in the capillary tube and firmly packed until a column of sample 1 to 2 mm is obtained. Occasionally it is difficult to introduce the sample into the capillary and get the sample firmly packed in the closed end of the capillary. Small quantities of material

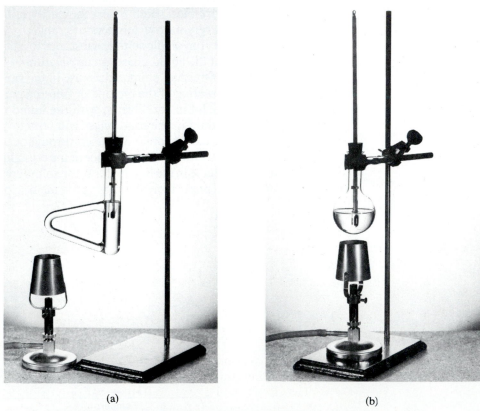

(a)   (b)

Figure 3.2   Setups for the determination of melting points. (a) Employing a Thiele tube. (b) Employing a small round-bottomed flask. (Photographs by Bruce Harlan, Notre Dame, Ind.)

can be forced into the open end of the capillary and then induced to move down the capillary to the sealed end by vibrations. This can be accomplished by firmly grasping the capillary tube and tapping against the bench or table top, by allowing the capillary to fall down a glass tube and bounce on the bench top, or by drawing a file across the side of the capillary.

The capillary is attached to a thermometer by means of a small rubber band cut from $\frac{1}{4}$-in. rubber tubing, and the thermometer and sample are placed in a heating medium. The sample must be maintained at the same level as the mercury bulb of the thermometer. Mineral oil is usually employed as the heating medium and can be safely used up to 225 to 250°C. For higher temperatures, a metal block device should be used.

A variety of liquid heating bath devices can be employed. Figure 3.2a illustrates the proper use of the Thiele tube. The thermometer and sample are placed in the Thiele tube so that the sample and sensing bulb of the thermometer are slightly below the bottom of the top side-arm neck of the tube. A Bunsen burner is used to heat the side arm of the Thiele tube gently, the circulation of the heating medium occurring by convection. More elaborate setups can be employed that contain internal resistance wire heaters and mechanical stirrers. A magnifying lens may be used to observe the sample during the melting process. Figure 3.2b illustrates the use of a small round-bottom flask as the heating bath vessel; however, maintaining a uniform temperature throughout the heating medium is quite difficult unless some means of stirring is provided, for example, a stream of air bubbles introduced by means of an air tube.

Figure 3.3 illustrates a popular commercial apparatus for melting point determinations. It shows two commonly used melting point apparatuses for recording melting points in capillary tubes. In both apparatuses the sample is viewed through a small magnifying eyepiece. The Fisher–Johns melting point apparatus (see Figure 3.4) is a typical hot stage melting point apparatus. A small portion of the sample is placed between two 18-mm microscope slide cover glasses, which are then placed in the depression of the electrically heated aluminum block. The rate of temperature rise is regulated by means of a rheostat and the sample is observed with the aid of a magnifying lens. The Fisher–Johns apparatus is useful for determining melting points between 30 and 300°C. This apparatus must be calibrated periodically using a set of melting point standards.

The rate of heating of a sample in any method of melting point determination should be approximately 1°C per minute while traversing the melting point range. Greater heating rates can be used to raise the temperature of the heating bath to about 10 to 15°C below the expected melting point temperature. Many compounds display melting ranges that depend on the rate of heating of the sample. Examples of compounds that display such behavior are those that undergo decomposition, chemical isomerization (thermally induced), or changes in crystal form.

The sample should be carefully observed during the melting process for changes in crystal form, decomposition with the formation of gaseous products, or decomposition with general charring. Such observations may be very useful in deriving structural information. Changes in crystal form may be due to a simple change of

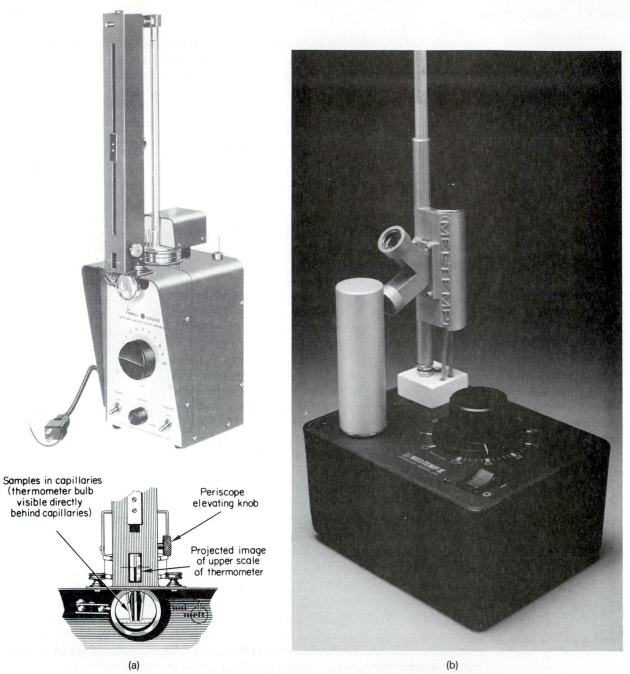

Samples in capillaries
(thermometer bulb
visible directly
behind capillaries)

**Periscope
elevating knob**

**Projected image
of upper scale
of thermometer**

(a)

(b)

Figure 3.3 (a) Thomas–Hoover melting point apparatus. Periscopic thermometer reader allows sample observation and temperature readings to be made at the same eye level. (Reproduced by courtesy of Arthur H. Thomas Company, Swedesboro, N.J.) (b) Mel-Temp melting point apparatus. (Reproduced by courtesy of Laboratory Devices, Holliston, Mass.)

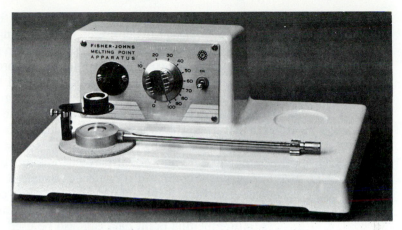

Figure 3.4 Fisher–Johns hot-stage melting point apparatus. (Reproduced by courtesy of Fisher Scientific Company.)

one allotropic crystal modification to another of the same substance, or they may be due to a thermally induced rearrangement to an entirely new compound, where the final melting point then corresponds to the rearranged material and not to the original sample. The evolution of a gas is generally the result of a thermal decarboxylation of β-carboxy carbonyl compounds, altough the loss of small molecules such as water, hydrogen halides, and hydrogen cyanide may occur with some compounds. Should such phenomena occur during the melting point determination, the thermally induced reaction should then be carried out on a larger portion of the sample and the product(s) identified. General decomposition with charring is typical of polyfunctional compounds, such as many sugars and very high melting compounds.

The melting points of substances that undergo decomposition will vary with the rate of temperature rise of the sample; hence, reliable comparisons of melting or decomposition points are not possible. Determination of the instantaneous melting or decomposition point, accomplished by dropping small quantities of the material on a previously heated hot stage and by increasing the temperature of the hot stage until the sample melts or decomposes upon striking the surface, gives a more reliable melting or decomposition temperature.

3.2.1 Mixture Melting Points

The phenomenon of melting point depression has many very useful applications in structure determinations. Qualitatively, melting point depressions, or really, the lack of such depression, can be used for the direct comparison of an unknown with possible knowns. For example, chemical and physical properties may limit the number of possibilities for the structure of an unknown to two or three. The melting point of the mixture of the unknown with a known that has the same structure as the unknown does not lead to a depression in the melting point, whereas mixtures of the unknown with knowns that differ in structure from the unknown show depressions in their mixture melting points. The mixtures are prepared by taking approximately equivalent amounts of unknown and known and then melting the mixture until homogeneous. The melt is allowed to crystallize, and the melting point of a small, finely ground portion of the mixture is determined. This procedure has limited utility in that it requires the availability of suitable known compounds. Data derived from such mixture melting point experiments must be used with caution, however, since some mixtures may not display a depression, or may display only a very small depression, due to compound or eutectic formation. Some mixtures may display an elevation of melting point. This can occur when compound formation occurs or when one has a racemic mixture (see following paragraph).

Mixture melting point data can also be used to distinguish between racemic mixtures [eutectics of equal amounts of crystals of $(-)$ and $(+)$ enantiomers], racemic compounds [crystals containing equal quantities of $(-)$ and $(+)$ molecules in a specified arrangement], and racemic solid solutions [crystals containing equal amounts of $(-)$ and $(+)$ molecules arranged in a random manner in the crystal].[1] The addition of a pure enantiomer: (1) to a racemic mixture results in an elevation of the melting point, (2) to a racemic compound results in a depression of the melting point, and (3) to a racemic solid solution results in no effective change in the melting point (see Section 3.6 for discussion of enantiomers).

The quantitative aspects of melting point or freezing point depression can be applied in the determination of molecular weights.

3.3 FREEZING POINTS

The freezing points of liquids can be used for purposes of characterization; however, freezing points are much more difficult to determine accurately and freezing point data for liquids are much more limited than are melting and boiling point data. In addition, the quantity of sample required to determine a freezing point is much greater than that required for a melting point determination. One to two milliliters of the sample is placed in a test tube and cooled by immersion in a cold bath (the temperature of the cold bath should not be too far below the freezing point of the sample in order to avoid extensive supercooling). The sample is continuously stirred, and the temperature of the sample is recorded periodically. The cooling curve will usually reach a minimum (supercooling) and rise to a constant temperature as crystal growth occurs. This temperature is the freezing point of the sample.

---

[1] E. L. Eliel, *Stereochemistry of Carbon Compounds* (New York: McGraw-Hill, 1962), pp. 43–46.

3.4 BOILING POINTS

The boiling point of a sample is generally determined during purification of the sample by distillation. Difficulties arise in comparison of boiling points, however, when a boiling point is determined at an applied pressure that is different from a literature value. Conversion of a boiling point at one pressure to another pressure can be accomplished by the use of a boiling point nomograph (see the Appendix).

The determination of a boiling point by distillation requires the availability of sufficient sample to attain temperature equilibrium in the distillation apparatus. Frequently, insufficient quantities of sample are available to determine boiling points by distillation techniques, and one must use semimicro or micro techniques. A semimicro technique requires only two or three drops of sample. The sample is placed in a micro test tube, and a very fine capillary, with one end sealed, is placed open end down into the sample. The test tube is suspended in an oil bath and gently heated until a steady stream of bubbles issues from the end to the capillary. The temperature of the oil bath is allowed to decrease slowly until the bubbling stops; this temperature is taken as the boiling point of the sample. The process should be repeated several times until a reproducible boiling point is obtained.

An ultramicro procedure uses a melting point capillary tube as the container vessel and requires only 3 to 4 μL of sample. The sample is injected into the top of the melting point capillary tube with a micro syringe, and the liquid is forced to the bottom of the tube by centrifugation. (*Micro syringes should be cleaned carefully and thoroughly immediately after each use.*) A very fine capillary is prepared by heating a piece of 1- to 3-mm soft glass tubing with a small burner and quickly drawing it out to a diameter small enough to fit into the melting point tube. A small length of the capillary is cut and one end is sealed. The capillary is placed open end down into the melting point tube containing the sample. The melting point tube is then placed in a melting point apparatus such as the Thomas–Hoover apparatus, shown in Figure 3.3a. The temperature of the melting point tube is raised until a fine stream of bubbles issues from the end of the small capillary. The temperature is then allowed to slowly decrease. The boiling point is the temperature at which the last bubble is formed. The sample should be reheated and the process repeated until a consistent temperature for the boiling point is achieved. Boiling points determined in this manner are quite reproducible and are generally accurate to $\pm 1\,°C$.

3.5 DENSITY MEASUREMENTS

The density of a substance is defined as its mass per unit volume and is expressed as grams per milliliter or grams per cubic centimeter. A convenient method for determining densities is a direct comparison of the weights of equal volumes of the sample

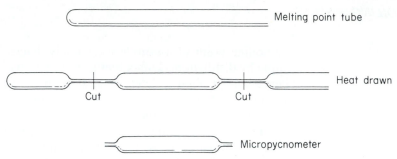

Figure 3.5 A micropycnometer prepared from a melting point tube. The length of the center section can be varied depending on the amount of sample available.

and water at a given temperature, t, and correcting to the density of water at 4°C as given in Equaton (3.1).

$$d_t = \frac{\text{weight of sample at } t}{\text{weight of water at } t} \times d_{H_2O} \qquad (3.1)$$

Densities are determined using vessels referred to as pycnometers. Pycnometers with capacities of 1 to 25 mL are commercially available; however, since quantities of material of this size are often not available, a pycnometer of suitable size must be prepared. A micropycnometer can be made from a melting point tube by carefully heating and drawing out the melting point tube to form fine capillary ends on a section of the tube. The capillary ends are carefully cut (see Figure 3.5) and the micropycnometer is weighted on an analytical balance. One end of the micropycnometer is placed in the sample and the tube normally fills itself by capillary action. When the tube is filled, it is again weighed on an analytical balance. (The evaporation of material is generally negligible when the capillary ends are very small.) The sample is removed from the micropycnometer and the pycnometer is rinsed with acetone and dried with a stream of air. The micropycnometer is then filled with water and once again weighed, and the density is calculated according to Equation (3.1). (Values for the density of water at various temperatures can be found in the *Chemical Rubber Company Handbook of Chemistry and Physics.*)

3.6 OPTICAL ROTATION

Molecules whose mirror image structures are not superimposable are called enantiomers. Enantiomers have identical physical properties (melting points, boiling points, infrared and nuclear magnetic resonance spectra, etc.) except their interaction with plane polarized light. On passage of plane polarized light through a sample of one enantiomer, the plane of polarization will be rotated in a clockwise direction, while the other enantiomer rotates the plane in a counterclockwise direction. (An equimolar mixture of the two enantiomers will not rotate the plane of plane polarized light.) When dealing with such substances, the determination of the optical rotation of the substance provides useful information on the absolute configuration of the chiral center and the enantiomeric purity of the sample.

A device known as a polarimeter is used to determine the optical rotation of a substance. The principle components of a polarimeter are a light source, a polarizing Nicol prism, a sample tube compartment, an analyzing Nicol prism, an observation lens, and a scale indicating the optical rotation of the sample (see Figure 3.6). The observation lens is used to observe the field through the sample as the analyzing prism is rotated manually until the two sides of the observation field are identical in shading. Modern instruments are available that give a direct readout of the rotation of the sample.

The purified sample, approximately 25 to 500 mg, is accurately weighed and dissolved in an appropriate volume of solvent, the volume depending on the volume of the sample tube to be used. If the rotation of the sample is low, higher concentrations may be used. Suitable solvents include water, methanol, ethanol, chloroform, or dioxane. The optical rotation of liquids can be determined using neat (no solvent) samples if sufficient sample is available. The blank, or zero, reading of the polarimeter should be determined with the pure solvent used to dissolve the sample. This blank, or zero value, is subtracted from the sample rotation. Care should be exercised to keep the samples free of small particles (dust, lint, etc.) which will interfere with the measurements. It is very difficult to measure the rotations of highly colored substances.

The specific rotation $[\alpha]$ is given by Equations (3.2) and (3.3),

$$\text{For solutions:} \qquad [\alpha]_D^T = \frac{\alpha \cdot 100}{l \cdot c} \qquad\qquad (3.2)$$

$$\text{For pure liquids:} \qquad [\alpha]_D^T = \frac{\alpha}{l \cdot d} \qquad\qquad (3.3)$$

where α is the observed rotation of the sample in degrees, l is the length of the sample tube in decimeters, c is the concentration of solute in g/100 mL of solution, d is the

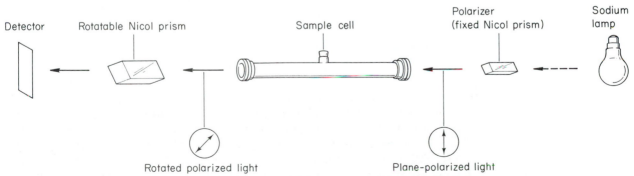

Figure 3.6   Components of a polarimeter.

density of the pure liquid, T is the temperature of the sample at which the rotation was measured, and D specifies the wavelength of the light used as that at the sodium D line (otherwise the wavelength is given in nanometers). The solvent, as well as the concentration, should be specified. The rotation of the sample is given, for example, as $[\alpha]_D^{20} + 65.2°$ (c 1.0, H_2O). Occasionally materials may possess very large specific rotations, and a single measurement will not allow one to distinguish between, for example, a moderate negative specific rotation and a large positive specific rotation. *Specific rotations can be compared only when using the same solvent, since specific rotations may vary with solvent.*

In quantitative comparisons of compounds of different molecular weights *molecular rotations,* [ϕ], as defined by Equation (3.4), are often used.

$$[\phi] = [\alpha]\frac{M}{100} \tag{3.4}$$

where M is the molecular weight.

3.7 REFERENCES

General

1. Cheronis, N. D., "Micro and Semimicro Methods," in *Technique of Organic Chemistry,* Vol. VI, Part I, A. Weissberger, ed. New York: Interscience, 1954.
2. Mayo, D. W., R. M. Pike, and S. S. Butcher, *Microscale Organic Laboratory.* New York: John Wiley, 1986.
3. Overton, K. H., "Isolation, Purification, and Preliminary Observations," in *Technique of Organic Chemistry,* Vol. XI, Part I, A. Weissberger, ed. New York: Interscience, 1963.

Compilations of Physical Properties of Orgainic Compounds

4. *Aldrich Handbook of Fine Chemicals,* Milwaukee, Wisc.: Aldrich Chemical, 1990 (updated version published every two years).
5. *CRC Handbook of Chemistry and Physics,* R. C. Weast, ed. Boca Raton, Fla.: CRC Press, 1989 (updated version published yearly).
6. *CRC Handbook of Tables for Organic Compound Identification,* 3rd ed., Z. Rappoport, ed. Cleveland, Ohio: Chemical Rubber, 1967.

4

Ultraviolet Spectroscopy

4.1 GENERAL INTRODUCTION TO SPECTROSCOPY

Perhaps the most useful physical properties for the characterization of a substance, or the determination of the structure of an unknown substance, are its spectral properties. The spectral properties of substances arise from the ability of materials to absorb light of very characteristic frequencies in the different regions of the electromagnetic spectrum. The absorption of radiant energy results in a transition between quantized energy levels within a molecule. The amount of energy required depends on the type of excitation involved and the types of functional groups present in the molecule. The regions of the electromagnetic spectrum that are of interest to the organic chemist are the ultraviolet and visible region in which the excitation of electrons from an occupied to an unoccupied orbital occurs, the infrared region in which bond deformation excitations occur, and the radiowave region in which excitation of nuclear spin states occurs. The wavelengths and the associated energies in these regions are illustrated in Table 4.1. The definitions of terms and equations are given in Table 4.2.

TABLE 4.1 Regions of the Electromagnetic Spectrum

Region	Wavelength	Energy (kcal mole^{-1})	Type of excitation
Ultraviolet	200–350 nm	143–82	Electronic
Visible	350–800 nm	82–36	Electronic
Infrared	2–16 μm	14–1.8	Bond deformation
Radiowave	meters	$\sim 10^{-6}$	Nuclear spin

TABLE 4.2   Definition of Terms and Equations*

Term	Symbol	Equation	Dimensions
Wavelength	λ	— — —	Å (Ångström)$(10^{-8}$cm) μm/micrometers nm/nanometers
Frequency	v	c/λ	Hz (cycles per second)
Wave number	n	$1/\lambda$	cm^{-1}
Energy	E	hv, hc/λ	Depends on the units of h

* c = velocity of light $(2.9979 \times 10^{10}$ cm-sec$^{-1})$: h = Planck's constant $(6.6237 \times 10^{-27}$ erg-sec)

4.2 INTRODUCTION TO ULTRAVIOLET AND VISIBLE SPECTROSCOPY

The absorption of energy in the ultraviolet and visible regions of the spectrum results in the excitation of an electron from an occupied molecular orbital to a higher energy unoccupied molecular orbital. The most common type involves the excitation of an electron from a π MO to a π^* MO, indicated as a $\pi \rightarrow \pi^*$ transition. These absorption bands are usually quite intense. Excitations of this type are observed with conjugated dienes, conjugated unsaturated carbonyl and aromatic compounds. The π to π^* transition of nonconjugated alkenes and carbonyl compounds occurs at wavelengths shorter than 200 nm, wavelengths at which air absorbs, thus requiring special techniques to record the absorption spectra.

The excitation of an electron from a nonbonded type MO to either a π^* or a σ^* MO (indicated as $n \rightarrow \pi^*$ or $n \rightarrow \sigma^*$ transitions) is also induced by absorption in the ultraviolet region. These bands are usually not very intense. Excitations of this type are observed with carbonyl compounds and saturated molecules containing atoms bearing a pair of nonbonding electrons. The wavelengths of maximum absorption (λ_{max}) of various types of functional groups are given in the tables in the following sections of this chapter. The absorption bands are usually quite broad, but with rigid molecules the bands occasionally show fine structure due to vibrational interactions as illustrated in Figure 4.1.

Ultraviolet and visible spectral data are obtained as plots of absorbance or extinction coefficient versus wavelength. The absorbance and extinction coefficients are calculated according to Equation (4.1), in which I_o is the intensity of the incident light, I is the intensity of the light transmitted through the sample, ϵ is the extinction coefficient, l is the cell length in centimeters, and c is the concentration in moles per liter. Extinction coefficients range from 1 to 100,000 depending on the nature of the absorbing chromophore. Unless direct comparison of features is desired, only the wavelength at maximum intensity of absorption is noted as $\lambda_{max}^{solvent}$,

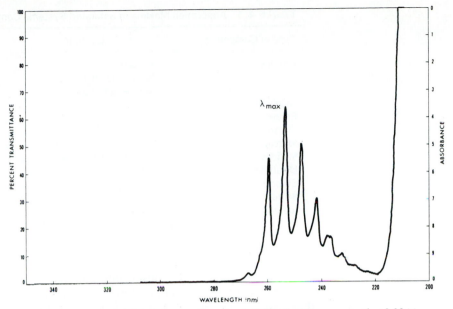

Figure 4.1   Ultraviolet spectrum of benzene in 95% ethanol: concentration 2.92 × 10^{-3} m/L.

followed by the extinction coefficient or log of the extinction coefficient in parentheses, that is, $\lambda_{max}^{C_2H_5OH}$ 227 nm (log ϵ 3.72).   Occasionally the wavelengths at which other features appear are also listed, such as inflections ($\lambda_{infl}^{solvent}$) and minima ($\lambda_{min}^{solvent}$).   The solvent used must always be specified, as in many cases the nature of the solvent has a profound effect on the position of maximum absorption.

$$\text{Absorbance} = \log\frac{I_o}{I} = \epsilon cl \tag{4.1}$$

4.3   SPECTRAL DATA AND STRUCTURE CORRELATIONS

4.3.1   Hydrocarbon Derivatives

Table 4.3 lists the wavelengths of typical $n \rightarrow \sigma^*$ absorption bands of saturated hydrocarbon derivatives.

4.3.2   Alkenes, Alkynes, and Polyenes

Nonconjugated alkenes and alkynes.   The $\pi \rightarrow \pi^*$ band of ethylene occurs at 172 nm.   Each sequential attachment of an alkyl group to the C=C produces a bathochromic shift (a shift to longer wavelength) of 3 to 5 nm.   As even the λ_{max} of tetraalkyl-substituted alkenes occurs below 200 nm, the λ_{max} of such chromophores

TABLE 4.3 Absorption Maxima of Saturated Systems Containing Heteroatoms

Type of Compound	λ_{max} (nm)	Approximate Log ϵ
Amines	190–200	3.4
	Also occasionally a longer wavelength shoulder	
Alcohols	180–185	2.5
Ethers	180–185	3.5
Epoxides	~170	3.6
Mercaptans	190–200 and 225–230 (shoulder)	3.2 and 2.2
Sulfides	210–215 and 235–240 (shoulder)	3.1 and 2.0
Disulfides	~250	2.6
Chlorides	170–175	2.5
Bromides	200–210	2.6
Iodides	255–260	2.7

cannot generally be observed. However, the absorption band widths are sufficiently broad that a portion of the absorption band extends beyond 200 nm (termed end absorption), and the presence of the C=C function can be detected in the absence of other strongly absorbing groups. Similarly, the intense $\pi \rightarrow \pi^*$ band of alkynes appears near 175 nm; however, in many cases a very weak band is also observed near 223 nm.

Conjugated polyenes.   Conjugated dienes and enynes have $\pi \rightarrow \pi^*$ absorptions that occur at wavelengths longer than 200 nm. Spectra-structure correlations have been developed that are useful in predicting the type and degree of substitution on the conjugated system. The empirical correlations initially proposed by R. B. Woodward and later extended by L. Fieser and M. Fieser are known as the Woodward rules or Woodward–Fieser rules.

A parent system is defined, and increments characteristic of the functions directly attached to the chromophore are added to the base absorption of the parent. The parent system is the acyclic diene, or heteroannular diene, system **1**, in which the two double bonds are not contained within a single ring. The base absorption of **1** is at 214 nm, while the homoannular diene, system **2**, has a base absorption at 253 nm. In addition to differences in the base absorption wavelengths, the heteroannular dienes absorb more intensely than do the homoannular diene systems. It should be noted that in the homoannular diene system the two carbon atoms of the six-mem-

1
λ_{max} 214 nm (log ϵ 3.9 – 4.3)

2
λ_{max} 253 nm (log ϵ 3.7 – 3.9)

TABLE 4.4 Substituent Constants for the Calculation of λ_{max}'s of Substituted Dienes

Function	Increment (nm)
Alkyl (per group)	+5
—OCOR*	0
—OR	+6
—SR	+30
—Cl, —Br	+5
—NR$_2$	+60
Extended conjugation (per double bond)	+30
Exocyclic double bond	+5

* R represents any alkyl group.

bered ring attached to the ends of the diene chromophore (boldface C's in **2**) are included in the base absorption value. Homoannular dienes contained in other size rings possess different base absorption values, for example, cyclopentadiene (228 nm) and cycloheptadiene (241 nm). Cross-conjugated polyenes, represented by the partial structure **3**, in general do not correlate well.

Table 4.4 lists the substituent increments. In general, the magnitude of the increment is not a function of the position of the substituent on the diene chromophore.

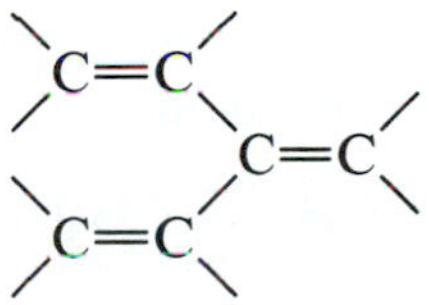

3

A few examples will illustrate the use of these rules. In structure **4**, the parent diene is of the homoannular type, with one of the double bonds exocyclic to ring A and one exocyclic to ring C, exocyclic being defined as having only one of the carbon atoms of the double bond belonging to a ring. In structure **5**, the parent diene is of the heteroannular type in which the double bond in ring B is exocyclic to ring A. Structure **6** is an example of a "through" conjugated triene system, and it presents us

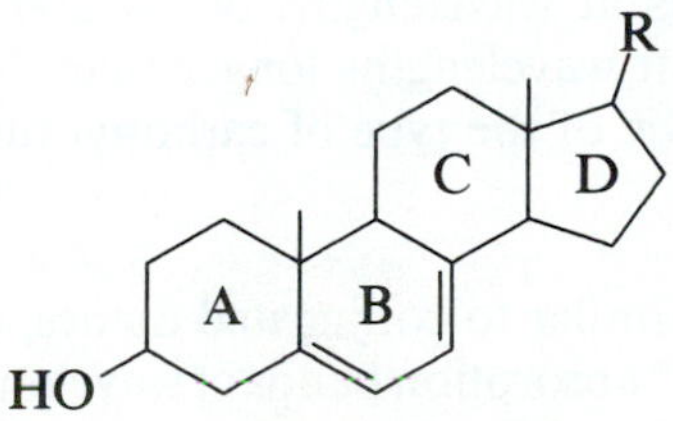

4

Parent diene	homoannular	253	nm
Substituents	(4 × 5)	20	nm
Exocyclic double bonds	(2 × 5)	10	nm
Calculated λ_{max}		283	nm
Observed		282	nm

Parent diene	heteroannular	214	nm
Substituents	(3 × 5)	15	nm
Exocyclic double bonds	(1 × 5)	5	nm
Calculated λ_{max}		234	nm
Observed		234	nm

5

with a choice of selecting either the homo- or heteroannular diene base. In general, one selects the longest wavelength base, that is, that of the homoannular diene. In counting the number of substituents on the chromophore in **6**, all substituents on the entire triene system are included; it should be pointed out that C_{10} (the circled carbon atom in **6**) of the steroid nucleus is a substituent on two different carbon atoms of the triene system and is counted as two alkyl groups.

Parent diene	homoannular	253	nm
Extended conjugation	(1 × 30)	30	nm
Substituents	(5 × 5)	25	nm
Exocyclic double bonds	(3 × 5)	15	nm
Calculated λ_{max}		323	nm
Observed		324	nm

6

Conjugated enynes and diynes. Conjugated enynes and diynes also possess $\pi \rightarrow \pi^*$ absorption bands that appear at wavelengths longer than 200 nm; however, no correlation of absorption band positions with structure has been published. Most compounds of these types display several closely spaced absorption bands due to vibrational interactions. The most intense peak of selected examples is given below the following examples.

CH_3CH_2 $C \equiv CH$ $CH_3CH = CH - C \equiv CCH_3$ $CH_3 - C \equiv C - C \equiv C - CH_3$
$\hspace{2cm}C = C$
$H \hspace{3cm} H$

λ_{max} 221 nm λ_{max} 223 nm λ_{max} 227 nm

4.3.3   Carbonyl Compounds

Nonconjugated carbonyl derivatives. The $\pi \rightarrow \pi^*$ absorption band of all nonconjugated carbonyl derivatives occurs at wavelengths below 200 nm. The $n \rightarrow \pi^*$ excitation band, however, occurs at wavelengths longer than 200 nm, the position of which is somewhat characteristic of the type of carbonyl function (see Table 4.5).

Conjugated carbonyl compounds. Similar to conjugated dienes, conjugated carbonyl compounds possess intense $\pi \rightarrow \pi^*$ absorption bands at wavelengths longer than 200 nm, and spectra-structure correlations have been developed. [Conjugated

TABLE 4.5 Absorption Maxima of Nonconjugated Carbonyl Compounds

Function	$\lambda_{max}(n \rightarrow \pi^*,$ nm$)$	Log ϵ
Aldehydes	290	0.9–1.4
Ketones	280	1.0–2.0
Carboxylic acids	206	1.6–2.0
Esters (in water)	206	1.6–2.0
(in nonprotic solvents)	212	1.6–2.0
Amides	210	1.0–2.0
Acid chlorides	235	1.0–2.0
Anhydrides	225	≈ 1.7

carbonyl compounds also show a weak $n \rightarrow \pi^*$ band in the 300 to 330 nm region (see Figure 4.2).] In contrast to the conjugated dienes, the substituent increment depends on the position of the substituent on the unsaturated carbonyl chromophore. The parent base absorptions for the various types of conjugated carbonyl compounds, along with the substituent parameters for the general chromophore **7**, are given in Table 4.6.

It should be noted that in contrast to the conjugated dienes, the position of maximum absorption of the conjugated carbonyl compounds is dependent on the

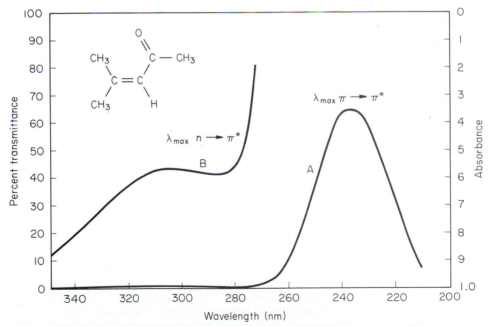

Figure 4.2 Ultraviolet spectrum of mesityl oxide in 95% ethanol. Trace A displays the $p \rightarrow p^*$ band (concentration 6.29×10^{-5} m/L), and trace B displays the $n \rightarrow p^*$ band (concentration 6.29×10^{-3} m/L).

TABLE 4.6 Base Absorptions and Substituent Parameters for
Calculation of λ_{max}'s of Conjugated Carbonyl Compounds
in Ethanol Solution

α,β-Unsaturated Parent	Base Absorption (nm)
Acyclic or six-membered or higher-ring ketone	215
Five-membered ring ketone	205
Aldehydes	210
Carboxylic acids and esters	195
Extended conjugation	+30
Homodienic component	+39
Exocyclic double bond	+5

Substituent	Substituent Parameters		
	α	β	γ
Alkyl	+10	+12	+18
Hydroxyl	+35	+30	+50
Alkoxyl	+35	+30	+17(+31 for δ)
Acetoxyl	+6	+6	+6
Dialkylamino		+95	
Chloro	+15	+12	
Bromo	+25	+30	
Alkylthio		+85	

Solvent	Solvent Correction (nm)
Water	−8
Methanol	0
Chloroform	+1
Dioxane	+5
Ether	+7
Hexane	+11
Cyclohexane	+11

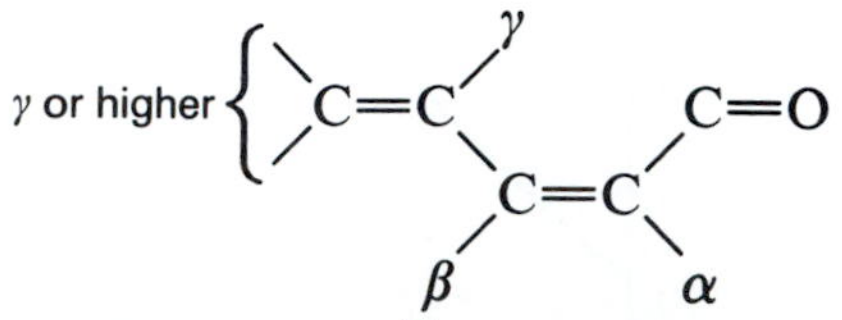

nature of the solvent. Also, in contrast to the change in position of the $\pi \rightarrow \pi^*$ band
with increasing solvent polarity, the $n \rightarrow \pi^*$ band shifts to lower wavelength with
increasing polarity of the solvent and in the presence of hydrogen bond donors.
The following examples illustrate the use of these rules.

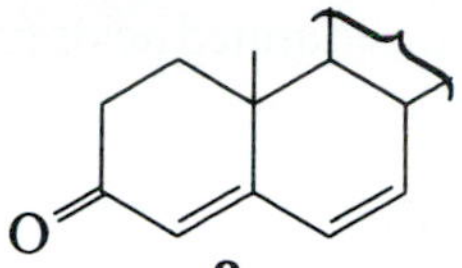

Parent base	215 nm
Extended conjugation	30 nm
Substituents	
beta	12 nm
delta	18 nm
Exocyclic double bond	5 nm
Calculated λ_{max}^{EtOH}	280 nm
Observed	284 nm

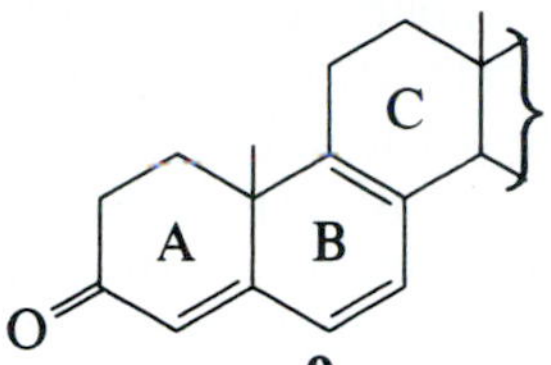

8

Parent base	215 nm
Extended conjugation (2×30)	60 nm
Homodienic component (ring B)	39 nm
Substituents	
beta	12 nm
3γ or higher	54 nm
Exocyclic double bond	5 nm
Calculated λ_{max}^{EtOH}	385 nm
Observed	388 nm

9

A homodienic component is included for compound **9** to account for the presence of two double bonds in the B ring. Compound **10** illustrates an application to a cyclopentenone derivative. Both the carbonyl group and the double bond must be in the five-membered ring:

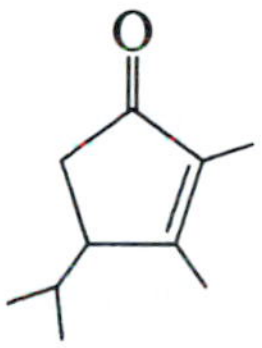

Parent base	215 nm
Substituents	
alpha	10 nm
2 beta	24 nm
Five-membered ring	−10 nm
Calculated λ_{max}^{EtOH}	239 nm
Observed	237.5 nm

10

$(CH_3)_2C=CHC$ (with O and H)

Parent base		215 nm
Substituents	2 beta	24 nm
Aldehyde		−5 nm
Calculated λ_{max}^{EtOH}		234 nm
Observed		235 nm

11

$(CH_3)_2C=CHC$ (with O and OH)

Parent base		215 nm
Substituents	2 beta	24 nm
Acid		−20 nm
Calculated λ_{max}^{EtOH}		219 nm
Observed		216 nm

12

Compounds **11** and **12** illustrate the application of these rules to aldehydes and acids. The correspondence between calculated and observed positions is much poorer in the application to unsaturated acids (± 5 nm), and certain caution must be exercised.

4.3.4 Aromatic Compounds

Substituted benzenes may display several absorption bands in the ultraviolet region. All substituted benzenes display what are commonly known as E and B bands in the 200 to 230, and 255 to 280 nm regions. Benzene rings bearing a conjugating group will display an intense band, referred to as a K band, in the 220 to 250 nm region. The presence of a group bearing a nonbonded pair of electrons will give rise to a weak band, referred to as an R band, in the 275 to 330 nm region. Examples are given in Tables 4.7 and 4.8. In Table 4.8 it should be noticed that the presence of an electron-donating group shifts the maxima to longer wavelengths, while the presence of an electron-withdrawing group shifts the maxima to shorter wavelengths. This is most notable in comparing the change in the positions of the maxima on deprotonation of phenol and the protonation of aniline. An empirical correlation has been developed that allows one to calculate the position of absorption of polysubstituted benzenes.[1]

The fusion of additional rings onto the benzene nucleus, for example as in naphthalene, anthracene, and so on, results in substantial shifts of the maxima to longer wavelengths. The B band of benzene appears at 255 nm, whereas in naphthalene and anthracene this type of band appears at 314 and 380 nm, respectively.

Heterocyclic aromatic compounds also absorb in the ultraviolet region. Table 4.9 lists the absorption maxima of several heterocyclic aromatic compounds. Note the similarity in position between the absorption bands of benzene and pyridine, and naphthalene and quinoline.

4.3.5 Miscellaneous Chromophores

There are many other types of organic functional groups that absorb in the ultraviolet region. Table 4.10 lists many such functional groups.

TABLE 4.7 Typical Absorption Maxima of Aromatic Compounds

Compound	Band λ_{max} nm(ϵ)			
	E	K	B	R
Benzene	198(8000)	—	255(230)	—
t-Butylbenzene	208(7800)	—	257(170)	—
Styrene		244(12,000)	282(450)	—
Acetophenone		240(13,000)	278(1100)	319(50)

[1] E. J. Cowles, *J. Org. Chem.*, **53**, 657 (1988).

TABLE 4.8　UV Absorption Maxima of Substituted Benzenes

Function	E Band (log ϵ)	B or K Band (log ϵ)	R Band (log ϵ)
Electron-donating substituents			
—CH$_3$	207　(3.85)	261(2.35)	
—C(CH$_3$)$_3$	207.5(3.78)	257(2.23)	
—OCH$_3$	217　(3.81)	269(3.17)	
—OH	210.5(3.79)	270(3.16)	
—O$^-$	235　(3.97)	287(3.42)	
—NH$_2$	230　(3.93)	280(3.16)	
—SH	236　(4.00)	269(2.85)	
Electron-withdrawing substituents			
—F	204(3.80)	254(3.00)	
—Cl	210(3.88)	257(2.23)	
—Br	210(3.88)	257(2.23)	
—I	207(3.85)	257(2.85)	
—NH$_3^+$	203(3.88)	254(2.20)	
π-Conjugating Substituents			
—CHO	244(4.18)	280(3.18)	328(1.30)
—COCH$_3$	240(4.11)	278(3.04)	319(1.70)
—CO$_2$H	230(4.00)	270(2.93)	
—CO$_2^-$	224(3.94)	268(2.75)	
—CN	224(4.11)	271(3.00)	
—NO$_2$	252(4.00)	270(2.90)	330(2.10)
—CH=CH$_2$	244(4.08)	282(2.65)	
—C$_6$H$_5$	246(4.30)		

TABLE 4.9　Absorption Maxima of Heterocyclic and Benzenoid Aromatics

		λ_{max} (nm)	ϵ
Benzene	(B band)	255	230
Pyridine	(B band)	252	2090
Pyridazine	(B band)	246	1150
Pyrimidine	(B band)	243	2950
Pyrazine	(B band)	261	6000
Naphthalene	(B band)	314	250
Quinoline	(B band)	313	2500
Anthracene	(B band)	380	9000
Furan		205	6300
Pyrrole		211	10000
Thiophene		235	4500
Imidazole		207	5000
Pyrazole		210	3100
Isoxazole		211	4000
Thiazole		204	4000

TABLE 4.10 Absorption Data for Miscellaneous Functional Groups*

Functional Group	λ_{max} nm(ϵ)
Allenes	175–185(~10,000)
Cumulenes (butatriene)†	241(~20,000)
Nitriles	~170(weak) and ~340(120)
Nitro	~210(~16,000)($\pi \rightarrow \pi^*$) and 270–280(~200)($n \rightarrow \pi^*$)
Nitrate	$\pi \rightarrow \pi^*$ end absorption and ~260–270(shoulder)(~150)
Nitrite	~350(~150)‡
Azo	$\pi \rightarrow \pi^*$ end absorption and ~350(weak)($n \rightarrow \pi^*$)
Diazo	~400(~3)‡
Sulfoxide	210–215(~1600)
Sulfone	no absorption above 208
Vinyl sulfone	~210(~300)

* Functional groups contained in aliphatic systems only.

† Cumulenes containing more double bonds absorb at correspondingly longer wavelengths.

‡ Composed of a number of bands for which the approximate center is given.

4.4 THE USE OF MODEL COMPOUNDS

Often the partial structure of a complex molecule can be deduced by comparison of the spectral properties of the compound with those of simpler molecules thought to contain the same chromophore. This is particularly true in cases in which the geometry of the molecule causes a distortion of the chromophore, resulting in shifts of the maxima to different wavelengths. Electronic interaction with a neighboring functional group may also lead to anomalous shifts in the band positions. The simpler molecules used for such comparisons are called model compounds. Model compounds are used extensively in natural products chemistry and for highly strained ring compounds.

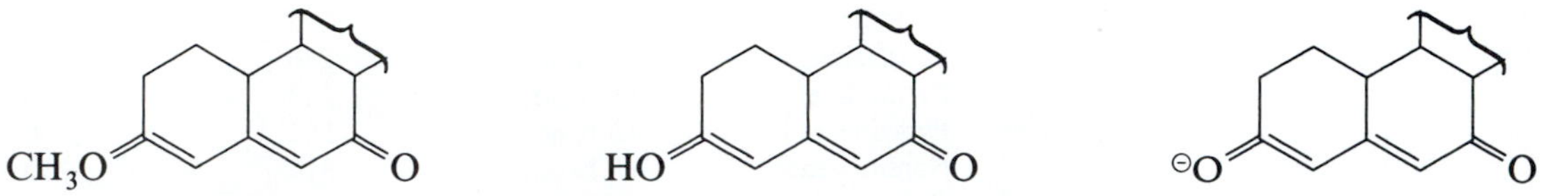

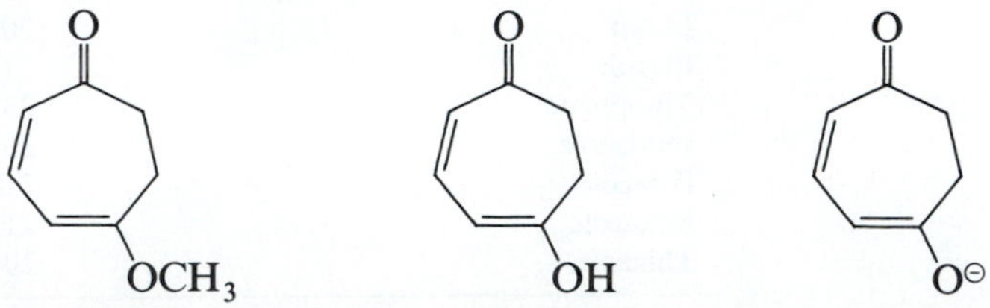

Although one generally uses the positions of the maxima for correlation purposes, similar changes in the positions of maxima throughout a series of compounds

308 nm $\Delta\lambda = 12$ nm 320 nm $\Delta\lambda = 73$ nm 393 nm

328 nm $\Delta\lambda = 12$ nm 340 nm $\Delta\lambda = 74$ nm 414 nm

may be very helpful. The chromophores may not be completely identical, but may be similar enough to allow a correlation to be made. The similarity in the shifts observed in the ultraviolet spectra of the following steroidal compounds and seven-membered ring compounds indicates a similarity in the chromophores.

4.5 ADDITIVITY OF CHROMOPHORES

Many molecules can contain more than one absorbing chromophore. In such cases, the observed ultraviolet absorption spectrum will be the sum of the absorption bands of the individual chromophores, provided there is no electronic interaction between the chromophores. If the absorption maxima of the two chromophores lie close to one another, resulting in a great deal of overlapping of the bands, we can subtract an expected absorption of one of the known chromophores to derive the absorption spectrum of the other chromophore. Experimentally this can be readily accomplished by placing a reference compound containing the desired chromophore into the reference cell of the spectrometer, with the sample in the sample cell. The resulting spectrum will be the difference spectrum between the sample and reference compound. The concentration of the reference compound may have to be adjusted so that the net absorptivity of the reference chromophore equals that of the same chromophore in the sample. A great deal of caution must be exercised in the selection of model compound because the chromophores must be identical.

4.6 PREPARATION OF SAMPLES

The primary consideration in the preparation of samples for recording their ultraviolet spectra is that of concentration. The concentration must be adjusted such that the absorption peak remains on the recording scale of the instrument and, furthermore, in the more accurate region on the recording scale. The region of greatest accuracy is between 0.2 and 0.7 absorbance units. If we specify that the maximum absorbance should be near 0.7 and we have a general idea of what type of chromophore is present and thus a general region for the value of the extinction coefficient, we can use Equation (4.1) to calculate the concentration required for use in a 1 cm cell. For conjugated diene chromophores with extinction coefficients in the general region of 8000 to 20,000, the concentration should be near 4×10^{-5} mol/L, whereas for the $n \rightarrow \pi^*$ transition of a carbonyl group (ϵ of 10 to 100), the concentration should be near 10^{-2} mol/L. Preparation of very dilute samples cannot, in general, be accomplished by a direct weighing technique. The quantity of material would be quite small, in the submilligram range, and large weighing errors would be introduced. In general, such sample concentrations are obtained by successive dilutions of more concentrated solutions.

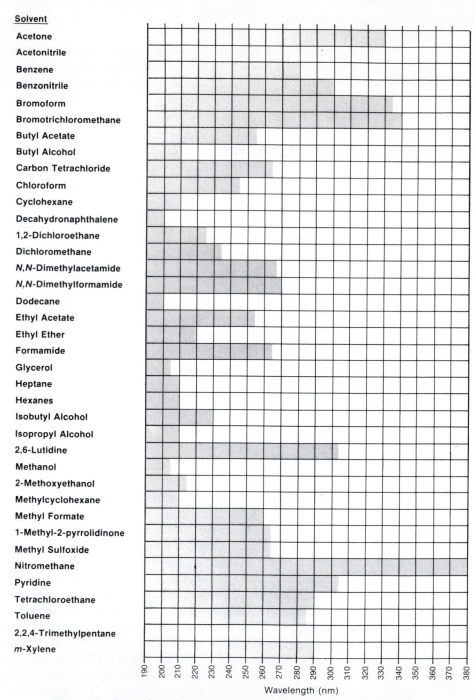

Figure 4.3 Chart of solvents for use in the ultraviolet region. The shaded areas represent absorbances greater than unity in a 1-cm pathlength cell, prohibiting use as a solvent in those regions. (Reproduced by courtesy of Eastman Kodak Company, Rochester, N.Y.)

The choice of solvent rests on several factors. The most important criterion is that the solvent be transparent in the region over which you wish to record the absorption spectrum. Figure 4.3 lists several solvents and the cutoff point below which the solvent itself absorbs.

4.7 PROBLEMS

1. Calculate the λ_{max}'s for each of the following compounds.

(a)

$$CH_3-\overset{\overset{\displaystyle O}{\|}}{C}-CH=CHN(CH_3)_2$$

(b)

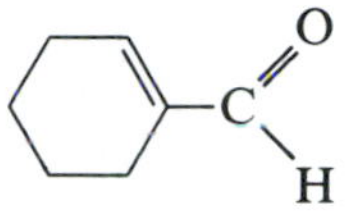

(c) $CH_2=C(CH_3)-CH=CH-CO_2CH_3$

(d)

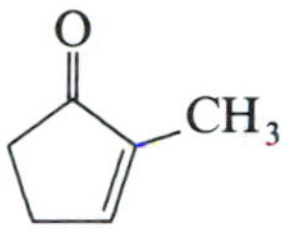

(e)

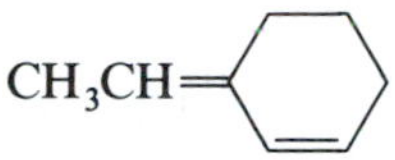

$CH_3CH=$

(f) $CH_2=CHCH_2CH_2\overset{\overset{\displaystyle O}{\|}}{C}CH_3$

(g)

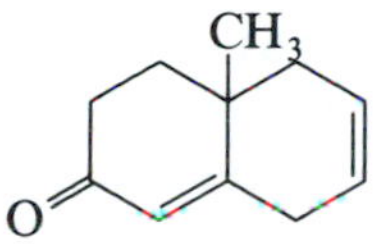

(h) $CH_3O_2CCH=C(CH_3)_2$

(i)

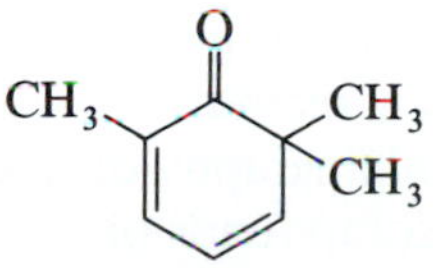

(j)

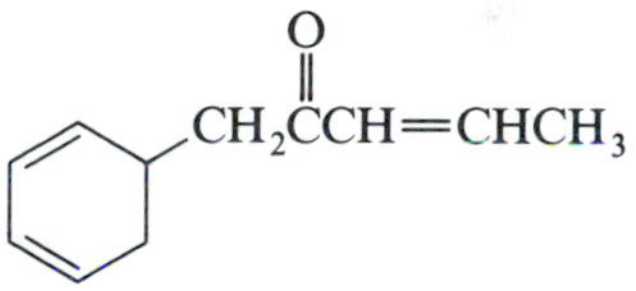

2. Suggest structures for each of the following that are compatible with the given spectral data (all in ethanol solution).
   (a) A $C_7H_{10}O$ compound (λ_{max} 259 nm), which undergoes acid-catalyzed hydrolysis to a C_6H_8O compound with λ_{max} 227 nm.
   (b) A C_7H_9N compound with λ_{max} 207 and 257 nm.
   (c) A C_4H_6O compound with λ_{max} 222 nm, which undergoes reduction with sodium borohydride to give a product that displays only end absorption.
   (d) A C_5H_8O compound with λ_{max} 290 (log ϵ 1.2) and end absorption, which does not readily add hydrogen in the presence of a catalyst.
   (e) A $C_8H_{10}O$ compound with λ_{max} 216 and 270 nm, which possesses the same spectral properties when dissolved in 0.1 M ethanolic sodium hydroxide.
   (f) A $C_9H_{12}O$ compound with λ_{max} 207 and 260 nm, which reacts with sodium dichromate in the presence of sulfuric acid to give a compound with λ_{max} 207, 261, and 280 (weak) nm.
   (g) A C_9H_{10} compound (λ_{max} 244 and 285 nm), which undergoes addition of hydrogen in the presence of a catalyst to produce a compound with λ_{max} 207 and 260 nm.

(h) A C_3H_8OS compound with λ_{max} 212 nm.

(i) A C_6H_{10} compound with λ_{max} 224 nm, which on reaction with one equivalent of ozone in methylene chloride-pyridine solution gives, as one of the products, a compound with λ_{max} 234 nm.

(j) A C_7H_9NO compound with λ_{max} 228 and 281 nm, $\lambda_{max}^{0.1\,M\,base}$ 228 and 281 nm, and $\lambda_{max}^{0.1\,M\,acid}$ 202 and 255 nm.

4.8 LOCATING ULTRAVIOLET AND VISIBLE ABSORPTION SPECTRAL DATA IN THE LITERATURE

Several compilations of ultraviolet and visible absorption data have been published. H. M. Hershenson has compiled literature references in two volumes entitled *Ultraviolet and Visible Absorption Spectra* (see reference 15) covering the years 1930 to 1954 and 1955 to 1959. Sixteen volumes entitled *Organic Electronic Spectral Data* (see reference 17) have been published that give spectral data and original references covering the years 1946 to 1974. The Chemical Rubber Company has published two volumes entitled *CRC Atlas of Spectral Data and Physical Constants for Organic Compounds* (see reference 13).

Reproductions of ultraviolet and visible spectra have been compiled by the American Petroleum Institute Research Project 44 (see reference 21), Sadtler Research Laboratories (see reference 19), as well as in the *UV Atlas of Organic Compounds* (see reference 22), the compilation of UV spectra of aromatic compounds (see reference 14), and *Absorption Spectra in the Ultraviolet and Visible Region* (see reference 12).

General references pertaining to the general theory and interpretation of ultraviolet and visible spectra are given in references 1 to 11. *The Handbook of Ultraviolet Methods* (see reference 10) is a bibliography of references of ultraviolet analytical procedures dealing with specific compounds or substances.

4.9 REFERENCES

General

1. Fieser, L. F., and M. Fieser, *Steroids.* New York: Reinhold, 1959, pp. 15–21.

2. Gillam, A. E., and E. S. Stern, *An Introduction to Electronic Absorption Spectroscopy in Organic Chemistry.* London: Edward Arnold, 1957.

3. Jaffe, H. H., and M. Orchin, *Theory and Applications of Ultraviolet Spectroscopy.* New York: John Wiley, 1962.

4. Mason, S. F., "Molecular Electronic Absorption Spectra," *Quart. Revs.,* **15,** 287 (1961).

5. Matsen, F. A., "Applications of the Theory of Electron Spectra," in *Technique of Organic Chemistry,* A. Weissberger, ed. New York: Interscience, 1956.

6. *Practical Absorption Spectrometry—UV Spectrometry Group,* A. Knowles and C. Burgess, eds. New York: Chapman and Hall, 1984.

7. Rao, C. N. R., *Ultra-Violet and Visible Spectroscopy,* London: Butterworths, 1961.

8. "Report on the Notation for the Spectra of Polyatomic Molecules," *J. Chem. Phys.,* **23,** 1997 (1955).

9. Sandorfy, C., *Electronic Spectra and Quantum Chemistry.* Englewood Cliffs, N.J.: Prentice-Hall, 1964.

10. White, R. G., *Handbook of Ultraviolet Methods.* New York: Plenum, 1965.

11. Woodward, R. B., *J. Am. Chem. Soc.,* **64,** 72 (1952).

Spectral Data and References

12. *Absorption Spectra in the Ultraviolet and Visible Region,* L. Lang, ed., Vols. 1–20, New York: Academic Press, 1961–1975; Vols. 21–24, New York: Robert E. Krieger, 1977–1984.

13. (a) CRC Atlas of Spectral Data and Physical Constants for Organic Compounds, J. G. Grasselli, ed. Cleveland, Ohio: Chemical Rubber, 1973. (b) *CRC Atlas of Spectral Data and Physical Constants for Organic Compounds,* 2nd ed., J. G. Grasselli and W. M. Ritchy, eds. Cleveland, Ohio: Chemical Rubber, 1975.

14. Friedel, R. A., and M. Orchin, *Ultraviolet Spectra of Aromatic Compounds.* New York: John Wiley, 1951.

15. Hershenson, H. M., *Ultraviolet and Visible Absorption Spectra.* New York: Academic Press, index for 1930–1954 published in 1956, index for 1955–1959 published in 1961.

16. Hirayama, K., *Handbook of Ultraviolet and Visible Absorption Spectra of Organic Compounds.* New York: Plenum Data Division, 1967.

17. *Organic Electronic Spectral Data.* New York: Interscience, nine volumes covering the years 1946–1973.

18. Scott, A. I., *Interpretation of the Ultraviolet Spectra of Natural Products.* New York: Pergamon, 1964.

19. *Standard Ultraviolet Spectra,* compiled by the Sadtler Research Laboratories, Philadelphia, Pa.

20. *Standards in Absorption Spectroscopy-UV Spectroscopy Group,* C. Burgess and A. Knowles, eds. New York: Chapman and Hall, 1981.

21. *Ultraviolet Spectral Data,* compiled by the American Petroleum Institute Research Project 44, Chemical Thermodynamics Properties Center, Department of Chemistry, Texas A & M University, College Station, Tex.

22. *UV Atlas of Organic Compounds.* London: Butterworths, five volumes published 1966–1971.

5

Infrared
Spectroscopy

5.1 INTRODUCTION

The positions of atoms within molecules do not remain constant, but undergo continual periodic movement (vibrations) relative to each other. For any given vibrational motion, only certain energies are possible for the vibrational energy states; that is, the energy levels are quantized. Transitions between the energy levels of the vibrational motions can be induced by the absorption of electromagnetic radiation in the infrared region. For a molecule containing n atoms there are $3n - 6$ independent vibrational motions called *normal modes* (linear molecules possess $3n - 5$ normal modes). Although these normal modes involve a vibrational motion of the entire molecule, which at first glance might make it appear that the infrared spectrum of a molecule containing a large number of atoms would contain a large number of randomly positioned absorption bands, and thus would be hopelessly complex, most of the normal modes are highly localized in a given bond or functional group. As a result, the absorption bands of individual functional groups appear in fairly well-defined and limited portions of the infrared region, allowing for the development of spectra-structure correlations.

The infrared region of the spectrum extends from the upper end of the visible region, approximately 0.75 μm, to the microwave region near 400 μm. The portion of this region generally used by the organic chemist for structural work is from 2.5 to 16 μm. This is primarily due to instrument design and cost, and to the fact that most of the useful information can be derived in this region. The region from 0.75 to

2.5 μm is referred to as the near-infrared region and contains absorption bands due to overtone and combination bands. The region extending from 16 to 400 μm is referred to as the far-infrared region. The normal and far-infrared regions contain absorptions due to fundamental, overtone, and combination bands.

The position of absorption in the infrared region can be expressed in the wavelength [micrometer (μm)] of the absorbed radiation, or in terms of the wave number (cm^{-1}, the reciprocal of the wavelength in centimeters). Prior to the early 1970s, wavelengths were commonly given in microns (μ); 1 μ = 1 μm. The wave number, or incorrectly termed the frequency, convention is the most commonly used. In the following portions of this text, absorption positions will be given in wave numbers followed by wavelength in μm in parentheses. In Chapter 9 only cm^{-1} will be given.

The vibrational motions within molecules fall into two categories: (1) those involving the vibration of a bond along the bond axis (a stretching deformation designated by v); and (2) those involving a vibration of a bond perpendicular to the bond axis (a bending deformation of the bond in a plane or perpendicular to an internal plane of symmetry in the function and designated by σ). If we view the

$$A \longleftrightarrow B \qquad\qquad A \underset{\longleftarrow}{\overset{\longrightarrow}{}} B$$

stretching deformation   bending deformation

vibrational motions of a bond as a classical harmonic oscillator (such as a spring), the vibrational frequency, v_{osc} in a system A—B, in which the mass of A is much larger than that of B, is given by Equation (5.1), in which k is the force constant (i.e., a measure of the energy required to deform the oscillator), and m is the mass of B.

$$v_{osc} = \frac{1}{2\pi}\sqrt{\frac{k}{m}} \tag{5.1}$$

In bonded systems k increases with increasing bond order, causing v_{osc} to increase. For example, the vibrational frequencies increase in the order $v_{C-C} < v_{C=C} < v_{C\equiv C}$. When the masses of A and B are comparable Equation (5.2) is applicable, in which μ is the reduced mass of the system ($m_A \times m_B / m_A + m_B$). If one applies Equation (5.2) to a C—H bond in a complex molecule of mol. wt. 100, increasing the mol. wt. to 114 (the next higher homolog) results in very little change in v_{osc}, which is consistent with the observation that bands arising from a given function tend to appear in narrow portions of the infrared region.

$$v_{osc} = \frac{1}{2\pi}\sqrt{\frac{k}{\mu}} \tag{5.2}$$

Not all of the $3n - 6$ normal modes of a molecule necessarily absorb in the infrared region. In order for a normal mode to absorb in the infrared, a bond must possess a dipole moment that changes in magnitude on vibrational excitation (vibrationally excited states generally have slightly longer bond lengths and, thus, larger

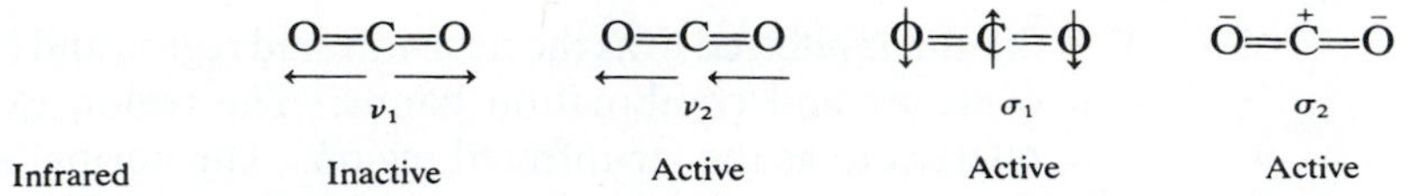

$$O=C=O \qquad O=C=O \qquad \phi=\overset{\ |}{C}=\phi \qquad \bar{O}=\overset{+}{\underset{\ }{C}}=\bar{O}$$
$$\overset{\longleftarrow \quad \longrightarrow}{\nu_1} \qquad \overset{\longleftarrow \quad \longleftarrow}{\nu_2} \qquad \sigma_1 \qquad \sigma_2$$

Infrared   Inactive   Active   Active   Active

Figure 5.1 Normal modes of carbon dioxide. Arrows indicate the direction of movement of the atoms in the plane of the page; (+) and (−) indicate movement forward and backward, respectively, out of the plane of the page.

dipole moments than the ground vibrational state).[1] As an example, let us consider the linear molecule CO_2, which possesses $3 \times 3 - 5 = 4$ normal modes of which two are stretching (ν_1 and ν_2) and two are bending (σ_1 and σ_2) deformations (see Figure 5.1). The stretching deformation ν_1 does not result in a change in the dipole moment on vibrational excitation, and thus does not give rise to an infrared absorption band. The deformations ν_2, σ_1, and σ_2 do result in a change in dipole moment on vibrational excitation; hence, they absorb infrared radiation.

Certain functions give rise to characteristic group deformation bands which are extremely useful in differentiating between related functions. In an A—B

system only a single stretching deformation is possible. In an $A\overset{\displaystyle B}{\underset{\displaystyle B}{\diagup}}$ system,

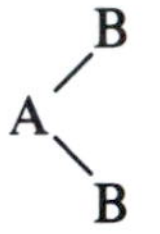

however, two stretching deformations are possible, one being a symmetric stretching deformation and the other being an antisymmetric deformation. Thus two bands are found in the infrared when such groups are present. For example, this difference provides a ready means for distinguishing between primary and

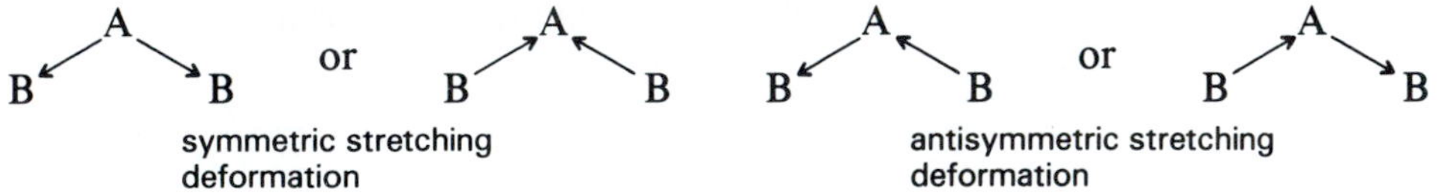

symmetric stretching deformation   or   antisymmetric stretching deformation

secondary amines, the former giving rise to two N—H bands and the latter to one N—H stretching band. Other functions readily distinguished are: primary and

secondary amides, nitroso (—N=O) and nitro $\left(-\overset{+}{N} \overset{\displaystyle \diagup\!\!\!O}{\diagdown_{O-}} \right)$, and sulfinyl

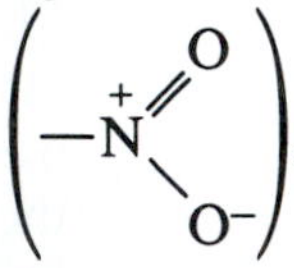

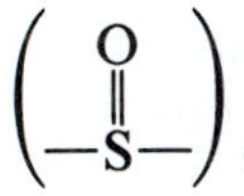

$\left(-\overset{O}{\underset{\|}{S}}- \right)$ and sulfonyl $\left(-\overset{O}{\underset{\underset{\displaystyle O}{\|}}{\overset{\|}{S}}}- \right)$ functions. AB_3 systems, such as the methyl

group, similarly give rise to two stretching deformations: one symmetric and one antisymmetric.

---

[1] This is an extremely simplified statement of the requirements for infrared activity. The reader is referred to F. A. Cotton, *Chemical Applications of Group Theory*, 3rd ed. (New York: Interscience, 1983), Chapter 9; and H. Jaffe and M. Orchin, *Symmetry in Chemistry* (New York: John Wiley, 1965) for more detailed and rigorous discussions of the requirements for infrared and Raman activity.

The intensities of the infrared absorption bands vary from strong to very weak (eventually disappearing if there is little or no change in dipole moment on excitation). The intensity of an infrared absorption band cannot be expressed as a unique constant as was possible in ultraviolet and visible spectroscopy. This is due to the fact that the slit widths (fine slits controlling the amount of infrared radiation passing through the instrument) used in most instruments are of the same order as the typical infrared bandwidths, which causes the measured optical density to be a function of the slit width. Despite this problem, infrared spectroscopy can be used as a quantitative analytical tool with certain precautions. Instead of determining the true molar extinction coefficient, one can determine an apparent molar extinction coefficient, ϵ_a, which will be a function of the slit width and even sample concentration. The apparent extinction coefficient is calculated as shown in Equation (5.3).

$$\epsilon_a = \frac{\text{absorbance}}{\text{concentration (mol/L)} \times \text{cell length (cm)}} \qquad (5.3)$$

A series of apparent extinction coefficients *vs.* concentration should be determined, and the data plotted for use in final analysis. The apparent extinction coefficients can be used only with the instrument they have been measured on. Analytical limits are approximately $\pm 5\%$ when these methods are used.

5.2 CHARACTERISTIC ABSORPTION BANDS

Not all the absorption bands appearing in an infrared spectrum will be useful in deriving structural information. Certain portions of the infrared region, for example, the region of 1300 to 1000 cm^{-1} (7.5 to 10 μm), are extremely difficult to interpret due to the variety and number of fundamental absorptions occurring in this region. Certain narrow regions of the infrared spectrum provide most of the important information. In the derivation of information from an infrared spectrum, the most prominent bands in these regions are noted and assigned first.

The characteristic bands of various individual functions will be discussed in the following sections. These discussions will include the effect of molecular structure and electronic effects on the more prominent absorption bands.

5.2.1 Carbon-Hydrogen Absorption Bands

Alkanes

Absorption by carbon-hydrogen bonds occurs in two regions: the C—H stretching region of 3300 to 2500 cm^{-1} (3 to 4 μm) and the bending region of 1550 to 650 cm^{-1} (6.5 to 15.4 μm). In a molecule that contains a large number of C—H bonds the interpretation of these regions becomes very difficult because of the large number of very close, sometimes overlapping, bands. An exception is for the methyl group, whose symmetric bending deformation appears near 1380 cm^{-1} (7.25 μm), a

TABLE 5.1 Selected Carbon-Hydrogen Absorption Bands*

Functional Group	Wave Number cm⁻¹	Wavelength μm	Assignment†	Remarks‡
Alkyl				
—CH₃	2960	3.38	ν_{as}	s, lu
	2870	3.48	ν_s	m, lu
	1460	6.85	σ_{as}	s, lu
	1380	7.25	σ_s	m, gu
—CH₂—	2925	3.42	ν_{as}	s, lu
	2850	3.51	ν_s	m, lu
	1470	6.83	σ	s, lu
	1250	~8.00	σ	s, nu
—C̶—H	2890	3.46	ν	w, nu
	1340	7.45	σ	w, nu
—OCOCH₃	1380–1365	7.25–7.33	σ_s	s, gu
—COCH₃	~1360	~7.35	σ_s	s, gu
—COOCH₃	~1440	~6.95	σ_{as}	s, gu
	~1360	~7.35	σ_s	s, gu
Vinyl				
=CH₂	3080	3.24	ν_{as}	m, lu
	2975	3.36	ν_s	m, lu
	~1420	~7.0–7.1	σ(in-plane)	m, lu
	~900	~11	σ(out-of-plane)	s, gu
C=C⟨H	3020	3.31	ν	m, lu
Monosubstituted	990	10.1	σ(out-of-plane)	s, gu
	900	11.0	σ(out-of-plane)	s, gu
Cis-disubstituted	730–675	13.7–14.7	σ(out-of-plane)	s, gu
Trans-disubstituted	965	10.4	σ(out-of-plane)	s, gu
Trisubstituted	840–800	11.9–12.4	σ(out-of-plane)	m-s, gu
Aromatic				
C—H	3070	3.30	ν	w, lu
5-adjacent H	770–730	13.0–13.7	σ(out-of-plane)	s, gu
	710–690	14.1–14.5	σ(out-of-plane)	s, gu
4-adjacent H	770–735	13.0–13.6	σ(out-of-plane)	s, gu
3-adjacent H	810–750	12.3–13.3	σ(out-of-plane)	s, gu
2-adjacent H	860–800	11.6–12.5	σ(out-of-plane)	s, gu
1-adjacent H	900–860	11.1–11.6	σ(out-of-plane)	m, gu
1,2-, 1,4-, and	1275–1175	7.85–8.5	σ(in-plane)	w, lu
1,2,4-substituted	1175–1125	8.5–8.9	(only with 1,2,4-)	
	1070–1000	9.35–10.0	(two bands)	w, lu
1-, 1,3-, 1,2,3-, and	1175–1125	8.5–8.9	σ(in-plane)	
1,3,5-substituted	1100–1070	9.1–9.35	(absent with 1,3,5-)	
	1070–1000	9.35–10.0		
1,2-, 1,2,3-, and	1000–960	10.0–10.4	σ(in-plane)	w, lu
1,2,4-substituted				

TABLE 5.1 (Continued)

Functional Group	Wave Number cm^{-1}	Wavelength μm	Assignment†	Remarks‡
Alkynyl				
$\equiv$C—H	3300	3.0	ν	m-s, gu
Aldehydic				
$-C{\overset{O}{\underset{H}{\Big\backslash}}}$	2820	3.55	ν	m, lu
	2720	~3.7~	Overtone or combination band	m, gu

* Values selected from tables presented in Bellamy (ref. 3) and Nakanishi (ref 13).

† Assignments are designated as follows: ν, stretching deformation; as, asymmetric; s, symmetric; σ, bending deformation.

‡ Intensities are designated as: s, strong; m, medium; w, weak. Band utilities for structural assignments are indicated as: gu, great utility; lu, limited utility (depends on the complexity of the structure); nu, no practical utility.

region that is almost devoid of other types of absorption bands. Table 5.1 lists the positions of absorption of a wide variety of C—H bonds.

Alkenes and Aromatic Compounds

The stretching absorption bands of vinyl and aromatic C—H bonds occur in the region of 3180 to 2980 cm^{-1} (3.15 to 3.36 μm), while the out-of-plane bending deformation (in which the C—H bond bends out of the plane of the molecule) bands appear in the 1000 to 650 cm^{-1} (10.0 to 15.4 μm), a region generally referred to as the *fingerprint region*. [The in-plane bending deformation bands appear in the 1400 to 1000 cm^{-1} (7.1 to 10.0 μm), a region that is highly congested and difficult to interpret and is generally ignored.] Although the higher frequency region is often difficult to interpret, the bands appearing in the fingerprint region are extremely useful for assigning the degree and stereochemistry of substitution on a C=C, and the degree and position of substitution on a benzene ring.

Vinyl groups (—CH=CH$_2$) give rise to two out-of-plane bending deformation bands, one band appearing near 900 cm^{-1} (11 μm), and the other near 990 cm^{-1} (10.1 μm). The terminal methylene of 1,1-disubstituted alkenes also appears near 900 cm^{-1} (11 μm). This band is quite intense and usually appears as one of the more prominent bands in a spectrum. *Cis-* and *trans*-disubstituted alkenes absorb near 685 and 965 cm^{-1} (14.2 and 10.4 μm), respectively. The out-of-plane bending deformation of a trisubstituted alkene appears near 820 cm^{-1} (12.1 μm). The position of this band is quite different compared with the corresponding band of the terminal vinyl group, which appears at 990 cm^{-1} (10.1 μm). Typical fingerprint regions of spectra of alkenes are illustrated in Figure 5.2.

A monosubstituted benzene displays two out-of-plane bending deformation bands near 750 and 710 cm^{-1} (13.4 and 14.3 μm). *Ortho*-disubstituted benzenes

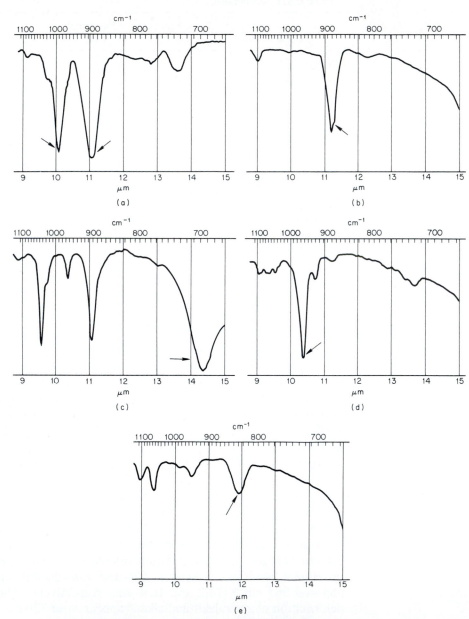

Figure 5.2   Olefinic C—H bending bands of a monosubstituted alkene (a), 1,1-di-substituted alkene (b), *cis*-di- (c), and *trans*-disubstituted (d) alkenes, and a trisubstituted alkene (e). Note that some of the spectra also contain other deformation bands in the low-frequency region that are not due to the alefinic C—H.

display a single band near 750 cm^{-1} (13.4 μm), characteristic of a four-adjacent hydrogen system; *para*-disubstituted benzenes display a single band near 830 cm^{-1} (12.1 μm), characteristic of a two-adjacent hydrogen system; *meta*-disubstituted benzenes display bands near 780 cm^{-1} (12.8 μm), characteristic of a three-adjacent hydrogen system, and 880 cm^{-1} (11.4 μm), characteristic of a one-adjacent hydrogen system. Several of these types of systems are illustrated at the end of this chapter. These characteristic absorption patterns for the various adjacent hydrogen systems are also observed with substituted pyridines and polycyclic benzenoid aromatics, for example, substituted naphthalenes, anthracenes, and phenanthrenes.

The 2000 to 1670 cm^{-1} (5 to 6 μm) region the spectra of aromatic compounds display several weak bands that are overtone and combination bands. These bands are usually quite weak, and a fairly concentrated solution of the sample must be used to record these bands (see Figure 5.3). These bands are also useful in deriving structural information; however, the positions of these bands are sensitive to the types of substituents present in the system, as was noted for the out-of-plane deformations discussed in preceding paragraphs. Figure 5.3 displays typical absorption patterns of substituted aromatics in the 2000 to 1670 cm^{-1} (5 to 6 μm) region.

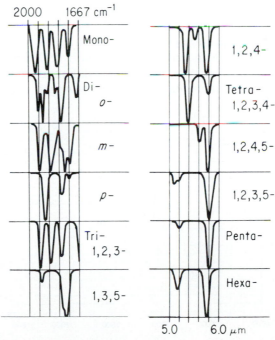

Figure 5.3   Typical absorption patterns of substituted aromatics in the 2000 to 1670 cm^{-1} (5.0 to 6.0 mm) region.

Acetylenic Compounds

Acetylenic carbon-hydrogen bonds absorb very close to 3330 cm⁻¹ (3.0 μm). The band is quite sharp and intense.

Aldehydic Carbon-Hydrogen

The aldehyde C—H gives rise to two bands at approximately 2820 and 2700 cm⁻¹ (3.55 and 3.7 μm). The lower-wave-number band is quite useful in detecting the presence of an aldehyde group.

5.2.2 Oxygen-Hydrogen Absorption Bands

The oxygen-hydrogen stretching deformation occurs in the 3700 to 2500 cm⁻¹ (2.7 to 4.0 μm) region. The position and the shape of the absorption bands vary greatly with structure and can be very useful in deriving information concerning the structure of the molecule. The position and shape of the absorption band are also quite sensitive to the type and extent of hydrogen bonding, either to another hydroxyl group or another acceptor group.

The nonbonded hydroxyl gives rise to a sharp absorption whose position varies slightly from primary to secondary to tertiary hydroxyl. The difference in band positions varies only about 10 cm⁻¹ (0.01 μm) between primary and secondary, and secondary and tertiary. This difference in position borders on the limits of resolution of spectrometers commonly available in laboratories, and it should not be relied on to decide the environment of the hydroxyl group. Nonbonded hydroxyl absorption can be observed only in dilute solutions in nonpolar solvents. At normal concentrations, approximately 5 to 10% by weight, extensive intermolecular hydrogen bonding

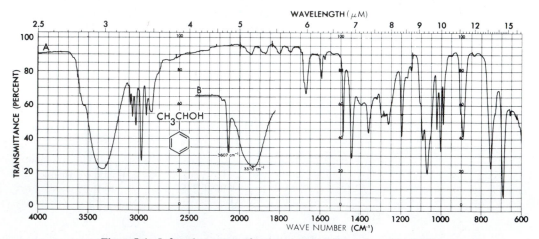

Figure 5.4 Infrared spectrum of 1-phenylethanol. Trace A, neat liquid film; trace B, 5% in carbon tetrachloride. Trace B is of the 3900 to 3150 cm⁻¹ region and has been displaced on the chart paper.

occurs, causing an additional broad absorption at lower wave number (see Figure 5.4). The position and shape of the band depend on the extent of hydrogen bonding, becoming more broad and moving to lower wave number as the strength of the hydrogen bond increases (see Table 5.2). Occasionally another functional group may be present in the same molecule as the hydroxyl group to act as an acceptor

TABLE 5.2   Selected X-H Absorption Bands

Functional Group	Wave Number cm^{-1}	Wavelength μm	Remarks
Alcohols			
Primary (nonbonded)	3640	2.72	m, gu, usually determined in dilute solution in nonpolar solvents.
Secondary	3630	2.73	
Tertiary	3620	2.74	
Phenols	3610	2.75	
Intermolecularly H-bonded	3600–3500	2.78–2.86	m, dimeric, rather sharp
	3400–3200	2.94–3.1	s, polymeric, usually quite broad
Intramolecularly H-bonded	3600–3500	2.78–2.86	m-s, much sharper than intermolecular hydrogen bonded OH; is not concentration dependent.
Amines			
RNH$_2$	~3500	~2.86	m, gu, ν_{as}
	~3400	~2.94	m, gu, ν_s
	1640–1560	6.1–6.4	m-s, gu, corresponds to scissoring deformation.
R$_2$NH	3500–3450	2.86–2.90	w-m, gu, ν
ArNHR	3450	2.90	w-m, gu
Pyrroles, indoles	3490	2.86	w-m, gu
Ammonium salts			
NH$_4^+$	3300–3030	3.0–3.3	s, gu
	1430–1390	7.0–7.2	s, gu
$-\overset{+}{\text{N}}\text{H}_3$	3000	~3.0	s, gu, usually quite broad
	1600–1575	6.25–6.35	s, gu, σ_{as}
	1490	~6.7	s, gu, σ_s
$\overset{+}{\diagdown}\text{N}\diagup\text{H}_2$	2700–2250	3.7–4.4	s, gu, ν_{as} and ν_s, usually broad or a group of bands
$\overset{+}{-}\text{N}\diagup\text{H}$	1600–1575	6.25–6.35	m, gu, σ
	2700–2250	3.7–4.4	s, gu, ν, σ_{NH^+}, band is weak and of no practical utility.
Thiols			
—SH	2600–2550	3.85–3.92	s, gu, band is often very weak and can be missed if care is not exercised.

functional group for intramolecular hydrogen bonding. The shape and intensity of the absorption band are not functions of the concentration of the sample and can be differentiated from intermolecular hydrogen bonding by dilution studies.

The oxygen-hydrogen stretching absorption bands of phenols occur at slightly lower frequency than those of alcohols, but usually not sufficiently different to allow for distinction between phenols and alcohols.

The oxygen-hydrogen stretching absorption of carboxylic acids appears as a very broad band with maximum absorption at approximately $2940 \, cm^{-1}$ ($3.4 \, \mu m$); it extends to nearly $2500 \, cm^{-1}$ ($4.0 \, \mu m$), due to the very strong intermolecular hydrogen bonding between carboxyl groups. Since acids exist as dimers in nonpolar solvents, the position and shape of this band are little affected by changes in concentration.

The oxygen-hydrogen bending absorption occurs in the 1500 to $1300 \, cm^{-1}$ (6.7 to $7.7 \, \mu m$) region and is of no practical value for analysis.

5.2.3 Nitrogen-Hydrogen Absorption Bands

Nitrogen-hydrogen stretching absorptions occur at slightly lower frequencies than the O—H stretching absorptions do. Primary amines display two bands corresponding to asymmetric and symmetric stretching deformations, whereas secondary amines display a single peak (Table 5.2). Imines, $\diagdown$C=NH, display a single peak in the same region. In general, it is easy to distinguish primary from secondary amines, but distinction between various types of $\diagdown$N—H is difficult.

Amines display N—H bending deformation absorption bands similar to those of —CH$_2$— and —C̣—H, except that they occur at slightly higher frequency. Protonation of amines, to give the corresponding amine salts, results in the formation of bands similar to those of —CH$_3$, —CH$_2$—, and —C̣—H, again appearing at higher wave numbers.

Primary amides display N—H stretching bands at slightly higher wave numbers than do amines. The N—H bending deformations of amides occur in the 1610 to $1490 \, cm^{-1}$ (6.2 to $6.7 \, \mu m$) region and are generally referred to as "amide II" bands. These bands are quite intense.

Hydrogen bonding involving N—H gives rise to similar effects noted with hydroxyl absorption bands, although N—H bonds show a lesser tendency to hydrogen-bond than do alcohols.

5.2.4 Sulfur-Hydrogen Absorption Bands

The sulfur-hydrogen stretching absorption occurs near 2630 to $2560 \, cm^{-1}$ (3.8 to $3.9 \, \mu m$), is usually very weak, and is not subject to large shifts due to hydrogen bonding.

5.2.5 *Carbon-Carbon Absorption Bands*

Carbon-carbon single-bond stretching bands appear from 1350 to 1150 cm^{-1} (7.4 to 8.7 μm), are extremely variable in position, and are usually quite weak in intensity. These absorption bands are of little practical use in structure determination.

Carbon-carbon double bond stretching absorption occurs in the 2000 to 1430 cm^{-1} (5 to 7 μm) region. The position and intensity of these bands are very sensitive to the degree and type of substitution and, in cyclic alkenes, to ring size. Alkyl substitution results in a shift of the absorption band to higher wave numbers. The shift to higher wave numbers with increasing substitution is least with acyclic alkenes and increases dramatically with decreasing ring size in cyclic alkenes (see Table 5.3). Decreasing the ring size of cyclic alkenes from a six- to a four-membered ring results in shifts to lower wave numbers; however, a further decrease in ring size to cyclopropene results in a dramatic shift to higher wave numbers. These shifts are probably due to changes in hybridization involving the carbon atoms that comprise the carbon-carbon double bond. The effect of conjugation with an aromatic nucleus results in a slight shift to lower wave number, and with another C=C or C=O, a shift to lower wave number of approximately 40 to 60 cm^{-1} (0.15 to 0.20 μm) with a substantial increase in intensity.

The intensity of the C=C stretching absorption band is a function of the symmetry of substitution on the double bond; mono- and trisubstituted alkenes produce more intense bands than do unsymmetrical di- and tetrasubstituted alkenes, with symmetrically substituted di- and tetrasubstituted alkenes generally showing no $\nu_{C=C}$ absorption. Enamines (C=C—NR$_2$), and the related enol ethers (C=C—OR), give rise to very intense absorption bands.

The skeletal C=C vibrations of aromatics give rise to a series of four bands in the 1660 to 1430 cm^{-1} (6 to 7 μm) region. These bands occur very close to 1600 cm^{-1} (6.24 μm), 1575 cm^{-1} (6.34 μm), 1500 cm^{-1} (6.67 μm), and 1450 cm^{-1} (6.9 μm). The first and third bands vary greatly in intensity, generally becoming more intense in the presence of electron-donating groups.

Bending deformation absorptions of C—C and C=C bonds occur below 670 cm^{-1} (15 μm) and are of little utility in structure identification.

The carbon-carbon double bonds of allenes (C=C=C) generally give rise to a rather intense, sharp band in the 1950 cm^{-1} (5.1 μm) region. This region is characteristic for the cumulatively bonded systems (X=Y=Z).

Carbon-carbon triple bond stretching absorption in terminal acetylenes occurs in the 2140 to 2080 cm^{-1} (4.6 to 4.8 μm) region. The absorption band is relatively weak, but quite sharp. A band in this region, along with the C—H stretching absorption band near 3300 cm^{-1} (3.0 μm) is indicative of the presence of a terminal acetylene.

Unsymmetrically disubstituted acetylenes absorb in the 2260 to 2190 cm^{-1} (4.42 to 4.57 μm) region. The intensity of this band is a function of the symmetry of the molecule; the more symmetrical the substitution, the less intense the absorption band.

TABLE 5.3 Selected C—C and C—N Absorption Bands

Functional Group	Wave Number cm^{-1}	Wavelength μm		Remarks
Acyclic C=C($\nu_{C=C}$)				
Monosubstituted	1645	6.08	m	These bands are of little utility in assigning
1,1-Disubstituted	1655	6.04	m	substitution and stereochemistry; the C—H
Cis-1,2-disubstituted	1660	6.02	m	out-of-plane bending bands in the fingerprint
Trans-1,2-disubstituted	1675	5.97	w	region are recommended for this purpose.
Trisubstituted	1670	5.99	w	
Tetrasubstituted	1670	5.99	w	
Allenes	1950	5.11	m, gu, $\nu_{C=C=C}$, terminal allenes display two bands and ν_{C-H} at 850 cm^{-1} (11.76 μm).	
Conjugated C=C($\nu_{C=C}$)				
Diene	1650 and 1600	6.06 and 6.25	s, gu	
With aromatic	1625	6.16	s, gu	
With C=O	1600	6.25	s, gu	
Cyclic C=C($\nu_{C=C}$)				
6-membered ring and larger	1646	6.08	m	Of limited utility due to closeness to the acyclic region.
Monosubstituted	1680–1665	5.95–6.00	m	
5-m. unsubstituted	1611	6.21	m	Can be of great utility although the ring size
Monosubstituted	1660	6.02	m	must be assigned first for those compounds
Disubstituted	1690	5.92	w	whose absorption falls into regions consistent
4-m. unsubstituted	1566	6.39	m	with other types of absorption bands (nuclear
3-m. unsubstituted	1640	6.10	m	magnetic resonance spectra can be very useful
Monosubstituted	1770	5.65	m	in this respect).
Disubstituted	1880	5.32	m	
Aromatic C=C	1600	6.24	w-s	In-plane skeletal vibrations; the intensities of the
	1580	6.34	s	1600 and 1500 cm^{-1} may be rather weak.
	1500	6.67	w-s	
	1450	6.9	s	
Acetylenes				
terminal	2140–2100	4.67–4.76	w, gu, $\nu_{C\equiv C}$, to be used in conjunction with the ν_{C-H} band.	
disubstituted	2260–2190	4.42–4.57	w, lu, $\nu_{C\equiv C}$, may be completely absent in symmetrical acetylenes.	
C—N				
Aliphatic amine	1220–1020	8.2–9.8	m-w, lu, ν_{C-N}	
Aromatic amine	1370–1250	7.3–8.0	s, gu-lu, ν_{C-N}	
C=N				
	1700–1615	5.9–6.2	s, gu, $\nu_{C=N}$	
Nitriles				
alkyl	2260–2240	4.42–4.47	m-s, gu, $\nu_{C\equiv N}$	
aryl	2240–2220	4.46–4.51	s, gu, $\nu_{C\equiv N}$	
α,β-unsaturated	2235–2215	4.47–4.52	s, gu, $\nu_{C\equiv N}$	

    Conjugation with a $C{=}C$ causes an increase in the band intensity and a shift to slightly lower wave numbers. The observed shifts are usually less than those observed with alkenes. Conjugation with carbonyl groups does not appreciably alter the position of absorption.

5.2.6   Carbon-Nitrogen Absorption Bands

Aliphatic amines display $C{-}N$ stretching absorptions in the 1220 to 1020 cm^{-1} (8.2 to 9.8 μm) region. These bands are of medium to weak intensity. Owing to the intensity and position of these bands, they are of little practical value for structural work, except in relatively simple molecules. Aromatic amines absorb in the 1370 to 1250 cm^{-1} (7.3 to 8.0 μm) region and give quite intense bands. Carbon-nitrogen double bond absorption occurs in the 1700 to 1615 cm^{-1} (5.9 to 6.2 μm) region and is subject to similar environmental effects as is the $C{=}O$ absorption bands (see following section).

    The carbon-nitrogen triple bond of nitriles absorbs in the 2260 to 2210 cm^{-1} (4.42 to 4.52 μm) region. Saturated nitriles absorb in the higher wave-number portion of the region, while unsaturated nitriles absorb in the lower wave-number end. The $C{\equiv}N$ absorption band is of much greater intensity than the $C{\equiv}C$ absorption bands.

5.2.7   Carbon-Oxygen Absorption Bands

Carbon-oxygen single bond absorption occurs in the 1250 to 1000 cm^{-1} (8 to 10 μm) region. Although both alcohols and ethers absorb in this region, distinction between these classes can be made from information gained from the hydroxyl region.

    The carbon-oxygen double bond absorbs in the 2000 to 1540 cm^{-1} (5 to 6.5 μm) region, except for ketenes, which absorb near 2200 cm^{-1} (4.65 μm). This is probably the most useful portion of the infrared region because the position of the carbonyl group is quite sensitive to substituent effects and the structure of the molecule. The band positions are solvent-sensitive due to the high polarity of the $C{=}O$ bond.

    An empirical correlation that can be used to adequately predict the positions of most carbonyl bands has been developed. Table 5.4 lists the band positions for a number of different types of compounds that contain the $C{=}O$ group. For ketones, aldehydes, acids, anhydrides, acid halides, and esters, the wave numbers (wavelengths) cited are for the normal, unstrained (acyclic or contained in a six-membered ring) *parent* compound in carbon tetrachloride solution. The introduction of a double bond or aryl group in conjugation with the $C{=}O$ results in a reasonably consistent 30 cm^{-1} (0.1 μm) shift to lower wave number. Introduction of a second double bond results in an additional shift of approximately 15 cm^{-1} (0.05 μm) in the same direction.

    The position of the $C{=}O$ band is very sensitive to changes in the C(CO)C bond angle. A decrease in ring size, resulting in a decrease in the C(CO)C bond angle, of ketones and esters results in a reasonably consistent shift of 30 cm^{-1} (0.1 μm) to

TABLE 5.4 Selected Carbon-Oxygen Absorption Bands

Functional Groups	Wave Number cm^{-1}	Wavelength μm	Remarks
C—O single bonds			
Primary C—O—H	1050	9.52	s, gu, ν_{C-O}
Secondary C—O—H	1100	9.08	s, gu, ν_{C-O}
Tertiary C—O—H	1150	8.68	s, gu, ν_{C-O}
Aromatic C—O—H	1200	8.33	s, gu, ν_{C-O}
Ethers-acyclic	1150–1070	8.7–9.35	s, gu, antisymmetric ν_{C-O-C}
C=C—O—C	1270–1200 and	7.9–8.3 and	s, gu, antisymmetric ν_{C-O-C}
	1070–1020	9.3–9.8	s, gu, symmetric ν_{C-O-C}
Cyclic ethers			
6-member and larger	1140–1170	8.77–9.35	s, gu
5-m.	1100–1075	9.1–9.3	s, gu
4-m.	980–970	10.2–10.3	s, gu
Epoxides	1250	8.0	s, gu
Cis-disubstituted	890	11.25	s, gu
Trans-disubstituted	830	12.05	s, gu
C—O double bonds			
Ketones	1715	5.83	s, gu, $\nu_{C=C}$, unstrained C=O group in acyclic and 6 m. ring compounds in carbon tetrachloride solution. (See discussion for effects of conjugation and ring size.)
α, β-unsaturated	1685	5.93	s, gu, $\nu_{C=O}$. For the s-*cis* conformation 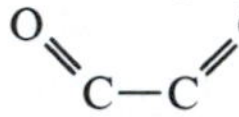 the $\nu_{C=C}$ may appear above 1600 cm^{-1} (below 6.26 μm) with an intensity approximately that of the $\nu_{C=O}$. The *trans*-conformation 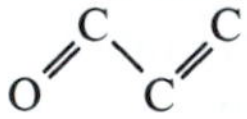 does not show this enhanced intensity of absorption of the C=C.
α- and β-diketones	1720	5.81	s, gu, $\nu_{C=O}$, two bands at higher frequency when in the s-*cis* conformation 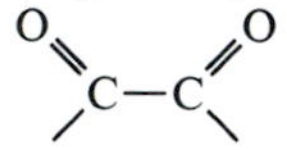
	1650	6.06	s, gu, $\nu_{C=C}$ if enolic. 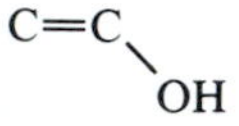
Ketenes	2150	4.65	s, gu
Quinones	1675	5.97	s, gu, $\nu_{C=O}$
Tropolones	1650	6.06	s, gu, $\nu_{C=O}$
	1600	6.26	s, gu, $\nu_{C=O}$ if intramolecularly hydrogen bonded as in α-tropolones.

TABLE 5.4 (Continued)

Functional Groups	Wave Number cm^{-1}	Wavelength μm	Remarks
Aldehydes $\left(-C\!\!\begin{array}{c} \nearrow O \\ \searrow H \end{array} \right)$	1725	5.80	s, gu, $\nu_{C=O}$ (See discussion for effects of conjugation.)
Carboxylic acids and derivatives			
$\left(-C\!\!\begin{array}{c} \nearrow O \\ \searrow OH \end{array} \right)$	1710	5.84	s, gu, $\nu_{C=O}$, usually as the dimer in nonpolar solvents, monomer absorbs at 1730 cm^{-1} (5.78 μm) and may appear as a shoulder in the spectrum of a carboxylic acid; for conjugation effects see the discussion.
Esters	1735	5.76	s, gu, $\nu_{C=O}$, acyclic and 6-m lactones; for conjugation and ring size effects see the discussion.
	1300–1050	7.7–9.5	s, lu, symmetric and antisymmetric ν_{C-O-C} giving 2 bands, indicative of type of ester, for example, formates: 1178 cm^{-1} (8.5 μm) acetates: 1242 cm^{-1} (8.05 μm) methyl esters: 1164 cm^{-1} (8.6 μm) others: 1192 cm^{-1} (8.4 μm) but the distinction generally is not great enough to be of diagnostic value.
Vinylesters $\left(-C\!\!\begin{array}{c} \nearrow O \\ \searrow O-C=C \end{array} \right)$	1755	5.70	s, gu, $\nu_{C=O}$
Anhydrides	1820 and 1760	5.48 and 5.68	s, gu, $\nu_{C=O}$, the intensity and separation of the bands may be quite variable. (See discussion for general effects of conjugation and ring size.)
Acid halides	1800	5.56	s, gu, $\nu_{C=O}$, acid chlorides and fluorides absorb at slightly higher wave number while the bromides and iodides absorb at slightly lower wave number. (See discussion for effects of conjugation.)
Amides	1650	6.06	s, gu, $\nu_{C=O}$, "Amide I" band, this frequency is for the associated amide (see COOH), free amide at 1686 cm^{-1} (5.93 μm) in dilute solution, cyclic amides shift to higher wave number as ring size decreases.
	1300	7.7	s, gu, ν_{C-N}, "Amide III" band, free amide at slightly higher wave number.
α, β-unsaturated	1665	6.01	s, gu, ν_{C-O}
Carboxylate $-C\!\!\begin{array}{c} \nearrow O \\ \searrow O^- \end{array}$	1610–1550 and 1400	6.2–6.45 and 7.15	s, gu, antisymmetric and symmetric stretching of $-C\!\!\begin{array}{c} \nearrow O \\ \searrow O^- \end{array}$

higher wave number per each decrease in ring size from the six-membered ring. In cyclic structures containing more than six atoms, slight shifts to lower wave numbers are observed. For example, a shift of -10 cm^{-1} ($+0.03$ μm) is observed with seven-membered ring compounds, and an approximate -5 cm^{-1} (0.01 to 0.02 μm) shift with eight- and nine-membered ring systems. Further increases in ring size result in a moderate increase in wave number back to the parent compound position. Highly strained bridge carbonyl compounds, for example, 7-oxobicyclo[2.2.1]heptanes and 6-oxobicyclo-[2.1.1]hexanes, absorb at relatively high wave numbers, in the general region of 1800 cm^{-1} (5.5 μm).

The effects of conjugation and ring size are cumulative and can be used to adequately predict the position of absorption in more complex molecules. For example, the calculated positions of absorption in compounds **1** and **2** in carbon tetrachloride solution are illustrated below and are compared with the observed absorptions.

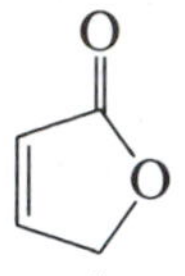

	Normal ester (from Table 5.4)	1735 cm^{-1}	5.76 μm
	Ring size correction	$+30$	-0.10
	Unsaturation	-30	$+0.10$
	Calculated $\nu_{C=O}$	1735 cm^{-1}	5.76 μm
1	Observed $\nu_{C=O}$	1740 cm^{-1}	5.74 μm

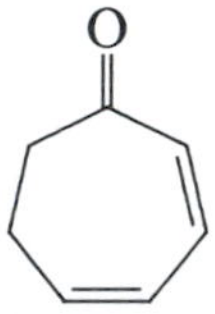

	Normal ketone	1715 cm^{-1}	5.83 μm
	Ring size correction	-10	$+0.03$
	Unsaturation (first C=C)	-30	$+0.10$
	(second C=C)	-15	$+0.05$
	Calculated $\nu_{C=C}$	1660 cm^{-1}	6.01 μm
2	Observed $\nu_{C=O}$	1666 cm^{-1}	6.00 μm

5.2.8 Absorption Bands of Other Functional Groups

The absorption bands of other selected functional groups occasionally encountered in organic compounds appear in Table 5.5.

5.3 INTERPRETATION OF REPRESENTATIVE INFRARED SPECTRA

In the analysis of the infrared spectrum of an unknown material certain guidelines are useful. As indicated in the previous sections, there are particular regions that provide unambiguous information concerning the nature of functions present in a molecule. We suggest the following steps in interpreting an infrared spectrum of an unknown.

1. Check the 3700 to 3300 cm^{-1} (2.7 to 3.0 μm) region for the possible presence of O—H or N—H containing functions.

TABLE 5.5 Absorption Bands of Selected Functional Groups

Functional Group	Wave Number cm^{-1}	Wavelength μm	Remarks
Nitro (R—NO$_2$)	1570–1500	6.37–6.67	s, gu, antisymmetric ν_{NO_2}
	1370–1300	7.30–7.69	s, gu, symmetric ν_{NO_2}, conjugated nitro absorbs in the lower-wave number portions of the cited regions.
Nitrate (RO—NO$_2$)	1650–1600	6.06–6.25	s, gu, antisymmetric ν_{NO_2}
	1300–1250	7.69–8.00	s, gu, symmetric ν_{NO_2}, additional bands at 870–855 cm^{-1} (11.5–11.7 μm) (ν_{O-N}), 763 cm^{-1} (13.1 μm) (out-of-plane bending) and 703 cm^{-1} (14.2 μm) NO$_2$ bending).
Nitramine (N—NO$_2$)	1630–1550	6.13–6.45	s, gu, antisymmetric ν_{NO_2}
	1300–1250	7.69–8.00	s, gu, symmetric ν_{NO_2}
Nitroso (R—N=O)	1600–1500	6.25–6.66	s, gu, $\nu_{N=O}$, conjugation shifts to lower wave numbers in the cited region
Nitrite (R—ONO)	1680–1610	5.95–6.21	s, gu, $\nu_{N=O}$, two bands usually present due to *cis* and *trans* forms.
Nitrone	1170–1280	7.8–8.6	s, gu, $\nu_{C=N}$, appears at 1580–1610 cm^{-1} (6.2–6.3 μm).
Amine Oxide aliphatic	970–950	10.3–10.5	v.s, gu, ν_{N-O}
aromatic	1300–1200	7.7–8.3	v.s, gu, ν_{N-O}
Azoxy	1310–1250	7.63–8.00	v.s, gu, ν_{N-O}
—S—O—	900–700	11.1–14.2	s, gu, ν_{S-O}
Sulfoxide	1090–1020	9.43–9.62	s, gu, $\nu_{S=O}$
Sulfone	1350–1310	7.42–7.63	s, gu, antisymmetric ν_{SO_2}
	1160–1120	8.62–8.93	s, gu, symmetric ν_{SO_2}
Sulfonic acid (—SO$_2$OH)	1260–1150	7.93–8.70	s, gu, antisymmetric ν_{SO_2}
	1080–1010	9.26–9.90	s, gu, symmetric ν_{SO_2}
	700–600	14.2–16.6	s, gu, ν_{S-O}, may appear outside the general infrared region; in addition, O—H absorption appears.
Sulfonyl chlorides (—SO$_2$Cl)	1385–1340	7.22–7.46	s, antisymmetric ν_{SO_2}
	1185–1160	8.44–8.62	s, symmetric ν_{SO_2}
Sulfonate (—SO$_2$OR)	1420–1330	7.04–7.52	s, gu, antisymmetric ν_{SO_2}
	1200–1145	8.33–8.73	s, gu, symmetric ν_{SO_2}
Sulfonamide (—SO$_2$NR$_2$)	1370–1330	7.30–7.52	s, gu, antisymmetric ν_{SO_2}
	1180–1160	8.47–8.62	s, gu, symmetric ν_{SO_2}

2. Check the 1750 to 1670 cm^{-1} (5.7 to 6.0 μm) region for the possible presence of C=O containing functions.
3. Check the 1660 to 1450 cm^{-1} (6.0 to 6.9 μm) region for the possible presence of aromatic and/or C=C functions. If sharp bands are noted in this region, analyze the fingerprint region 1000 to 650 cm^{-1} (10.0 to 15.4 μm) for C—H bending deformation bands for possible assignment of degree and position of substitution on aromatic rings and double bonds.
4. Finally, look for the most intense band, or bands, in other regions, and attempt assignments based on the molecular formula and chemical properties, and by reference to Tables 5.1 to 5.5.

In the following, an interpretation of representative infrared spectra is outlined.

The infrared spectrum of 1-phenylethanol is shown in Figure 5.4. The complete trace was taken of a neat liquid film between sodium chloride plates (see discussion on the preparation of samples, Section 5.4). The intense, broad band near 3360 cm^{-1} (3 μm) represents the hydrogen bonded O—H stretch. The nonbonded absorption peak is barely perceptible as a shoulder near 3550 cm^{-1} (2.81 μm). Trace B is of the hydroxyl region of a carbon tetrachloride solution of 1-phenylethanol, which has been displaced on the chart paper for greater clarity, showing the sharp nonbonded band and the broad bonded band. The C—H stretching region shows the aromatic C—H stretch as three bands just above 3000 cm^{-1} (3.33 μm). The band at 2970 cm^{-1} (3.36 μm) represents the asymmetric stretch of the methyl. The bending deformation bands of the methyl group appear at 1450 and 1365 cm^{-1} (6.9 and 7.3 μm), the former overlapping one of the aromatic skeletal stretching bands. The out-of-plane C—H bending deformation bands occur at 755 and 685 cm^{-1} (13.1 and 14.8 μm), characteristic of a monosubstituted aromatic. The in-plane C—H bending deformation bands appear in the 1100 to 1000 cm^{-1} (9 to 10 μm) region, along with the C—O stretch and other possible skeletal deformation bands, thus making definite assignments rather tenuous. The overtone and combination bands of the out-of-plane and in-plane C—H deformations appear in the 2000 to 1670 cm^{-1} (5 to 6 μm) region. These bands are perceptible in trace A. Note the similarity of these band shapes and positions with those predicted in Figure 5.3. Only two of the aromatic skeletal stretching bands are readily visible in the 1650 to 1430 cm^{-1} (6 to 7 μm) region, those appearing at 1590 and 1485 cm^{-1} (6.3 and 6.7 μm). The immediate information that a reader should derive from this spectrum, if given to the person as an unknown, is the presence of O—H, a monosubstituted benzene ring, a methyl group, and probably very little other aliphatic C—H.

Figure 5.5 displays the spectrum of phenylacetylene. The acetylenic C—H band appears at 3290 cm^{-1} (3.03 μm) as a very sharp, intense band. The C≡C stretch, on the other hand, appears as the relatively weak band at 2100 cm^{-1} (4.76 μm). The typical out-of-plane C—H bending bands appear at 745 and 680 cm^{-1} (13.4 and 14.6 μm), with the in-plane C—H bending bands appearing in the 1100 to 900 cm^{-1} (9 to 11 μm) region. The typical pattern of a monosubstituted benzene ring is clearly seen in the 2000 to 1650 cm^{-1} (5 to 6 μm) region. The 1590 and 1570 cm^{-1}

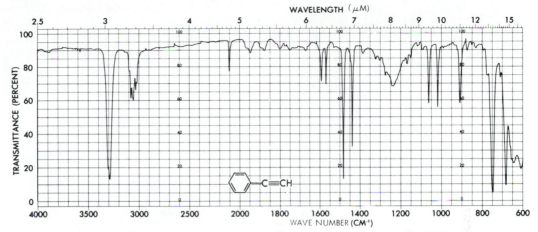

Figure 5.5   Infrared spectrum of phenylacetylene as a neat liquid film.

(6.28 and 6.37 μm) bands due to the aromatic system are clearly more distinct than in Figure 5.4, along with the 1480 and 1440 cm^{-1} (6.74 and 6.94 μm) bands.

Figure 5.6 shows the infrared spectrum of *trans-β*-bromostyrene. The olefinic C—H out-of-plane deformation appears at 930 cm^{-1} (10.7 μm), corresponding to a *trans*-disubstituted alkene. The C=C double-bond stretch overlaps the 1600 cm^{-1} (6.2 μm) aromatic band. The interpretation of the remainder of the spectrum is left to the reader.

Figure 5.7 shows the infrared spectrum of *m*-nitroaniline in chloroform solution. The N—H absorption of the primary amine appears as two bands at 3500 and 3400 cm^{-1} (2.86 and 2.93 μm). The aromatic C—H stretch is barely resolved near 3030 cm^{-1} (3.3 μm). The aromatic C—H out-of-plane deformations occur at 860 cm^{-1} (11.7 μm), characteristic of a single isolated C—H, and at 820 cm^{-1} (12.2 μm).

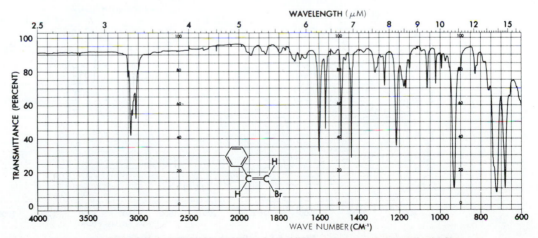

Figure 5.6   Infrared spectrum of *trans-β*-bromostryene as a neat liquid film.

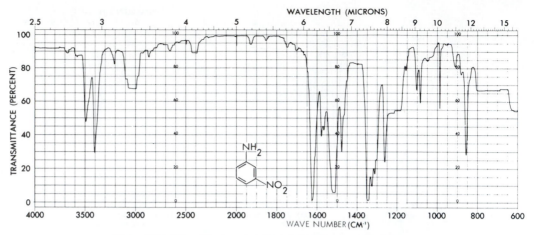

Figure 5.7   Infrared spectrum of *m*-nitroaniline in chloroform solution.

This latter band appears only as a slight shoulder where the solvent (chloroform) absorbs very strongly. The characteristically very intense nitro group bands appear at 1515 and 1340 cm^{-1} (6.6 and 7.4 μm). Note the intense band at 1625 cm^{-1} (6.16 μm) due to in-plane bending of the —NH$_2$.

The infrared spectrum of 5-hexen-2-one, Figure 5.8, shows the olefinic C—H stretch at 3080 cm^{-1} (3.24 μm), with C—H out-of-plane deformation bands characteristic of the —CH=CH$_2$ group appearing at 985 and 905 cm^{-1} (10.1 and 11.0 μm). The methyl C—H bending deformation appears at 1355 cm^{-1} (7.39 μm). The C=O stretching band appears at 1720 cm^{-1} (5.83 μm), and the bending band appears at 1155 cm^{-1} (8.65 μm). The terminal C=C stretching deformation band appears at 1635 cm^{-1} (6.11 μm).

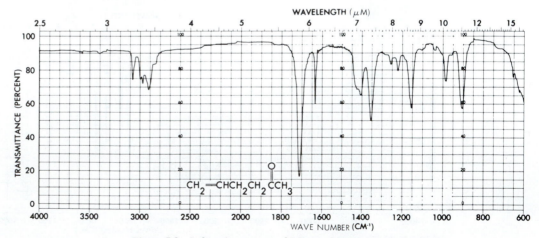

Figure 5.8   Infrared spectrum of 5-hexen-2-one as a neat liquid film.

5.4   PREPARATION OF THE SAMPLE

The method of handling the sample for recording the infrared spectrum is generally dictated by the physical state of the sample and the region one wishes to record.

Gas samples are placed in gas cells whose internal path length can be greatly multiplied by internal mirrors. The effective length of gas cells may extend from a few centimeters up to several meters (generally in multiples of meters). The handling of gases usually requires a vacuum line for storage and transfer to the cell.

Liquid samples can be run either as a neat (pure liquid) sample or in solution. Neat samples can be prepared by placing one or two small drops of the sample between two highly polished pieces of cell material (see Figure 5.9). The thickness of the capillary film is very difficult to control and reproduce, giving spectra with varying absorption intensities. Thin metal foil spacers with thicknesses accurately measured down to 0.001 mm can be used to control the thickness of the sample. A few small drops of the sample are placed in the open area of the spacer on one plate, and the second plate is firmly pressed to the spacer in a cell holder (see Figure 5.9a).

Solution spectra are obtained by dissolving the sample in an appropriate solvent (see subsequent paragraphs) and placing it in a cell (see Figure 5.9b). The concentration range normally employed is 2 to 10% by weight. A variety of cells are available for holding the sample solution. In addition to the sample cell, a reference cell of the same thickness as the sample cell is filled with pure solvent and placed in the reference beam of the instrument. The weak absorption bands occurring in the reference beam offset similar weak absorption bands of the solvent occurring in the sample beam; thus interfering extraneous absorption bands are removed from the spectrum of the compound. Major absorption bands of the solvent absorb so

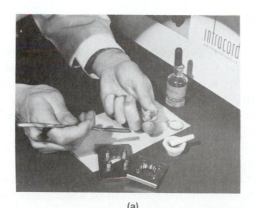

(a)   (b)

Figure 5.9   (a) Preparation of a neat sample for recording the infrared spectrum of a liquid. A drop of the liquid, or mull, is placed between sodium chloride plates, and the plates are secured between metal retainers of the cell (bottom of the photograph). (b) The filling of a solution cell. (Courtesy of the Perkin-Elmer Corporation, Norwalk, Conn.)

strongly that no effective infrared energy passes through the cell, and no differential absorption due to the sample can be detected.

Solid samples can be run as solutions, mulls, or as a solid dispersion in potassium bromide. The solutions are prepared as described above. Mulls are prepared by suspending finely ground sample particles in Nujol (paraffin or mineral oil) and then recording the spectrum of the mull as a neat sample. Fluorolube can also be used in the preparation of mulls. Not all solids can be mulled successfully. In addition, the Nujol displays intense C—H absorption and renders these regions useless for identification purposes. In general, mulls should be used only if no other method is available.

Solid dispersions of samples, approximately 1% by weight in potassium bromide, are prepared by carefully grinding a mixture of the sample and potassium bromide until composed of very finely ground particles. The finely ground mixture is then pressed into a transparent disk under several tons of pressure. The disk is mounted on a holder, and the spectrum is then recorded. Since potassium bromide is transparent in the infrared region, only bands corresponding to the sample appear in the spectrum. The spectrum obtained thus, as well as with Nujol mull dispersions, will be of the material in the solid phase and may differ from solution spectra owing to restrictions of molecular configurations (in microcrystals) or increased functional group interactions. Broad hydroxyl absorption near 3300 cm^{-1} (3 μm) is usually present, due to moisture absorbed by the potassium bromide, unless one is careful when preparing the disk.

The sample container is placed in the sample beam of the instrument along with the appropriate reference beam blank (pure solvent), and the spectrum is recorded. Instructions for operating the instrument are provided in the manufacturer's instruction manual and should be thoroughly understood before individual operation.

5.4.1 Solvents

The choice of solvent depends on the solubility of the sample and the absorption characteristics of the sample and solvent. Since almost all solvents employed in infrared spectroscopy are organic molecules themselves, they will give absorption bands characteristic of the types of bonds present. It is always desirable to use a solvent having the least amount of absorption in the infrared region. For example, carbon tetrachloride absorbs strongly only in the 830 to 670 cm^{-1} (12 to 15 μm) region, whereas chloroform absorbs strongly near 3030, 1220, and 830 to 670 cm^{-1} (3.3, 8.2, and 12 to 15 μm); thus carbon tetrachloride is more useful than chloroform, provided suitable concentrations can be achieved in both solvents. Frequently it may be necessary to record the spectrum of a sample in two different solvents to derive all of the available spectral information. For example, the use of carbon tetrachloride as the solvent for alkenes and aromatics allows one to observe all absorption bands out to 900 cm^{-1} (11 μm), but the fingerprint region will be obscured. Recording the spectra of these compounds in carbon disulfide, or acetonitrile, allows one to record the absorption bands appearing in the fingerprint region. Figure 5.10 lists several solvents suitable in the infrared region. The darkened areas

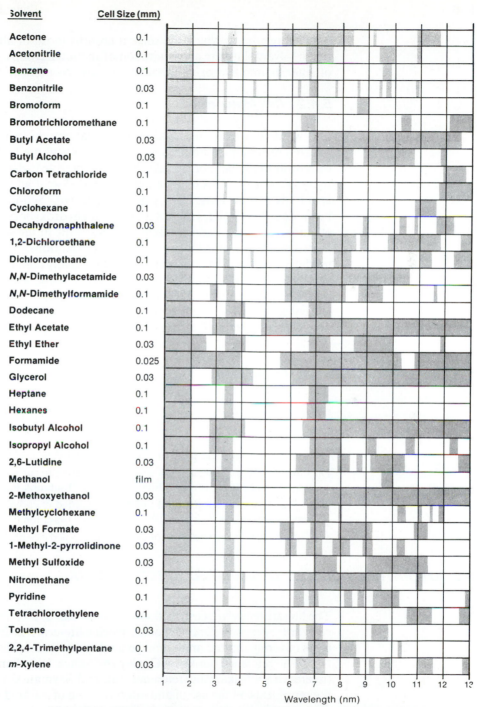

Solvent	Cell Size (mm)
Acetone	0.1
Acetonitrile	0.1
Benzene	0.1
Benzonitrile	0.03
Bromoform	0.1
Bromotrichloromethane	0.1
Butyl Acetate	0.03
Butyl Alcohol	0.03
Carbon Tetrachloride	0.1
Chloroform	0.1
Cyclohexane	0.1
Decahydronaphthalene	0.03
1,2-Dichloroethane	0.1
Dichloromethane	0.1
N,N-Dimethylacetamide	0.03
N,N-Dimethylformamide	0.1
Dodecane	0.1
Ethyl Acetate	0.1
Ethyl Ether	0.03
Formamide	0.025
Glycerol	0.03
Heptane	0.1
Hexanes	0.1
Isobutyl Alcohol	0.1
Isopropyl Alcohol	0.1
2,6-Lutidine	0.03
Methanol	film
2-Methoxyethanol	0.03
Methylcyclohexane	0.1
Methyl Formate	0.03
1-Methyl-2-pyrrolidinone	0.03
Methyl Sulfoxide	0.03
Nitromethane	0.1
Pyridine	0.1
Tetrachloroethylene	0.1
Toluene	0.03
2,2,4-Trimethylpentane	0.1
m-Xylene	0.03

Wavelength (nm)

Figure 5.10   Chart of solvents for use in the infrared region. The unshaded areas represent useful regions in which transmittance of the pure solvents is > 60%. (Reproduced by courtesy of Eastman Kodak Company, Rochester, N.Y.)

indicate regions in which the solvent absorbs most, or all, of the available energy. Slight solvent shifts are noted with some functional groups, particularly those capable of entering into hydrogen bonding either as acceptors or donors.

5.4.2 Cell Materials

Cell window or plate material must be transparent in the desired portion of the infrared region, must not be soluble in the solvent used, and must not react chemically with the solvent or sample. Sodium chloride is the most commonly used cell material because of its relative low cost and ease of handling. Potassium bromide also finds extensive use, but is more expensive. Sodium chloride and potassium bromide cells and plates cannot be used with hydroxylic solvents such as methanol or ethanol due to the slight solubility of these salts in the solvents. Several other cell materials are available for use with aqueous solutions. The reader is advised to look up the properties of the cell materials if there is any doubt as to whether that material can be used for the intended purpose. When using sodium chloride or potassium bromide cells or plates, care must be exercised to make sure that the sample and solvent are free of water. After use, the cells or plates should be carefully washed with dry acetone, dried with a stream of dry air or nitrogen, and stored in a dessicator.

5.5 INFRARED SPECTRAL PROBLEMS

Identify each of the following compounds from their infrared spectra.

1. $C_5H_8O_2$; infrared spectrum of unknown recorded as a neat liquid film (see Figure 5.11).
2. $C_8H_{10}O$; infrared spectrum of unknown recorded in carbon tetrachloride (4000 to 1000 cm^{-1}) and carbon disulfide (1000 to 600 cm^{-1}) solution (see Figure 5.12).
3. $C_9H_{10}O$; infrared spectrum of unknown recorded in carbon tetrachloride (4000 to 1000 cm^{-1}) and carbon disulfide (1000 to 600 cm^{-1}) solution (see Figure 5.13).
4. $C_7H_7NO_3$; infrared spectrum of unknown recorded as a neat liquid film (see Figure 5.14).

5.6 LITERATURE SOURCES OF INFRARED SPECTRAL DATA

The literature covering infrared spectroscopy can be conveniently grouped into three categories: general reference books, specific literature reviews and spectral compilations covering limited areas, and general spectral compilations.

The reference books by Conley (reference 5), Bellamy (reference 3), Lawson (reference 9), Nakanishi (reference 13), and Szymanski (reference 19) are general reference texts on the use of and interpretation of infrared spectroscopy. *Infrared-A Bibliography* by Brown et al. (reference 4) is a compilation of original literature references on instrumentation, techniques, and general compound classifications

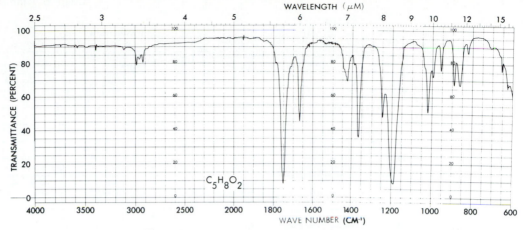

Figure 5.11   Infrared spectrum of unknown in problem 1.

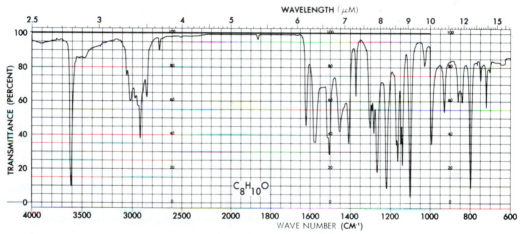

Figure 5.12   Infrared spectrum of unknown in problem 2.

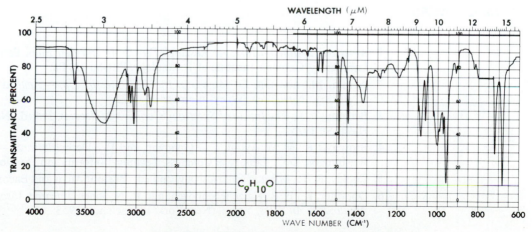

Figure 5.13   Infrared spectrum of unknown in problem 3.

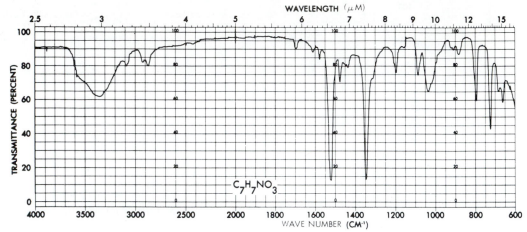

Figure 5.14 Infrared spectrum of unknown in problem 4.

covering the years 1935 to 1951. Specific literature reviews covering limited areas are too numerous to cite; only a very few are cited in this text. The reader is referred to abstracting and current literature services to locate specific reviews, for example, reference 8.

Literature references for specific compounds can be found in references 7, 8, 11, and 16. Published spectra of compounds can be found in the *Aldrich Library of Infrared Spectra* (reference 14), the *Sadtler Standard Spectra* files (reference 16), the *American Petroleum Institute Infrared Spectra Files* (reference 1), the *Manufacturing Chemists' Association Research Project Infrared Files* (reference 10), and the atlases of *Infrared Absorption Spectra of Alkaloids* and *Steroids* (references 6 and 15, respectively). Instructions on the use and applications of the various compilations of literature references and spectra accompany each compilation.

5.7 REFERENCES

1. *American Petroleum Institute Research Project 44 Infrared Files,* Petroleum Research Laboratory, Carnegie Institute of Technology, Pittsburgh, Pa.
2. Avram, M., and Gh. D. Mateescu, *Infrared Spectroscopy—Applications in Organic Chemistry.* New York: Wiley-Interscience, 1972.
3. Bellamy, L. J., *The Infra-red Spectra of Complex Molecules,* 3rd ed. New York: Wiley/Halstead, 1975.
4. Brown, C. R., M. W. Ayton, T. C. Goodwin, and T. J. Derby, *Infrared—A Bibliography.* Washington, D.C.: Library of Congress, Technical Information Division, 1954.
5. Conley, R. T., *Infrared Spectroscopy,* 2nd ed. Boston: Allyn and Bacon, 1972.
6. Dobriner, K., E. R. Katzenellenbogen, and R. N. Jones, *Infrared Absorption Spectra of Alkaloids—An Atlas,* Vol. 1. New York: Interscience, 1953.

7. Hershenson, H. M., *Infrared Absorption Spectra Index.* New York: Academic Press, 1959.

8. *IR, Raman, Microwave Current Literature Service.* London: Butterworths; and Weinheim, Germany: Verlag Chemie, Gmbh.

9. Lawson, K. E., *Infrared Absorption of Inorganic Substances.* New York: Reinhold, 1961.

10. *Manufacturing Chemists' Association Research Project Infrared Files,* Chemical Thermodynamics Properties Center, Department of Chemistry, Texas A & M University, College Station, Tex.

11. Ministry of Aviation Technical Information and Library Services, ed., *An Index of Published Infrared Spectra,* Vols. 1 and 2. London: Her Majesty's Stationery Office, 1960.

12. Nakamoto, K., *Infrared and Raman Spectra of Inorganic and Coordination Compounds,* 3rd ed. New York: John Wiley, 1978.

13. Nakanishi, K., and P. H. Solomon, *Infrared Absorption Spectroscopy,* 2nd ed. San Francisco: Holden Day, 1977.

14. Pouchert, C. J., *Aldrich Library of Infrared Spectra.* Milwaukee, Wisc.: Aldrich Chemical, 1977.

15. Roberts, G., B. S. Gallagher, and R. N. Jones, *Infrared Absorption Spectra of Steroids—An Atlas,* Vol. 2. New York: Interscience, 1958.

16. *Sadtler Standard Spectra,* Sadtler Research Laboratories, Philadelphia, Pa.

17. Socrates, G., *Infrared Characteristic Group Frequencies.* New York: John Wiley, 1980.

18. Szymanski, H. A., *Infrared Band Handbook* (with annual supplements). New York: Plenum Press, 1963, supplements, 1964 and 1966.

19. Szymanski, H. A., *Interpreted Infrared Spectra,* Vols. 1 to 3. New York: Plenum Press, 1964, 1966, and 1967.

6

Nuclear Magnetic Resonance

6.1 INTRODUCTION

Nuclear magnetic resonance (NMR) spectroscopy involves transitions between the spin states of nuclei. Nuclei possess mechanical spins, which, in conjunction with the charge of the nuclei, produce magnetic fields whose axes are directed along the spin axes of the nuclei. When a nucleus is placed in a magnetic field of strength H_0 (during the past few years the symbol B_o has also been used for magnetic field strength), the nucleus will assume $(2I + 1)$ spin orientations (spin states) with respect to the direction of the applied magnetic field, where I is the spin quantum number of the nucleus. The possible spin states of a nucleus are indicated as $-I, (-I + 1), \ldots, +I$; for example with $I = \frac{1}{2}$, spin states of $-\frac{1}{2}$ and $+\frac{1}{2}$ are possible, and with $I = 1$, spin states of $-1, 0,$ and $+1$ are possible. Spin transitions occur only between adjacent spin states. Nuclei having $I = 0$ have only one spin state and cannot undergo spin excitation. The difference in energy between the spin states is very small ($\sim 10^{-6}$ kcal per mole), requiring electromagnetic radiation in the radiowave region of the spectrum.

In the absence of a magnetic field at room temperature, the energies of the spin states of a nucleus are degenerate. As a magnetic field is impressed on the nucleus, the spin states lose their degeneracy and become separated by an energy difference (ΔE), which is directly proportional to the strength of the applied field (see Figure 6.1). The relationship between ΔE and the field strength (H_0) for a nucleus is given by Equation (6.1),

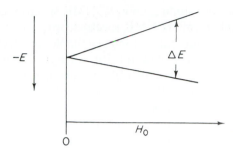

Figure 6.1 Spin energy level separation for a nucleus with I of $\frac{1}{2}$ as a function of the applied field H_0.

$$\Delta E = h\nu = \frac{\mu \beta_N H_0}{I} \tag{6.1}$$

where h is Planck's constant, ν is the frequency of the exciting radiation, μ is the magnetic moment of the nucleus, and β_N is a constant called the nuclear magneton constant. Because various nuclei differ in their values of μ and I, they undergo nuclear spin transitions at different frequencies in the same applied magnetic field. Table 6.1 lists the more common nuclei encountered in organic and biochemical applications of NMR spectroscopy, along with their nuclear spin quantum number I, μ, resonance frequency in a 10,000-gauss field, and natural abundance. Since the frequency and field strength are directly proportional, the data given in Table 6.1 can be scaled up or down to correspond to other applied field strengths.

Inspection of the data in Table 6.1 reveals that certain nuclei are present in very low natural abundance, for example, ^2H, ^{13}C, ^{15}N, and ^{17}O. Furthermore, the sensitivities, or signal intensities, of equal numbers of nuclei of different kinds are not the same, most of the low-abundance nuclei having sensitivities much lower than that of ^1H. In the past it was necessary to increase the atom concentration of low-abundance nuclei by chemical synthesis in order to record their NMR spectra. Such procedures are very tedious and costly. The use of fast Fourier transform techniques (see later paragraphs of this section) has made it possible to record NMR spectra of low abundance nuclei such as ^{13}C. Until recently the organic chemist utilized essentially

TABLE 6.1 Resonance Frequencies at 10,000 Gauss and Related Nuclear Properties

Nucleus	Spin Quantum Number, I	Magnetic Moment, μm	Resonance Frequency, MHz	Natural Abundance, %
^1H	$\frac{1}{2}$	2.79268	42.5759	99.9844
^2H	1	0.857386	6.53566	0.0156
^{13}C	$\frac{1}{2}$	0.70220	10.705	1.108
^{14}N	1	0.40358	3.076	99.635
^{15}N	$\frac{1}{2}$	-0.28304	4.315	0.365
^{17}O	$\frac{5}{2}$	-1.8930	5.772	0.037
^{19}F	$\frac{1}{2}$	2.6273	40.055	100.0
^{31}P	$\frac{1}{2}$	1.1305	17.236	100.0

only ^{1}H NMR in structural studies; now ^{13}C NMR spectroscopy routinely complements the data available from ^{1}H NMR spectroscopy.

There are two very different methods available for recording an NMR spectrum. If one has 10 mg or more of sample the NMR spectrum of the sample can be recorded with a single pass (sweep) through the appropriate radio frequency region. This is called *continuous wave* (CW) spectroscopy. However, if one has only a very limited amount of sample or wishes to record the NMR spectrum of a very low-abundant nucleus such as ^{13}C, this form of NMR spectroscopy is not applicable. It is then necessary to use what is known as *fast Fourier transform* (FT) spectroscopy.

In FT NMR spectroscopy all the nuclei of a given type are excited with a single pulse covering a broad range in the radiofrequency region. The nuclei return to their ground spin states (relax) with half-lives on the order of a second or so. In doing so, each relaxing nucleus emits a characteristic radio frequency signal which is recorded by the spectrometer. The sum of the emitted frequencies of all the nuclei is called the *free induction decay* (FID) signal. This summed FID is then converted to the normal spectrum by a mathematical process known as *Fourier transformation*. This FT technique allows one to pulse, or scan, the entire resonance region of a given type of nucleus repeatedly every second or less. The FIDs derived from each pulse are summed, thus greatly increasing the signal-to-noise ratio. In this way one can record the NMR spectra of extremely small quantities of samples, even those containing a low-abundance nucleus. Essentially all the modern NMR spectrometers use this technique.

6.2 ORIGIN OF THE CHEMICAL SHIFT

Let us now consider a nucleus contained in a molecule that is in the vicinity of various other types of atoms, or functional groups. The magnetic field "felt" at the nucleus of the atom will not be equal to that of the applied field H_0, but may be less than or more than H_0 due to the interactions of the neighboring atoms and functional groups with the applied field, and will be represented by H_{net} (the net magnetic field felt at the nucleus under consideration). H_{net} is the function of the strength of the applied field H_0 and can be represented by Equation (6.2), in which σ_A is the shielding constant for a particular nucleus. (σ_A is comprised of several contributions which will not be discussed here.) In a molecule of any degree of complexity the environments of the various nuclei will be different, resulting in different σ_A's, thus the different nuclei will undergo resonance at different frequencies and appear at different positions in the NMR spectrum.

$$H_{net} = H_0(1 - \sigma_A) \tag{6.2}$$

The shielding constant σ_A may result in an increase *(shielding)* or a decrease *(deshielding)* in the resultant field felt by the nucleus. The presence of an electron-withdrawing group results in deshielding, while the presence of an electron-donating group results in shielding. The orientation of a functional group with respect to the nucleus being observed may also result in the shielding or deshielding of that nucleus. It is beyond the scope of this text to discuss in detail these effects.

6.3 DETERMINATION OF CHEMICAL SHIFT

Equation (6.1) contains two potential variables, the frequency v and the magnetic field strength H_0. In the CW spectroscopy, experimentally we might keep one of these variables constant while varying the other. The frequencies required for resonance in magnetic fields of 10,000 gauss or higher are in the region of 10^6 Hz. With proton magnetic resonance, differences in chemical shifts may be of the order of a hertz (Hz) or less that demand an extremely fine control and measurement of the frequency, if frequency is to be used as the variable. Electronically it is easier to maintain a highly constant radio frequency signal and vary the strength of the magnetic field. This is accomplished by summing a constant high-intensity magnetic field with a weak varying field controlled by the flow of a small current through the "sweep" coils (see Figure 6.2). The relaxing nuclei induce a signal in the receiver coil that is amplified and sent to a recorder.

In FT NMR spectroscopy the experimental setup is quite different. The magnetic field is kept constant and the sample is irradiated with broad-band pulses of radio frequency radiation. The resulting FID signals are summed and the spectrum

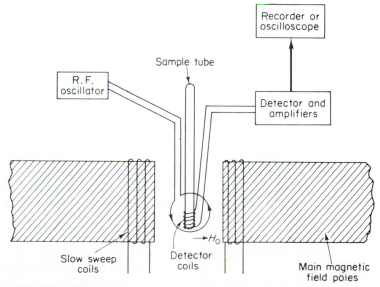

Figure 6.2 Nuclear induction system of a nuclear magnetic resonance spectrometer.

is obtained by Fourier transformation of the FID. In FT NMR spectroscopy the magnet is almost always a high-field, superconducting magnet capable of generating magnetic fields of over 10^5 gauss.

The difficulty in specifying a resonance frequency to the degree of precision required to correlate NMR spectral data (~ 1 part in 10^6), such as is done in ultraviolet and infrared spectroscopy, is that it requires the use of a standard that is added to the sample. It is assumed that the standard undergoes resonance at a constant frequency regardless of the nature of the solvent and the sample. The standard chosen should have a single resonance that is well displaced from the resonances of the sample, does not chemically or physically interact with the solvent or the sample, and is easily removed so that the sample can be recovered. Tetramethylsilane [$(CH_3)_4Si$, TMS, bp $26°C$] is ideal in all respects and is the most commonly used *internal* standard in ^1H and ^{13}C NMR spectroscopy. In some cases it may not be possible to use an internal standard that is added directly to the sample because of solubility or chemical reactivity problems. In such cases a standard is placed inside a small sealed capillary that is placed inside the NMR tube. Such standards are referred to as *external* standards. In both cases the standard is assigned a resonance position, or *chemical shift,* of 0. When standards other than TMS are used, corrections can be applied to change the chemical shifts relative to TMS.

Several chemical shift designation conventions have been employed to indicate the chemical shifts of nuclei. The position of resonance may be given in Hz from the standard used. As the chemical shift is field-strength dependent, the frequency of the spectrometer must also be specified. In ^1H and ^{13}C NMR, peaks appearing at lower field than the standard are assigned a positive sign, whereas peaks appearing at higher field are assigned a negative sign.

As NMR spectrometers of many different frequencies are currently in use (e.g., 60, 90, 100, 200, 300, and 500 MHz, to cite a few), it is more convenient to use a convention that is independent of the frequency of the spectrometer. Such is the δ system that is defined in Equation (6.3),

$$\delta = \frac{\text{chemical shift in Hz} \times 10^6}{\text{spectrometer frequency}} \qquad (6.3)$$

in which the chemical shift in Hz is the difference between the resonance frequency of a particular nucleus and TMS. For example, a proton resonance appearing at $+72$ Hz on a spectrometer with a frequency of 60 MHz has a δ value of $+1.20$. (Note that the sign of δ is the same as that of the chemical shift in Hz, and that the spectrometer frequency and standard need not be specified.)

Whereas the δ convention is now predominantly used, the *tau* (τ) scale was used extensively in the past. The τ scale implies the use of TMS as the standard, with τ being calculated according to Equation (6.4). The scale can be used only for ^1H chemical shifts.

$$\tau = 10 - \frac{\text{chemical shift in Hz relative to TMS} \times 10^6}{\text{spectrometer frequency}} \qquad (6.4)$$

6.4 HYDROGEN MAGNETIC RESONANCE

The chemical shift of a proton is highly dependent on the electron density at the proton nucleus, or associated with the atom to which it is bonded. As the electron density decreases, the proton experiences a greater deshielding and absorbs at lower field positions (higher δ values). The position of resonance shifts to lower field as one proceeds from left to right across the periodic table, and from bottom to top. In fact, the chemical shifts of the methyl hydrogens of CH_3—X-type compounds show a nearly linear correspondence with the electronegativity of the X atom from lithium to oxygen. The chemical shifts for a variety of acyclic compounds are given in Table 6.2. The shielding effect of a functional group decreases by about 90% on going from the α- to the β-position, and another 90% on going from the β- to the γ-position.

Cyclic hydrocarbons absorb at only slightly different frequencies relative to a CH_2 group in an acyclic hydrocarbon (δ 1.3–1.6), except for cyclopropane, which appears at very high field (δ 0.222). Cyclobutane appears at δ 1.96, cyclopentane at 1.51, cyclohexane at 1.44, and cycloheptane at 1.53.

The presence of a second substituent Y on a carbon to give X—CH_2—Y-type compounds results in further deshielding, or shielding, of the CH_2 hydrogens. Shoolery has developed an empirical method for calculating the chemical shifts of the hydrogens in CH_3—X and X—CH_2—Y compounds using Equation 6.5,

$$\delta = 0.233 + \sum \sigma_{i_{eff}} \tag{6.5}$$

in which $\sigma_{i_{eff}}$ is the effective shielding constant of the substituent(s) and 0.233 is the chemical shift of methane. Although the correlation works well for CH_3—X and X—CH_2—Y compounds, the correlation does not always work well for calculating the chemical shifts of methyne hydrogens in CHXZY-type compounds. The $\sigma_{i_{eff}}$'s for various functional groups are given in Table 6.3. Several comparisons are calculated with experimentally observed chemical shifts in Table 6.4.

Compounds that contain double and triple bonds and aromatic rings possess more polarizable and delocalized electronic structures than do compounds that contain only single bonds. Thus, they produce significantly greater induced magnetic fields when placed in a magnetic field. The magnitude and the sign of the induced fields are highly directional with respect to the axes of the π systems. It is beyond the scope of this text to discuss these effects in detail. The general ranges of chemical shifts of hydrogen atoms bonded to π-containing systems are given in Table 6.5.

Within the chemical shift regions given in Table 6.5, the chemical shift of a hydrogen attached to a C=C or a benzene ring is dependent on the type of functions attached to the C=C or the ring. By considering resonance and inductive interactions of functions with the C=C or the aromatic ring, one can qualitatively predict the shielding effect exerted by the group. For example, the terminal methylene hydrogens of **1** appear at lower field than in **2** due to the contributions of resonance

TABLE 6.2   Average Chemical Shifts (δ) of α-Hydrogens in Substituted Alkanes*

Functional Group X	CH_3X	$-CH_2X$	$>CHX$
H	0.233	0.9	1.25
CH_3 or CH_2	0.9	1.25	1.5
F	4.26	4.4	—
Cl	3.05	3.4	4.0
Br	2.68	3.3	4.1
I	2.16	3.2	4.2
OH	3.47	3.6	3.6
O—Alkyl	3.3	3.4	—
O—Aryl	3.7	3.9	—
OCO—Alkyl	3.6	4.1	5.0
OCO—Aryl	3.8	4.2	5.1
SH	2.44	2.7	—
SR†	2.1	2.5	—
SOR	2.5	—	2.8
SO_2R	2.8	2.9	3.1
NR_2	2.2	2.6	2.9
NR—Aryl	2.9	—	—
NCOR	2.8	—	3.2
NO_2	4.28	4.4	4.7
CHO	2.20	2.3	2.4
CO—Alkyl	2.1	2.4	2.5
CO—Aryl	2.6	3.0	3.4
COOH	2.07	2.3	2.6
CO_2R	2.1	2.3	2.6
$CONH_2$‡	2.02	2.2	—
$CR=CR^1CR^2$	2.0–1.6	2.3	2.6
Phenyl	2.3	2.7	2.9
Aryl§	3.0–2.5	—	—
$C≡CR$	2.0	—	—
$C≡N$	2.0	2.3	2.7

* The data appearing in this table were taken from various published compilations and original literature references. The tabulated values are average values for compounds that do not contain another functional group within two carbon atoms from the indicated hydrogens. For methylene and methine hydrogens most values fall within ± 0.15 ppm of the tabulated values.

† S-aryl derivatives absorb at somewhat lower fields.

‡ Replacement of NH_2 by N-alkyl$_2$ results in a slight shift upfield.

§ Includes polycyclic and many heterocyclic aromatics. The values for $-CH_2-X$ and $>CH-X$ probably appear at lower fields also.

TABLE 6.3 Shoolery's Effective Shielding Constants

Functional Group	$\sigma_{i_{eff}}$	Functional Group	$\sigma_{i_{eff}}$
—Cl	2.53	—CR^1=CR^2R^3	1.32
—Br	2.33	—C_6H_5	1.85
—I	1.82	—C≡C—R	1.44
—OH	2.56	—RC=O	1.70
—O—Alkyl	2.36	$\overset{\displaystyle O}{\underset{\displaystyle \parallel}{}}$ —C—OR	1.55
—O—Aryl	3.23		
$\overset{\displaystyle O}{\underset{\displaystyle \parallel}{}}$ —OCR	3.13		
—SR	1.64	$\overset{\displaystyle O}{\underset{\displaystyle \parallel}{}}$ —C—NR_2	1.59
—N(Alkyl)$_2$	1.57	—CF_3	1.14
—CH_3	0.47	—C≡N	1.70

$$CH_2{=}CH{-}\overset{\displaystyle O}{\overset{\displaystyle \parallel}{C}}{-}OR \longleftrightarrow \overset{+}{C}H_2{-}CH{=}\overset{\displaystyle O^-}{\overset{\displaystyle |}{C}}OR$$

(a)   (b)

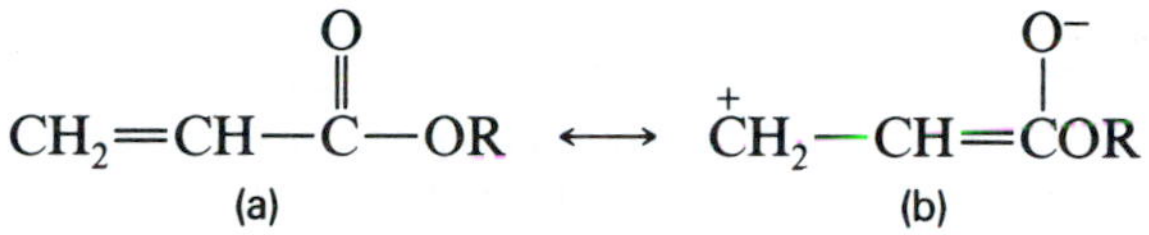

1

structures (b), which results in deshielding of the terminal hydrogens (δ 6.20 and 6.38), and to **2**, which results in shielding of the similar hydrogens (δ 4.55 and 4.85).

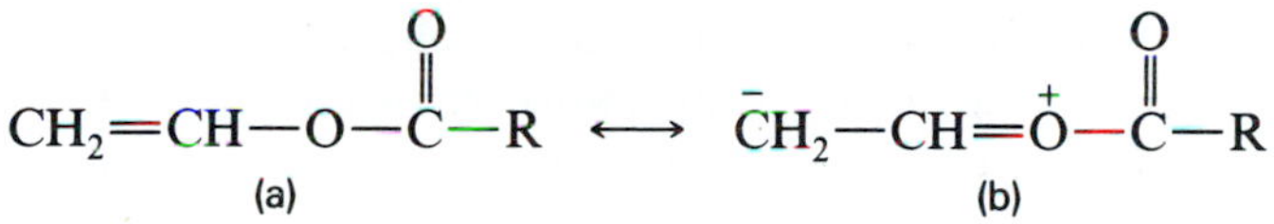

$$CH_2{=}CH{-}O{-}\overset{\displaystyle O}{\overset{\displaystyle \parallel}{C}}{-}R \longleftrightarrow \overset{-}{C}H_2{-}CH{=}\overset{+}{O}{-}\overset{\displaystyle O}{\overset{\displaystyle \parallel}{C}}{-}R$$

(a)   (b)

2

TABLE 6.4 Comparison of Observed and Predicted Chemical Shifts Using Shoolery's Rules

Compound	Calculated δ	Observed δ
$BrCH_2Cl$	5.09	5.16
ICH_2I	3.87	4.09
$C_6H_5CH_2OR$	4.44	4.41
$C_6H_5CH_2CH_3$	2.52	2.55
$C_6H_5CH_2C_6H_5$	3.91	3.92
C=C—CH_2OH	3.92	3.91
C=C—CH_2—C=C	2.87	2.91
$CH_3CH_2\overset{\displaystyle O}{\overset{\displaystyle \parallel}{C}}{-}R$	2.40	2.47
$(C_2H_5O)_3CH$	7.31	4.96
$(CH_3)_2CHI$	2.99	4.24

TABLE 6.5 Chemical Shifts of Hydrogens Bonded to Unsaturated Centers

Type	Unconjugated, δ	Conjugated, δ
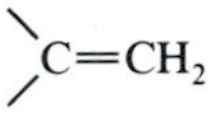 $\diagdown$ $\diagup C{=}CH_2$	4.6–5.0	5.4–7.0*
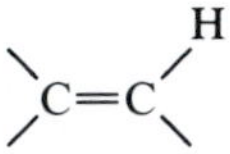 $\diagdown$ H $\diagup C{=}C\diagup$ $\diagdown$	5.0–5.7	5.7–7.3*
Aromatic	6.5–8.3†	—
Nonbenzenoid aromatic	6.2–9.0	—
Acetylenic	2.3–2.7	2.7–3.2
Aldehydic	9.5–9.8	9.5–10.1*
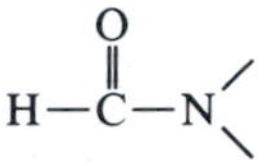 $\overset{\text{O}}{\underset{\parallel}{}}$ H—C—N$\diagup$ $\diagdown$	7.9–8.1	—
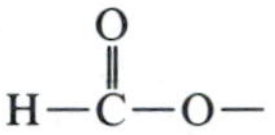 $\overset{\text{O}}{\underset{\parallel}{}}$ H—C—O—	8.0–8.2	—

* The position depends on the type of functional group in conjugation with the unsaturated group.

† The position of aromatic hydrogen resonance depends on the type of substituent attached to the aromatic ring (Table 6.6).

In an approach similar to that of Shoolery, an empirical correlation has been developed to calculate the chemical shifts of vinyl hydrogens. The chemical shift of ethylenic hydrogens in carbon tetrachloride solution is calculated by use of Equation (6.6),

$$\delta = \delta_{\text{ethylene}} + \sum_i Z_i \qquad (6.6)$$

where δ_{ethylene} is the chemical shift of the hydrogens of ethylene and is taken as $\delta = 5.28$, and Z_i is the $\Delta\delta$ that results on substitution of an ith functional group for hydrogen in ethylene. The values of Z_i depend on the stereochemical relationship of the substituent with respect to the hydrogen in question in the substituted ethylene.

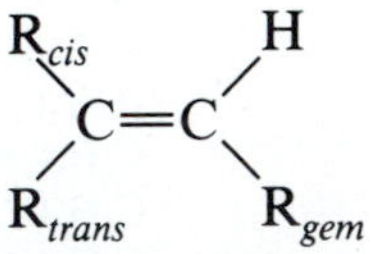

$$R_{cis}\diagdown \qquad \diagup H$$
$$C{=}C$$
$$R_{trans}\diagup \qquad \diagdown R_{gem}$$

The values for Z_i are given in Table 6.6; they give values of δ within 0.15 of the experimental values in 73.3% of the 1070 compounds correlated. Serious deviations sometimes occur in the more highly complex systems.[1] An example of the use of

---

[1] U. E. Matter, C. Pascual, E. Pretsch, A. Pross, W. Simon and S. Sternhill, *Tetrahedron*, **25**, 2023 (1969).

Equation (6.6) is illustrated with *cis-* and *trans-*stilbene using the Z_i values given in Table 6.6.

$$\delta_{cis} = 5.28 + Z_{i_{gem}} + Z_{i_{trans}} = 5.28 + 1.35 + (-0.10) = 6.53$$

$$\text{Observed } \delta = 6.55$$

$$\delta_{trans} = 5.28 + Z_{i_{gem}} + Z_{i_{cis}} = 5.28 + 1.35 + 0.37 = 7.00$$

$$\text{Observed } \delta = 6.99$$

The chemical shift of hydrogens attached to a benzene ring is similarly quite sensitive to the type of substituent attached to the ring. Electron-withdrawing groups

TABLE 6.6 Substituent Constants for Substituted Ethylenes*

Substituent	Gem	Cis	Trans	Substituent	Gem	Cis	Trans
—H	0	0	0	—CHO	1.03	0.97	1.21
—Alkyl	0.44	−0.26	−0.29	$-\overset{\overset{\text{O}}{\|}}{\text{C}}-\text{N}\!\big\langle$	1.37	0.93	0.35
—Cycloalkyl	0.71	−0.33	−0.30				
—CH₂O, —CH₂I	0.67	−0.02	−0.07	$-\overset{\overset{\text{O}}{\|}}{\text{C}}-\text{Cl}$	1.10	1.41	0.99
—CH₂S	0.53	−0.15	−0.15	—OR†	1.18	−1.06	−1.28
—CH₂Cl, —CH₂Br	0.72	0.12	0.07	—OR‡	1.14	−0.65	−1.05
—CH₂N	0.66	−0.05	−0.23	—OCOR	2.09	−0.04	−0.67
—C≡C	0.50	0.35	0.10	—Aryl	1.35	0.37	−0.10
—C≡N	0.23	0.78	0.58	—Cl	1.00	0.19	0.03
—C=C§	0.98	−0.04	−0.21	—Br	1.04	0.40	0.55
—C=C‖	1.26	0.08	−0.01				
—C=O§	1.10	1.13	0.81	$-\text{N}\!\big\langle\!\!\begin{smallmatrix}\text{R†}\\ \text{R}\end{smallmatrix}$	0.69	−1.19	−1.31
—C=O‖	1.06	1.01	0.95				
—COOH§	1.00	1.35	0.74	$-\text{N}\!\big\langle\!\!\begin{smallmatrix}\text{R‡}\\ \text{R}\end{smallmatrix}$	2.30	−0.73	−0.81
—COOH‖	0.69	0.97	0.39				
—COOR§	0.84	1.15	0.56	—SR	1.00	−0.24	−0.04
—COOR‖	0.68	1.02	0.33	—SO₂R	1.58	1.15	0.95

* For carbon tetrachloride solutions. Taken from C. Pascual, J. Meier, and W. Simon, *Helv. Chim. Acta,* **49,** 164 (1966).

† R = aliphatic.

‡ R = unsaturated group.

§ For a single functional group in conjugation with the first C=C.

‖ For a functional group that is also conjugated to a further substituent (e.g., 1,3,5-hexatriene).

TABLE 6.7 Chemical Shifts of Hydrogens
in Monosubstituted Benzenes

Functional Group	Chemical Shifts (δ) Relative to Benzene (δ 7.27)		
	Ortho	*Meta*	*Para*
—NO$_2$	−0.97	−0.30	−0.42
—CO$_2$R	−0.93	−0.20	−0.27
—COCH$_3$	−0.63	−0.27	−0.27
—COOH	−0.63	−0.10	−0.17
—CCl$_3$	−0.80	−0.17	−0.23
—CH$_3$	0.10	0.10	0.10
—Cl	0.00	0.00	0.00
—OCH$_3$	0.23	0.23	0.23
—OH	0.37	0.37	0.37
—NH$_2$	0.77	0.13	0.40
—N(CH$_3$)$_2$	0.50	0.20	0.50

result in deshielding of the ring hydrogens, particularly the *ortho* hydrogens, while electron-donating functions result in shielding of the ring hydrogens, again primarily the *ortho* hydrogens. Table 6.7 lists the chemical shifts of *ortho,* meta, and *para* hydrogens in a number of substituted benzenes.

6.4.1 Chemical Shifts of Hydrogen Attached to Atoms Other than Carbon

Hydrogen Bonded to Oxygen

The chemical shifts of hydrogens bonded to oxygen are extremely sensitive to structural and environmental changes. The hydroxyl hydrogens of alcohols generally absorb in the δ 0.5 to 5.0 region (see Table 6.8), the position being dependent on the concentration. This is due, in general, to intermolecular hydrogen bonding and rapid exchange, and results in a deshielding of the hydroxyl hydrogen. Distinction between intermolecular and intramolecular hydrogen bonding can be made by use of dilution studies. On dilution, a plot of the chemical shift versus concentration results in a line with a significant slope for intermolecularly bonded hydrogen and a very small slope for intramolecularly bonded hydrogen (theoretically, one would predict a zero slope). The resonances of monomeric hydroxyl hydrogens are near δ 0.5 in carbon tetrachloride. Phenolic hydroxyl hydrogens absorb at lower fields, approximately δ 4.5 for monomeric species. Their chemical shift is also a function of the concentration.

The acidic hydrogens of enols and carboxylic acids appear at very low fields. Enolic hydroxyl hydrogens absorb near δ 15.5, the enhanced deshielding being due to the positive nature of the hydrogen and strong intramolecular hydrogen bonding.

TABLE 6.8 Chemical Shifts of Hydrogen Bonded to Oxygen, Nitrogen, and Sulfur

	Functional Group	Chemical Shift, δ	Remarks
OH	Aliphatic alcohols	0.5	(Monomeric)
		0.5–5	(Associated)
	Phenols	4.5	(Monomeric)
		4.5–8	(Associated)
	Enols	15.5	
	Carboxylic acids	9–12	(Dimeric)
	Hydrogen bonded to carbonyl systems	13–16	
NH_2	Alkylamine	0.6–1.6	
	Arylamine	2.7–4.0	
	Amide	7.8	
NH	Alkylamine	0.3–0.5	
	Arylamine	2.7–2.8	
$R_3\overset{+}{N}H$	Ammonium salts	7.1–7.7	(In trifluoroacetic acid)
SH	Aliphatic	1.3–1.7	
	Aromatic	2.5–4	

Carboxylic acid hydrogens absorb in the δ 9 to 12 region and are not affected by concentration changes. This is due to the fact that carboxylic acids normally exist in dimeric form in nonpolar solvents.

The resonance signal of hydroxyl hydrogens is further complicated by the possibility of chemical exchange of hydrogens between different molecules. The peak shape, both in broadness and splitting by spin-spin interaction with the carbinol proton (see discussions on spin-spin coupling, Section 6.5), of the hydrogen resonance peaks depends on the extent of chemical exchange and will be discussed in greater detail under time-averaging of chemical shifts (see Section 6.9). Since the hydrogen exchange is acid catalyzed, it can be substantially reduced by careful purification of the compound and solvent, or by recording the spectrum in dimethyl sulfoxide solution.[2]

An alternative method of characterizing the hydroxyl hydrogen resonance is to remove the hydrogen and replace it with deuterium. This is accomplished by dissolving a small portion of the sample in ether or chloroform and shaking the solution several times with deuterium oxide. The organic phase is dried over a drying agent, and the organic compound is recovered by evaporation. Quite often, addition of one drop of deuterium oxide to the NMR tube followed by vigorous shaking is significant to exchange the hydroxyl protons.

---

[2] The exchange process is slowed considerably owing to hydrogen bonding between the hydroxyl hydrogen and the sulfoxide group. The hydroxyl hydrogen of a primary alcohol appears as a triplet (see Section 6.5 on spin-spin coupling), of a secondary alcohol as a doublet, and of a tertiary alcohol as a singlet [O. L. Chapman and R. W. King, *J. Am. Chem. Soc.*, **86**, 1257 (1964)].

Hydrogen Bonded to Nitrogen

The chemical shifts of hydrogens bonded to nitrogen are also quite sensitive to structural and environmental changes. The resonance signals are generally very broad, and they may not be readily apparent. Hydrogen exchange also occurs that may lead to line broadening (see latter discussion). Table 6.8 lists the resonance ranges for various types of N—H containing compounds.

Hydrogen Bonded to Sulfur

The chemical shifts of hydrogens bonded to sulfur in aliphatic thiols appear in the δ 1.2 to 1.6 region. Aromatic thiols absorb at a somewhat lower field, near δ 3.0 to 3.5 (Table 6.8). Hydrogen bonding effects are generally minimal.

6.5 SPIN-SPIN COUPLING

The correlation of chemical shifts with structural features provides very useful information for assigning the structures of organic molecules. However, the value of NMR is greatly enhanced by the presence of another phenomenon, that involving nuclear spin-spin interactions that may result in the appearance of a number of resonance lines instead of only a single resonance line. In order to illustrate this, let us consider the molecular fragment **3** in which the two hydrogens H_A and H_X have

$$Y-\underset{\underset{H_X}{|}}{\overset{\overset{Z}{|}}{C}}-\underset{\underset{H_A}{|}}{\overset{\overset{A}{|}}{C}}-B$$

3

different chemical environments, and thus have different chemical shifts. If H_A were not present, H_X would give rise to a single resonance line. In **3**, H_A will exist in the $+\frac{1}{2}$ and $-\frac{1}{2}$ spin states, which in a magnetic field will act as tiny bar magnets, that will affect the relative energies at which H_X will undergo resonance. One of the spin states of H_A will result in a decrease in H_{net} felt at H_X, the other an increase in H_{net}. Thus the resonance of H_X will appear as two resonance lines termed a *doublet*. The difference in frequency between the two resonance lines is called the *coupling constant, J. J* has the dimensions of Hz and is *independent of the strength of the applied field H_0*. The difference in the behavior of the chemical shift and the coupling constant as a function of the field strength H_0 results in substantial changes in the appearance of the NMR spectrum of a compound when recorded at different field strengths. A similar analysis of the effect of the spin states of H_X on the appearance of the resonance of H_A leads to the prediction that two resonance lines will be present for H_A. The separation between the two resonance lines for H_A will be the *same* as between the two resonance lines for H_X (i.e., the coupling constant J for the interaction between the two nuclei is the same in both directions). The chemical shift of H_A and H_X is taken as the center of gravity of the resonance lines representing the two nuclei (see Figure 6.3). (The intensities of the two lines will, in general, never be the same. The reason for this will be discussed later.)

The spin interactions between nuclei are transmitted through the bonding electrons. The magnitude of coupling constants generally decreases as the number of intervening bonds increases. The number of bonds through which the spin interaction is transmitted is often indicated as a superscript before J, while the identities of the nuclei involved are indicated as postsubscripts. For example, in **3** it would be designated as $^3J_{HH}$. The coupling in **3** is also referred to as *vicinal* coupling. When the coupling occurs through more than three bonds it is generally referred to as *long-range* coupling. In theory, all the nuclei in a molecule with $I \neq 0$ are spin coupled; however, most of the long-range coupling constants are so small that they cannot be resolved by most routinely available NMR spectrometers.

Let us now return to our discussion of spin interactions between nuclei and consider the fragments **4** and **5** in Figure 6.3. In **4** the spin states of the two identical H_X nuclei will combine to form three spin states with total spin of -1, 0, and $+1$ (see Figure 6.3). There are only single combinations for the -1 and $+1$ spin states. For the 0 spin state, however, there are two different combinations of the H_X nuclei. Thus, there are four different combination spin states, which have essentially equal populations. The three spin states will interact with H_A to produce a three-line resonance pattern, with relative intensities of $1:2:1$ reflecting the relative populations of the individual spin states. Such a pattern is referred to as a *triplet*. The

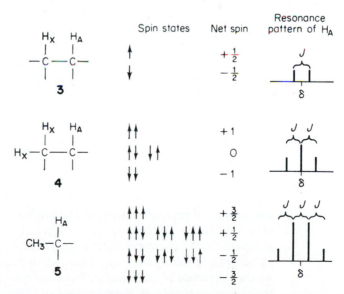

Figure 6.3 Diagram of spin-spin interactions between adjacent hydrogens.

interaction of the single H_A with the two H_X's produces a doublet resonance pattern for H_X. The separations between the three lines of the triplet and the two lines of the doublet are the same: the value of the coupling constant J.

In fragment **5**, the three methyl hydrogens are chemically equivalent and represent four separate spin states with total spins of $-\frac{3}{2}, -\frac{1}{2}, +\frac{1}{2}$, and $+\frac{3}{2}$, for which there is only one spin combination for the $-\frac{3}{2}$ and $+\frac{3}{2}$ states and three combinations for the $-\frac{1}{2}$ and $+\frac{1}{2}$ states. All eight spin-state combinations are essentially equally populated. The resulting interaction of these spin-state combinations with H_A results in the appearance of the resonance of H_A as a four-line pattern *(quartet)* with relative intensities of $1:3:3:1$ with an equal spacing of J. H_X will appear as a doublet with a separation of J.

In general, the multiplicity of a resonance pattern can be represented by Equation (6.7), in which I is the spin quantum number of the coupled nucleus and n is the number of chemically equivalent coupled nuclei.

$$\text{no. of peaks} = 2In + 1 \qquad (6.7)$$

The relative intensities of the peaks produced by spin-spin coupling between hydrogens that have widely differing chemical shifts approaches the coefficients of r in the expanded form of Equation (6.8), in which n is the number of adjacent, chemically equivalent hydrogens.

$$(r + 1)^n \qquad (6.8)$$

For example, the use of Equations (6.7) and (6.8) predicts a four-line pattern with relative intensities $1:3:3:1$ [coefficients of r in the expanded form of $(r = 1)^3 = r^3 + 3r^2 + 3r + r^0$] for an interaction with a methyl group. The multiplicity of hydrogen absorption patterns can be referred to as doublets, triplets, quartets, and so on, if the relative intensities of the peaks follow the coefficients of the expanded form of Equation (6.8).

In most complex molecules, a particular hydrogen may be coupled to two or more sets of hydrogens of different chemical shift, as in the fragments **6** and **7**. In **6**,

$$-CH_n-\underset{\underset{H_A}{|}}{\overset{|}{C}}-CH_m \qquad\qquad -CH_n-\underset{\underset{H_A}{|}}{\overset{\overset{|}{CH_p}}{\underset{|}{C}}}-CH_m$$

$$\textbf{6} \qquad\qquad\qquad\qquad \textbf{7}$$

the number of resonance lines for H_A will be $(2In + 1)(2Im + 1)$, and in **7** it will be $(2In + 1)(2Im + 1)(2Ip + 1)$. In addition, the assumption that the hydrogens have greatly different chemical shifts (implying $\Delta\delta \gg J$), as employed in the previous paragraphs, is hardly ever applicable when dealing with typical organic molecules. As the difference in chemical shifts of the coupled nuclei decreases with respect to their coupling constants, the resonance patterns described in the previous paragraphs become highly distorted and a simple interpretation of the patterns is not always possible.

In addition to the effect of the number of intervening bonds on the magnitude of J, the type of intervening bonds, the type of functions present in the molecule, and the geometrical relationship between the bonds also affect the magnitude of the coupling constant. Table 6.9 lists typical ranges for proton-proton coupling constants. (It will be noted that signs are associated with most of the constants given in Table 6.9. Although the mathematical signs are important in the detailed theory of spin-spin coupling, differences in sign do not affect the appearance of NMR spectra.

TABLE 6.9 Hydrogen-Hydrogen Coupling Constants

Structure	J_{HH} (Hz)
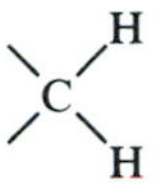	$-9--20*$
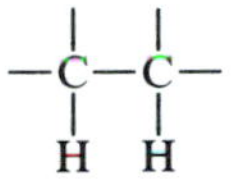	$0-+9†$
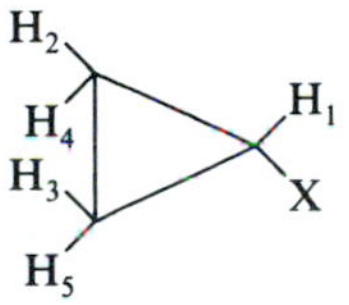	$J_{12}+6-+8.5‡$ $J_{25}+5.5-+8.3$ $J_{14}+2.9-+5.2$ $J_{23}+8.8-+10.9$ $J_{24}-3.3--6.3$ $J_{cis}\,7-13$ $J_{trans}\,4-9.5$
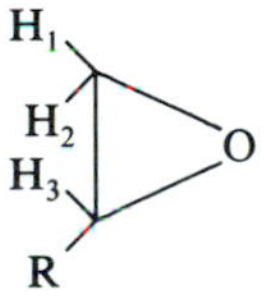	$J_{12}\sim+5.5$ $J_{23}\,3-5$ $J_{13}\,2-3.5$
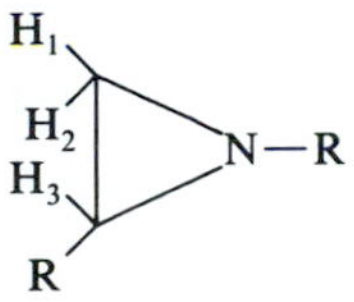	$J_{12}\sim+2.0$
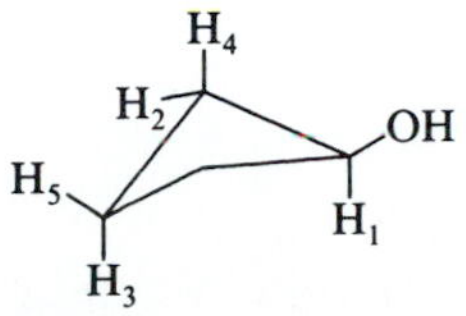	$J_{12}\,6.9§$ $J_{23}\,7.9$ $J_{45}\,9.6$ $J_{14}\,8.1$ $J_{25}\,2.3$ $J_{43}\,10.4$
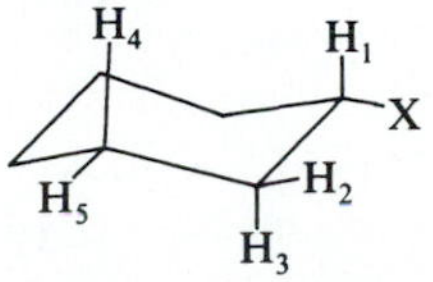	J_{13}, J_{34} (*trans*-diaxial) $10-13$ J_{12}, J_{24} (*cis*-axial-equatorial) $4-7$ J_{25} (*trans*-diequatorial) $2-5$ $J_{23}, J_{45}\,11-13$

TABLE 6.9 (Continued)

Structure	J_{HH} (Hz)

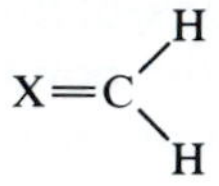

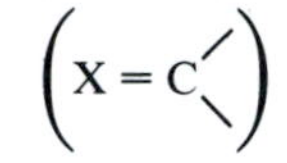

$$X=C\diagup_{\diagdown}{}^{H}_{H}$$

$$\left(X=C\diagup_{\diagdown}\right)$$
                                                         $-4.5-+3.5\|$

(X = N—R) ~ 16
(X = NOH) $\sim 8.5\P$
(X = C=O, ketene) -15.8
(X = O) $+40-+42\P$

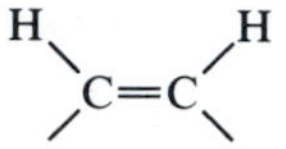

$\overset{H}{\diagup}C=C\overset{H}{\diagdown}$ $6-14\|$

$\overset{H}{\diagup}C=C\diagup^{}{}_{H}$ $11-18\|$

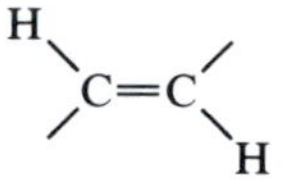

1.4

H_3 H_2
H_4
H_5
H_6 H_1
                  $J_{23}+1.00$ $J_{34}-12.0$ $J_{12}\,2.85$
                  $J_{35}+4.65$ $J_{36}+1.75$

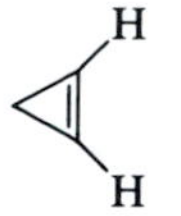

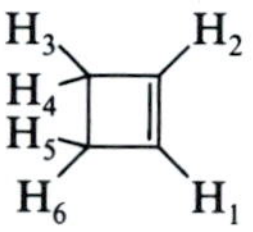

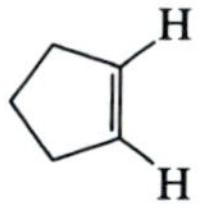

~ 5.5

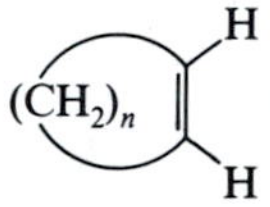

$(CH_2)_n$ $(n = 4-6)$ $10-11$

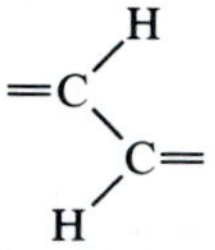

$10-11$

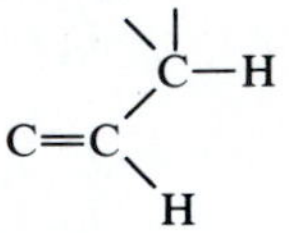

$1-7\dagger$

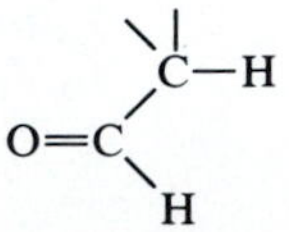

$1-3$

TABLE 6.9 (Continued)

Structure	J_{HH} (Hz)

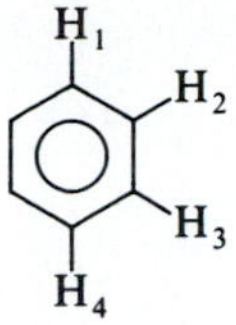

J_{12} 7–10
J_{13} 1–3
J_{14} ~1

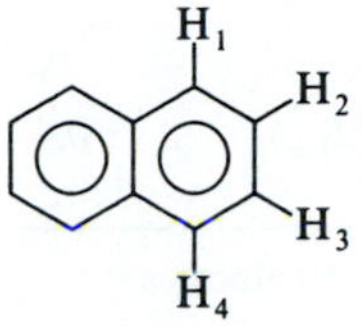

J_{12} 8.30 J_{14} 0.74
J_{13} 1.20 J_{23} 6.83

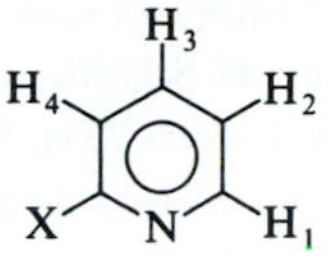

J_{12} 4.6–5.1 J_{23} 7.1–7.9
J_{13} 1.7–2.1 J_{24} 0.9–1.3
J_{14} 0.8–1.0 J_{34} 7.7–8.4

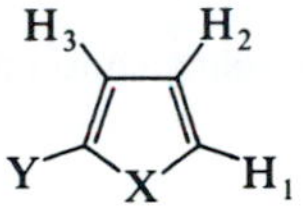

(X = O) J_{12} 1–2 J_{23} ~3.6 J_{13} 1–2
(X = NH) J_{12} ~2.6 J_{23} ~3.7 J_{13} 1–2
(X = S) J_{12} ~5.5 J_{23} ~3.5 J_{13} 1–2

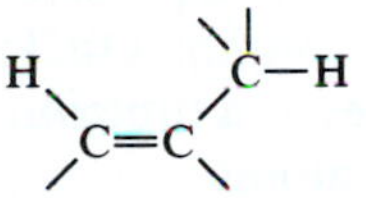

$(^4J_{HH})$** -1.0–-1.9†

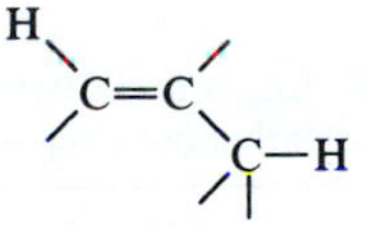

-1.5–-2.5†

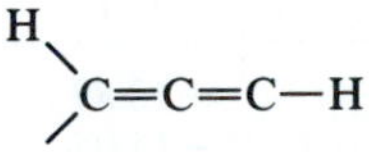

5–6

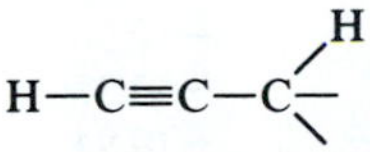

2–4

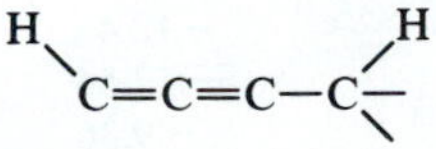

$(^5J_{HH})$**   2–3

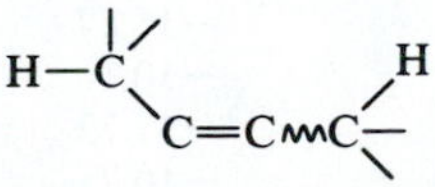

0–3.0†

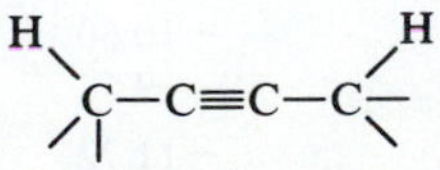

2–3†

(table continues)

TABLE 6.9 (Continued)

Structure	J_{HH} (H_Z)

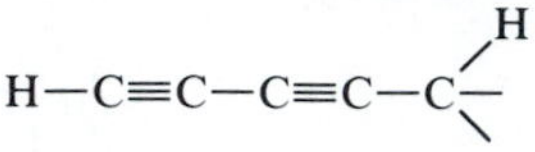

$$H-C\equiv C-C\equiv C-C\begin{smallmatrix}H\\ \\ \end{smallmatrix} \quad (^6J_{HH}) \quad 1-2$$

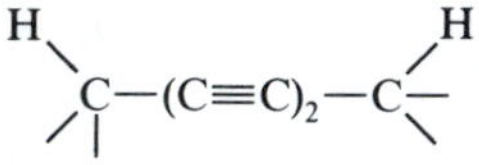

$$\begin{smallmatrix}H\\ \\ \end{smallmatrix}C-(C\equiv C)_2-C\begin{smallmatrix}H\\ \\ \end{smallmatrix} \quad (^7J_{HH}) \quad \sim 1$$

$$\begin{smallmatrix}H\\ \\ \end{smallmatrix}C-(C\equiv C)_3-C\begin{smallmatrix}H\\ \\ \end{smallmatrix} \quad (^9J_{HH}) \quad \sim 0.4$$

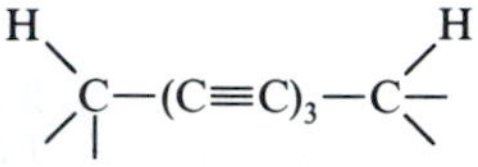

* See Table 6.10 for individual values.

† J is a function of the dihedral angle and substituents (see discussion).

‡ K. B. Wiberg, D. E. Barth and P. H. Schertler, *J. Org. Chem.,* **38**, 378 (1973).

§ K. B. Wiberg and D. E. Barth, *J. Am. Chem. Soc.,* **91**, 5128 (1969).

‖ J is a function of the H—C—H bond angle and substituents (see discussion).

¶ Value of J depends on the solvent.

** For a review of long-range coupling constants see S. Sternhall, *Rev. Pure Appl. Chem.,* **14**, 15 (1964).

The interested reader is referred to the more comprehensive NMR texts cited at the end of the chapter.)

The geminal coupling constant,$^2J_{HH}$, varies greatly: from approximately -17 to $+42$ Hz, as is illustrated in the examples in Tables 6.9 and 6.10.

The magnitude of $^3J_{HH}$ varies as a function of the dihedral angle between the C—H bonds on adjacent carbon atoms.

TABLE 6.10 $^2J_{HH}$'s and Chemical Shifts in Substituted Methanes, CH_3-X

X	$^2J_{HH}$	δ
Li	-12.89	-1.74
$Si(CH_3)_3$	-14.05	0.0
$Ge(CH_3)_3$	-12.96	0.13
$Sn(CH_3)_3$	-12.37	0.06
$Pb(CH_3)_3$	-10.94	0.71
H	-12.4	0.23
CH_3	-12.56	0.90
I	-9.36	2.16
SCH_3	-11.87	2.08
Br	-10.1	2.68
$N(CH_3)_2$	-11.73	2.12
Cl	-10.7	3.05
OCH_3	-10.60	3.24
F	-9.5	4.26
$\overset{+}{N}H(CH_3)_2$	-11.72	3.05

For ethane the relationship of $^3J_{HH}$ with ϕ is given by Equation (6.9),

$$^3J_{HH} = 4.22 - 0.5 \cos \phi + 4.2 \cos^2 \phi \qquad (6.9)$$

a plot of which is illustrated in Figure 6.4. Experimental values of $^3J_{HH}$ as a function of ϕ have been measured in conformationally rigid systems in which the bond angles are reasonably well known, as in conformationally biased cyclohexane derivatives (see entries in Table 6.9). These values are in reasonable agreement with the theoretically calculated values.

The magnitudes of $^3J_{HH}$'s are also affected by the substituents attached to the spin system. This effect has been evaluated by several groups,[3] and is apparent in the data given in Table 6.11.

The magnitude of $^3J_{HH}$ between vinyl hydrogens depends on their stereochemical relationship, electronegativity of functions attached to the C=C, and bond angle strain in the C=C. As might be suggested by Equation (6.9), it is observed that *trans* $^3J_{HH}$ is greater than *cis* $^3J_{HH}$ (see entries in Table 6.9). The effect of the electronegativities of groups attached to the C=C on $^3J_{HH}$, as well as on $^2J_{HH}$, is given by Equations (6.10) to (6.12), in which E_X and E_Y are the Pauling electronegativities of the func-

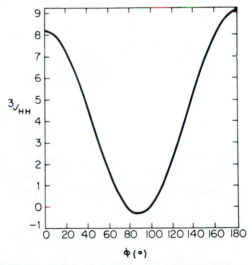

Figure 6.4 Variation of $^3J_{HH}$ with dihedral angle, ϕ.

[3] M. Karplus, *J. Am. Chem. Soc.,* **85,** 2870 (1963); D. H. Williams and N. S. Bhacca, *J. Am. Chem. Soc.,* **86,** 2742 (1964); H. Booth, *Tetrahedron Lett.,* 411 (1965); H. R. Buys, *Recl. Trav. Chim. Pays-Bas,* **88,** 1003 (1969).

TABLE 6.11 Hydrogen Coupling Constants (Hz) as a Function of the Substituent X in **8**

X	J_{AX}	J_{BX}	J_{AB}
CN	4.6	9.3	−12.6
COOH	4.4	8.5	−12.6
C_6H_5	4.2	8.9	−12.7
Cl	3.2	8.0	−13.2
OH	2.4	7.4	−12.6
$OCOCH_3$	2.5	7.6	−13.3

tions X and Y, or the attachment atoms of the functions (such as C of CH_3) in $CHX{=}CHY$ (either X or Y can be H, with $E_H = 2.1$).[4]

$$^3J_{HH}\ (cis) = \frac{91.2}{E_X + E_Y} - 9.7 \tag{6.10}$$

$$^3J_{HH}\ (trans) = \frac{75.0}{E_X + E_Y} \tag{6.11}$$

$$^2J_{HH}\ (geminal) = \frac{61.6}{E_X + E_Y} - 12.9 \tag{6.12}$$

The effect of bond angle strain in the $C{=}C$ on $^3J_{HH}$ is reflected in the values of the coupling constants for cyclopropene, cyclobutene, and the larger ring cycloalkenes given in Table 6.9. As the ring strain decreases, the coupling constant increases. Comparisons between other cyclic and acyclic alkenes are given in Table 6.12. Note that in the acyclic alkenes, as the strain increases the coupling constant increases, the *cis* coupling constant increasing more rapidly than the *trans* coupling constant.

Long-range coupling, a term used to indicate coupling that is transmitted through more than three bonds, is generally more readily observed in unsaturated systems such as aromatic compounds, alkenes, and alkynes. Typical examples of long-range coupling are given in Table 6.9. Long-range coupling through saturated bonds is less frequently observed, and is highly dependent on the conformation of the system. When the bonded system exists in the **W**, or **M**, conformation, that is,

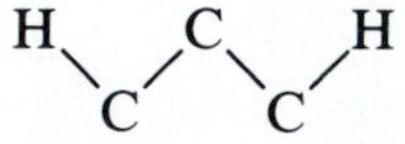

appreciable long-range coupling can be observed, as is illustrated in the structures **9** and **10**.

---

[4] T. Schaefer and H. M. Hutton, *Can. J. Chem.,* **45,** 3153 (1967).

TABLE 6.12　Comparison of $^3J_{HH}$'s in Strained and Unstrained Systems*

Strained Compound	$^3J_{HH}$	Less Strained Compound	$^3J_{HH}$
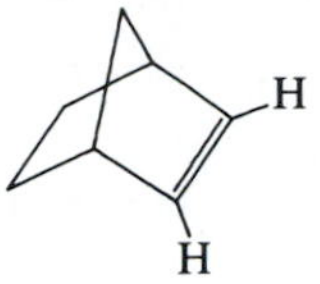	5.68	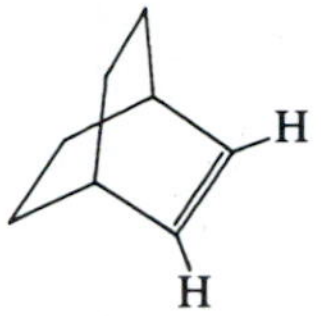	8.24
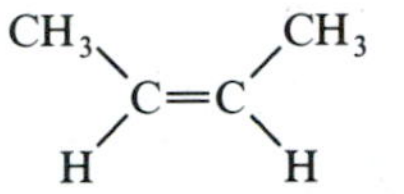	10.88 (11.0)†	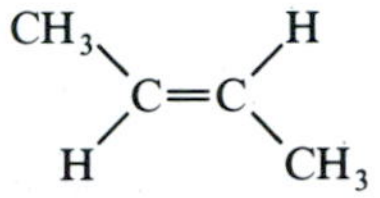	15.09 (17.0)†
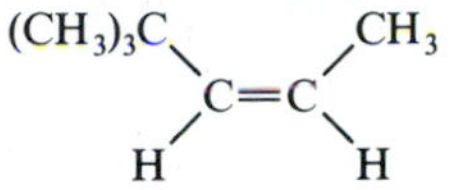	12.02	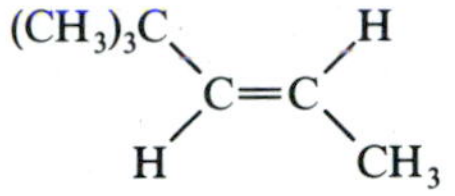	15.50
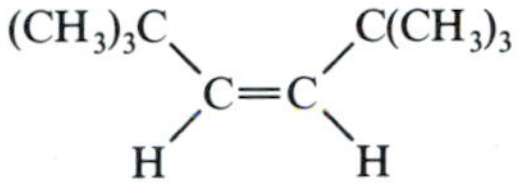	14.2	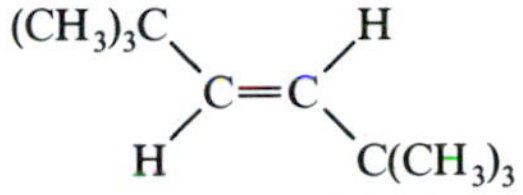	16.10

* Taken from M. A. Cooper and S. L. Manatt, *Org. Magnetic Res.*, **2**, 511 (1970).
† Parenthesized values are the calculated values, using 2.2 for E_C of the attached methyl and *t*-butyl groups.

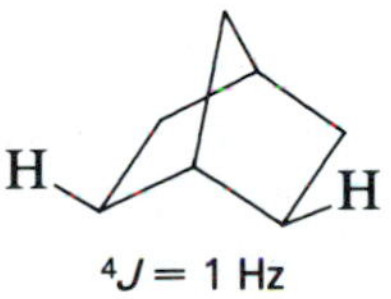

$^4J = 1$ Hz　　　　　$^4J = 7$ Hz
9　　　　　　　　　**10**

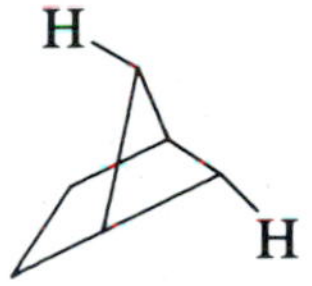

6.5.1　Spin-Spin Coupling Between Other Nuclear Combinations

As indicated in the opening paragraphs of this chapter, spin-spin coupling can exist between nuclei having I not equal to zero. It should now be obvious that the organic chemist is extremely fortunate that the predominant isotopes of carbon and oxygen, ^{12}C and ^{16}O, possess I values of zero and do not spin-spin couple with other nuclei. If this were not true, the complexity of hydrogen magnetic resonance spectra would be formidable. However, there are many other nuclei encountered in organic molecules that possess I values not equal to zero. Table 6.13 lists typical ranges of coupling constants for a variety of different nuclear combinations. These coupling constants also vary with bond angle, dihedral angle, the number and type of intervening bonds, and the electronegativity of the attached atoms or functions in the

TABLE 6.13 Coupling Constants for Various Nuclear Combinations

Coupling Constant	System	Coupling Constant Range (H_z)
J_{HD}	General	$0.154 J_{HH}$
$J_{^{10}BH}$	^{10}B—H	~30–50
$J_{^{11}BH}$	^{11}B—H	75–182
	^{11}B———H———^{11}B(bridged)	30–50
	^{11}B—C—H	~0
$J_{^{13}CH}$*	$^{13}CH_3X$	110–155
	$=^{13}C$—H	150–170
	$\equiv^{13}C$—H	~250
	^{13}C—C—H	4–6
	^{13}C—C—C—H	3–4
$J_{^{14}NH}$	NH_4^+	52.6
$J_{^{15}NH}$	NH_4^+	73.7
$J_{^{19}FH}$	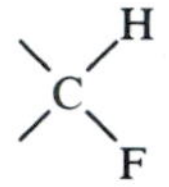	44–81
	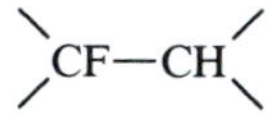 CF—CH	0–30
	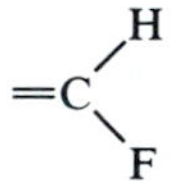	70–80
	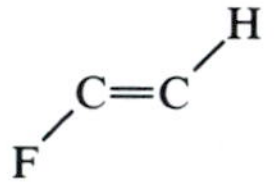 C=C H (*trans*)	30–50
	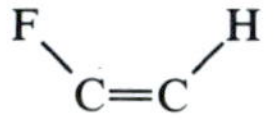 C=C H (*cis*)	2–20
	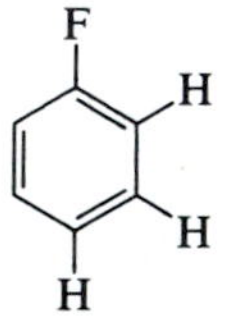 (*ortho*) (*meta*) (*para*)	8–10 5–8 ~2
$J_{^{13}C^{19}F}$	^{13}C—^{19}F	250–300
	^{13}C—C—^{19}F	30–40
$J_{^{19}F^{19}F}$	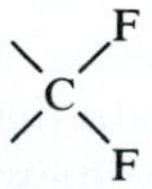	155–225
	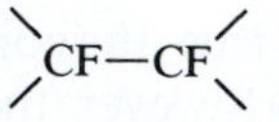 CF—CF (*trans*) (*gauche*)	≈ 16 ≈ 18
	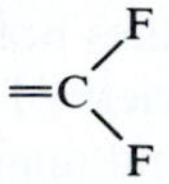	28–87

TABLE 6.13 (Continued)

Coupling Constant	System	Coupling Constant Range (H_z)
	F, F (cis) $C=C$	20–58
	F (trans) $C=C$ F	95–120
	F (ortho) (meta) (para) F	~20 2–4 11–15
$J_{^{29}SiH}$	^{29}Si—H	190–380
$J_{^{31}PH}$	^{31}P—H	179–700
	^{31}P—C—H	~4
	^{31}P—C—C—H	~12–14
$J_{^{199}HgH}$	^{199}Hg—C—H	80–235
	^{199}Hg—C—C—H	115–200

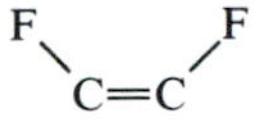

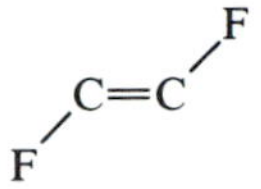

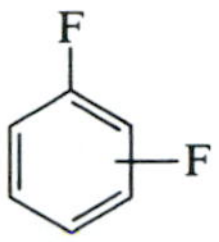

* Trends in the magnitude of ^{13}C coupling constants with structure are discussed in Section 6.15.

same manner as described for hydrogen-hydrogen coupling. Spin-spin coupling interaction of ^{13}C will be discussed in more detail in a later section.

6.5.2 Spin Decoupling

In spectra of relatively complex molecules it may be quite difficult to determine which nuclei are spin-coupled leading to the observed patterns. This is particularly true if several of the coupling constants are of similar size, if one or more of the nuclei are buried in a complex multiplet, or if long-range coupling is present. In such cases it would be desirable to destroy the spin-spin coupling interaction between specific nuclei in the system. Spin-spin splitting is observed when the nucleus that is responsible for the splitting remains in a given spin state for a sufficient period of time that is long in comparison with the reciprocal of the difference in the chemical shifts, in Hz, of the coupled nuclei. If it is possible to cause the lifetime of a spin state to decrease sufficiently so that the absorbing nucleus "sees" only a time-averaged spin state for the coupled nucleus, a single resonance line will be observed. Experimentally, irradiating the nucleus (nuclei) that is (are) coupled to the nucleus being observed at its (their) resonance frequency destroys the coupling interaction. This is referred to as *double resonance*. When the type of nucleus being decoupled is the same as that being observed, the process is termed *homonuclear double resonance;* when the nucleus being decoupled is not the same as that being observed, it is termed *heteronuclear double resonance.* An example of homonuclear double resonance is shown in Figure

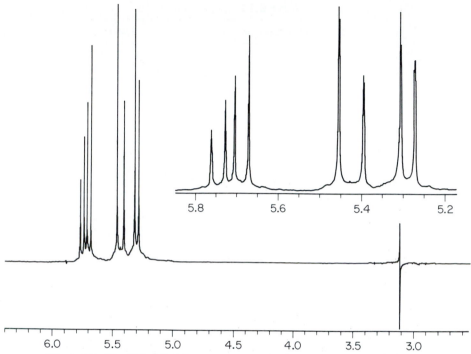

Figure 6.5 The 300 mHz ^{1}H NMR spectrum of allyl cyanide in which the CH_2 hydrogens have been decoupled from the other nuclei. The small "blip" at ~δ 3.1 is where the CH_2 hydrogens would have appeared had they not been saturated.

6.5 in which the methylene hydrogens of allyl cyanide (H_2C=$CHCH_2CN$) have been irradiated, destroying the *vicinal* coupling to the =CH hydrogen and the long-range coupling to the H_2C= hydrogens (see Figure 6.14 for the nondecoupled NMR spectrum of allyl cyanide).

6.6 DESIGNATION OF SPIN SYSTEMS

In a spectrum containing the absorption peaks of a variety of differing nuclei, a system of designating the types of spin systems present is desirable. By a spin system, we mean a set of nuclei that are spin-spin coupled giving rise to recognizable resonance patterns. A single molecule may contain several such distinguishable spin systems.

The designation of the various nuclei in a spin system is based on the relative chemical shifts and the size of the chemical-shift difference with respect to the coupling constant J. The symbols used to designate individual nuclei and spin systems are included in Table 6.14. Examples of the use of these symbols appear in the following paragraphs.

TABLE 6.14   Designation of Individual Nuclei and Spin Systems

Symbol	Description*
A	H_A with δ_A.
A_n	nH_A of same δ_A and magnetically identical.
AA′	Hydrogens H_A and H_A, with same chemical shift, but magnetically different with respect to other nuclei.
AB	Hydrogens H_A and H_B with δ_A and δ_B, where $J_{AB} \geq (\delta_A - \delta_B)$.
AX	Hydrogens H_A and H_X with δ_A and δ_X, where $J_{AX} < (\delta_A - \delta_X)$.
AMX	Hydrogens H_A, H_M, and H_X with δ_A, δ_M, and δ_X, where $J_{AM} < (\delta_A - \delta_M)$ and $J_{XM} < (\delta_M - \delta_X)$.
ABX	Hydrogens H_A, H_B, and H_X with δ_A, δ_B, and δ_X, where $J_{AB} \geq (\delta_A - \delta_B)$, $J_{AX} < (\delta_A - \delta_X)$, and $J_{BX} < (\delta_B - \delta_X)$.
ABC	Hydrogens H_A, H_B, and H_C with δ_A, δ_B, and δ_C, where all $\Delta\delta$'s $< J$'s.

* Although all entries in this table utilize the hydrogen atom, these systems can be used with any other nuclear system or mixed nuclear system.

Most of the entries in Table 6.14 are self-explanatory and will be illustrated by the use of actual resonance spectra. In general, spectra are presented in the literature with field strength increasing from left to right. The symbols designating the spin systems are employed such that the first letters of the alphabet (A, B, C, etc.) are used to represent nuclei appearing at lowest fields, while the latter letters (M, X, Y, etc.) represent nuclei appearing at progressively higher fields. For example, in an ABX system, the A nucleus would appear at the lowest field strength, the X nucleus appearing at the highest field strength.

The use of primed symbols, for example, in AA′, also requires further explanation. Up to now we have considered only the chemical identity, or nonidentity, of nuclei in the resonance region. The term *magnetic identity* implies an equal spin-spin interaction, that is, an equal coupling constant between each nucleus of a given set of coupled nuclei and the observed nucleus. This condition is not always met. In many instances, two chemically identical nuclei interact with different coupling constants with another nucleus and, thus, are not magnetically identical with respect to that nucleus. The prime is used to indicate this magnetic nonidentity. For example, in the cyclopropane derivative **11**, we have three chemically different types of hydrogens, H_A, H_B, and H_X. The hydrogens on C_2 and C_3 *cis* to the X function are chemically identical and are designated as H_A's; similarly, the two hydrogens on C_2 and C_3 *trans* to the X function are chemically identical (H_B's), but are not chemically identical with the H_A's. Although the H_A's are chemically identical, they are not magnetically identical. H_A on C_2 is coupled to H_B on C_2 with a geminal coupling constant (^2J), but is coupled to the H_B on C_3 with a *trans* vicinal coupling constant (^3J). On the other hand, the H_A on C_3 is coupled to the H_B on C_2 with a ^3J and to the H_B on C_3 with a ^2J. Thus, although both H_A's are chemically identical, they are not equally coupled to the two H_B's on C_2 and C_3, they are therefore magnetically nonequivalent and are designated as H_A and $H_{A'}$. A similar analysis shows that the two H_B's are not magnetically identical with respect to their interactions with the two H_A's. It should be pointed out, however, that H_A and $H_{A'}$, as well as H_B and $H_{B'}$ are magnetically identical in their spin coupling to H_X, that is, $J_{AX} = J_{A'X}$ and $J_{BX} = J_{B'X}$.

A similar situation exists in 1,3-butadiene (12). It is obvious that hydrogens H_A and $H_{A'}$ are chemically identical, as are H_B and $H_{B'}$, and H_C and $H_{C'}$; however, the spin-spin interactions between C' and A', and C' and B' hydrogens are different from those between the C' and A, and C' and B hydrogens. Again magnetic nonidentity is indicated. Magnetic nonidentity may be obvious in some spectra, whereas in others, the complexity of the absorption patterns may not allow a straightforward analysis.

11   12

Certain ambiguities in the designation of spin systems may still arise when applying the nomenclature rules outlined in the foregoing paragraphs. For example, if we were to designate a system as AA'BB'CC' (butadiene), we would not be clearly indicating the sequence of nuclei with respect to the carbon-atom framework of the molecule. To avoid such confusion, the symbols designating the types of nuclei bonded to one carbon atom can be separated by a hyphen from the symbols designating the types of nuclei on the adjacent carbon atom. Thus the preferred designation for butadiene would be BC—A—A'—B'C'.

6.7 EXAMPLES OF SIMPLE SPIN SYSTEMS

6.7.1 The Two-Spin System

We shall begin our discussion of simple spin systems by considering the two-spin system, the one-spin system being trivial and giving rise to a single resonance line. The simplest two-spin system is the A_2 system, which also gives rise to a single resonance line. If we allow the chemical shifts of the two nuclei to become slightly different, and then introduce coupling between the nuclei such that $J_{AB} > \Delta\delta_{AB}$, a number of lines appear in the resonance spectrum. If we were to spin-couple the resonance lines of nuclei A and B in a symmetrical fashion about their respective chemical shifts, the lower field line of the B nucleus doublet would appear at lower field than the higher field line of the A nucleus (crossing of energy states). However, the energy states of one nucleus produced by spin-spin interaction with a neighboring nucleus can not overlap (cross) the energy states of that nucleus. To avoid the crossing of states in the AB system, the inner lines of the A and B doublets become

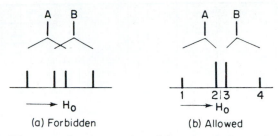

Figure 6.6 Diagrammatic representation of the spin-spin interaction in the AB system.

more intense than the outer lines so that the centers of gravity of the A and B doublets are at their respective chemical shifts. The spin-spin coupling interaction is diagrammed in Figure 6.6. As the chemical shift difference ($\Delta\delta$) between the nuclei A and B increases relative to J_{AB}, the outer lines increase in intensity at the expense of the inner lines. Finally, as the system graduates to the AX system, the lines become more equal in intensity.

The relative intensities of the outer-to-inner lines in an AB or AX system and the difference in chemical shifts ($\Delta\delta$) are calculated using Equations (6.13) and (6.14), respectively.

$$\text{Relative intensity} \left(\frac{\text{line 1}}{\text{line 2}} \right) = \frac{Q - J}{Q + J} \tag{6.13}$$

$$\Delta\delta = (Q^2 - J^2)^{1/2} \tag{6.14}$$

The coupling constant J is the distance between lines 1 and 2, or 3 and 4, and the quantity Q is the distance between lines 1 and 3, and 2 and 4 expressed in Hz.

Examples of spectra containing AB and AX systems are illustrated in Figure 6.7. The hydrogen resonance spectrum of *cis-β*-phenylmercaptostyrene, Figure 6.7a, shows an AB pattern centered at δ 6.48 for the two ethylenic hydrogens. The coupling constant J_{AB} is 10.7 Hz, and $\Delta\delta_{AB}$ is 0.095 ppm. The observed intensity ratio is 0.056; compared to 0.061 as calculated from Equation (6.13). The complex resonance pattern at lower field is due to the aromatic hydrogens.

Figure 6.7b shows the hydrogen resonance spectrum of *cis-β*-ethoxystyrene, in which the ethylenic hydrogens appear as an AX system, with δ_A 5.18 and δ_X 6.06 and $J = 7.2$ Hz. The observed intensity ratio is 0.77, with a calculated value of 0.77. The triplet at δ 1.21 and the quartet at δ 3.77 represent the A_2X_3 pattern of the CH_3CH_2— group, while the aromatic hydrogens give rise to the complex multiplet centered at δ 7.3.

6.7.2 Three-Spin Systems

The simplest three-spin system is the A_3 system, for example, an isolated methyl group, which gives rise to a single resonance line. If one of the nuclei becomes slightly different in chemical shift relative to the other two nuclei such that $\Delta\delta$ is less than the coupling constant, we have an A_2B or AB_2 system. The spectra of such systems do

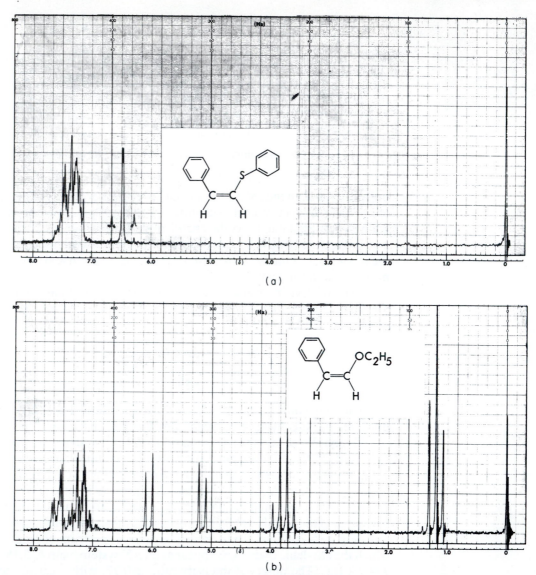

Figure 6.7   (a) Hydrogen resonance spectrum of *cis-β*-phenylmercaptostyrene.
(b) Hydrogen resonance spectrum of *cis-β*-ethoxystyrene.

not possess any symmetry and cannot be readily analyzed as described for the AB and AX systems. A similar situation is encountered with the ABC system, in which all three nuclei are slightly different in chemical shift. In spectra where there appears to be a three-spin system of the A_2B or ABC type, the reader is referred to Wiberg and Nist's *The Interpretation of NMR Spectra* (reference 19) in order to determine the chemical shifts and coupling constants. Wiberg and Nist have calculated the absorp-

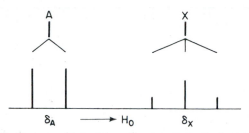

Figure 6.8 Representation of the spin-spin splitting in the A_2X system.

tion patterns for various spin systems, maintaining a constant chemical shift difference (6 Hz) while varying the coupling constant. The student selects the calculated spectrum that most closely resembles the portion of the spectrum of his sample in peak shape, relative positions, and intensities. The difference in chemical shifts and coupling constants can then be scaled up to correspond to the observed pattern. Numerous computer programs are available for the calculation of spectra and for comparison with observed spectra.

The A_2X or AX_2 system, in which one nucleus is of a greatly different chemical shift, can be represented in a straightforward fashion as indicated in Figure 6.8. The resonance of the A hydrogens is split into a $1:1$ doublet by the single X hydrogen, while the X resonance is split into a $1:2:1$ triplet by the two A hydrogens. The diagram in Figure 6.8 is a splitting diagram that illustrates how the multiplicity of the resonance patterns is generated. The theoretical intensities of the lines in the A and X portions of the spectrum, as calculated from Equation (6.8), apply only when $\Delta\delta_{AX}/J_{AX} \to \infty$. As the ratio of $\Delta\delta_{AX}$ to J_{AX} decreases, the absorption patterns become distorted, with the inner lines increasing in intensity at the expense of the outer lines, as in the $AX \to AB$ transformation.

Another three-spin system is the ABX system where $\Delta\delta_{AB} < J_{AB}$, and $\Delta\delta_{AX}$ and $\Delta\delta_{BX} > J_{AX}$ and J_{BX}. The spin-spin interaction is diagrammed in Figure 6.9. If we first couple the A and B nuclei independently of the X nucleus, we derive the typical AB pattern. Spin-spin interaction between the A and X nuclei doubles the A and X resonance lines, and interaction between the B and X nuclei doubles the B and again the X lines. It should be pointed out that coupling of the A and B nuclei with the X

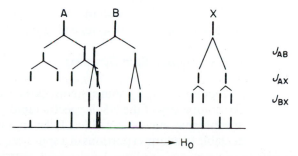

Figure 6.9 Representation of the spin-spin splitting in the ABX system.

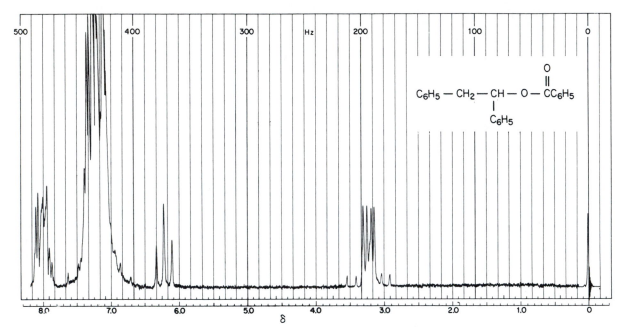

Figure 6.10 Hydrogen resonance spectrum of 1,2-diphenylethyl benzoate.

nucleus leads to an overlapping of the A and B resonance lines. (This is permissible except in cases when direct coupling between two nuclei might lead to crossing.)

Occasionally the X portion of the ABX spectrum may appear as an apparent triplet if $J_{AX} \approx J_{BX}$, which results in an overlap of the two central lines of the X portion. In such cases the term triplet should not be used; the pattern should be described as "the X portion of an ABX system." A typical example of a spectrum containing an ABX system is shown in Figure 6.10 for 1,2-diphenylethyl benzoate. The AB portion arises from the dissimilarity of the diastereotopic methylene hydrogens.

The final three-spin system is the AMX system. The resonance pattern is predicted in the straightforward manner as illustrated in Figure 6.11. The relative sizes of the coupling constants will vary, depending on whether we are dealing with an AM-X or an A-M-X system. A typical example of an AMX pattern is illustrated in the NMR spectrum of styrene (Figure 6.12).

6.7.3 Higher-Spin Systems

Except for the $A_n X_m$ spin systems, as the number of nuclei in a spin system increases the complexity of the NMR spectra rapidly increases, and first-order analyses of the spectra may not be possible. Even in such complex cases, it may, however, be possible to interpret portions of a spin system. For example, in the NMR spectrum of

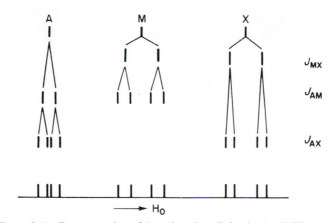

Figure 6.11 Representation of the spin-spin splitting in the AMX system.

1-chlorobutane (Figure 6.13) we can readily recognize a low-field, slightly distorted triplet representing the —CH_2Cl hydrogens, which are split by the adjacent methylene hydrogens. At higher field we can also make out a highly distorted triplet representing the methyl group [the distortion arises from the fact that the difference between the chemical shifts of the methyl and the adjacent methylene hydrogens is

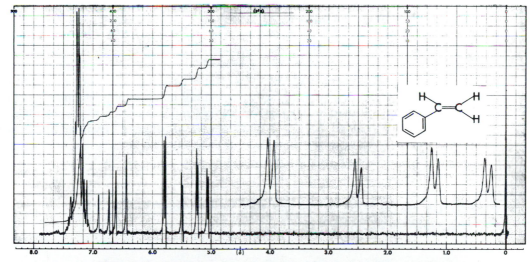

Figure 6.12 Hydrogen resonance spectrum of styrene showing an AMX spin system. H_A appears as a double doublet at δ 6.62 with $J_{AM} = 17.5$ Hz and $J_{AX} = 11.7$ Hz, H_M at δ 5.68 and H_X at δ 5.16 with $J_{MX} = 1.7$ Hz. The tracing at the upper right is that of the H_X and H_M resonance region, which has been expanded to 100 Hz total sweep width with an offset of 296 Hz (to keep the resonance region on the chart paper). The vertical-stepped tracing is the integral line (see discussion in Section 6.9).

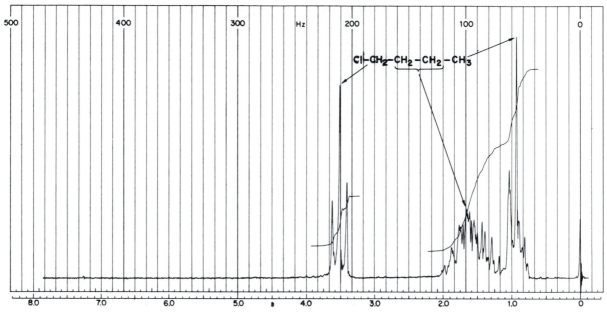

Figure 6.13 ^{1}H NMR spectrum of 1-chlorobutane.

relatively small and only slightly larger than the coupling constant (~ 6.7 Hz)]. The resonance pattern centered at δ 1.6 represents the two methylene groups, which cannot be resolved and analyzed. The assignment that the multiplet at δ 1.6 represents two methylene groups, that is, four hydrogens, can be made on the basis of the spectrum integral (see Section 6.10). Therefore, had the spectrum in Figure 6.13 been that of an unknown, we would have been able to assign a unique structure to the unknown even though we could not make a complete analysis of the spectrum.

Even with much larger and more complex molecules than 1-chlorobutane, partial structures of a molecule can be assigned. By integrating chemical and other spectral data for the molecule, the complete assignment of structures of such molecules can usually be made.

6.8 EFFECT OF FIELD STRENGTH ON THE APPEARANCE OF THE NMR SPECTRUM

It has been noted in Section 6.5 that the chemical shift is field-strength dependent, while the magnitude of the coupling constant is not. This difference in behavior results in significant differences in the appearance of the spectra when recorded at different field strengths. For example, at lower field strengths a two-spin system will appear as AB doublets. At higher field strengths, however, the difference in chemical shift, in Hz, increases and the spectrum will have the appearance of an AX spin system.

In more complex molecules where several nuclei of similar chemical shift are spin coupled, the NMR spectrum recorded at low field strength may show only a complicated overlapping set of resonances that cannot be easily interpreted. At higher field strengths the individual nuclei are sufficiently separated in chemical shift that a complete analysis of the spectrum is easy to carry out. Figure 6.14 shows the differences in the appearance of the NMR spectrum of allyl cyanide recorded at 90 and 300 MHz.

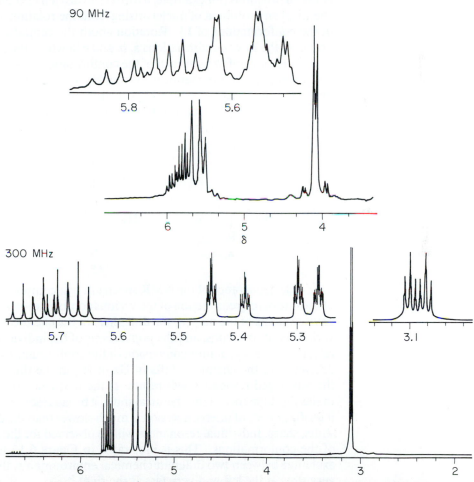

Figure 6.14 The ^{1}H NMR spectra of allyl cyanide recorded at 90 (top) and 300 (bottom) mHz. Inserts are of expansions of regions of the spectra. Note the overlap of the two =CH$_2$ multiple doublets at δ 5.28 and 5.42 that occurs in the 90 mHz spectrum, thus making the interpretation of the 300 mHz spectrum much easier.

6.9 EFFECTS OF THE EXCHANGE OF THE CHEMICAL ENVIRONMENT OF NUCLEI

Our discussion of chemical shift and spin-spin coupling phenomena has thus far been quite general. We have not stopped to consider in detail the dynamic processes that may be occurring within or between molecules and the effect that these dynamic processes have on the appearance of the NMR spectrum of compound. Three dynamic processes that affect the appearance of the NMR spectra of compounds are (1) rotations about bond axes, (2) inversion of an atom (for example, the inversion of the nitrogen atom in amines), and (3) intra- and intermolecular exchange of nuclei between functions (for example, hydroxyl-hydrogen exchange). The exchange of the chemical environment of nuclei arising from the rotation about a bond is illustrated in the conformations of **13**. Rotation about the central carbon-carbon bond of **13** produces the three conformations **a**, **b**, and **c** in which the chemical environments of H_A and H_B are different in each of the conformations. As the conformations are not equally populated, the chemical shifts of H_A and H_B will be different. Also, because of the different H_X—C—C—H_A dihedral angles in the three conformations, the H_X—H_A coupling constant in **13a** will be much larger than in **13b** and **c**, while the H_X—H_B coupling constant will be much larger in **13b** than in **13a** and **c**. This will result in $J_{AX} \neq J_{BX}$.

13

The appearance of the NMR spectrum of a compound such as **13** is a function of the rate of interconversion of the various conformations (the exchange of chemical environments of the nuclei), the difference in the chemical shifts of the nuclei in the various conformations, and the populations of the individual conformations. If the rate, or frequency, of interconversion of the conformations is much greater than the difference in the chemical shifts in Hz of H_A in the three conformations, a sharp, time-averaged resonance will result. If the frequency of interconversion approximates the difference in the chemical shifts, a broad resonance line will result. Finally, if the frequency of interconversion is much slower than the difference in the chemical shifts, sharp, individual resonances will be observed for the individual nuclei in each of the conformations. This is illustrated in Figure 6.15 for a nucleus undergoing exchange between two different chemical environments, the rate of exchange being very slow at the left and very fast at the right.[5]

---

[5] The NMR spectrometer can be compared to a camera equipped with a relatively slow shutter. A clear picture will be obtained of a nearly stationary object, whereas a blurred picture will be obtained of a rapidly moving object.

The chemical shift of a nucleus in a rapidly interconverting system is given by Equation (6.15),

$$\delta_{A_{av}} = \sum_i N_i \delta_{A_i} \qquad (6.15)$$

in which N_i is the mole fraction of the ith conformer and δ_{i_A} is the chemical shift of H_A in the ith conformation. Similarly, the observed coupling constant is given by Equation (6.16),

$$J_{AX_{av}} = \sum_i N_i J_{AX} \qquad (6.16)$$

in which J_{AX_i} is the coupling constant between H_A and H_X in the ith conformation.

It is instructive to consider in greater detail the chemical shifts of H_A and H_B in structure **13**. Applying Equation (6.15) to H_A and H_B gives

$$\delta_A = N_a \delta_{A_a} + N_b \delta_{A_b} + N_c \delta_{A_c}$$

and

$$\delta_B = N_a \delta_{B_a} + N_b \delta_{B_b} + N_c \delta_{B_c}$$

H_A and H_B are diastereotopic in all three of the rotamers **13a**, **b**, and **c**. (A *diastereotopic* relationship exists between two atoms or groups when replacement of each atom or group separately produces a pair of diastereomers.) In structure **13**, the diastereomeric environments are provided by the asymmetric carbon $-CXYH_X$. H_A and H_B are *not* chemically equivalent despite the fact that both hydrogens are bonded to the same carbon! The chemical shifts of H_A and H_B are consequently different in each of the three rotamers although the difference may be quite small, that is, $\delta_A \neq \delta_B$ under any conditions of rapid rotation and rotamer populations. Similarly, $J_{AX} \neq J_{BX}$. Occasionally it is useful to modify a function to increase the difference between the chemical shifts of diastereotopic hydrogens. For example, the difference in the chemical shifts of the diastereotopic hydrogens of 1,2-diphenylethyl alcohol is very small and is barely discernible, but is greatly enhanced on converting the alcohol (see Figure 6.16) to the benzoate (see Figure 6.10), in which the diastereotopic hydrogens appear at δ 3.17 and δ 3.35.

Just as a pair of diastereotopic hydrogens or groups in the same molecule may have different chemical shifts, corresponding hydrogens or groups in two diastereomers, in principle, will have different chemical shifts. When the chemical shifts of hydrogens in diastereomers are of sufficient difference, nuclear magnetic resonance

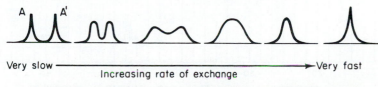

Figure 6.15   Line shapes of resonance peaks of an equilibrating system.

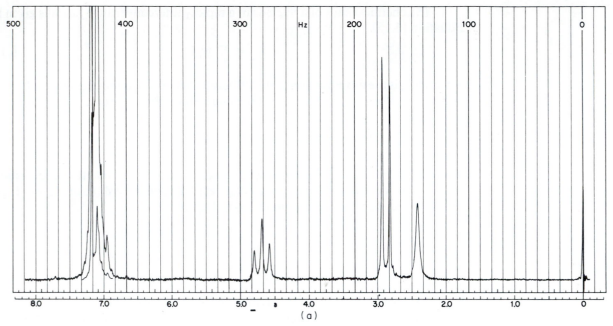

Figure 6.16   ¹H NMR spectrum of 1,2-diphenylethanol. See Figure 6.10 for the ¹H NMR spectrum of the corresponding benzoate.

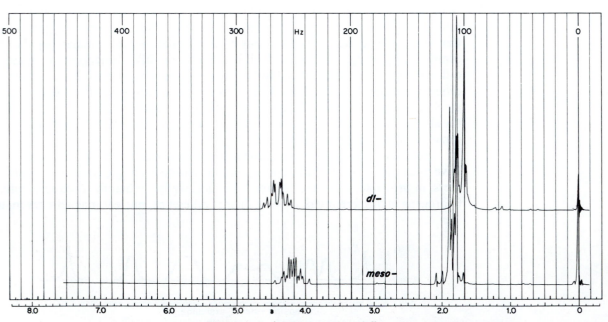

Figure 6.17   ¹H NMR spectra of *dl,* and *meso*-2,3-dibromobutane.

provides a very accurate method for the analysis of mixtures of diastereomers. For example, the NMR spectra in Figure 6.17 are those of *dl*- and *meso*-2,3-dibromobutane, which are considerably different from each other.

The effect of the exchange of nuclei between functions, either inter- or intramolecular, is illustrated in Figure 6.18. The top spectrum is that of very pure ethanol, in which the coupling between the hydroxyl hydrogen and the methylene hydrogens is readily apparent. In this case the rate of exchange of the hydroxyl hydrogen between different molecules of ethanol in the sample is slow enough so that the hydroxyl and methylene hydrogens have sufficient time to "see" each other.

The lower spectrum is that of ethanol containing a trace of acid (which greatly accelerates the exchange of the hydroxyl hydrogens). In this case the rate of exchange

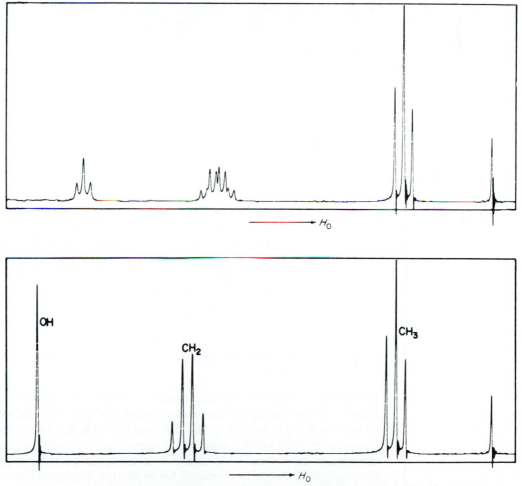

Figure 6.18   ¹H NMR spectra of pure ethanol (top) and acidified ethanol (bottom) illustrating the loss of coupling under the rapid exchange conditions.

is so fast that the hydroxyl and methylene hydrogens do not have time to "see" each other, but only a time-averaged picture of all of the exchanging hydrogens in the sample. The rate of exchange of hydroxyl hydrogens is slowed down on dilution of the sample, and by using certain solvents, such as dimethylsulfoxide-d_6, which also slow down the rate of exchange.

6.10 QUANTITATIVE APPLICATIONS OF NMR SPECTROSCOPY

In continuous wave NMR spectroscopy the response of equal numbers of the same type of nuclei is the same regardless of their environment. Thus the measurement of the areas underneath the individual resonance patterns, termed the integral of the spectrum, allows one to derive the relative ratios of the nuclei present in the different chemical environments in the molecule. If a definite structural assignment can be made for any one of the resonances in terms of the number of nuclei present, the relative ratios can be converted into the absolute numbers of each nuclei present. The integration of an NMR spectrum is done electronically by the NMR instrument. Examples are shown in Figures 6.12 and 6.13. The line appearing above the resonance patterns is the integral line, and the height of the vertical steps is proportional to the number of nuclei represented by the resonance patterns.

In Fourier transform (FT) spectroscopy, however, the integrals may not be directly related to the absolute numbers of each nuclei present. The relative response is related to the rate of spin relaxation and the time of and duration of the acquisition of the FID signal. If the spin relaxation has reached completion during the acquisition time, and if there is no delay between the pulse and the start of the acquisition of the FID, the integral will be directly related to the relative numbers of each type of nuclei present. Such is the case with ^{1}H FT NMR spectroscopy. With ^{13}C FT NMR spectroscopy, however, the rate of spin relaxation of the ^{13}C nuclei is much slower and the rates vary over a wide range. Thus, in general, the integral obtained from a ^{13}C FT NMR spectrum will not accurately represent the numbers of the individual nuclei present. This will be discussed again later.

6.11 THE NUCLEAR OVERHAUSER EFFECT (NOE)

One of the ways in which a spin-excited nucleus can undergo spin relaxation is to transfer its spin energy to that of an adjacent nucleus. The efficiency of this energy transfer is directly related to the distance between the two nuclei. The Nuclear Overhauser Effect (NOE) takes advantage of this spin energy transfer. In such an experiment the normal NMR spectrum of the sample is recorded and the integral is determined. If one now irradiates (as in double resonance) one of the nuclei and records the NMR spectrum and its integral, the integrals of the nuclei closest in space to the irradiated nucleus will be enhanced in intensity relative to those for more remotely positioned nuclei. For example, if one irradiates the vinyl hydrogen in **14** and records the integral while doing so, the integral of the *cis* methyl group will be larger than that of the *trans* methyl group. This technique is very useful in helping to determine stereochemical relationships in molecules that may be difficult to do otherwise. This is nicely illustrated with **14**. On irradiation of the *cis* methyl group (a), H(c) shows a 17% increase in its integral intensity. Irradiation of the more remote *trans* methyl group (b) causes no change in the intensity of the integral of H(c).

6.12 ¹³C NMR SPECTROSCOPY

The development of the application of Fourier transform (FT) techniques to NMR spectroscopy and advances in instrumentation have made it possible to record routinely natural abundance ¹³C NMR spectra. ¹³C chemical shifts span slightly over 200 ppm in contrast to the typical 8 to 9 ppm range in ¹H NMR; thus considerably more structural information is generally available from ¹³C NMR chemical shift data. A second, very important difference between ¹H and ¹³C NMR spectroscopy is that whereas diamagnetic effects are dominant in the shielding of the hydrogen nucleus, paramagnetic effects are the dominant contributors to the shielding of the ¹³C nucleus. Long-range shielding effects that were important in ¹H NMR are less important in ¹³C NMR. As a result, ¹³C chemical shifts generally do not parallel ¹H chemical shifts.

In ¹H NMR spectroscopy, the responses of different nuclei are directly proportional to the numbers of the nuclei present. In ¹³C FT NMR spectroscopy this is not so. The computer sums the free-induction decay (FID) signal for a short period of time (on the order of a few tenths of a second up to 1 to 2 sec), during which time varying percentages of the different nuclei have undergone relaxation. If the FID is recorded very shortly after the initial pulse, the nuclei undergoing rapid relaxation contribute more to the FID than do the more slowly relaxing nuclei. Delaying the recording of the FID results in proportionately greater contributions from the more slowly relaxing nuclei, the population of the rapidly relaxing nuclei having approached the equilibrium distribution; however, incorporating a substantial delay between the initial pulse and the recording of the FID greatly increases the instrument use time, due not only to the delay period but also to the lower intensity of the FID signal per scan, and a reasonable compromise must be achieved. In general, the rate of relaxation of a ¹³C nucleus is directly proportional to the number of hydrogens directly attached to that carbon atom, that is, methyl carbons undergo relaxation more rapidly than do methylene carbons, which in turn relax more rapidly than do methine carbons. Quaternary and carbonyl carbons generally possess long relaxation times. Examples of the variation in response is evident in the various ^{13}C spectra presented later.

^{13}C NMR spectra are generally recorded under double resonance conditions in which the coupling of ^{1}H to ^{13}C is destroyed. (Coupling of ^{13}C with ^{13}C is generally not observed because of the low probability of having adjacent ^{13}C's.) Complete ^{1}H coupling is accomplished by irradiating the ^{1}H resonance region with a broad band width radiofrequency radiation, termed "noise," sufficient to cover the entire ^{1}H resonance region. The ^{13}C NMR spectra thus obtained contain only singlet resonances. Although this greatly simplifies the appearance of the spectra, it does result in the loss of information, that is, the number of hydrogens attached to each carbon atom cannot be determined from the multiplicity of each ^{13}C resonance signal according to Equation (6.7). Two benefits accrue, however, by using the double resonance technique. As all lines in each multiplet are summed to give a single line in the decoupled spectra, the signal-to-noise ratio (noise here is the electronic background level variations in the spectrometer system) is considerably greater than in the nondecoupled recordings involving the same number of scans. The second benefit leading to increased signal-to-noise ratios arises from the operation of the Nuclear Overhauser Effect (NOE) (see Section 6.11) under the double resonance conditions. Saturation of the hydrogens attached to the ^{13}C nuclei results in enhanced populations of the excited spin states of ^{13}C and induces more rapid relaxation. This results in a proportionately greater contribution by such a nucleus to the FID signal. The NOE enhancements are directly related to the number of attached hydrogens. Carbon atoms not bearing hydrogens, for example, quaternary and carbonyl carbons, are unaffected and are always of weaker intensity. In order to observe carbonyl carbon resonances, it is often necessary to incorporate a much longer delay period after the initial pulse before recording the FID signal. The emissions from the less rapidly relaxing nuclei are thus proportionately greater. The need for a greatly increased number of scans, with a resultant increase in instrument use time in recording a nondecoupled spectrum, is illustrated in Figure 6.19. Figure 6.19b is the ^{1}H decoupled spectrum of 4-methyl-1-pentene, which required 128 scans. Figure 6.19a is the proton coupled spectrum, which required 1244 scans! The signal-to-noise ratio in this spectrum is very low, some of the resonances being barely detectable for the CH_2 and CH carbon atoms, with no ^{3}J or ^{4}J couplings being observable.

A more recently developed, and more efficient, technique for recording nondecoupled ^{13}C spectra involves alternate recording of the decoupled and coupled spectra in which some NOE effect still persists during the nondecoupled acquisition period, thus increasing the relative intensity of the nondecoupled spectrum. The decoupled and coupled FIDs are alternately added and subtracted, resulting in one spectrum being oriented upward and the other downward. Figure 6.20 shows the decoupled (up) and proton coupled (down) spectra of allyl cyanide obtained with 250 scans up and 250 scans down.

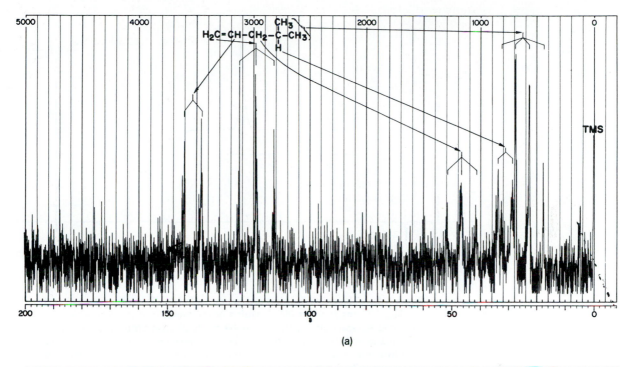

(a)

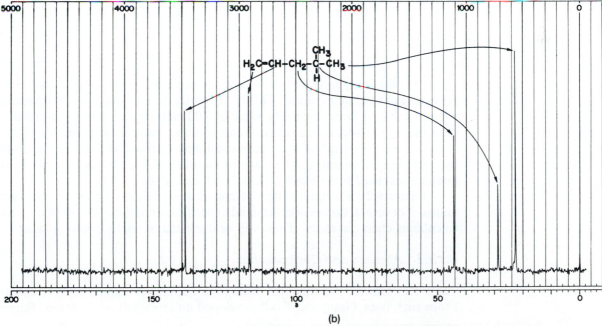

(b)

Figure 6.19   Proton coupled (a) and decoupled (b) ¹³C NMR spectra of 4-methyl-1-pentene.

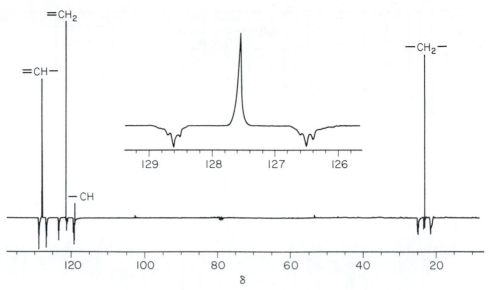

Figure 6.20 The decoupled (up) and coupled (down) ^{13}C NMR spectra of allyl cyanide. The insert shows the expansion of the lowest field signal in which the nondecoupled spectrum appears as a double multiplet arising from a large coupling with the attached hydrogen and from smaller two- and three-bond couplings with the other hydrogen atoms (see Table 6.13). The small inverted 1 : 1 : 1 triplet near δ 79 is the solvent ^{13}CHCl$_3$ resonance.

6.13 ^{13}C CHEMICAL SHIFTS

As in ^{1}H NMR, a standard must be used to correlate ^{13}C chemical shifts. The ^{13}C resonance of tetramethylsilane (TMS) is currently the most widely used standard, its chemical shift, δ, being assigned a value of 0.00, and in general being at higher field than most ^{13}C resonances in organic compounds. Early studies used the ^{13}C resonance of benzene or carbon disulfide as standards. Corrections to the δ scale (implied use of TMS as standard) can be made by adding 128.7 to the chemical shifts when benzene was used as standard, or 193.7 with carbon disulfide.

Figure 6.21 illustrates ranges of chemical shifts for various types of carbon atoms. For several classes of compounds, specific correlations have been derived for calculating chemical shifts. These will be discussed in the following sections.

6.13.1 Alkanes and Cycloalkanes

Table 6.15 lists the ^{13}C chemical shifts for a number of alkanes and cycloalkanes. From such data, Grant and Paul[6] developed an empirical relationship [Equation

---

[6] D. M. Grant and E. G. Paul, *J. Am. Chem. Soc.,* **86**, 2984 (1964).

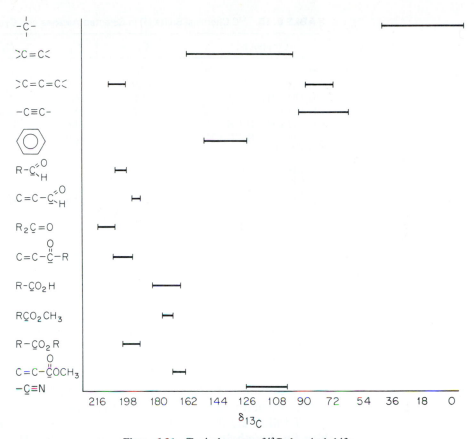

Figure 6.21 Typical ranges of ^{13}C chemical shifts.

(6.17)] to calculate ^{13}C chemical shifts in which $B = -2.5$ for alkanes, A_j is the

$$\delta_{^{13}C} = B + \sum_i A_j n_{ij} \tag{6.17}$$

value for the appropriate shift parameter in Table 6.16, and n_{ij} is the number of each group present. The shift parameters are defined in the following manner. The α parameter, or more commonly termed the α "effect," is for the structural change $-C-H \rightarrow -C-CH_3$, in which the bold face **C** is the ^{13}C under consideration. The β effect is for the structural change $-C-C-H \rightarrow -C-C-CH_3$, and similarly, the γ, δ, and ε effects are for the substitution of a methyl carbon for hydrogen at the γ, δ, and ε positions from the ^{13}C under consideration. The reader should note that the γ effect is opposite that of the α, β, δ, and ε effects. This has been attributed to a steric effect in a synclinal, or *gauche*, conformation (this steric effect will be discussed in greater detail later). The last eight shift parameters in Table 6.16 are corrections that consider the degree of substitution at the α carbon. For example, $1°(3°)$ is the correction factor for a methyl, or primary carbon, when it is attached to a

TABLE 6.15 ^{13}C Chemical Shifts (δ) in Selected Alkanes and Cycloalkanes

Compound	C_1	C_2	C_3	C_4	C_5
CH_4	−2.1				
CH_3CH_3	5.9	5.9			
$CH_3CH_2CH_3$	15.6	16.1	15.6		
$CH_3CH_2CH_2CH_3$	13.2	25.0	25.0	13.2	
$CH_3CH_2CH_2CH_2CH_3$	13.7	22.6	34.5	22.6	13.7
$CH_3CH_2CH_2CH_2CH_2CH_3$	13.9	22.9	32.0	32.0	22.9
CH_3CHCH_3 $\quad\mid$ CH_3	24.3	25.2	24.3		
CH_3CCH_3 $\quad\mid$ CH_3	31.5	27.9	31.5		
$\quad\quad CH_3$ $\quad\quad\mid$ $CH_3CH_2CCH_3$ $\quad\quad\mid$ $\quad\quad CH_3$	28.9	30.4	36.7	8.7	
Cyclopropane	−2.6				
Cyclobutane	23.3				
Cyclopentane	26.5				
Cyclohexane	27.8				
Cycloheptane	29.4				
Cyclooctane	27.8				

tertiary carbon, and 2°(4°) is the correction factor for a methylene, or secondary carbon, when it is attached to a quaternary carbon.

To illustrate the use of Equation (6.17), we shall calculate the ^{13}C chemical shifts for 3-methylpentane: $CH_3CH_2CH(CH_3)CH_2CH_3$. For C_1, there are one α (+9.43), one β (+8.81), two γ (−2.5), and one δ (+0.3) contributions resulting in a value of $\Sigma_j A_j n_{ij} = 13.5$ and $\delta_{C_1} = 11.0$, compared to the observed value of δ 11.3. The correction for the degree of substitution at C_1, a 2°(1°) effect, does not appear in

TABLE 6.16 Shielding Parameters for Calculations of ^{13}C Chemical Shifts in Acyclic Hydrocarbons*

Shift Parameter	$A_i (\Delta\delta)$	Shift Parameter	$A_i (\Delta\delta)$
α	+9.43	2°(3°)	−2.5
β	+8.81	2°(4°)	−7.2
γ	−2.5	3°(2°)	−3.7
δ	+0.3	3°(3°)	−9.5
ε	+0.1	4°(1°)	−1.5
1°(3°)	−1.1	4°(2°)	−8.4
1°(4°)	−3.4		

* Taken from reference 5, p. 58, except the α and β shift parameters, which are taken from D. K. Dalling and D. M. Grant, *J. Am. Chem. Soc.*, **96**, 1827 (1974).

Table 6.16 and is assigned a value of zero. Similar calculations for C_2, C_3, and the methyl carbon attached to C_3 are detailed as follows:

$$
\begin{array}{llr}
C_2: & 2\alpha & +18.9 \\
 & 2\beta & +17.6 \\
 & 1\gamma & -2.5 \\
 & 2^\circ(3^\circ) & -2.5 \\
 & \Sigma_j A_j n_{ij} & +31.5 \\
 & \text{Base} & -2.5 \\
 & \delta_{C_2} & +29.0 \\
 & \text{Observed} & +29.3
\end{array}
\qquad
\begin{array}{llr}
C_3: & 3\alpha & +28.3 \\
 & 2\beta & +17.6 \\
 & 2\ 3^\circ(2^\circ) & -7.4 \\
 & \Sigma_j A_j n_{ij} & +38.5 \\
 & & -2.5 \\
 & \delta_{C_3} & +36.0 \\
 & & +36.7
\end{array}
\qquad
\begin{array}{llr}
C_3-\underline{C}H_3: & 1\alpha & +9.4 \\
 & 2\beta & +17.6 \\
 & 2\gamma & +5.0 \\
 & 1^\circ(3^\circ) & -1.1 \\
 & \Sigma_j A_j n_{ij} & +20.9 \\
 & & -2.5 \\
 & \delta_{\underline{C}H_3} & +18.4 \\
 & & +18.6
\end{array}
$$

Equation (6.17) cannot be used to calculate the ^{13}C chemical shifts for cyclic compounds due to the limited number of conformations possible with cyclic compounds compared to the large number of molecular conformations possible with acyclic structures that contribute to the A_j's. Separate empirical correlations are required for the cycloalkanes (see reference in footnote to Table 6.16).

6.13.2 Alkenes

The ^{13}C chemical shifts for a number of acyclic and cyclic alkenes appear in Table 6.17. A number of trends are obvious even in the limited data presented in Table 6.17, which apply in general. In straight-chain alkenes, the difference in chemical shift ($\Delta\delta_C$) between the two vinyl carbon atoms in a terminal alkene is ~ 24 ppm, that in Δ^2 alkenes is 7 to 10 ppm, and in Δ^3 alkenes is 1 to 2 ppm. Introduction of a methyl group onto one of the vinyl carbon atoms, that is,

$$
C_\beta{=}C_\alpha\!\!\begin{array}{c}H\\ \diagup\end{array} \longrightarrow C_\beta{=}C_\alpha\!\!\begin{array}{c}CH_3\\ \diagup\end{array}
$$

TABLE 6.17 ^{13}C Chemical Shifts (δ) in Alkenes

Compound	C_1	C_2	C_3	C_4
$H_2C{=}CH_2$	122.8	122.8		
$CH_3CH{=}CH_2$	115.4	135.7		
$CH_3CH_2CH{=}CH_2$	112.8	140.2		
$CH_3CH_2CH_2CH{=}CH_2$	113.5	137.6		
$CH_3CH_2C{=}CH_2$ 　　　　$\mid$ 　　　CH_3	107.7	146.4		
$(CH_3CH_2)_2C{=}CH_2$	105.5	151.7		
$CH_3CH{=}CHCH_3$		123.3	123.3	
$CH_3CH_2CH{=}CHCH_3$		122.8	132.4	
$CH_3(CH_2)_3CH{=}CHCH_2CH_3$			130.5	129.3
Cyclopentene	130.6	130.6		
Cyclohexene	127.2	127.2		

TABLE 6.18 Shielding Parameters for Calculation of Vinyl ^{13}C Chemical Shifts in Alkenes*

Shift Parameter	$A_j (\Delta\delta)$
α	$+11.0 \pm 0.4$
β^{π}	-7.1 ± 0.4
β^{σ}	$+6.0 \pm 0.3$
γ^{π}	-1.9 ± 0.3
γ^{σ}	-1.0 ± 0.3
δ^{π}	$+1.1 \pm 0.3$
δ^{σ}	$+0.7 \pm 0.4$
ε	$+0.2 \pm 0.3$
cis	-1.2 ± 0.3
gem^{α}	-4.9 ± 0.3
gem^{β}	$+1.2 \pm 0.3$
$mult^{\sigma}$	$+1.3 \pm 0.3$
$mult^{\pi}$	-0.7 ± 0.3

* Taken from p. 75 of reference 18 at the end of this chapter. Superscript π is for when the substituent is attached to the C=C, and σ is for when it is attached to a saturated carbon of the alkene. *Cis* is the shielding effect on the vinyl carbon atoms for *cis,* tri-, or tetrasubstituted alkenes. *Gem*$^{\alpha}$ is the shift effect on the vinyl carbon bearing two groups, and *gem*$^{\beta}$ is the effect on the other vinyl carbon atom. *Mult*$^{\sigma}$ and *mult*$^{\pi}$ are effects caused by branching at a carbon directly attached to the C=C.

causes a deshielding of C_{α} by 5 to 7 ppm, but the C_{β} experiences a *shielding* of 5 to 7 ppm. This β effect is in the opposite direction of that observed in alkanes and is indicated as a β^{π} effect. When the methyl group is not attached directly to one of the vinyl carbon atoms, the β^{σ} effect is similar to that observed in alkanes, a β^{σ} effect being one in which the β carbon is attached to a saturation carbon atom. A correlation similar to that for alkanes has been derived in which $B = 122.1$, and the A_j's have the values given in Table 6.18 (definitions of the shielding parameter are given in the footnotes to the table). Correspondence between calculated and observed ^{13}C chemical shifts is not as good as with the alkanes. Ring and steric strain cause a deshielding of the vinyl carbons, as is illustrated by the examples given in Table 6.18 and in the following structures.

In contrast to ^1H NMR, in which the C=C has a profound effect on the chemical shift of allylic hydrogens, the C=C has only a minor effect on the chemical shifts of carbon atoms separated from the C=C by only one or two bonds, with no effect observed on carbon atoms further removed. This is illustrated in the following comparison.

An interesting feature of the above data is the fact that the carbon atoms attached to the C=C in the *cis* alkenes are *shielded* by 6.0 ± 0.6 ppm relative to the *trans* isomer (a steric shift). This difference is often useful in distinguishing between conformational isomers of alkenes, particularly with trisubstituted alkenes.

Another contrast in the trends of ^{1}H and ^{13}C chemical shifts is that the introduction of a second double bond has relatively little effect on the ^{13}C chemical shifts of the vinyl carbon atoms, as illustrated by the comparison of the ^{13}C chemical shifts of C_1 and C_2 in 1-butene and 1,3-butadiene.

$$CH_3CH_2CH{=}CH_2 \qquad H_2C{=}CH{-}CH{=}CH_2$$
$$140.2 \qquad 112.8 \qquad\qquad 137.2 \qquad 116.6$$

6.13.3 Alkynes

Acetylenic carbon atoms undergo resonance in a region intermediate between alkane and vinyl carbons (see Table 6.19). The effect of substituents on the chemical shifts of the acetylenic carbons is very similar to that observed with the alkenes; replacement of the acetylenic hydrogen by a methyl deshields that carbon by ~ 6.5 ppm, but shields the β carbon by ~ 5.0 ppm (a β^{π} effect). The β^{σ} effect is $\sim +7$ ppm, while the γ^{σ} effect is ~ -1 to 2 ppm, and that of γ^{π} is $+0.2$ to $+3.0$ ppm. The C≡C appreciably affects only the shielding of the attached carbon atoms (~ -9 to -13 ppm), the further removed carbon atoms having chemical shifts comparable to those in the corresponding alkane.

6.13.4 Aromatic Hydrocarbons

Aromatic ring carbon atoms undergo resonance in the δ 125 to 150 region (see Table 6.20). The introduction of a methyl group causes a deshielding of the *ipso* carbon (the ring carbon bearing the function) by $+9.2$ ppm, a slight deshielding of the *ortho* carbon, essentially no effect on the *meta* carbon, and a shielding of the *para* carbon. Again an empirical correlation has been developed for calculation of aromatic carbon chemical shifts using Equation (6.17), in which $B = 128.3$ and the A_j's are those given in Table 6.21.

TABLE 6.19 ^{13}C Chemical Shifts (δ) in Alkynes

Compound	C_1	C_2	C_3
$CH_3CH_2C{\equiv}CH$	67.3	85.0	
$CH_3C{\equiv}CCH_3$		73.9	73.9
$CH_3(CH_2)_3C{\equiv}CH$	68.6	84.0	
$CH_3(CH_2)_2C{\equiv}CCH_3$		74.9	78.1
$CH_3CH_2C{\equiv}CCH_2CH_3$			81.1

TABLE 6.20 ^{13}C Chemical Shifts (δ) of Aromatic Ring Carbon Atoms

Compound	C_1	C_2	C_3	C_4	C_5	C_6
Benzene	128.7	128.7	128.7	128.7	128.7	128.7
Methylbenzene	137.8	129.3	128.5	125.6	128.5	129.3
1,2-Dimethylbenzene	136.4	136.4	129.9	126.1	126.1	129.9
1,3-Dimethylbenzene	137.5	130.1	137.5	126.4	128.3	126.4
1,4-Dimethylbenzene	134.5	129.1	129.1	134.5	129.1	129.1
Ethylbenzene	144.1	128.1	128.5	125.9	128.5	128.1
n-Propylbenzene	142.5	128.7	128.4	125.9	128.4	128.7
t-Butylbenzene	149.2	125.4	128.4	125.9	128.4	125.4

TABLE 6.21 Chemical Shift Parameters for Calculation of ^{13}C Chemical Shifts of Aromatic Ring Carbon Atoms

Shift Parameter*	$A_j\,(\Delta\delta)$
σ_α	$+9.2 \pm 0.2$
σ_o	$+0.8 \pm 0.2$
σ_p	-2.5 ± 0.2
σ_q	-2.3 ± 0.2
σ_t	$+1.2 \pm 0.2$

* σ_α, σ_o, and σ_p are the shielding parameters for the *ipso, ortho,* and *para* carbon atoms when a hydrogen on the *ipso* carbon atom is substituted by methyl. σ_q is a shielding contribution to the carbon bearing a hindered methyl group (as in 1,2-dimethylbenzene), and σ_t is the contribution to the adjacent *tertiary* ring carbon atom (not the adjacent carbon atom bearing the other methyl group!).

TABLE 6.22 ^{13}C Chemical Shifts (δ) of Aryl Carbon Atoms in Monosubstituted Benzenes, C_6H_5—X*

X	C_1	C_2	C_3	C_4
O^-	168.3	120.5	130.6	115.1
OH	155.6	116.1	130.5	120.8
OCH_3	158.9	113.2	128.7	119.8
$OCOCH_3$	151.7	122.3	130.0	126.4
NH_2	147.9	116.3	130.0	119.2
$N(CH_3)_2$	151.3	113.1	129.7	117.2
$NHCOCH_3$	139.8	118.8	128.9	123.1
F	163.9	114.6	130.3	124.3
Cl	135.1	128.9	129.7	126.7
BR	123.3	132.0	130.9	127.7
I	96.7	138.9	131.6	129.7
$CH{=}CH_2$	138.2	126.7	128.9	128.2
CO_2CH_3	130.0	128.2	128.2	132.2
COCl	134.5	131.3	129.9	136.1
CHO	137.7	129.9	129.9	134.7
$COCH_3$	136.6	128.4	128.4	131.6
CN	109.7	130.1	127.2	130.1
NO_2	148.3	123.4	129.5	134.7

* Taken from Table 5.58 on p. 197 of reference 18 at the end of this chapter.

TABLE 6.23 ^{13}C Chemical Shifts (δ) in Heterocyclic Compounds

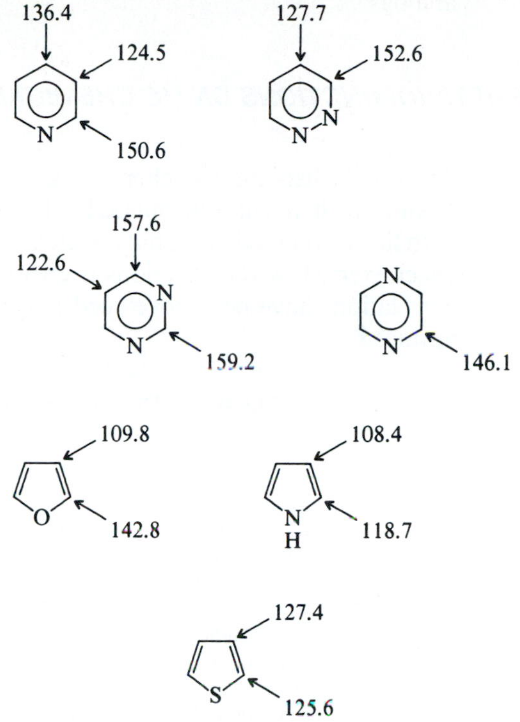

The effect of an aromatic ring on the ^{13}C chemical shifts of an attached alkyl group is to deshield the α carbon (the carbon attached to the ring) by ~ 23 ppm and the β carbon by ~ 9.5 ppm, and to shield the γ carbon by ~ 2 ppm. Carbon atoms further removed are unaffected.

Tables 6.22 and 6.23 list ^{13}C chemical shifts of ring carbon atoms in monosubstituted benzenes and in several selected monocyclic and heterocyclic aromatic compounds.

6.13.5   Carbonyl Compounds

The ranges of ^{13}C chemical shifts of the carbonyl carbon atoms in various carbonyl compounds are illustrated in Figure 6.21. Changes in substituents have little effect on their chemical shifts (β effect of $\sim +2$, γ effect of ~ -1). α,β-Unsaturated compounds undergo resonance at ~ 7 to 10 ppm higher field than do their saturated counterparts. Thioesters,

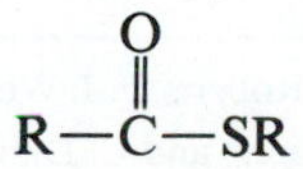

$$R-\overset{\overset{\textstyle O}{\|}}{C}-SR$$

undergo resonance at 15 to 20 ppm downfield relative to the corresponding ester analogs.

6.14 EFFECT OF HETEROFUNCTIONS ON ^{13}C CHEMICAL SHIFTS

Table 6.24 lists the ^{13}C chemical shifts of a series of 1-substituted butanes. The change in chemical shift caused by the functions in other systems is fairly consistent with those observed in the butyl system. Note that in the case of all polar groups, the γ effect exerted by the functions is greater than that for a methyl group. Empirical correlations have been developed to calculate ^{13}C chemical shifts in alcohols[7] and amines.[8]

TABLE 6.24 ^{13}C Chemical Shifts (δ) in 1-Substituted Butanes

Compound	C_1	C_2	C_3	C_4
Butane	13.2	25.0	25.0	13.2
1-Chlorobutane	44.6	35.2	20.4	13.4
1-Bromobutane	33.2	35.4	21.7	13.5
1-Iodobutane	7.1	36.1	24.2	13.7
1-Butanol	61.7	35.3	19.4	13.9
Dibutyl ether	71.2	33.1	20.3	14.6
Butyl acetate	63.8	31.2	19.7	14.1
Butylamine	42.3	36.7	20.4	14.0
2-Hexanone*	43.5	31.9	23.8	14.0
Pentanoic acid†	34.1	27.0	22.0	13.5

* Considered to be 1-acetylbutane, i.e., $CH_3^4CH_2^3CH_2^2CH_2^1COCH_3$.
† Considered to be 1-carboxybutane, i.e., $CH_3^4CH_2^3CH_2^2CH_2^1CO_2H$. ^{13}C chemical shifts in esters are very similar to those in the corresponding carboxylic acids.

6.15 SPIN COUPLING INVOLVING ^{13}C

The most extensively studied coupling interaction of ^{13}C is that with ^1H. The magnitude of $^1J_{^{13}C-H}$ has been found to be very sensitive to the electronegativity of the function(s) attached to the carbon atom under consideration. Table 6.25 lists $^1J_{^{13}C-H}$'s for a number of substituted methanes in which, in general, the coupling constant is observed to increase with increasing electronegativity of the attached group. This effect is rather large, as exemplified in the series CH_3Cl (148.6 Hz), CH_2Cl_2 (176.5 Hz), and $CHCl_3$ (208.1 Hz).

[7] J. D. Roberts, F. J. Weigert, J. I. Kroschwitz, and H. J. Reich, *J. Am. Chem. Soc.,* **92**, 1338 (1970).
[8] H. Eggert and C. Djerassi, *J. Am. Soc.,* **95**, 3710 (1973).

TABLE 6.25   Values of $^1J_{^{13}C-H}$ in Substituted Methanes, CH_3-X*

X	$^1J_{^{13}C-H}$ (Hz)	X	$^1J_{^{13}C-H}$ (Hz)
Li	98	CN	133
$MgCH_3$	105.5	SCH_3	138
$C(CH_3)_3$	124	OCH_3	140
H	125	OH	142
CH_3	125	NO_2	147
C_6H_5	126	F	149
$COCH_3$	127	Cl	150
$C\equiv CH$	131	Br	151
NH_2	133	I	151

* Taken from Chapter 10 of reference 18 at the end of this chapter.

In addition to the effect of substituent electronegativity, $^1J_{^{13}C-H}$ is also sensitive to the degree of hybridization of the orbitals on carbon involved in the C—H bond, increasing markedly as the s-character of the orbital on the carbon increases. For example, with sp^3-hybrid orbitals (25% s-character), $^1J_{^{13}C-H}$ is ~ 125 Hz, with sp^2-hybrid orbitals (33% s-character) ~ 160 Hz, and for sp-hybrid orbitals (50% s-character) ~ 250 Hz; a nearly linear relationship of percent s-character with the magnitude of $^1J_{^{13}C-H}$! The values of $^1J_{^{13}C-H}$ in strained ring systems have been used to calculate the degree of hybridization in such systems. Table 6.26 lists values of $^1J_{^{13}C-H}$'s for a variety of systems.[9]

Values for $^2J_{^{13}C-H}$ (−2 to −6 Hz) and $^3J_{^{13}C-H}$ (4 to 6 Hz) have been measured and trends evaluated in terms of electronegativity, bond angles, and dihedral angles. Remarkably, the trends observed parallel those observed with $^2J_{HH}$'s and $^3J_{HH}$'s.

Values for ^{13}C-^{13}C coupling constants have also been determined. Because of the low abundance of ^{13}C, ^{13}C-^{13}C coupling constants are difficult to measure even using FT techniques. Such determinations generally require the preparation of ^{13}C enriched compounds. Values of the various ^{13}C-^{13}C coupling constants are ^1J, 35 to 80 Hz; ^2J, 0 to 2 Hz; ^3J, 0 to 5.2 Hz, these values depending on the dihedral angle between the C-C bonds as in $^3J_{HH}$, and ^4J, 0 to 0.6 Hz. In general, $J_{^{13}C-^{13}C}$ = 0.27 $J_{^{13}C-H}$.

6.16   EXAMPLES OF ^{13}C NMR SPECTRA

Let us begin by considering the proton nondecoupled and decoupled spectra of 4-methyl-1-pentene shown earlier in Figure 6.19. The proton decoupled spectrum shows three peaks in the saturated carbon region at δ 22.5, 28.7, and 43.9, and two peaks in the vinyl carbon region at δ 115.6 and 137.8. Although we could reasonably

---

[9] For a more detailed discussion of ^{13}C coupling constants, see Chapter 10 in reference 18 at the end of this chapter.

TABLE 6.26   Values of $^1J_{^{13}C-H}$ in Various Systems*

Compound	J (H$_z$)	Compound	J (H$_z$)
CH$_3$—CH$_3$	125	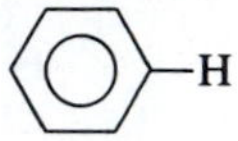 ⌬—H	159
△	160.5	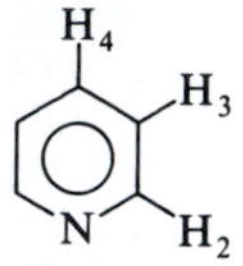 pyridine (H$_4$, H$_3$, H$_2$)	(2) 180 (3) 157 (4) 160
□	136	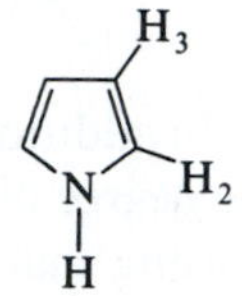 pyrrole (H$_3$, H$_2$)	(2) 184 (3) 170
 ⬠	128	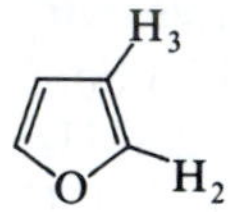 furan (H$_3$, H$_2$)	(2) 201 (3) 175
 ⬡	125	HC≡CH	248.7
H$_2$C=CH$_2$	156.2	CH$_3$C≡CH*	247.6
H$_2$C=CH—CH=CH$_2$*	158	CH$_3$C(=O)H	172.4
H$_2$C=C=CH$_2$	168	 (oxirane)	176
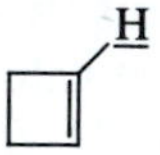 cyclobutene —H	170	 (aziridine)	171
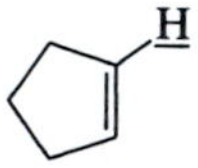 cyclopentene —H	160	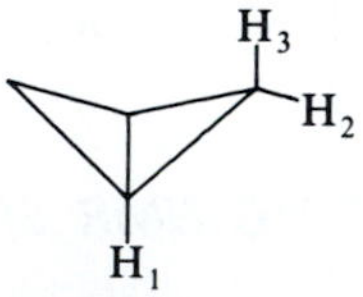 (H$_3$, H$_2$, H$_1$)	(1) 169 (2) 153 (3) 205
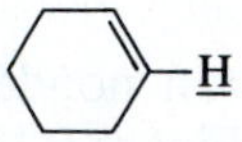 cyclohexene —H	157	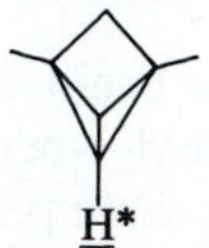 (H*)	212

* Values are for the underlined CH bond.

assign each of the resonances to the individual carbon atoms in the molecule by the use of Equation 6.17 and the substituent parameters given in Tables 6.16 and 6.18, the multiplicities of the resonances in the nondecoupled spectrum allow for immediate assignment. [In FT NMR spectroscopy, the relative intensities of the components of a multiplet do not always conform to those calculated by Equation (6.8).] In the nondecoupled spectrum, the resonance at δ 22.5 appears as a quartet (J_{CH} = 123 Hz), and thus must represent the two identical methyl carbon atoms. The resonances at δ 28.7 and 43.9 appear as a doublet (J = 126 Hz) and a triplet (J = 128 Hz) and must represent the methine and methylene carbon atoms respectively. The vinyl resonances at δ 115.6 and 137.8 appear as a triplet (J = 151 Hz) and doublet (J = 148 Hz) representing the $=CH_2$ and $=CH$ carbon atoms respectively. It should be noted that the relative intensities of the resonances in the decoupled spectrum do not accurately reflect the relative numbers of the carbon atoms in the molecule. Of the many C_6H_{12} possible isomeric structures, only 4-methyl-1-pentene can give rise to the ^{13}C NMR spectra that are shown in Figure 6.19.

The proton decoupled ^{13}C spectrum of allyl cyanide (Figure 6.20) shows one peak in the saturated carbon region at δ 23.0 and two peaks in the vinyl carbon region at δ 122 and 128. The lowest field peak in the nondecoupled spectrum appears as a doublet that immediately identifies that resonance as belonging to the $=CH-$ carbon atom. The nondecoupled resonances for the $=CH_2$ and $-CH_2-$ carbon atoms appear as distorted triplets. This is often common when this technique is used and when the center resonance line of a multiplet coincides in chemical shift with the decoupled resonance line. The ^{13}C resonance of the carbon atom of the cyano group appears as the very weak resonance just upfield of the downward resonance line of the $=CH_2$ carbon atom at approximately δ 188. The very low intensity of the carbon resonance is because this carbon atom is not bonded to a hydrogen atom and its rate of relaxation is extremely slow and is not as effectively detected as those carbon atoms bearing hydrogen atoms. This is typical of carbon atoms such as those in carbonyl groups of aldehydes, ketones, and carbonylic acid derivatives.

6.17 CHEMICAL SHIFT REAGENTS

Certain paramagnetic lanthanide metal complexes produce changes in the chemical shifts of nuclei when added to solutions of organic compounds that contain a Lewis basic functional group. Such complexes are called *chemical shift reagents*. The lanthanide complex forms a Lewis acid-base complex with the Lewis basic functional group. Chemical shifts reagents are used to "spread" out a spectrum when several resonances appear very close to each other resulting in the overlap of peaks. The magnitude of the induced shifts depend on the ability of the organic compound to form a Lewis acid-base complex. The largest induced shifts occur with amines and alcohols, with moderate shifts occurring with carbonyl compounds, and with very small shifts occurring with ethers, sulfides and alkenes, alkynes, and aromatics. The magnitude of the induced shift also depends on the distance and orientation from the lanthanide metal of the nucleus being observed. In general, the induced shift decreases with increasing distance from the functional group that undergoes complexation with the lanthanide reagent.

The magnitude, and the direction, of the induced shift also depends on the nature of the lanthanide metal and the ligands attached to the metal, the amount of the complex added to the sample, and the type of functional group and structural features present in the organic molecule. The most commonly used complexes are the *tris-β*-diketonate complexes possessing the general structure shown in **15**.

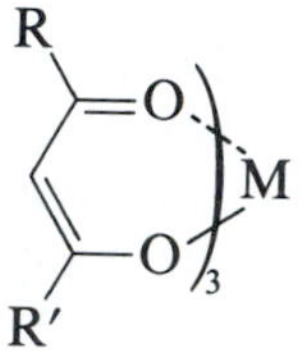

15

The commonly used *β*-diketonate ligands and their abbreviations are listed in Table 6.27.[10] The complexes of erbium (Er), europium (Eu), ytterbium (Yb), and thulium (Tm) induce shifts to lower field, the magnitude of the induced shifts in-

TABLE 6.27 Structures and Abbreviations of Ligands Commonly Used in NMR Chemical Shift Reagents

Structure of β-Diketonate Anion	Abbreviation
$(CH_3)_3C-C-CH-C-C(CH_3)_3$	thd, tmhd, or dpm
$CF_3CF_2CF_2-C-CH-C-C(CH_3)_3$	fod
$CF_3CF_2CF_2-C-CH-C-CF_3$	dfhd
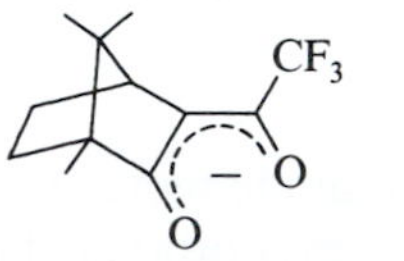 (structure with CF_3)	facam
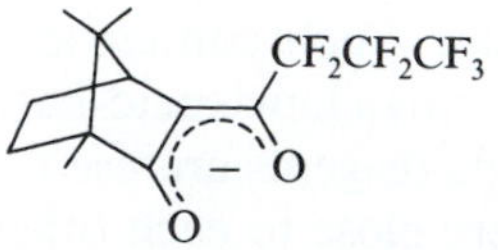 (structure with $CF_2CF_2CF_3$)	hfbc

---

[10] For a review of chemical shift reagents see K. A. Kime and R. E. Sievers, *Aldrich Chemica Acta,* **10** (4), 54 (1977).

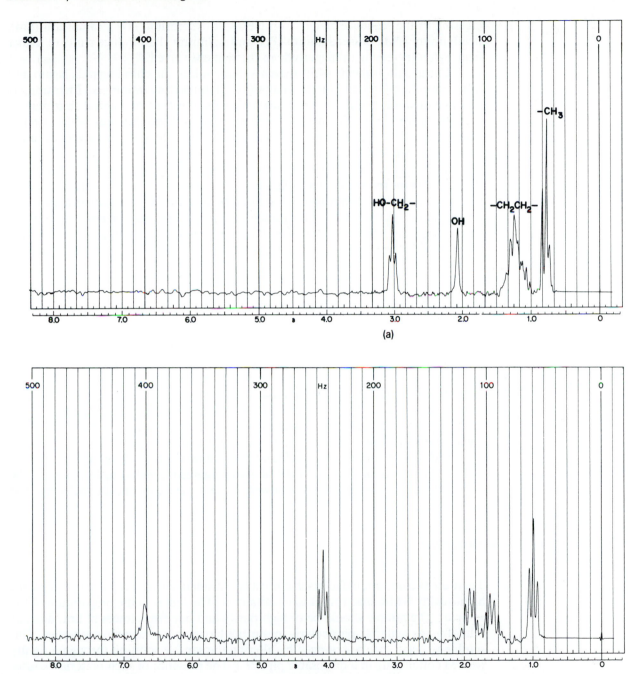

Figure 6.22 (a) Normal ^{1}H NMR spectrum of 1-butanol. (b) 1H NMR spectrum of a 1:0.5 molar ratio of 1-butanol and Eu(fod)$_3$.

creasing in the sequence given. Complexes of dysprosium (Dy), samarium (Sm), neodymium (Nd), praseodymium (Pr), and holmium (Ho) induce shifts to higher field, again increasing in magnitude in the sequence given.

Chiral shift reagents (shift reagents that contain the chiral ligands facam and hfbc) are extremely useful for the analysis of mixtures of enantiomers. Enantiomers produce identical NMR spectra; however, in the presence of a chiral shift reagent complexes are formed that are diastereomeric and possess different NMR spectra. Integration of the diastereomerically related sets of nuclei provides a very accurate method for determining optical purities.

The effect of $Eu(fod)_3$ (see Table 6.27) on the NMR spectrum of 1-butanol is illustrated in the spectra shown in Figure 6.22. The top spectrum is that of 1-butanol in the absence of the shift reagent. Note that both of the CH_2 resonances appear as a complex multiplet at δ 1.2. In the presence of the shift reagent the resonances of the two CH_2 groups have separated, the $HO-CH_2$-methylene hydrogens have shifted significantly downfield, and the hydroxyl hydrogen has shifted downfield the greatest extent.

6.18 ADVANCED APPLICATIONS OF NMR SPECTROSCOPY

Although the chemical shift and coupling constant data described above provide a great deal of information concerning the structure of a molecule, it may not be sufficient to define the entire structure of a complex molecule. Many other very sophisticated applications of NMR have been developed that allow correlating the ^1H and ^{13}C chemical shift and coupling relationships, for example, two-dimensional NMR spectroscopy. These methods are too advanced to be introduced here and the reader should consult more advanced texts on the subject.

NMR spectroscopy has also found applications in the biological and medical sciences. For example, the ^1H NMR of living organisms, including humans, can be recorded that has developed into a powerful, noninvasive diagnostic tool known as NMR imaging.

6.19 SAMPLE PREPARATION

There are two important factors that must be considered when preparing a sample if a high-quality NMR spectrum is to be obtained. The first is in the preparation of the sample. The sample should be of high purity and free from traces of lint, dust, or other foreign material, particularly impurities having paramagnetic properties. (Insoluble material can be removed by filtering the sample solution through a tightly packed cotton plug in a disposable pipette.) The presence of such materials causes extensive line broadening that decreases the resolution of the spectrum. The second factor is the quality of the NMR tube. There are a wide variety of sizes and quality of NMR tubes that are commercially available. (The size of the NMR tube will depend on the requirements of the spectrometer for a given application.) For top-quality

NMR spectra, high-quality NMR tubes must be used. For qualitative applications, such as the analysis of chromatographic factions, lower-quality (and lower-priced) NMR tubes can be used.

The quantity of sample required depends on whether a continuous wave or FT NMR spectroscopy will be employed. For continuous wave spectroscopy, at least 10 mg of sample is required. For FT NMR spectroscopy, the maximum sample size should be no more than a few mg and may be as small as a few hundredths of a mg.

The choice of solvent will depend on the type of spectrometer used. Some spectrometers require the use of a deuterium labeled solvent (to act as "lock" signal), while others do not. Regardless of the type of instrument used, the solvent should not contain the same nucleus being observed (except carbon, which can hardly be avoided). If a deuterium labeled solvent is not required, the best choice of a solvent is one of the fully chlorinated hydrocarbons such as CCl_4, C_2Cl_6, or $CCl_2{=}CCl_2$. There are many deuterium labeled solvents that are commercially available that can be used when such a solvent is required, the choice then being made on the basis of the solubility of the sample in the solvent.

6.20 NMR SPECTRAL PROBLEMS

From the data given and the NMR spectrum, assign structures to the compounds in each of the following problems.

1. C_6H_{12}. NMR spectrum is shown in Figure 6.23.

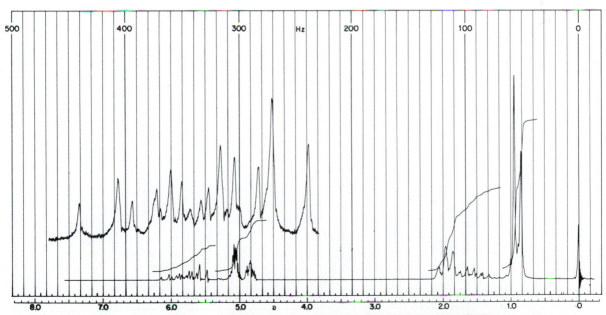

Figure 6.23 ^{1}H NMR spectrum of unknown in problem 1.

2. $C_{10}H_{12}O_2$. Absorbs in the IR at 1740 cm^{-1} and in the UV at 255 nm. NMR spectrum is shown in Figure 6.24.

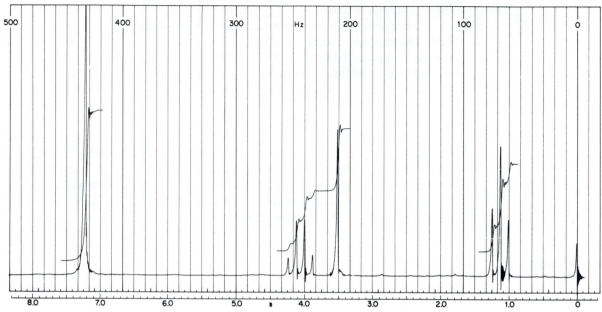

Figure 6.24 ^{1}H NMR spectrum of unknown in problem 2.

3. C_5H_8O. Transparent in the 3600 and 2000 to 1600 cm^{-1} regions. NMR spectrum is shown in Figure 6.25.

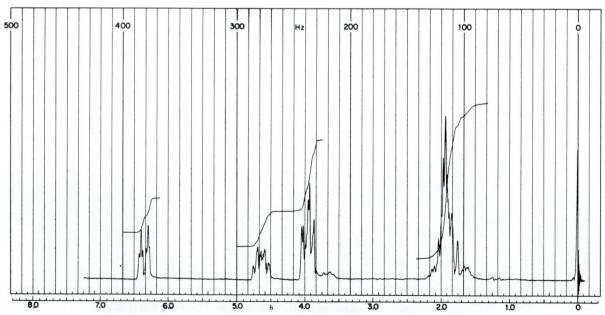

Figure 6.25 ^{1}H NMR spectrum of unknown in problem 3.

4. C_9H_8O. Absorbs in the IR at 1685 cm^{-1}. NMR spectrum is shown in Figure 6.26.

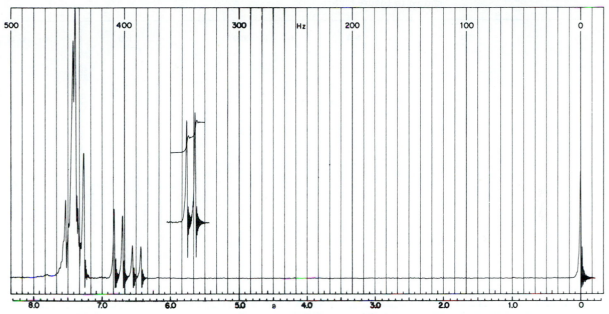

Figure 6.26   60 mHz ^{1}H NMR spectrum of unknown in problem 4. Insert is offset 250 Hz.

5. C_4H_8O. Absorbs in the IR at 3500 cm^{-1} (broad). NMR spectrum is shown in Figure 6.27.

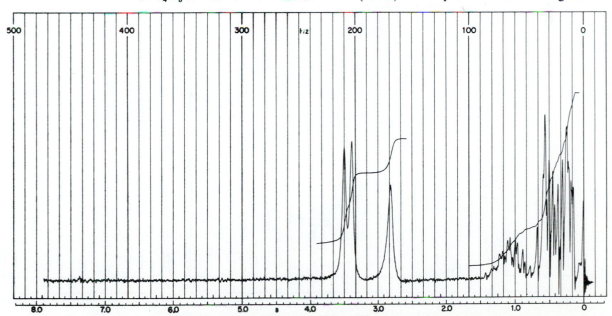

Figure 6.27   60 mHz ^{1}H NMR spectrum of unknown in problem 5 recorded at 250 Hz total sweep width.

6. $C_9H_{14}O$. Absorbs in the IR at 1680 cm^{-1} and at 239 nm in the UV. NMR spectrum is shown in Figure 6.28.

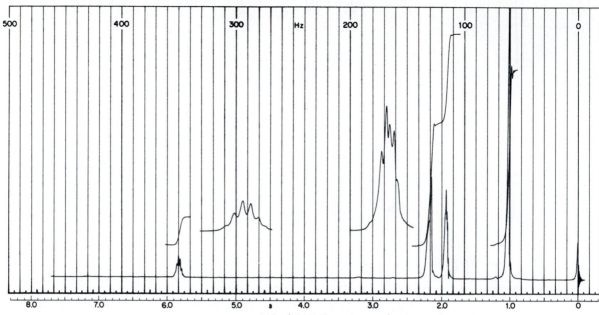

Figure 6.28 ^{1}H NMR spectrum of unknown in problem 6.

7. $C_8H_{11}N$. Absorbs in the IR at 3450 cm^{-1}. NMR spectrum is shown in Figure 6.29.

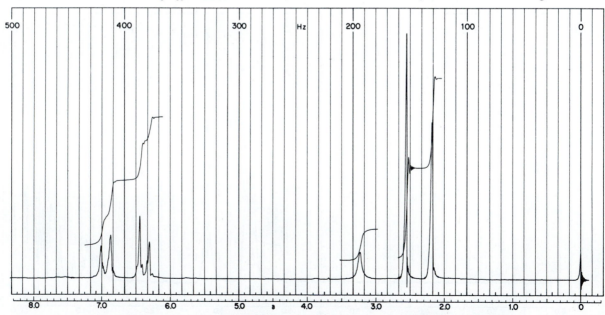

Figure 6.29 ^{1}H NMR spectrum of unknown in problem 7.

8. $C_6H_{10}O$. Absorbs in the IR at 3600 and 3300 cm^{-1}. NMR spectrum is shown in Figure 6.30.

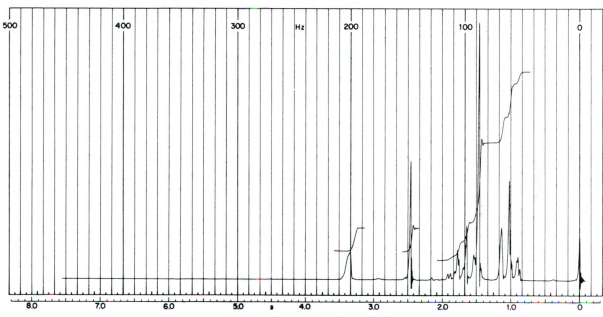

Figure 6.30 ^{1}H NMR spectrum of unknown in problem 8.

9. $C_8H_{11}N$. Absorbs in the IR at 3500 and 3400 cm^{-1}. NMR spectrum is shown in Figure 6.31.

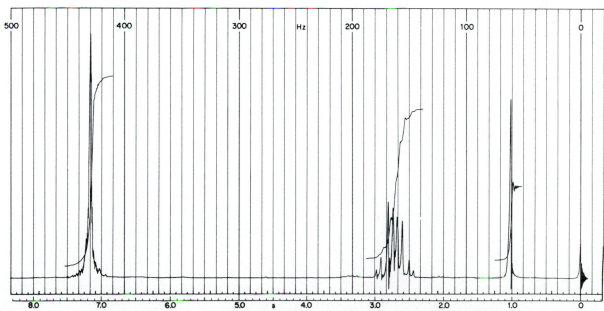

Figure 6.31 ^{1}H NMR spectrum of unknown in problem 9.

10. C_3H_6O. Absorbs in the IR at 3600 cm⁻¹. ¹³C NMR spectrum is shown in Figure 6.32.

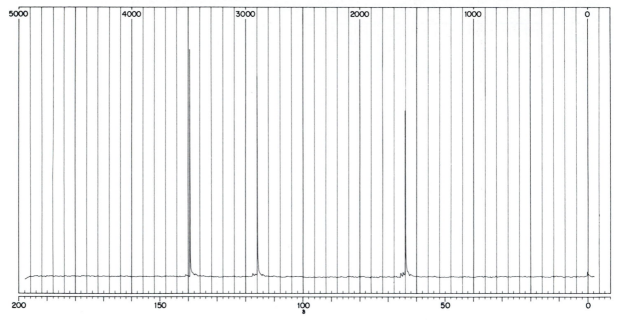

Figure 6.32 ¹³C NMR spectrum of unknown in problem 10.

11. $C_5H_{10}O$. Absorbs in the IR at 1710 cm⁻¹. ¹³C NMR spectrum is shown in Figure 6.33.

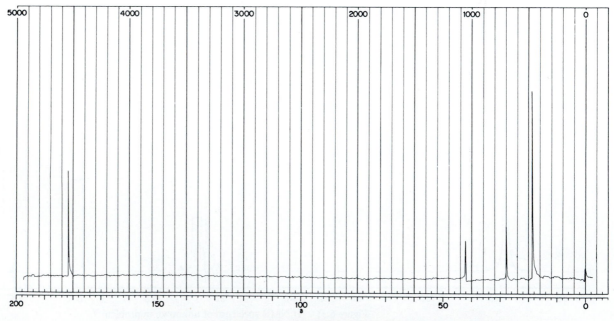

Figure 6.33 ¹³C NMR spectrum of unknown in problem 11.

12. $C_7H_{12}O$. Absorbs in the IR at 3600 and 3300 cm^{-1}. ^{13}C NMR spectrum is shown in Figure 6.34. Calculate the $\delta_{^{13}C}$'s for your answer using the correlations discussed earlier and compare with the observed values in the spectrum.

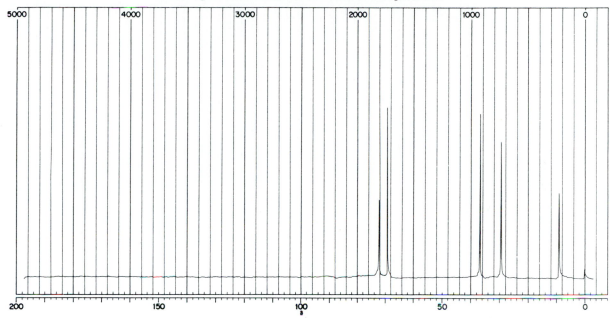

Figure 6.34 ^{13}C NMR spectrum of unknown in problem 12.

13. $C_{10}H_{12}O$. Absorbs in the IR at 1685 cm^{-1}. ^{13}C NMR spectrum is shown in Figure 6.35.

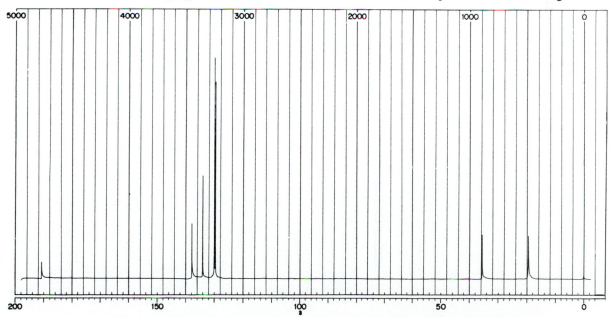

Figure 6.35 ^{13}C NMR spectrum of unknown in problem 13.

14. $C_6H_{14}O$. Absorbs in the IR at 3600 cm^{-1}. ^{13}C NMR spectrum is shown in Figure 6.36. Calculate the $\delta_{^{13}C}$'s for your answer using the correlations discussed earlier and compare with the observed values in the spectrum.

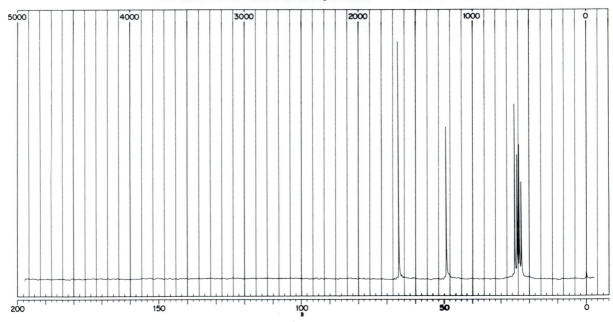

Figure 6.36 ^{13}C NMR spectrum of unknown in problem 14.

15. $C_{10}H_{12}O_2$. Absorbs in the IR at 1740 cm^{-1}. ^{13}C NMR spectrum is shown in Figure 6.37.

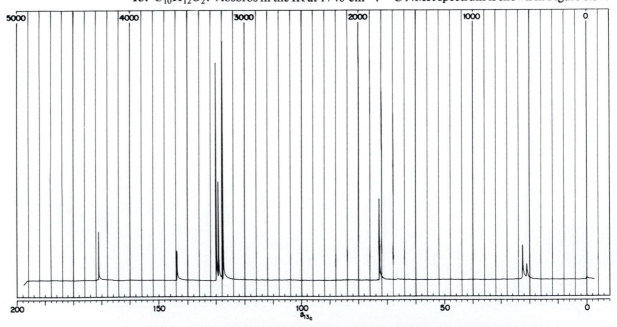

Figure 6.37 ^{13}C NMR spectrum of unknown in problem 15.

16. $C_8H_{19}N$. ^{13}C NMR spectrum is shown in Figure 6.38.

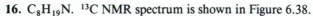

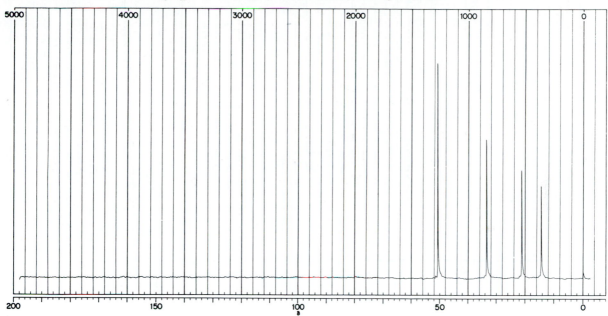

Figure 6.38 ^{13}C NMR spectrum of unknown in problem 16.

17. $C_6H_{10}O$. ^{13}C proton nondecoupled spectrum is shown in Figure 6.39a, and the proton decoupled spectrum is shown in Figure 6.39b.

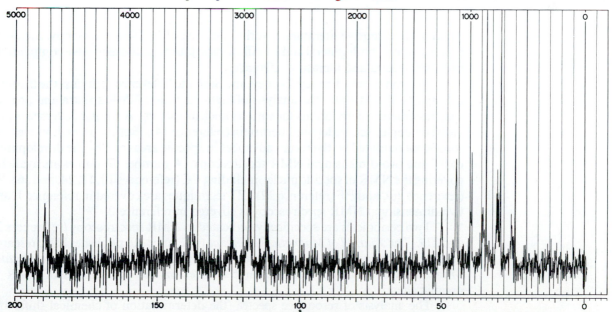

(a)

Figure 6.39 (a) Proton nondecoupled ^{13}C NMR spectrum of unknown in problem 17.

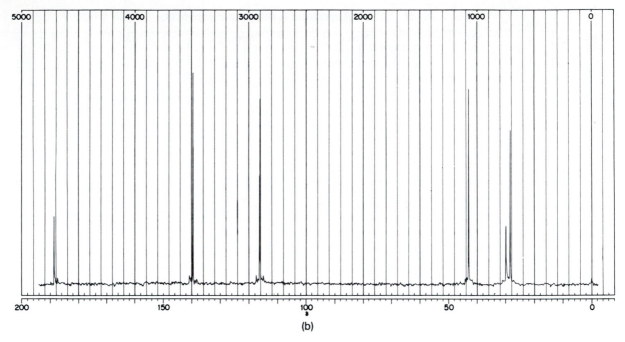

Figure 6.39   (b) Proton decoupled ^{13}C NMR spectrum.

6.21 NMR LITERATURE

There is an extensive literature covering the area of NMR spectroscopy, including books covering both general and specific topics, journals, and extensive compilations of NMR spectra and literature references.

The volumes by Emsley, Feeney, and Sutcliffe (reference 7) discuss the theoretical aspects of NMR and the analysis of high-resolution NMR spectra and provide spectra-structure correlations for various nuclear species. A similar coverage is provided by Pople, Schneider, and Bernstein (reference 14). Both sources, however, are somewhat outdated by the recent advances in specific areas and by the development of FT and two-dimensional NMR spectroscopy. A number of sources are available covering the NMR of specific nuclei, including ^{13}C (references 10, 18, 24, and 36), ^{31}P (reference 26), and ^{19}F (reference 22). FT techniques are covered in reference 1, and two-dimensional NMR spectroscopy is covered in reference 2.

Two periodicals are specifically dedicated to NMR spectroscopy (references 20 and 21). The *Journal of Magnetic Resonance* (reference 20) publishes basic research articles on theory and new techniques and applications of NMR, while *Organic Magnetic Resonance* publishes research articles on the applications of NMR in organic chemistry.

The analysis of resonance spectra (the determination of chemical shifts and coupling constants) in complex spin systems in greatly facilitated by Wiberg and

Nist's *The Interpretation of NMR Spectra* (reference 19), in which line positions and intensities have been calculated as functions of the coupling constant and the difference in chemical shifts.

Yearly reviews and reports on various aspects of NMR spectroscopy appear in references 28 and 29.

Numerous compilations of NMR spectra have appeared — the most extensive being the Sadtler collection (reference 41) and the second edition of the *Aldrich Library of NMR Spectra,* which contains over 8500 NMR spectra in two volumes (reference 40).

6.22 REFERENCES

General

1. Akitt, J. W., *NMR and Chemistry: An Introduction to the Fourier Transform-Multinuclear Era.* London: Chapman and Hall, 1981.

2. Bax, A., *Two-Dimensional Nuclear Magnetic Resonance in Liquids.* Boston: Reidel, 1982.

3. Becker, E. D., *High Resolution NMR: Theory and Chemical Applications.* New York: Academic Press, 1969.

4. Bhacca, N. S., and D. H. Williams, *Applications of NMR Spectroscopy in Organic Chemistry.* San Francisco: Holden-Day, 1964.

5. Bible, R. H., *Interpretation of NMR Spectra: An Empirical Approach.* New York: Plenum Press, 1965.

6. Bovey, F. A., *Nuclear Magnetic Resonance Spectroscopy.* New York: Academic Press, 1969.

7. Emsley, J. W., J. Feeney, L. H. Sutcliffe, *High Resolution Nuclear Magnetic Resonance Spectroscopy,* Vols. 1 and 2. New York: Pergamon Press, 1965.

8. Farrar, T. C., and E. D. Becker, *Pulse and Fourier Transform NMR: An Introduction to Theory and Methods.* New York: Academic Press, 1971.

9. Jackman, L. M., and S. Sternhell, *Applications of Nuclear Magnetic Resonance Spectroscopy in Organic Chemistry.* New York, Pergamon Press, 1969.

10. Levy, G. C., and G. L. Nelson, *Carbon-13 NMR in Organic Chemistry.* New York: Wiley-Interscience, 1972.

11. Marchand, A. P., *Stereochemical Applications of NMR Studies in Rigid Bicyclic Systems.* Deerfield Beach, Fla.: Verlag Chemie International, 1982.

12. Marshall, J. L., *Carbon-Carbon and Carbon-Proton NMR Couplings: Applications to Organic Stereochemistry and Conformational Analysis.* Deerfield Beach, Fla.: Verlag Chemie International, 1983.

13. *Nuclear Magnetic Resonance for Organic Chemists.* D. W. Mathieson, ed. New York: Academic Press, 1967.

14. Pople, J. A., W. G. Schneider, and H. J. Bernstein, *High Resolution Nuclear Magnetic Resonance.* New York: McGraw-Hill, 1959.

15. Roberts, J. D., *An Introduction to the Analysis of Spin-Spin Splitting in High-Resolution Nuclear Magnetic Resonance Spectra.* New York: Benjamin, 1962.

16. Roberts, J. D., *Nuclear Magnetic Resonance: Application to Organic Chemistry.* New York: McGraw-Hill, 1959.

17. Shaw, D., *Fourier Transform N.M.R. Spectroscopy.* New York: Elsevier, 1984.

18. Stothers, J. B., *Carbon-13 NMR Spectroscopy.* New York: Academic Press, 1972.

19. Wiberg, K. B., and B. J. Nist, *The Interpretation of NMR Spectra.* New York: Benjamin, 1962.

Journals

20. *Journal of Magnetic Resonance.* New York: Academic Press, continuous from Vol. 1, 1969.

21. *Organic Magnetic Resonance.* London: Heyden, continuous from Vol. 1, 1969.

Specific Topics

22. *An Introduction to ^{19}F NMR Spectroscopy.* E. F. Mooney, ed. London: Heyden, 1970.

23. Foster, M. A., *Magnetic Resonance in Medicine and Biology.* New York: Pergamon Press, 1984.

24. Mann, B. E., *^{13}C NMR Data for Organometallic Compounds.* New York: Academic Press, 1981.

25. Memory, J. D., *NMR of Aromatic Compounds.* New York: John Wiley, 1982.

26. *Phosphorus-31 NMR: Principles and Applications,* D. J. Gorenstein, ed. New York: Academic Press, 1984.

27. Sandstrom, J., *Dynamic NMR Spectroscopy.* New York: Academic Press, 1982.

Yearly Reviews

28. *Annual Review (Reports) on NMR Spectroscopy.* E. F. Mooney, ed. New York: Academic Press, continuous from 1968.

29. *Progress in Nuclear Magnetic Resonance,* J. W. Emsley, J. Feeney, and L. H. Sutcliffe, eds. New York: Pergamon Press, continuous from 1966.

Compilations of NMR Spectra and Literature References

30. *A Catalogue of the Nuclear Magnetic Resonance Spectra of Hydrogen in Hydrocarbons and Their Derivatives.* Baytown, Tex.: Humble Oil and Refining.

31. Bovey, F. A., *NMR Data Tables for Organic Compounds.* Vol. 1. New York: Interscience, 1967.

32. Brugel, W., *Nuclear Magnetic Resonance Spectra and Chemical Structures.* New York: Academic Press, 1967.

33. Dugan, C. H., and J. R. Van Wazer, *Compilation of Reported F ^{19} NMR Chemical Shifts (1951–1967).* New York: Wiley-Interscience, 1970.

34. *Formula Index to NMR Literature Data,* Howell, M. G., A. S. Kende, and J. S. Webb, eds. New York: Plenum Press, 1965.

35. *High Resolution NMR Spectra Catalogue,* Vol. 1 (1962) and Vol. 2 (1963). Palo Alto, Calif.: Varian Associates.

36. Johnson, L. F., and W. C. Jankowski, *Carbon-13 NMR Spectra; A Collection of Assigned, Coded and Indexed Spectra.* New York: Wiley-Interscience, 1972.

37. *NMR, EPR, and NQR Current Literature Service.* Washington, D.C.: Butterworths.

38. *Nuclear Magnetic Resonance Spectra Data* (American Petroleum Institute, Project No. 44). College Station, Tex.: Chemical Thermodynamic Properties Center, Agricultural and Mechanical College of Texas, 1962.

39. Pham, Q. T., R. Petiaud, M.-F. Llaure, and H. Waton, *Proton and Carbon NMR Spectra of Polymers.* New York: John Wiley, 1984.

40. Pouchert, C. J., and J. R. Campbell, *The Aldrich Library of NMR Spectra,* Vols. I–XI, (1974–1975). Milwaukee, Wisc.: Aldrich Chemical. Pouchert, C. J., *The Aldrich Library of NMR Spectra,* 2nd ed., Vols. 1 and 2.

41. *Sadtler NMR Spectra.* Philadelphia, Pa.: Sadtler Research Laboratories.

42. Szymanski, H. A., *NMR Band Handbook.* New York: Plenum Press, 1968.

43. *The Sadtler Guide to Carbon-13 NMR Spectra.* Philadelphia, Pa.: Sadtler Research Laboratories.

7

Mass Spectrometry

7.1 INTRODUCTION

Most of the spectroscopic and physical methods employed by the chemist in structure determination are concerned only with the physics of molecules; mass spectrometry (MS) deals with both the chemistry and the physics of molecules, particularly with gaseous ions. In conventional mass spectrometry, the ions of interest are positively charged ions. The mass spectrometer has three functions:

1. To produce ions from the molecules under investigation.
2. To separate these ions according to their mass-to-charge ratio.
3. To measure the relative abundances of each ion.

The demonstration of the basic principles of mass spectrometry preceded that of most of the other physical methods currently used for structure determination. As early as 1898, Wein showed that positive rays could be deflected by means of electric and magnetic fields; in 1912, J. J. Thompson recorded the first mass spectra of simple low molecular weight molecules. The earliest of the prototypes of today's instruments were the mass spectrometer of Dempster (1918) and the mass spectrograph of Aston (1918). By the early 1940s, instruments suitable for the examination of the spectra of organic molecules of moderate molecular weight were commercially available. The earliest extensive studies of organic systems were made by the petroleum industry on hydrocarbons. Although mass spectrometry was used for the quantitative analysis of hydrocarbon mixtures, there seemed to be little, if any, recognizable relationship between structure and spectra. This situation led to a dormant period in the utility of mass spectrometry for studies of organic molecules. In the late 1950s, Beynon, Biemann, and McLafferty clearly demonstrated the role of functional groups in directing fragmentation, and the power of mass spectrometry for organic structure determination began to develop. At that time the technology employed was essentially that available since 1940. Today mass spectrometry has achieved status as one of the primary spectroscopic methods to which a chemist faced with a structural problem turns; great advantage is found in the extensive structural information that can be obtained from submilligram or even submicrogram quantities of material. *Mass spectrometry is the most sensitive method of molecular analysis available to the chemist.*

7.2 INSTRUMENTATION

A rather large variety of commercial mass spectrometers is currently available. There is considerable variation in the mechanics by which the various mass spectrometers accomplish their tasks of production, separation, and measurement of ions. It is entirely beyond the scope and purpose of this book to provide the reader with a detailed discussion of instrumentation. However, so that the reader can at least have general ideas of the principles by which most mass spectrometers operate, representative examples will be discussed briefly.

Three methods are generally used for the production of ions from organic molecules: *electron ionization (EI), chemical ionization (CI),* and *desorption ionization (DI).* Before proceeding with a more detailed examination of these methods we will look briefly at the overall operation of mass spectrometers functioning in the EI mode, the most common of the ionization methods.

Throughout this text the term *low resolution* will refer to instruments capable of distinguishing only between ions of differing nominal mass, that is, ions whose mass-to-charge ratios differ by one mass unit. *High resolution* (HRMS) will refer to spectrometers capable of distinguishing between ions whose masses differ in the third decimal place or less, for example, $C_2H_6^+$, CH_2O^+, and CH_4N^+, with masses (based on carbon 12.0000) of 30.0469, 30.0105, and 30.0344, respectively.

7.2.1 Magnetic Focusing EI Instruments

A schematic diagram of a representative single-focusing low-resolution mass spectrometer operating in the EI mode is shown in Figure 7.1. The sample is introduced into the ionization chamber; the vapor of the sample in the ionization chamber is bombarded with an electron beam of variable energy [usually 50 to 70 electron volts (eV), the latter value being most commonly employed]. A small percentage of the

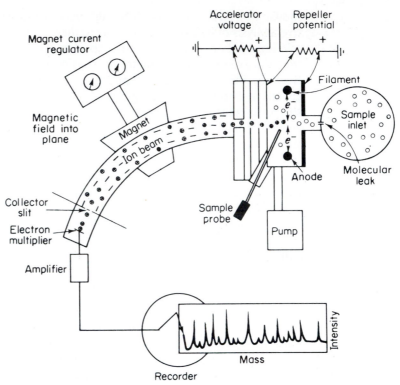

Figure 7.1 Schematic diagram of a magnetic single-focusing mass spectrometer with a 60° sector magnet.

molecules are ionized, by electron impact, into positively charged ions that subsequently form fragment ions. Negative ions are also formed to a small extent. The small repeller potential between the back wall of the ionization chamber and the first accelerator plate attracts the negative ions to the back wall and discharges them. At the same time, this potential pushes the positive ions toward the accelerating region. There they are accelerated by a potential of approximately 2000 V between plates that have a slit in them. The ions are focused on an exit slit by means of subsidiary accelerating plates and slits. The positive ions are accelerated by this potential according to Equation (7.1),

$$\tfrac{1}{2}mv^2 = zV \tag{7.1}$$

where m is the mass of the ion, z the electronic charge, V the potential of the ion accelerating plate, and v the velocity of the particle. The accelerated ions then pass into the magnetic field H generated between the two poles of an electromagnet. In the magnetic field, the ions are deflected along a circular path according to Equation (7.2),

$$r = \frac{mv}{zH} \tag{7.2}$$

where r is the radius of the path. Elimination of v from the preceding two equations yields Equation (7.3).

$$\frac{m}{z} = \frac{H^2 r^2}{2V} \tag{7.3}$$

From this equation one can conclude that, at given values of H and V, only particles with a particular mass-to-charge ratio will arrive at the collector slit placed along the fixed path r. It can be further seen from the equation that particles with mass-to-charge ratios of $m/1, 2m/2$, and $3m/3$ would all follow the same path and be detected and recorded simultaneously. Fortunately, under the conditions usually employed in recording mass spectra of organic compounds, most of the particles are singly charged, ions of higher charge generally being produced in insignificant quantities.

Equation (7.3) also indicates that a spectrum of ions could be obtained at the collector slit by variation of either the magnetic field strength or the accelerating voltage. In instruments with a 180° magnetic arc path, the scanning is accomplished by decreasing the accelerating voltage at a suitable rate. In the 60° and 90° arc path instruments, the accelerating voltage is fixed and the magnetic field is varied.

The collector assembly consists of a series of collimating slits and an ion current detector. Modern electron multipliers are capable of detecting single ions. The current that is amplified and measured is directly related to the abundance of the ions at the mass being examined. Modern instruments are equipped with computerized data acquisition systems that provide both a printout of the spectrum and a tabulation of m/z's and intensities. Using the data acquisition systems, spectra can be scanned within a few tenths of a second or less, with the ability to scan repeatedly the complete m/z region every few tenths of a second.

The high-resolution, double-focusing instruments (e.g., Figure 7.2) incorporate most of the principles and methods, with additional refinements, outlined in the foregoing discussion of low-resolution instruments. It will be seen in Equation (7.3) that, at a constant magnetic field H, any spread in the magnitude of velocity (equivalent to the spread of V) will result in a spread in r for any given value of m/z. The peaks obtained in low-resolution mass spectrometry are indeed broadened by such a spread, which is caused by the contributions of initial kinetic or thermal energy to the kinetic energy gained by the particle during acceleration in the ionization chamber.

A several thousand-fold increase in the resolving power of an instrument can be achieved by elimination of this energy spread in the ion beam before it enters the magnetic field. In the double-focusing instrument (see Figure 7.2), ions are passed through a radial electrostatic field that sorts them into monoenergetic paths (velocity focusing) before they are passed into the magnetic field where they are refocused according to their mass-to-charge ratio. According to the design of the instrument, the spectrum can be scanned by the conventional sweeping of the magnetic field or the accelerating potential, the ions being recorded, as in the case of the low-resolution instrument, as they come into focus on the collector (see Figure 7.2). Some double-focusing instruments are designed in reverse order with magnetic focusing preceding velocity focusing.

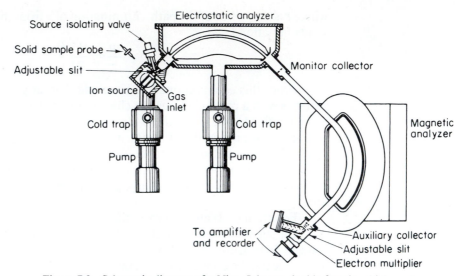

Figure 7.2   Schematic diagram of a Nier–Johnson double-focusing, high-resolution mass spectrometer showing the radial electric analyzer and the magnetic analyzer.

In addition to the commonly used magnet sector instruments discussed above, mass analyzers based on quadrupole mass filter, ion cyclotron resonance, and time-of-flight technology are available but discussion of such instruments is beyond the scope of this text.

7.3 PRODUCTION OF IONS IN THE MASS SPECTROMETER; SAMPLE HANDLING

7.3.1 Electron Ionization (EI)

In the ionization chamber of the mass spectrometer, high energy electrons emitted from a heated filament collide with the molecules of the sample. If the energy of the electrons is greater than the ionization potential of the molecule, an electron will be ejected from the molecule, producing a molecular radical cation (see the example reaction scheme below). The ionization potentials of most molecules fall in the range of 8 to 11 electron volts (eV); however, most mass spectrometers generally induce ionization with electrons having an average energy of 70 eV. When such an energetic electron collides with a molecule, the resulting molecular ion that is formed contains considerable excess internal energy, which is dissipated through subsequent fragmentation and/or rearrangement reactions.

Ionization

$$\text{C}_6\text{H}_5-\overset{\displaystyle O}{\overset{\|}{\text{C}}}\text{CH}_2\text{CH}_2\text{CH}_3 + e^- \longrightarrow \text{C}_6\text{H}_5-\overset{\displaystyle O^{\overset{+}{\cdot}}}{\overset{\|}{\text{C}}}\text{CH}_2\text{CH}_2\text{CH}_3 + 2e^-$$

radical ion

Fragmentations

$$\text{C}_6\text{H}_5-\overset{\displaystyle O^{\overset{+}{\cdot}}}{\overset{\|}{\text{C}}}\text{CH}_2\text{CH}_2\text{CH}_3 \longrightarrow \text{C}_6\text{H}_5-\text{C}\equiv\text{O}^+ + \overset{\displaystyle \cdot}{\text{C}}\text{H}_2\text{CH}_2\text{CH}_3$$

ion   radical

ion   +   +   CO

ion   neutral

Fragmentation with rearrangement

$$\left[\text{C}_6\text{H}_5-\overset{\displaystyle O}{\overset{\|}{\text{C}}}\begin{matrix}\text{H}\\ \text{CH}_2\\ \text{CH}_2\\ \text{CH}_2\end{matrix}\right] \longrightarrow \left[\text{C}_6\text{H}_5-\text{C}\overset{\displaystyle \text{O}-\text{H}}{\underset{\text{CH}_2}{=}}\right]^{\overset{+}{\cdot}} + \text{CH}_2{=}\text{CH}_2$$

radical ion   neutral

Due to this great excess of energy imparted to the parent molecular ion, in many cases the intensity of the parent ion is very weak, and in some cases is not detectable. To circumvent this problem, the energy of the ionizing electrons can be lowered; however, this also results in a general lowering in intensity of all of the peaks in the mass spectrum. The alternative techniques of chemical and desorption ionization (discussed below) provide molecular ions with little excess internal energy resulting in relatively less fragmentations with the molecular ions being more intense.

To obtain an EI mass spectrum, the sample must be introduced into the ionization chamber in the vapor state in sufficient quantity so that the vapor pressure exerted by the sample is in the range of 10^{-5} to 10^{-9} mm Hg. Most spectrometers have several sample introduction systems available; the method of choice is governed by the volatility of the sample.

For most organic compounds, direct insertion of the sample into the highly evacuated ionization chamber is advantageous. In a commonly employed system, the sample is placed in a small crucible on the end of a probe that is inserted into the ionization chamber through a vacuum lock system; if necessary to produce sufficient vapor pressure, the sample can be heated either by heating the entire ionization chamber or by heating elements associated with the probe. With probe heating, temperatures as high as 1000°C can be achieved; the method is limited only by the thermal stability of the compound.

For gases and liquids or solids of high volatility, the *inlet system* is often used for the introduction of the sample into the ion source. The inlet system consists of a series of vacuum locks, a reservoir, a sensitive pressure-measuring device, and a molecular leak into the ion source (see Figure 7.3). The volume of the reservoir on commercial instruments varies from 1 to 5 L; these relatively large volumes require proportionately larger samples compared with direct insertion. Sufficient sample is introduced into the reservoir to produce a pressure of 10^{-2} mm Hg. The entire system is in an oven that can be heated if necessary to produce sufficient pressure. The sample in the reservoir, under a pressure higher than that required in the ion source, is allowed to leak through a very small hole, producing a constant stream of the sample into the ionization region. The large volume of the reservoir minimizes depletion of the sample during the run; this is necessary if reproducible spectra are to be obtained at slow scan speeds.

An extremely useful and versatile method for the direct introduction of samples into the ionization chamber involves the direct hookup of a mass spectrometer and a gas chromatograph (GCMS). Fortunately, the sensitivities of mass spectrometer and gas chromatographic detectors are comparable; the quantities of material usually eluted from a gas chromatography column are much too small for analysis by infrared spectroscopy and most other identification techniques. The eluent from a column can be monitored by direct introduction into the ionization chamber. The dilution of the sample by the carrier gas can be reduced by means of special separators

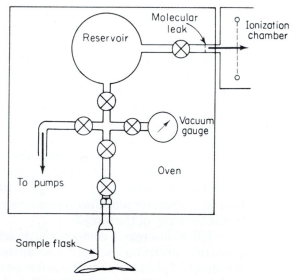

Figure 7.3 Schematic diagram of an inlet system.

that pump off the carrier gas preferentially. Since the concentration of any eluent is constantly and rapidly changing, this direct monitoring can be achieved only with spectrometers having rapid recording systems such as those spectrometers equipped with computerized data acquisition systems.

Interfacing of liquid chromatographs with mass spectrometers (LCMS) presents problems associated with solvent separation, involatile and thermally labile samples, and, in some cases with reverse phase columns, the presence of buffer salts. Ingenious solutions to these problems have been provided by techniques known as moving belt interfaces and thermospray, and LCMS is now a practical and rapidly growing procedure.

Even with the variety of sample handling techniques, mass spectrometry cannot be applied routinely to all compounds; the limiting factor is often thermal stability of the sample at the temperature necessary to produce the minimum pressure. It is often possible to convert thermally unstable compounds to more stable and/or more volatile derivatives, for example, acids to esters, and alcohols to acetates or trimethylsilyl ethers.

EI is the most common and universal technique of ionization. The advantages are (1) simplicity of operation, (2) generality (most organic molecules of moderate molecular weight give usable EI mass spectra), (3) highly reproducible spectra, and (4) availability of extensive compilations. The disadvantages are (1) sample must be capable of being volatilized without thermal decomposition, and (2) ionization energy often results in extensive and difficult-to-interpret fragmentations and low-to-negligible abundance molecular ions (from which information on molecular weight and molecular formulas are obtained).

7.3.2 Chemical Ionization (CI)

In chemical ionization MS, a reagent gas is used in great excess ($> 10^3$ times) over the sample so that initial ionization by the 70 eV electron beam results in essentially only ionization of the reagent. The sample molecules react with the reagent ions to produce "quasimolecular ions," for example $(M + H)^+$ (see Table 7.1), with little excess internal energy. The result, compared with conventional EI, is relatively less fragmentation and more intense molecular or quasimolecular ions. The composition and fragmentation tendency of these ions are dependent on the nature of the

TABLE 7.1 Commonly Used Reagent Gases in CIMS

Reagent Gas	Reagent Ion	Quasimolecular Ion	Fragmentation Tendency (internal energy)
H_2	H_3^+	$(M + H)^+$	High
CH_4	CH_5^+	$(M + H)^+$	High
i-C_4H_{10}	$C_4H_9^+$	$(M + H)^+$	Low
NH_3	NH_4^+	$(M + NH_4)^+$	Low
Ar	Ar^+	M^+	High

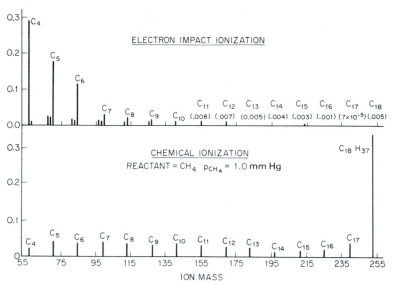

Figure 7.4   Electron impact and chemical ionization spectra of n-$C_{18}H_{38}$.

reagent gas. Reagent ions can be selected to result in either large or small internal energies in the sample molecular ions. Some common reagent gases, their resulting reagent ions, and their characteristics are given in Table 7.1.

  With CIMS the sample is usually introduced with a direct insertion probe and the reagent gas with an inlet system. Figure 7.4 illustrates the differences in relative intensities of ions when the same sample is subjected to EI and CI mass spectrometry.

7.3.3 Desorption Ionization (DI)

In desorption ionization MS, the sample in a condensed phase (solid, neat liquid, or solution) is bombarded by a particle or photon beam. In the simplest cases, preformed ions, for example, ammonium ions, are sputtered from a matrix (solid or liquid) into the gas phase. More typically neutral molecules are sputtered from surface layers as a dense gas containing positive and negative ions as well as neutral molecules. The latter may be ionized in the plasma immediately above the sample surface. The resulting gas phase ions are swept from the ion source, mass separated, and detected. DI is especially useful for nonvolatile or thermally labile samples. The three most common methods of DIMS and the characteristic beams are:

Method	*Beam*
Laser desorption MS (LDMS)	Photons from laser
Secondary ion MS (SIMS)	Xe^+, Cs^+, Ar^+
Fast atom bombardment MS (FABMS)	$Ar°$, $Xe°$

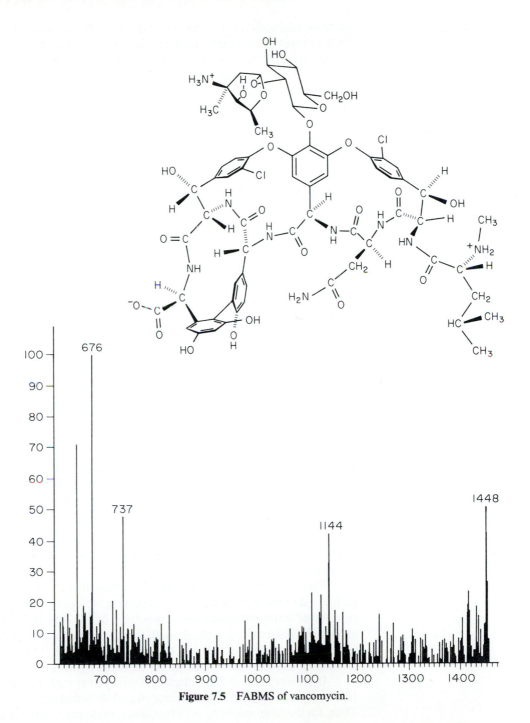

Figure 7.5 FABMS of vancomycin.

All these methods use a direct probe insertion technique. In sample preparation for FAB analysis a clean metal probe tip is wetted with a matrix material, often glycerol, and approximately 1 μL of a solution of the sample in a solvent such as methanol or water is added. The sample solution may also contain certain additives such as acids or "cationizing" salts to facilitate production of quasimolecular ions such as (M + H)$^+$, (M + Na)$^+$, and so on. Using FAB techniques the (M + Na)$^+$ ion at m/z 1678 of the peptide antimoebin II and the (M + H)$^+$ ion at m/z 1448 of the antibiotic vancomycin (see Figure 7.5) can be observed.

7.4 DETERMINATION OF MOLECULAR WEIGHTS AND FORMULAS

The mass spectrum of an organic compound indicates that the compound is composed of several different molecular species of different combinations of naturally occurring isotopes. The masses that we should calculate for mass spectrometry should not be based on the chemical scale, but should be the sum of the masses of the isotopes occurring in the species under consideration. For example, the molecular weight of methyl bromide is 94.95 on the chemical scale. This chemical scale weight is based on the weighted averages of the naturally occurring isotopes

$$C \quad 12.011$$

$$H \quad 1.008$$

$$Br \quad 79.909$$

However, in the mass spectrum, we, of course, do not see the average molecular species but rather the molecular species corresponding to each possible isotopic combination according to the following naturally occurring abundances:

| ^{12}C 98.89% | ^{1}H 99.985% | ^{79}Br 50.54% |
| ^{13}C 1.11% | ^{2}H 0.015% | ^{81}Br 49.46% |

The low-resolution mass spectrum of methyl bromide in the region of the molecular ion is shown in Figure 7.6.

The radical cation produced by the removal of a single electron from a molecule composed of the lightest naturally occurring isotopes of the elements present in the compound is the *molecular ion,* symbolized by M. The terms M + 1, M + 2, and so on signify peaks at 1, 2, and so on, of nominal mass units higher than M. Likewise terms such as (M − 1), (M − H), (M + H), (M − 29), (M − C$_2$H$_5$), and so on, are often used to relate other ions to the molecular ion.

If a molecular ion has sufficient stability to accord it a lifetime of approximately 10^{-5} seconds, it will be fully accelerated and recorded at its corresponding m/z value. Thus, for the large majority of compounds, mass spectrometry provides an exact and

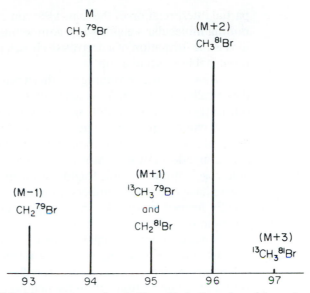

Figure 7.6   Mass spectrum of methyl bromide in the region of the molecular ion. Small contributions by low abundance ^2H are not indicated. Contributions by species containing both ^{13}C and ^2H are negligible.

unambiguous method for the determination of the molecular weight of a molecule. The molecular ion and related isotopic species correspond to the peaks of highest mass in the spectrum in the absence of collision processes (see below). However, in a number of cases, compounds give molecular ions of very low or negligible intensity, and care must be taken not to confuse peaks due to impurities or fragments with molecular ions.

The stabilization of the positive charge in the molecular ion and the tendency toward fragmentation influence the intensity of the molecular ion peak. Biemann has suggested the following approximate order of decreasing stability of the molecular ion:

> Aromatic compounds — conjugated olefins — alicyclic compounds — sulfides — straight chain hydrocarbons — mercaptans — ketones — amines — esters — ethers — carboxylic acids — branched hydrocarbons — alcohols

Cyclic compounds naturally tend to give rise to correspondingly more intense molecular ion peaks than their acyclic analogs because, in the cyclic structures, cleavage of one bond does not lead to the splitting off of a fragment of lower mass. It should be emphasized that one should always check to see that the fragmentation pattern exhibited in the rest of the spectrum can be accommodated by the suggested molecular ion and that the intensity of that peak is of the magnitude expected for the proposed structure. In some cases, even though a compound does not exhibit a molecular ion, it is possible to arrive at the molecular weight of the compound by the partial interpretation of the mass spectrum. Another approach often used for determining molecular weights when compounds do not exhibit recognizable molecular ions is the formation of a derivative chosen to give a predicted mode of cleavage or a more stable molecular ion.

As an aid in determining whether a peak is due to a molecular ion or a fragment, the so-called *nitrogen rule,* which holds for all compounds containing carbon, hydrogen, oxygen, nitrogen, sulfur, and/or the halogens, as well as phosphorus, arsenic, and silicon, can be employed. The rule states that a neutral molecule of even-numbered molecular weight must contain either no nitrogens or an even number of nitrogen atoms; an odd-numbered molecular weight compound must contain an odd number of nitrogen atoms. Simple fragments formed from even-numbered compounds will always have odd mass numbers, with the exception of rearrangement peaks that are usually formed by the elimination of a neutral molecule with the formation of a fragment of even mass number.

Since only a very small fraction of the molecules in the ionization chamber are actually ionized, those ions that are formed are present in a high population of un-ionized molecules. Ion-molecule collisions, which result in the abstraction of an atom or a group of atoms from the neutral molecule by the positive ion, may occur, resulting in a particle of mass higher than the expected molecular ion. At the pressures usually employed in EI mass spectrometry, the only significant reaction of this type is the abstraction of a hydrogen atom by the molecular ion resulting in a peak at $M + 1$. This $M + 1$ peak becomes particularly important in those cases (ethers, esters, amines, aminoesters, nitriles) in which the molecular ion is relatively unstable, whereas the corresponding protonated species is quite stable. Since this $M + 1$ peak is formed in a bimolecular process, it can be recognized as such by the proportionality of its intensity to the square of the sample pressure. As the sample pressure is increased, the intensity of the $M + 1$ peaks will increase relative to the intensity of the other peaks in the spectrum. An $M + 1$ peak can also be recognized by its sensitivity to the repeller potential; an increase in the repeller potential will decrease the residence time of ions in the ionization chamber and thus decrease the likelihood of collision processes. The importance of quasimolecular ions such as $(M + H)$, $(M + Na)$ in CIMS and DIMS has already been noted (see Sections 7.3.2 and 7.3.3).

As indicated earlier in this chapter, with high-resolution mass spectrometry it is possible to measure masses to four decimal places. Table 7.2 gives the exact masses of the nuclides frequently found in organic compounds.

It can be readily seen that each formulation of elements will have a unique mass associated with it. Measurement of the masses with sufficient accuracy allows one to determine the molecular formula. An excellent example of this approach is found in the work of Djerassi on the alkaloid vobtusine. For this alkaloid, earlier work based on microanalysis and other classical data had suggested possible molecular formulas of $C_{45}H_{54}N_4O_8$ (778), $C_{42}H_{48}N_4O_6$ (704), and $C_{42}H_{50}N_4O_7$ (722). Low-resolution mass spectrometry showed the molecular weight to be 718, which is inconsistent with all of the foregoing, but is consistent with the formulations of $C_{43}H_{50}N_4O_6$ (718.3730) and $C_{42}H_{46}N_4O_7$ (718.3366). A high-resolution mass

TABLE 7.2 Masses and Natural Abundances of Nuclides of Importance in High-Resolution Mass Spectrometry of Organic Compounds

Nuclide	Mass	Natural Abundance (%)
^{1}H	1.007825	99.985
^{2}H	2.01400	0.015
^{12}C	12.00000	98.9
^{13}C	13.00335	1.11
^{14}N	14.00307	99.63
^{16}O	15.994914	99.80
^{19}F	18.998405	100.00
^{28}Si	27.976927	92.21
^{31}P	30.973763	100.00
^{32}S	31.972074	94.62
^{34}S	33.96786	4.2
^{35}Cl	34.968855	24.5
^{37}Cl	36.9658	75.5
^{79}Br	78.918348	49.5
^{81}Br	80.983	50.5
^{127}I	126.904352	100.00

spectrum indicated a molecular weight of 718.3743, clearly showing that the formula $C_{43}H_{50}N_4O_6$ was the correct one.

Extensive tables (see Table 7.3 and reference 1) as well as computer programs for relating elemental composition to exact masses for use with HRMS are available. From high-resolution mass spectra the elemental composition not only of the molecular ion but of each fragment can be determined. The spectrum can then be interpreted directly according to the elemental formula of the molecule and its fragments.

From low-resolution spectra, it is sometimes possible to determine a unique molecular formula, or at least to limit severely the number of possibilities, by consideration of the relative intensities of the various isotope peaks (isotope clusters). To determine the molecular formula, one measures the intensity of the molecular ion M^{+} and expresses the intensity of the $M^{+} + 1$ peak and the $M^{+} + 2$ peak as percentages of the molecular ion peak intensity. For any possible molecular formula, the observed values of the $M^{+} + 1$ and $M^{+} + 2$ peaks can be compared with the theoreti-

TABLE 7.3 Sample Table: Mass versus Elemental Composition at Nominal Mass 43

Mass	Elemental Composition
43.0058	CHNO
43.0184	C_2H_3O
43.0269	CH_3N_2
43.0421	C_2H_5N
43.0547	C_3H_7

cal relative intensities; tables for various combinations of C, H, N, and O have been published and appear in reference 1 by Beynon.

An example of the use of this method follows. From a mass spectrum we obtain the following data:

m/z	Percent
100(M)	(100)
101(M + 1)	5.64
102(M + 2)	0.60

We can readily eliminate from further consideration formulas containing sulfur, chlorine, or bromine from the relatively small contribution of the M + 2 peak (these elements are usually easy to detect because of their high contribution to the M + 2 peak) (see Table 7.2). Iodine need not be considered because its mass alone is 127. The relatively large contribution of the M + 1 peak suggests that fluorine is also not present. If we look in the Beynon table of C, H, N, O compound formulas of mass 100, we find the following data:

100	M + 1	M + 2	100	M + 1	M + 2
$C_2H_2N_3O_2$	3.42	0.45	$C_4H_{10}N_3$	5.63	0.13
$C_2H_4N_4O$	3.79	0.26	$C_5H_8O_2$	5.61	0.53
$C_3H_2NO_3$	3.77	0.65	$C_5H_{10}NO$	5.98	0.35
$C_3H_4N_2O_2$	4.15	0.47	$C_5H_{12}N_2$	6.36	0.17
$C_3H_6N_3O$	4.52	0.28	$C_6H_{12}O$	6.72	0.39
$C_3H_8N_4$	4.90	0.10	$C_6H_{14}N$	7.09	0.22
$C_4H_4O_3$	4.50	0.68	C_7H_2N	7.98	0.28
$C_4H_6NO_2$	4.88	0.50	C_7H_{16}	7.82	0.26
$C_4H_8N_2O$	5.25	0.31	C_8H_4	8.71	0.33

On the basis of the nitrogen rule, we need not consider further any formulas containing an odd number of nitrogens. From the M + 1 and M + 2 values, the formula $C_5H_8O_2$ is found to be the best fit. The measured isotope peaks are usually slightly higher than the calculated ones because of small contributions from bimolecular collisions, impurities, M − 1 peaks, background, and so on. Moreover, it should be emphasized that the method is limited to compounds that produce relatively intense molecular ions and to compounds of low-to-moderate molecular weight. Relatively intense molecular ion peaks are necessary if the isotope peaks are to be measurable. In compounds that contain large numbers of atoms, the isotopic distribution functions are too complex for accurate analysis, and the distinction between possible formulas is not possible owing to experimental uncertainties in the peak intensities.

7.5 REACTIONS OF IONS IN THE GAS PHASE

In the formation of a molecular ion by electron bombardment, removal of an electron from a nonbonded electron pair on a heteroatom appears easier than removal of a pi electron, which in turn is easier to remove than a sigma electron. It is therefore possible and often desirable to depict fragmentation processes with the positive charge and free valence localized on a specific atom, for example, $CH_3CH_2-\overset{+}{\overset{\cdot}{O}}-CH_2CH_3$. In diagrams in this text, we shall occasionally choose to show charges and radicals localized, but more often we shall represent ions and radical ions by enclosing the structural formulas in brackets, for example, $[CH_3CH_2-O-CH_2CH_3]^{+\cdot}$.

7.5.1 Fragmentations

The mass spectrum of a compound provides the structural chemist with two types of information: the molecular weight of the compound and the mass of various fragments produced from the molecular ion. To obtain structural information, the chemist attempts on paper to reassemble these fragments in a way not unlike the assembling of a jigsaw puzzle. Hypothetical molecules containing these fragments are constructed, and the fragmentation patterns predicted for these models are checked against the spectrum obtained.

Fragmentation is a chemical process that results in bond breaking; the energetic considerations that are applicable to classical chemical reactions are also applicable to these fragmentation processes. There are two major differences between conventional ion chemistry and chemistry in the mass spectrometer. First, we are dealing with particles in excited states, and the energies involved are considerably higher than those from typical chemical reactions. Second, at the very low operating pressures ($\approx$ low concentration) of the mass spectrometer, we are dealing with unimolecular reactions, and energy is not dissipated to any appreciable extent by collisions with other molecules of the compound or solvent as might occur in solution chemistry; thus, further fragmentation of the initially formed ions occurs extensively. It must be pointed out, therefore, that the intensity of a given ion peak depends not only on its rate of formation, but also on the rate of its subsequent fragmentation.

Fragmentations are best interpreted based on the known stability of carbocations and free radicals. In the two possible modes of fragmentation shown in the following equation, the dominant process in general will be to generate the most stable cation, although the stability of the neutral radical product can be very important. In cases where both A^+ and B^+ cations are of similar stability, the more abundant will be the higher molecular weight cationic fragment.

$$[A-B]^{+\cdot} \Big\langle \begin{array}{l} A^+ + B\cdot \\ A\cdot + B^+ \end{array}$$

For examples of simple bond fragmentation the EI mass spectrum of *sec*-butyl ethyl ether is illustrative. The molecular ion (m/z 102) fragments to produce the following ions:

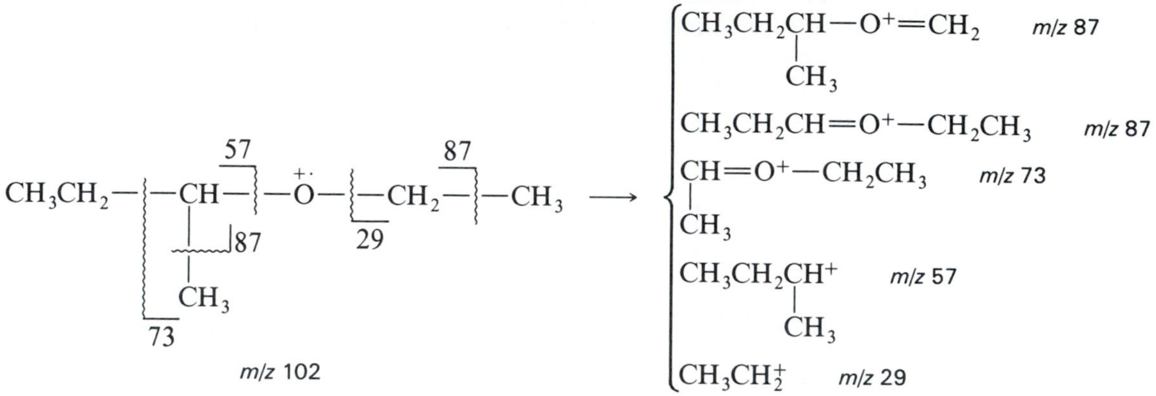

$$CH_3CH_2\overset{57}{\underset{\underset{\underset{73}{CH_3}}{\overset{|}{\underset{|87}{|}}}}{-\!\{-CH-\}-}}\overset{+\cdot}{O}\overset{87}{-\!\{-CH_2-\}-}\overset{}{\underset{29}{}}CH_3 \longrightarrow$$

m/z 102

$$\begin{cases} CH_3CH_2\underset{\underset{CH_3}{|}}{CH}\!-\!O^+\!\!=\!\!CH_2 & m/z\ 87 \\ CH_3CH_2CH\!=\!O^+\!\!-\!CH_2CH_3 & m/z\ 87 \\ \underset{\underset{CH_3}{|}}{CH}\!=\!O^+\!\!-\!CH_2CH_3 & m/z\ 73 \\ CH_3CH_2\underset{\underset{CH_3}{|}}{CH^+} & m/z\ 57 \\ CH_3CH_2^{+} & m/z\ 29 \end{cases}$$

Observe the notation of using a squiggly line through a bond with the mass given on the side corresponding to the observed fragment.

As an aid in the visualization of fragmentation and rearrangement processes, arrows are often used to symbolize bond breaking and bond making, as shown in Equation (7.4).

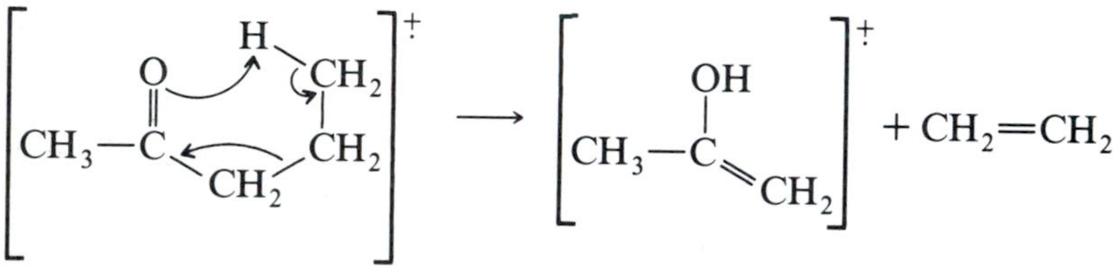

$$\left[CH_3-\overset{\overset{O}{\|}}{C}\underset{CH_2^{\cdot}}{\overset{H\diagdown CH_2}{\diagup\ \ \diagdown}}CH_2 \right]^{+} \longrightarrow \left[CH_3-\overset{OH}{C}\underset{CH_2}{\diagup\!\!=} \right]^{+} + CH_2\!=\!CH_2 \quad (7.4)$$

Since in conventional chemical notation doubly-barbed arrows indicate two electron shifts, the "fish hook" or "half arrow" is often used in connection with mass spectrometry to indicate a single electron shift, particularly with structures where charge and radical sites are designated, for example,

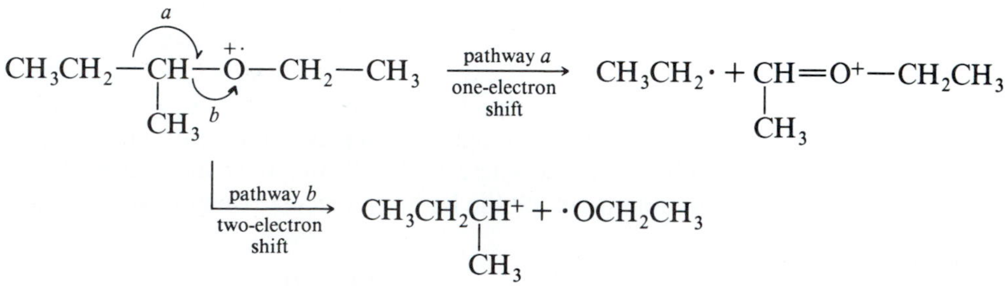

$$CH_3CH_2\overset{a}{\underset{\underset{CH_3}{|}\ b}{-CH-}}\overset{+\cdot}{O}-CH_2-CH_3 \quad\xrightarrow[\substack{\text{one-electron}\\\text{shift}}]{\text{pathway } a}\quad CH_3CH_2\cdot + \underset{\underset{CH_3}{|}}{CH}\!=\!O^+\!\!-\!CH_2CH_3$$

$$\xrightarrow[\substack{\text{two-electron}\\\text{shift}}]{\text{pathway } b}\quad CH_3CH_2\underset{\underset{CH_3}{|}}{CH^+} + \cdot OCH_2CH_3$$

Most of the important types of fragmentation are summarized in general form below.

1. Simple carbon-carbon bond cleavages

$$\text{(a)} \quad \left[R-\overset{|}{\underset{|}{C}}- \right]^{\ddagger} \longrightarrow R^+ + \cdot\overset{|}{\underset{|}{C}}-$$

$$\text{(b)} \quad R-\overset{|}{\underset{|}{C}}-\overset{|}{\underset{|}{C}}{}^+ \longrightarrow R^+ + \overset{\diagdown}{\diagup}C{=}C\overset{\diagup}{\diagdown}$$

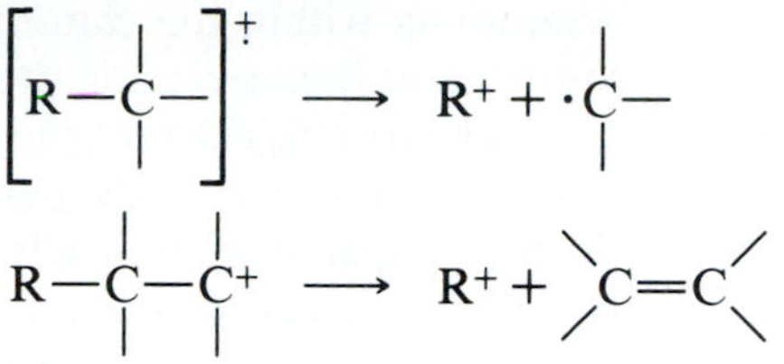

The relative importance of R^+ increases in the order

$$CH_3 < CH_2R' < CHR_2' < CR_3' < CH_2{-}CH{=}CH_2 < CH_2{-}\!\!\bigcirc\!\!\!\bigcirc$$

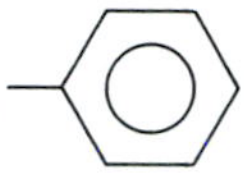

$$\text{(c)} \quad \left[R-\overset{O}{\overset{\|}{C}}-R \right]^{\ddagger} \longrightarrow R\cdot + R\overset{+}{C}{=}O$$

$$\text{(d)} \quad R-\overset{+}{C}{=}O \longrightarrow R^+ + C{=}O$$

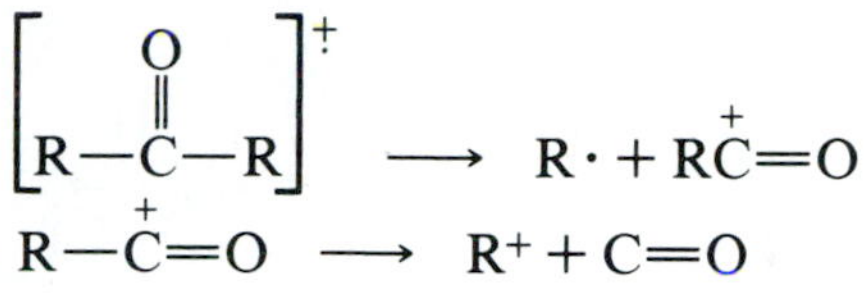

2. Cleavages involving heteroatoms

$$X = \text{halogen, } OR', SR', NR_2' \; (R' = H, \text{ Alkyl, Aryl})$$

$$\text{(a)} \quad \left[-\overset{|}{\underset{|}{C}}-X \right]^{\ddagger} \longrightarrow -\overset{|}{\underset{|}{C}}{}^+ + X\cdot$$

$$\text{(b)} \quad \left[R-\overset{|}{\underset{|}{C}}-X \right]^{\ddagger} \longrightarrow R\cdot + \overset{\diagdown}{\diagup}\overset{+}{C}-X$$

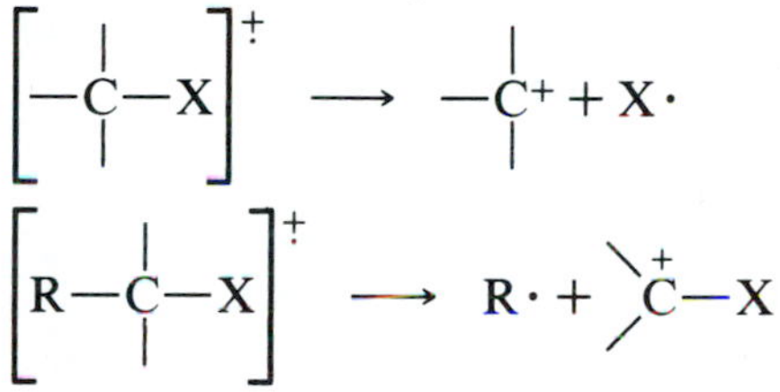

3. Converted cleavages

$$\text{(a)} \quad \left[\bigcirc \right]^{\ddagger} \longrightarrow \left[\bigcirc \right]^{\ddagger} + \overset{\diagdown}{\diagup}\overset{C}{\underset{C}{\|}}\overset{\diagup}{\diagdown}$$

$$\text{(b)} \quad \left[\bigcirc \right]^{\ddagger} \longrightarrow \left[\overset{C}{\underset{C}{\|}} \right]^{\ddagger} + 2\overset{\diagdown}{\diagup}C{=}C\overset{\diagup}{\diagdown}$$

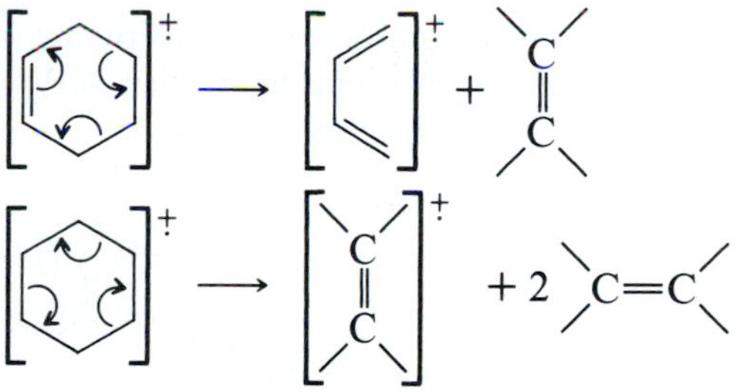

7.5.2 Rearrangements

In the mass spectra of almost all compounds, there are fragment peaks whose presence cannot be accounted for by the simple cleavage of bonds in the parent compound. These fragments are the result of rearrangement processes. Some rearrangements occur during a fragmentation process and are classified as specific rearrangements. Other rearrangements occur, leading to hydrogen and carbon atom scrambling within the cationic species, and are termed random rearrangements. Subsequent fragmentation produces ions not directly related to the parent ion.

The specific rearrangement processes often involve the migration of a hydrogen atom from one part of the parent ion to another during the fragmentation process. Such rearrangements typically involve the migration of a specific hydrogen atom that is sterically accessible to a radical cation site. (Recall that in the ionization, electrons are most easily removed from π electron systems and nonbonded pairs of electrons on heteroatoms.) Perhaps the most common rearrangement of this type is the McLafferty rearrangement involving the migration of a hydrogen atom to a π electron system via a six-membered ring transition state. This type of rearrangement is illustrated in general terms as follows:

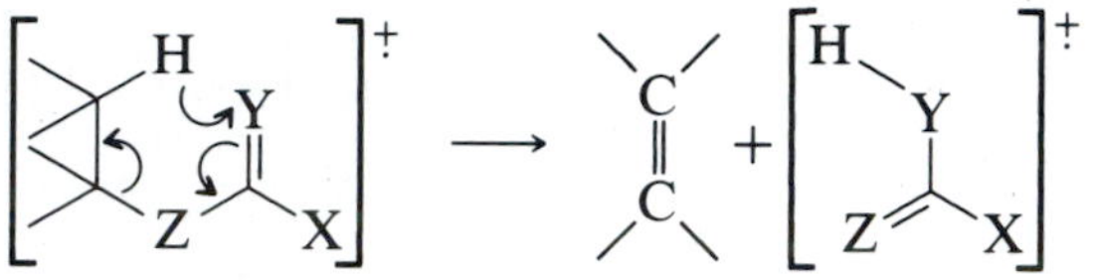

X, Y, and Z can be O, N, C, S, and any combinations thereof.

A specific example of a McLafferty rearrangement is illustrated below for the case of 2-hexanone-5,5-d_2.

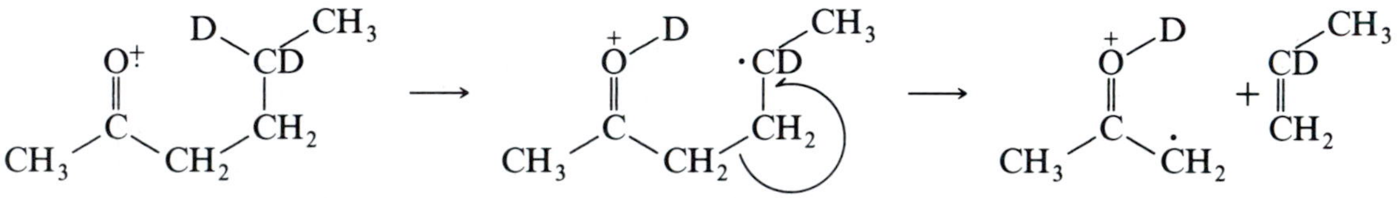

Other common rearrangements involving hydrogen atom migration include elimination reactions in which 1,4-elimination via a six-centered transition state is generally the dominant process.

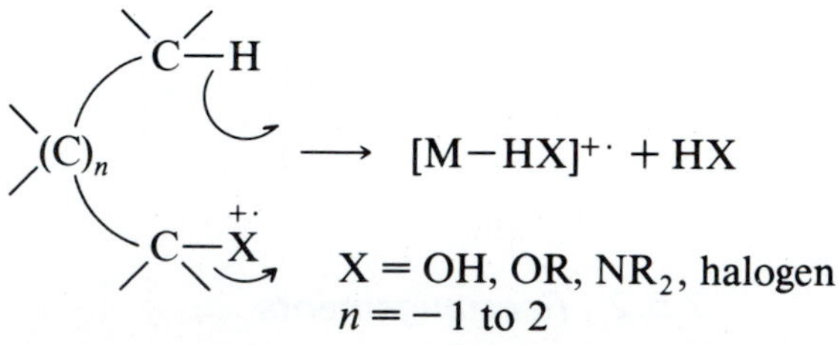

$$[M-HX]^{+\cdot} + HX$$

X = OH, OR, NR_2, halogen
$n = -1$ to 2

Similar migrations involving three-, four-, five-, and seven-centered transition states are also observed. Interestingly, a 1,4-elimination of water is most common with long chain alcohols whereas a 1,3-elimination of hydrogen halide dominates with halides.

Carbocations undergo 1,3-migrations with elimination of an alkene.

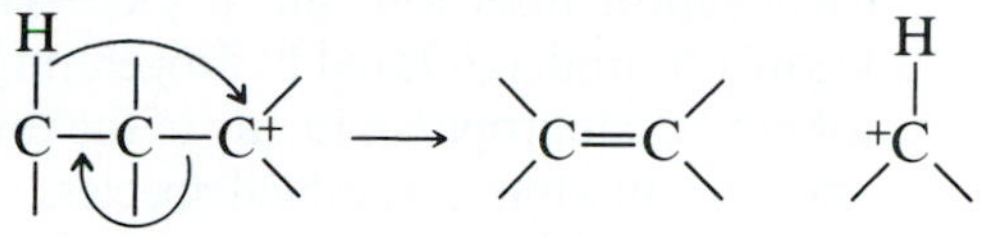

Another type of rearrangement process is that involving 1,2-shifts. As in ordinary chemical reactions, the driving force for this type of rearrangement is the formation of a more stable species.

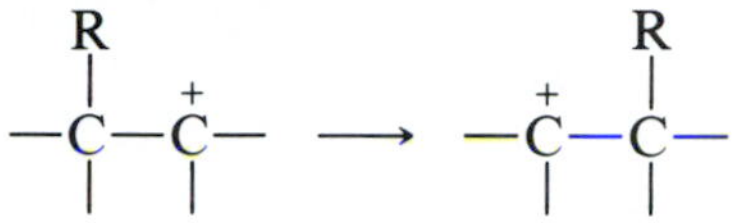

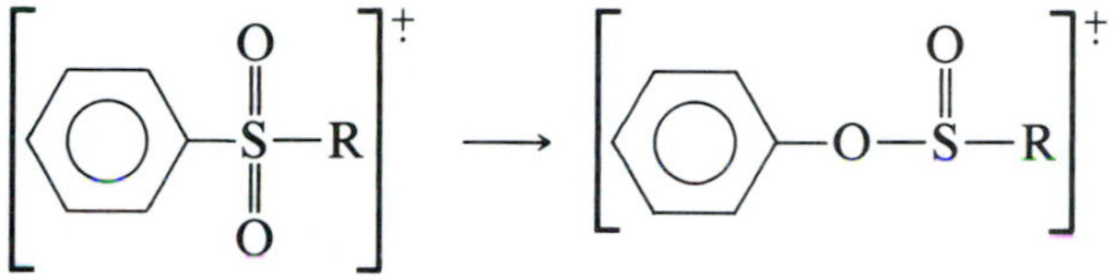

One can readily envision how 1,2-shifts might ultimately lead to fragments that seemingly have little to do with the initial structure of the compound, for example, aryl sulfones lose carbon monoxide during fragmentation. Perhaps the most extensively studied rearrangement of this type is the formation of tropylium ion from toluene. In this case it appears that ring expansion occurs in the parent ion before loss of a hydrogen atom. It has been demonstrated that in the intermediate $C_7H_8^+$ all hydrogens are equivalent. The spectra of 13 nonaromatic isomers of toluene have been studied. In all cases, the spectra are remarkably similar and, in every example, the most abundant ion is mass 91 ($C_7H_7^+$), apparently formed from a common $C_7H_8^+$ intermediate.

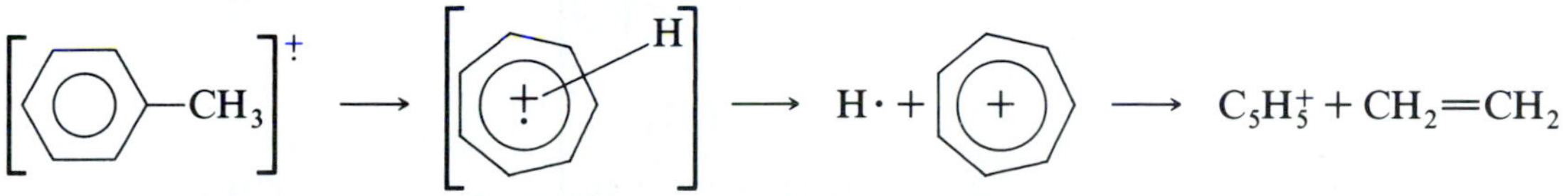

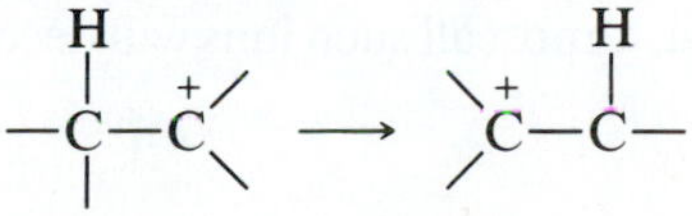

Mass spectral studies with ^{13}C and ^{2}H labeled molecules have revealed that in most cases the carbon and hydrogen atoms become essentially randomly distributed in the carbocation species formed after fragmentation of the parent radical cation, and in the succeeding carbocation species. Two processes appear to be responsible; one involving the well-known typical 1,2-hydride and alkyl shifts in carbocations (outlined above), and the other involving formation of protonated cyclopropanes by carbocation insertion into a γ-C—H bond. Protonated cyclopropanes are well known to undergo rapid hydrogen migration around the ring. Opening of the protonated cyclopropanes in any of the three possible manners results in both hydrogen and carbon atom scrambled species.

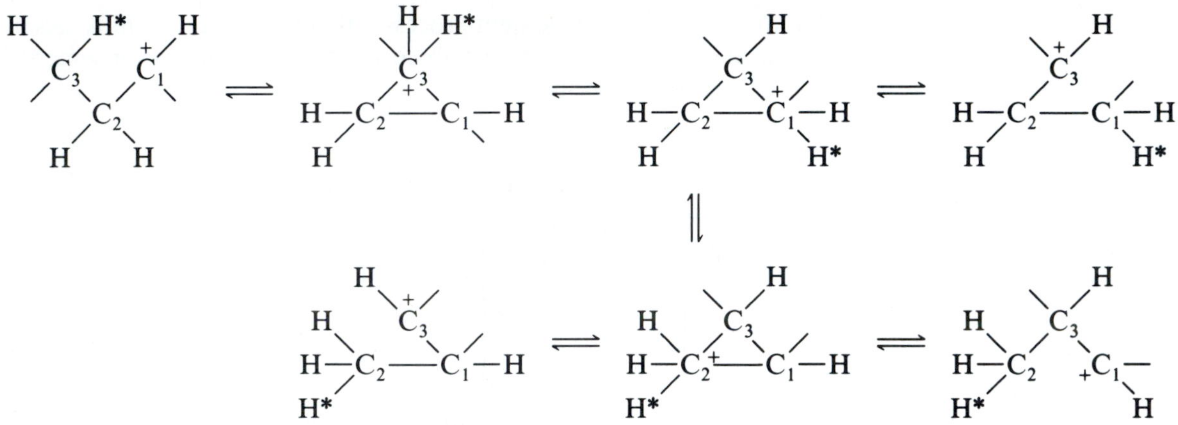

Obvious examples in which such processes must be occurring are illustrated in the following equations.

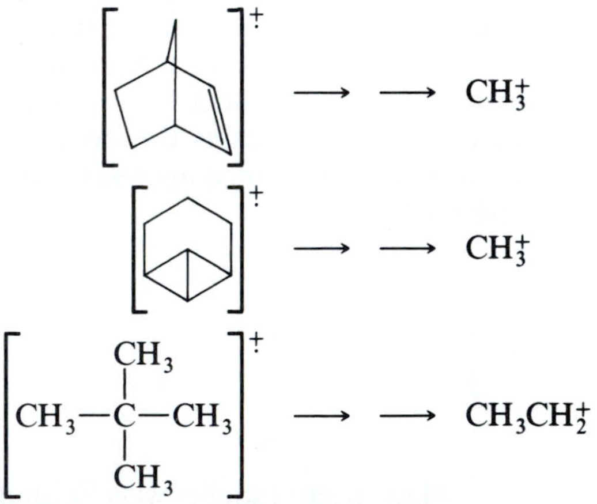

7.5.3 Metastable Peaks

If the rate of decomposition of an ion m_1 formed in the ionization chamber is very fast, almost all such ions will decompose before reaching the acceleration region, and

$$m_1^+ \longrightarrow m_2^+ + \text{neutral fragment}$$

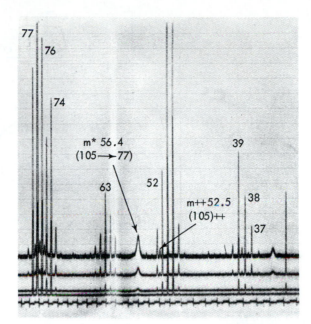

Figure 7.7   A portion of a mass spectrum traced by a four-element galvanometer illustrating normal peaks, metastable peaks, and peaks due to doubly charged ions.

only the fragments m_2 will be accelerated, deflected, and detected in any significant quantity. If the original ion (m_1) is very stable, it will be observed as an intense peak relative to any daughter ions (m_2). Ions that decompose at an intermediate rate should show considerable intensity for both the primary (m_1) and the daughter (m_2) ions. With this intermediate rate of decomposition, some of the primary ions will decompose into fragments while traveling through the accelerating region of the instrument. Such ions will at first be accelerated as mass m_1, decompose with loss of some kinetic energy to the neutral fragment, and then continue to be accelerated and deflected as mass m_2. Such an ion will be recorded as a broad peak of low intensity (generally 1% or less) at mass m^*, as shown in Equation (7.5):

$$m^* = \left[\frac{(m_2)^2}{m_1} \right] \tag{7.5}$$

The metastable peaks (m^*) are usually found at 0.1 to 0.4 mass units higher than calculated from this equation. An example of a characteristic metastable peak is shown in Figure 7.7.

   Metastable peaks are very important in the deduction of fragmentation mechanisms, for they indicate that the fragment of mass m_2 is formed in a one-step process from mass m_1. For example, in the mass spectrum of phthalic anhydride the following principal fragmentation pattern is indicated:

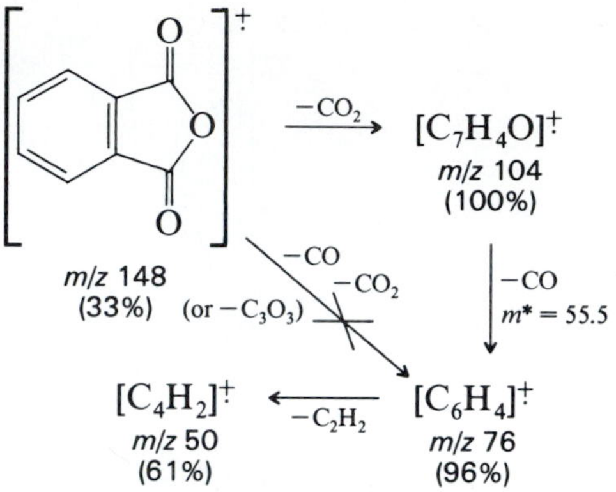

The metastable peak at 55.5 (calculated 55.5) tells us that the ion at m/z 76 (m_2) (benzyne?) is formed from the m/z 104 (m_1) ion and not by concerted loss of carbon monoxide and carbon dioxide from the molecular ion.

7.6 ISOTOPIC LABELING

Using isotopic labeling at strategically located points in a molecule and then follow-ing this label through to the products is a well-established method of studying chemi-cal and biological reaction mechanisms. For the detection of radioisotopes, various counting techniques are employed; for the detection of deuterium and sometimes ^{13}C, NMR techniques can be used (deuterium can also be measured by combustion of the sample to water and measurement of the density of the water). However, mass spectrometry is the only generally applicable method for the detection of stable isotopes. The mass spectrometric technique has the advantages of examining the intact molecule and often pinpointing the location of the label.

Isotopic labeling is also a very useful method of establishing reaction pathways that occur in the mass spectrometer. The method has special pertinence to low-resolution spectrometry, which cannot distinguish between ions of the same nominal mass as illustrated by the following: The mass spectrum of cyclopentanone exhibits a prominent peak at mass 28 that could be attributed to either carbon monoxide or ethylene. Since this fragment retains two deuterium atoms (producing a peak at m/z 30) in the spectrum of the tetradeuterio derivative, the peak is due to ethylene.

7.7 *HANDLING AND REPORTING OF DATA*

Because of space limitations in journals and books, just as in other forms of spectroscopy, it is impractical to present complete mass spectral data. Likewise, as is characteristic of other forms of spectroscopy, the methods of reporting data are considerably varied.

Data are frequently presented in tabular or linear form, in which the mass numbers and the intensities of the peaks are listed. The simplest and most frequently used method expresses the intensities in relation to the most intense peak in the spectrum. This peak, known as the *base peak*, is arbitrarily assigned a value of 100. In the experimental section of a journal the low-resolution mass spectral data on a compound would appear as follows:

EIMS m/z (relative intensity) 176 (M^+, 3), 111 (47), 93 (7), 91 (12), 82 (8), 79 (7), 77 (11), 67 (10), 66 (100), 65 (13), 55 (7), 54 (6), 53 (7), 51 (7).

In many instances, especially with high-resolution mass spectra, only the molecular ion or quasimolecular ion is reported, for example,

HRMS (CI, M + H) calcd for $C_{14}H_{24}O_4(H)$ 257.1750, found 257.1756.

The observation of a molecular ion in HRMS with m/z in agreement ($\pm$ 10 to 20 ppm) to that calculated is a generally accepted criterion for the establishment of the composition of a new compound (isolated or synthesized). It should be noted, however, that the detection of an ion with the expected composition has no bearing

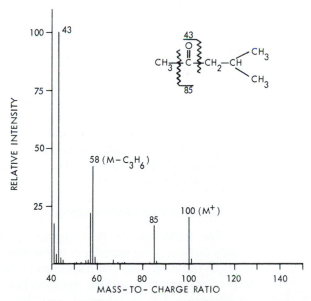

Figure 7.8 Mass spectrum of an aliphatic ketone.

on the purity of the sample under investigation other than demonstrating that there are no impurities detected with higher molecular weight!

Chemists, especially organic chemists, immediately associate the term spectrum with a picture with lines on it. Mass spectral data are often presented in the form of a bar graph, the ordinate indicating the relative abundances (intensity) as percent of the base peak and the abscissa indicating the mass-to-charge ratios. This method of presentation has the advantage that the overall characteristics of the spectrum are available at a glance. An example of a typical bar graph mass spectrum is shown in Figure 7.8.

7.8   MASS SPECTRA OF SOME REPRESENTATIVE COMPOUNDS

The important fragmentation patterns for a number of compounds representative of the most common chemical classes are described in this section. For a more comprehensive treatment, the general references cited at the end of this chapter should be consulted. *Mass Spectrometry of Organic Compounds* by Budzikiewicz, Djerassi, and Williams is a very thorough and suitable treatment for chemists interested in organic structure determination. In many cases, the structural formulas shown in this section for mass spectral fragments should be considered only as probable representations of the observed elemental compositions of the fragments.

7.8.1   Hydrocarbons

Saturated Hydrocarbons

The utility of mass spectra of hydrocarbons is greatly enhanced by the availability of a large number of reference spectra. Straight-chain hydrocarbons exhibit clusters of peaks, 14 mass units (CH_2) apart, of decreasing intensity with increasing fragment weight (see Figure 7.9). The molecular ion peak is almost always present, albeit of low intensity. Branching causes a preferential cleavage at the point of the branch, resulting in the formation of a secondary carbocation, as shown in Figure 7.10. Low-intensity ions from random rearrangements are common.

Saturated cyclic hydrocarbons give rise to complex spectra. The molecular ion peak is usually rather intense. Characteristic peaks are usually formed from the loss of ethylene and side chains.

Unsaturated Hydrocarbons

The molecular ion peak, apparently formed by removal of a π electron, is usually distinct. The prominent peaks from monoalkenes have the general formula C_nH_{2n-1} (allyl cations), but fragments of the composition C_nH_{2n} formed in McLafferty rearrangements are also common (see Figure 7.11).

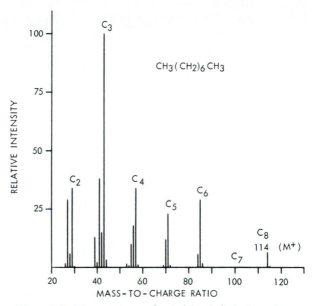

Figure 7.9 Mass spectrum of a straight-chain hydrocarbon.

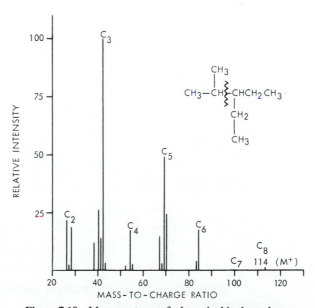

Figure 7.10 Mass spectrum of a branched hydrocarbon.

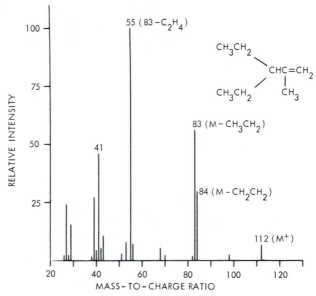

Figure 7.11   Mass spectrum of a typical alkene.

Aromatic Hydrocarbons

Hydrocarbons containing an aromatic ring usually give readily interpretable mass spectra, which exhibit strong molecular ion peaks. The molecular ion of benzene $[C_6H_6]^+$ accounts for 43% of the total ion current at 70 eV in the mass spectrum of benzene. The most characteristic cleavage of the alkyl benzenes occurs at the bonds beta to the aromatic ring. In the usual case, substituents are preferentially lost resulting in the formation of tropylium or substituted tropylium ions.

If side chains of propyl or larger are present, McLafferty rearrangements may also occur (see Figure 7.12).

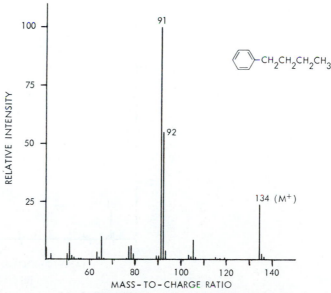

7.8.2 Halides

Iodine and fluorine are monoisotopic, whereas chlorine and bromine occur naturally as mixtures of two principal isotopes; hence a molecular ion or fragment ion containing a chlorine will show an M + 2 isotope peak amounting to about a third of the intensity, whereas a bromine-containing ion will be accompanied by an M + 2 peak of almost equal intensity (see Figures 7.6 and 7.13). For organic halides, the abundance of the molecular ion in a given series of compounds increases in the order F < Cl < Br < I. The intensity of the molecular ion peak decreases with increasing size of the alkyl group and α-branching. The following fragmentation processes are listed in approximately decreasing order of importance.

1. $R—X^{\ddagger} \rightarrow R^{+} + X \cdot$. Most important for X = I or Br.
2. $R—CH_2CH_2X^{\ddagger} \rightarrow [RCH{=}CH_2]^{\ddagger} + HX$. More important for X = Cl and F than for I or Br.

Figure 7.12   Mass spectrum of an aromatic hydrocarbon.

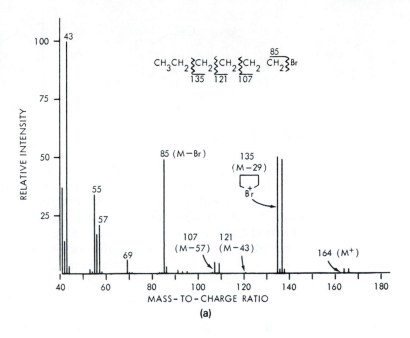

(a)

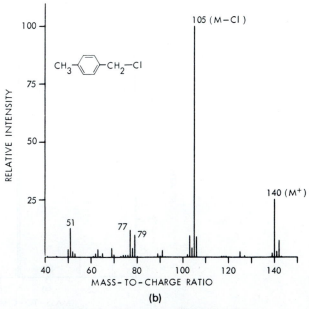

(b)

Figure 7.13 Mass spectra of organic halides.

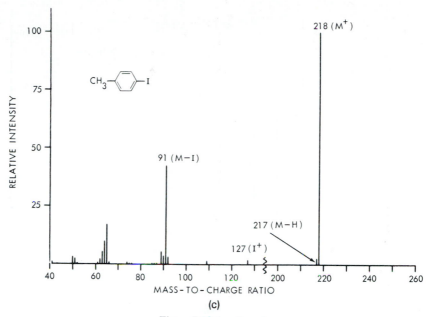

Figure 7.13 *(continued)*

3. $R-CH_2X^+ \rightarrow R\cdot + CH_2{=}X^+$. Heaviest group preferentially lost.

4.

$$X = Br \text{ or } Cl$$

The molecular ion peak of aromatic halides is usually abundant. An $M - X$ peak is usually observed (see Figure 7.13c).

7.8.3 Alcohols and Phenols

The molecular ion peak of primary and secondary alcohols is usually weak; that of tertiary alcohols is usually not detectable. The most important general fragmentation process involves cleavage of the bond beta to the oxygen atom. The largest group is lost most readily.

$$R_2-\underset{\underset{R_3}{|}}{\overset{\overset{R_1}{|}}{C}}-\overset{+\cdot}{O}H \longrightarrow R_1^\cdot + \underset{R_3}{\overset{R_2}{\diagdown}}C{=}\overset{+}{O}H$$

Loss of water generally occurs by a cyclic mechanism where $n = 2$ is the dominant process.

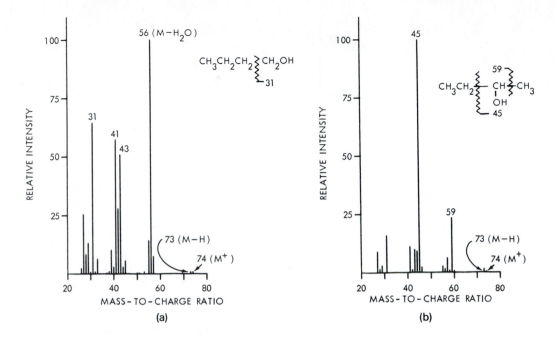

(a)

(b)

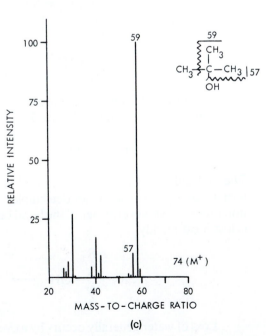

(c)

Figure 7.14 Mass spectra of isomeric butanols.

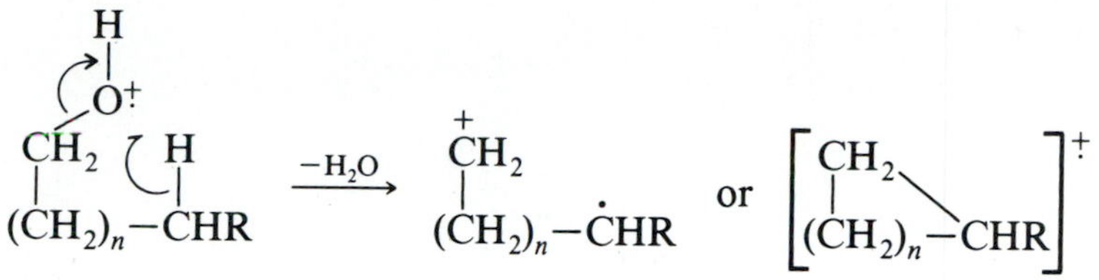

In addition a concurrent elimination of water and an alkene is usually observed with alcohols of four or longer carbon chains.

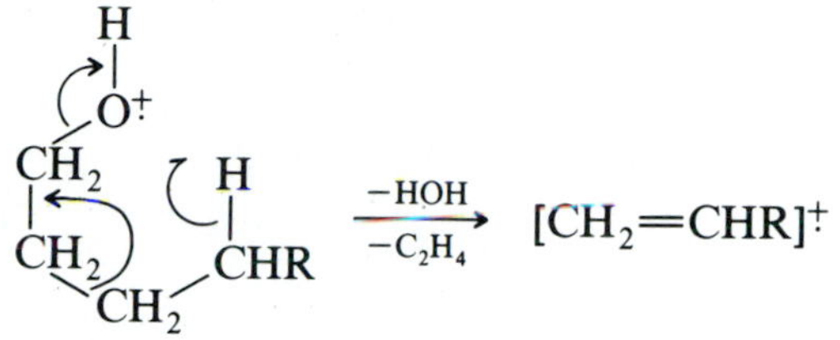

Examples of mass spectra of alcohols are shown in Figure 7.14. The fragmentation of a typical cyclic alcohol is illustrated below.

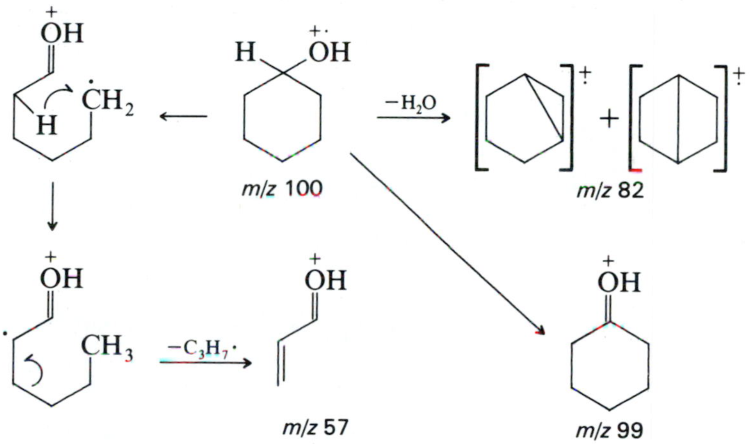

Benzylic alcohols typically exhibit intense molecular ion peaks, as well as M − 17 (loss of ·OH) peaks and rearrangement fragment peaks.

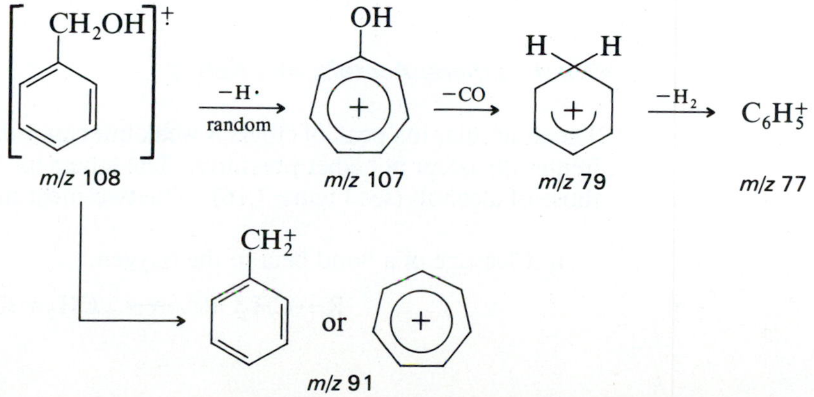

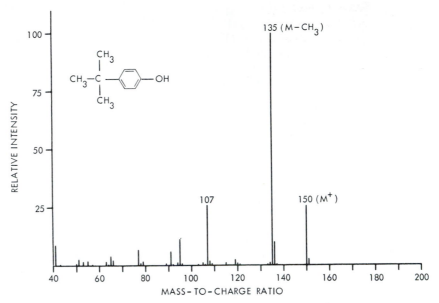

Figure 7.15 Mass spectrum of *p-t*-butylphenol.

Phenols also exhibit intense molecular ion peaks as illustrated in Figure 7.15.

Trimethylsilyl ethers of alcohols are widely employed in mass spectrometry and gas liquid chromatography because of their higher volatility compared with the parent alcohol. Although the molecular ion peak of these ethers is often of low intensity, the molecular weight can usually be inferred from the strong $M - CH_3$ peak.

$$CH_2=\overset{+}{O}Si(CH_3)_3 \quad \overset{-R\cdot}{\longleftarrow} \quad RCH_2\overset{+\cdot}{O}Si(CH_3)_3 \quad \overset{-CH_3\cdot}{\longrightarrow} \quad RCH_2\overset{+}{O}Si(CH_3)_2$$

$m/z\ 93$ $\hspace{9cm}$ $M - 15$

$$(CH_3)_3Si^+ \hspace{3cm} (CH_3)_3SiO^+ \hspace{3cm} HO-\overset{+}{Si}(CH_3)_2$$

$m/z\ 73$ $\hspace{3cm}$ $m/z\ 89$ $\hspace{3cm}$ $m/z\ 75$

7.8.4 Ethers, Acetals, and Ketals

The molecular ion peak of ethers is weak but can usually be observed. $M + 1$ peaks frequently occur at higher pressures. The ethers undergo fragmentations similar to those of alcohols (see Figure 7.16). The two most important pathways are:

1. Cleavage of a bond beta to the oxygen.

$$R-CH_2\overset{+\cdot}{O}R \quad \longrightarrow \quad CH_2=\overset{+}{O}R + R\cdot$$

In the case of branching at the α-carbon, the largest R group is lost preferentially.

2. Cleavage of a C—O bond.

$$R-\overset{+\cdot}{O}-R \longrightarrow RO\cdot + R^+$$

Acetals and ketals behave in a similar fashion.

Important fragmentations of aromatic ethers are illustrated below and in Figure 7.17.

7.8.5 Aldehydes

Aliphatic aldehydes generally exhibit weak molecular ion peaks. α-Cleavage occurs extensively to produce M − 1 and M − 29 peaks (see Figure 7.18).

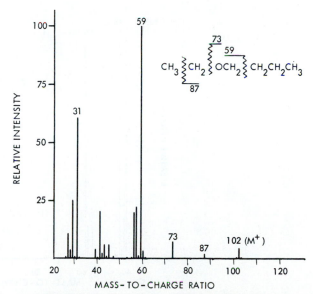

Figure 7.16 Mass spectrum of butyl ethyl ether.

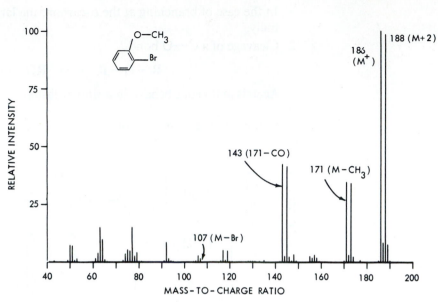

Figure 7.17 Mass spectrum of an aryl ether.

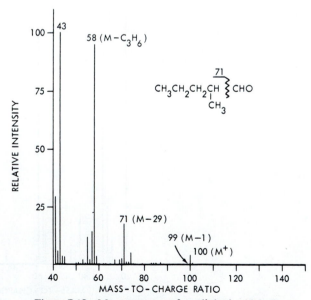

Figure 7.18 Mass spectrum of an aliphatic aldehyde.

α-Cleavage $\quad [R-CHO]^{\ddot{+}} \quad \longrightarrow \quad R-\overset{+}{C}{=}O + H\cdot$

$\qquad\qquad\quad [R-CH_2-CHO]^{\ddot{+}} \quad \longrightarrow \quad R-CH_2^+ + H\dot{C}O$

β-Cleavage $\quad [R-CH_2-CHO]^{\ddot{+}} \quad \longrightarrow \quad R^+ + H_2C{=}CH-O\cdot$

McLafferty rearrangements generally result in the formation of the dominant fragment peaks.

$$\begin{array}{c}\text{structure}\end{array} \longrightarrow \left[\begin{array}{c}\text{structure}\end{array}\right]^{\ddot{+}} + CH_2{=}CH-OH$$

M − 44

Aromatic aldehydes typically exhibit intense molecular ion and M − 1 peaks.

$$\left[\text{structure}\right]^{\ddot{+}} \xrightarrow{-H\cdot} \text{structure}-C{\equiv}\overset{+}{O} \xrightarrow{-CO} \text{structure}^+$$

7.8.6 Ketones

The molecular ion of aliphatic ketones is generally intense. Fragmentation pathways are similar to those of aldehydes (see Figure 7.19). The peak arising from α-cleavage

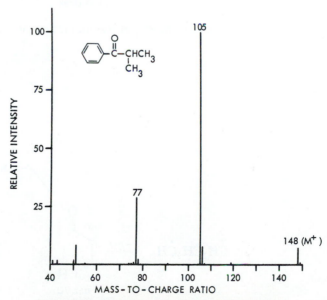

Figure 7.19 Mass spectrum of an aromatic ketone.

with loss of the larger alkyl group is more intense than that from loss of the smaller one (see Figure 7.8). Thus the base peak for most methyl ketones is m/z 43 (CH_3CO^+) (intense peaks may also occur at m/z 43 with propyl or isopropyl ketones representing loss of $C_3H_7^+$ from the molecule ion).

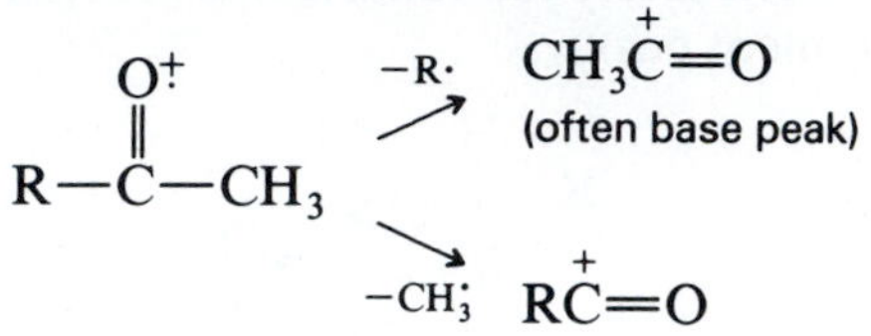

With ketones containing γ-hydrogens, McLafferty rearrangements (see Section 7.5.2) are prevalent.

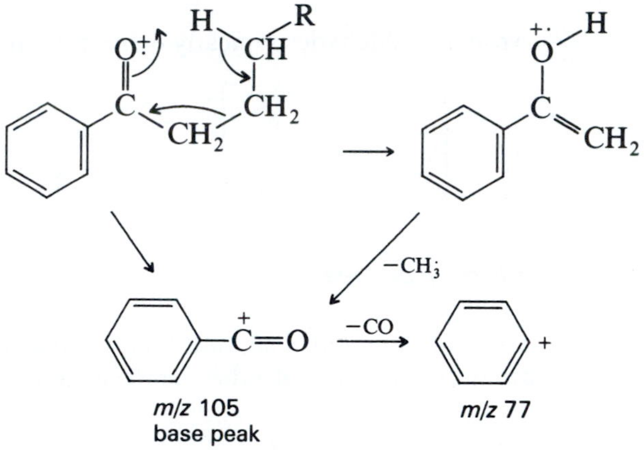

m/z 105
base peak

m/z 77

Important modes of fragmentation of cyclic ketones are illustrated below.

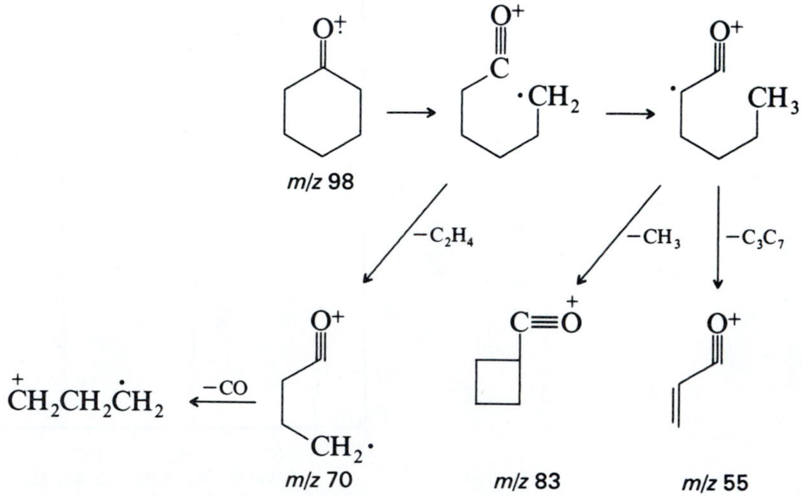

m/z 98

m/z 70

m/z 83

m/z 55

7.8.7 Carboxylic Acids

Acids are more frequently and often more conveniently examined as their methyl esters. The free acids exhibit weak but discernible molecular ion peaks. Normal aliphatic acids in which a γ-hydrogen is available for transfer have their base peak at m/z 60.

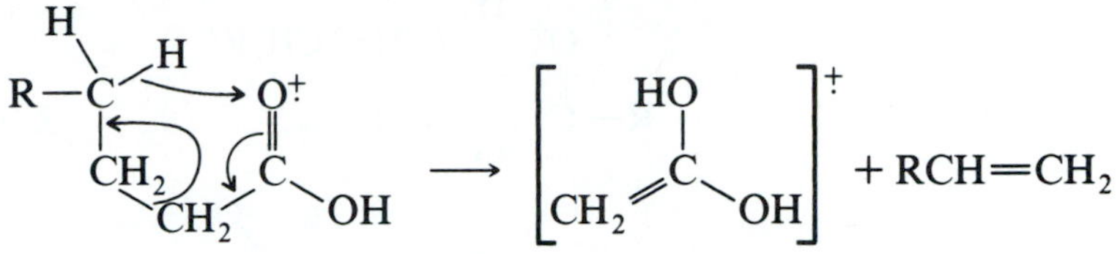

$$
\begin{array}{c}
\text{H} \\
\text{R}-\text{C} \cdots \rightarrow \text{O}^+ \\
\text{CH}_2 \quad \text{C} \\
\text{CH}_2 \quad \text{OH}
\end{array}
\longrightarrow
\left[
\begin{array}{c}
\text{HO} \\
\text{C} \\
\text{CH}_2 \quad \text{OH}
\end{array}
\right]^{\ddagger}
+ \text{RCH}=\text{CH}_2
$$

m/z 60

Both aliphatic and aromatic acids exhibit a peak due to [COOH] at m/z 45.

$$
[\text{R}-\text{COOH}]^{\ddagger} \xrightarrow{-\text{R}\cdot} {}^+\text{COOH}
$$

m/z 45

In the spectra of aromatic acids, the M − OH peak is prominent.

$$
\left[
\begin{array}{c}
\text{O} \\
\text{C} \\
\text{OH}
\end{array}
\right]^{\ddagger}
\xrightarrow{-\text{OH}\cdot}
\text{C}\equiv\text{O}^+
\xrightarrow{-\text{CO}}
+
$$

m/z 105 　　　　　 m/z 77

7.8.8 Esters and Lactones

Methyl Esters

The molecular ion is usually discernible. Fission of bonds adjacent to the carbonyl group may occur to yield four ions.

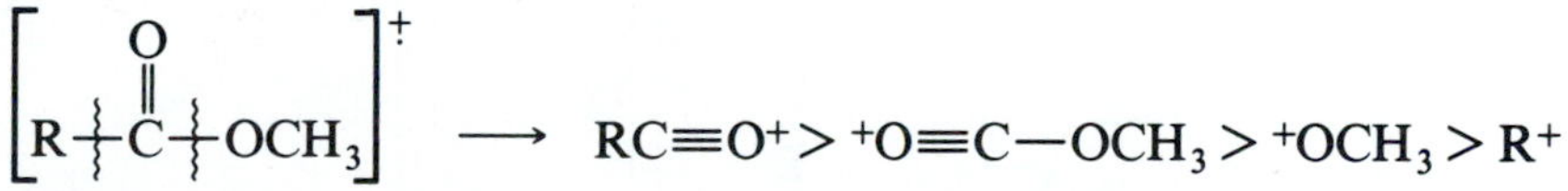

$$
\left[
\begin{array}{c}
\text{O} \\
\| \\
\text{R}\!+\!\text{C}\!+\!\text{OCH}_3
\end{array}
\right]^{\ddagger}
\longrightarrow
\text{RC}\equiv\text{O}^+ > {}^+\text{O}\equiv\text{C}-\text{OCH}_3 > {}^+\text{OCH}_3 > \text{R}^+
$$

The McLafferty rearrangement is a most important process in the longer chain methyl esters. The base peak in the mass spectrum of methyl esters of C_6 to C_{26} fatty acids occurs at m/z 74.

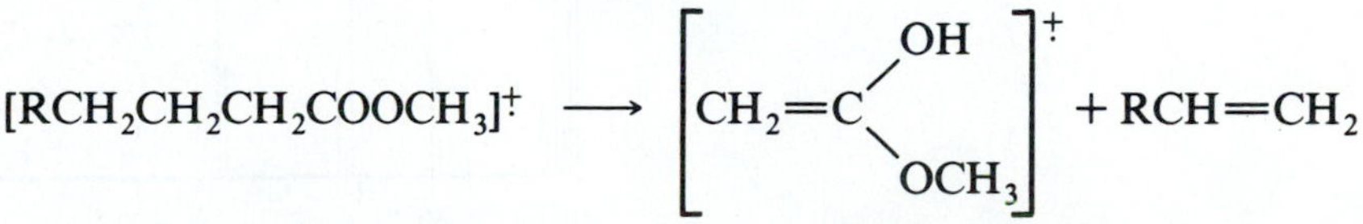

$$
[\text{RCH}_2\text{CH}_2\text{CH}_2\text{COOCH}_3]^{\ddagger}
\longrightarrow
\left[
\begin{array}{c}
\text{OH} \\
\text{CH}_2=\text{C} \\
\text{OCH}_3
\end{array}
\right]^{\ddagger}
+ \text{RCH}=\text{CH}_2
$$

m/z 74

Higher Esters

The higher esters undergo fragmentations similar to those of the methyl esters, but the overall spectrum is complicated by additional fragmentation involving the alkoxy group (see Figure 7.20). Such esters usually exhibit a peak corresponding to a protonated acid that may arise by the following pathway.

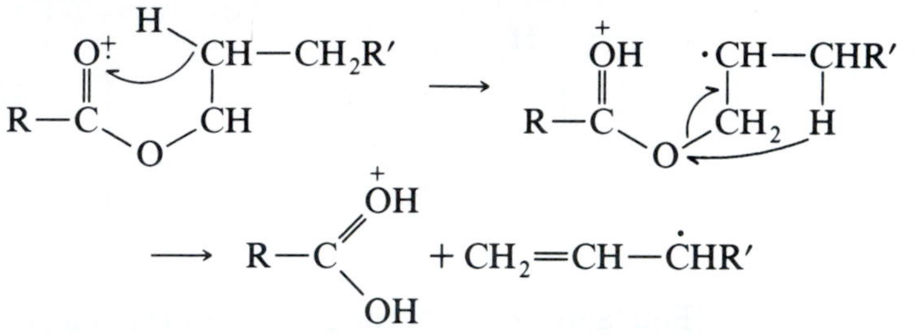

Rearrangement ions occur at m/z 60 and 61, corresponding to acetic acid and protonated acetic acid in the spectrum of ethyl or higher esters of butyric or higher acids.

Benzyl esters fragment with the loss of ketene.

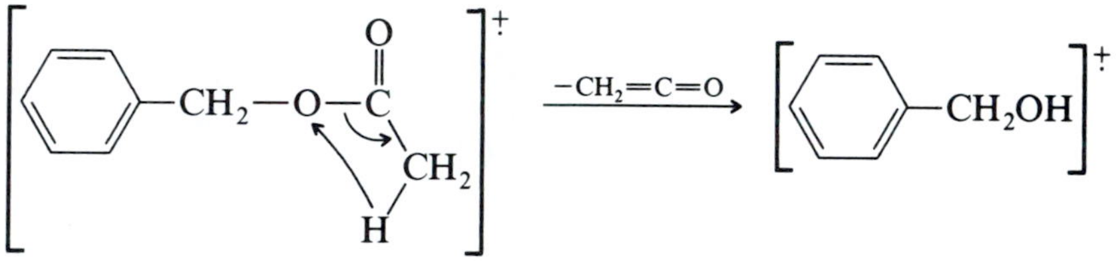

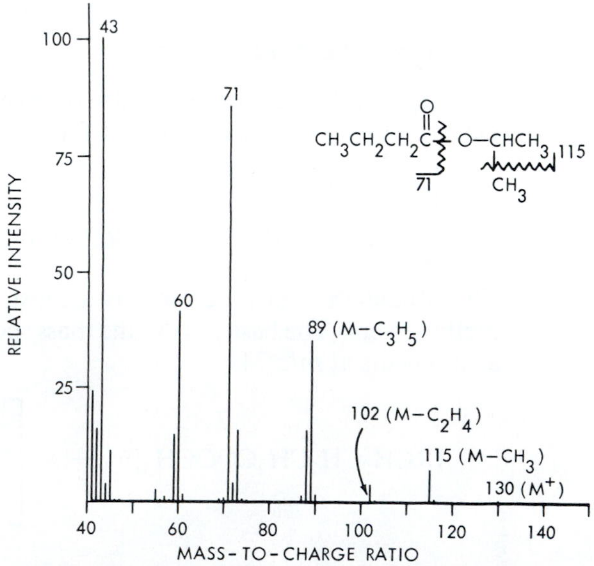

Figure 7.20 Mass spectrum of isopropyl benzoate.

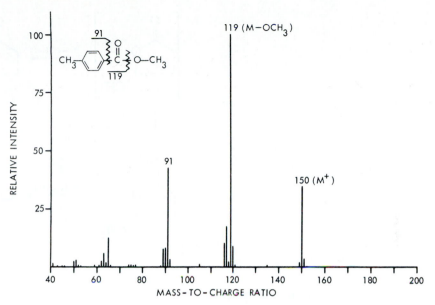

Figure 7.21   Mass spectrum of methyl *p*-toluate.

The loss of the alkoxy group is a very important fragmentation process for alkyl benzoates (see Figure 7.21).

$$\left[\underset{\overset{|}{OR}}{\overset{O}{\underset{\|}{C}}}\right]^{\overset{+}{\cdot}} \xrightarrow{-RO\cdot} C{\equiv}\overset{+}{O} \longrightarrow + + CO$$

It is typical that *ortho* groups exert a marked influence on fragmentation processes.

$$\left[\overset{O}{\underset{\overset{|}{CH_2}}{\underset{\|}{C}}}\overset{}{\underset{}{O{-}R}}\overset{H}{}\right]^{\overset{+}{\cdot}} \xrightarrow{-ROH} \left[\underset{CH_2}{}\overset{}{C{=}\overset{}{O}}\right]^{\overset{+}{\cdot}}$$

Some of the important ions recorded in the spectrum of γ-valerolactone are illustrated below.

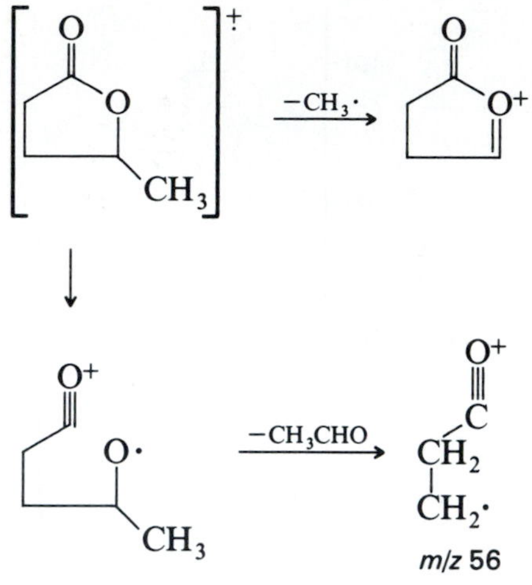

7.8.9 Amines

Amines and other nitrogen compounds containing an odd number of nitrogen atoms have an odd-numbered molecular weight.

Aliphatic and Aromatic Amines

The molecular ion peak of aliphatic amines is weak to absent. The most intense peak results from β-cleavage (see below and in Figure 7.22).

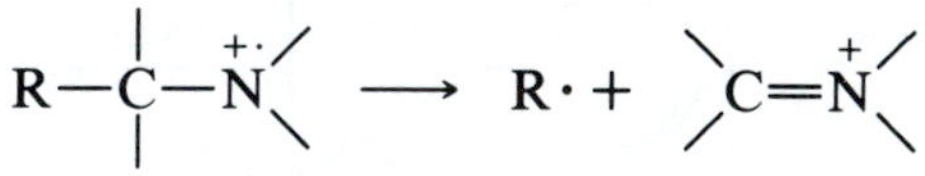

Again the largest alkyl group is preferentially lost. The base peak for all primary amines unbranched at the α-carbon is at m/z 30 [$CH_2=\overset{+}{N}H_2$]. The presence of this peak is good but not conclusive evidence for a primary amine since this fragment may arise from sequential fragmentation from secondary and tertiary amines. Cyclic fragments apparently occur during the fragmentation of longer chain amines.

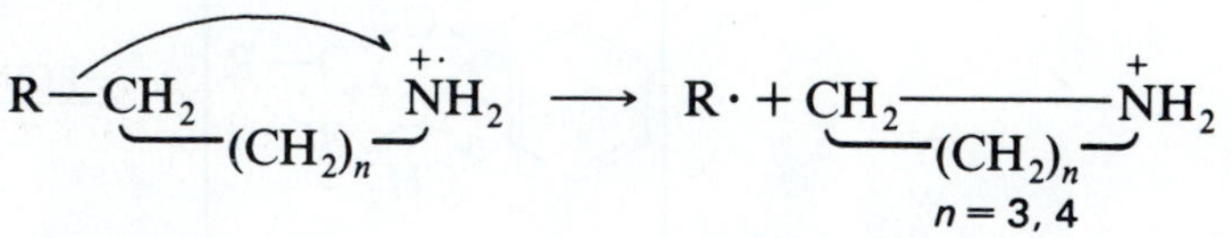

Cyclic amines give rise to strong molecular ion peaks; primary fragmentation occurs at α-carbons.

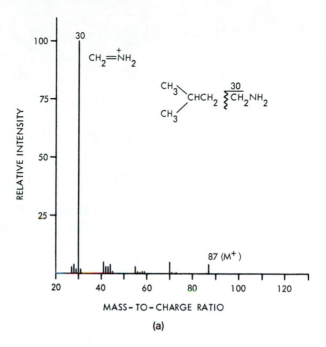

(a)

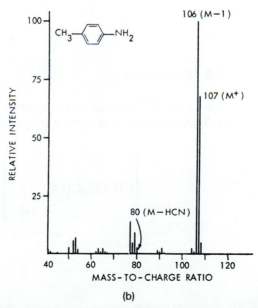

(b)

Figure 7.22   Mass spectra of (a) an aliphatic and (b) an aromatic amine.

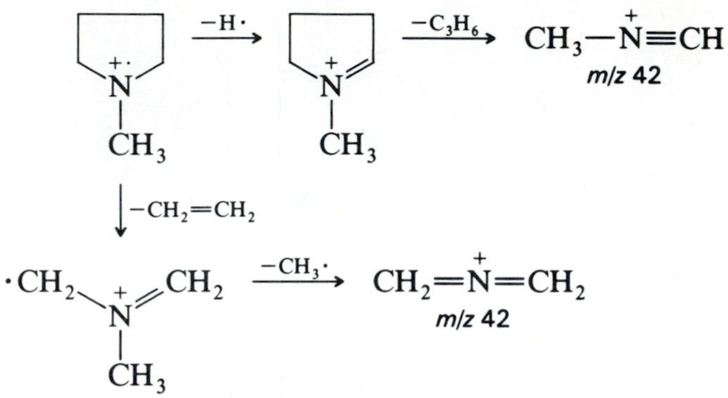

$$\text{(pyrrolidine)} \xrightarrow{-H\cdot} \text{(dihydropyrrole)} \xrightarrow{-C_3H_6} CH_3-\overset{+}{N}\equiv CH$$
$$m/z\ 42$$

$$\downarrow -CH_2=CH_2$$

$$\cdot CH_2 \diagdown \overset{+}{\underset{CH_3}{N}} \diagup CH_2 \xrightarrow{-CH_3\cdot} CH_2=\overset{+}{N}=CH_2$$
$$m/z\ 42$$

Aromatic amines give very intense molecular ion peaks accompanied by a moderate M − 1 peak (see below and in Figure 7.22).

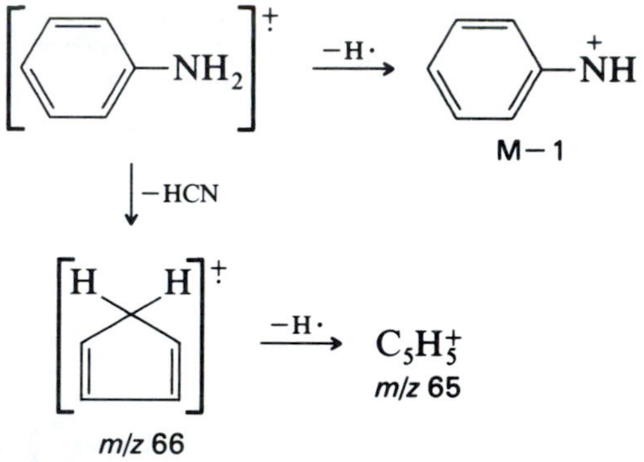

$$\left[\bigcirc\!\!-NH_2\right]^{\dagger} \xrightarrow{-H\cdot} \bigcirc\!\!-\overset{+}{N}H$$
$$M-1$$

$$\downarrow -HCN$$

$$\left[\overset{H\quad H}{\bigcirc}\right]^{\dagger} \xrightarrow{-H\cdot} C_5H_5^{+}$$
$$m/z\ 66 \qquad\qquad m/z\ 65$$

7.8.10 Amides

The molecular ion peak is usually observed. McLafferty rearrangements are important. A strong peak at *m/z* 44 is indicative of a primary amide.

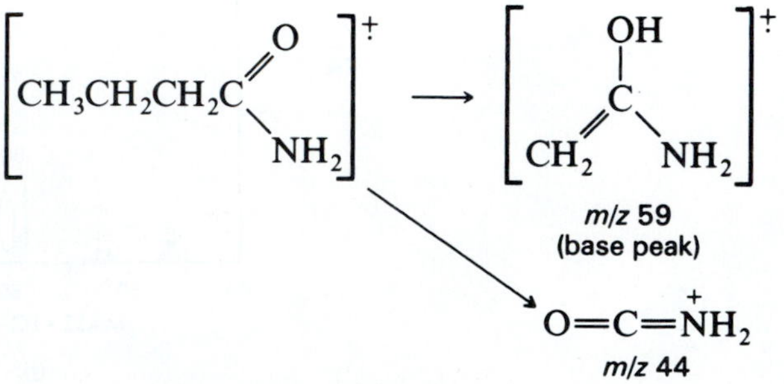

$$\left[CH_3CH_2CH_2C\!\!\diagup^{O}_{\diagdown NH_2}\right]^{\dagger} \longrightarrow \left[\overset{OH}{\underset{CH_2\quad NH_2}{C}}\right]^{\dagger}$$
$$m/z\ 59$$
$$\text{(base peak)}$$

$$O=C=\overset{+}{N}H_2$$
$$m/z\ 44$$

7.8.11 Nitriles

The molecular ion of aliphatic nitriles is usually not observed. At higher pressures an M + 1 ion may appear. A relatively weak but useful M − 1 ion due to R—CH=C=N$^+$ is often found. The base peak generally results from a McLafferty rearrangement and is found at m/z 41 in the spectrum of C_4 to C_9 straight-chain nitriles.

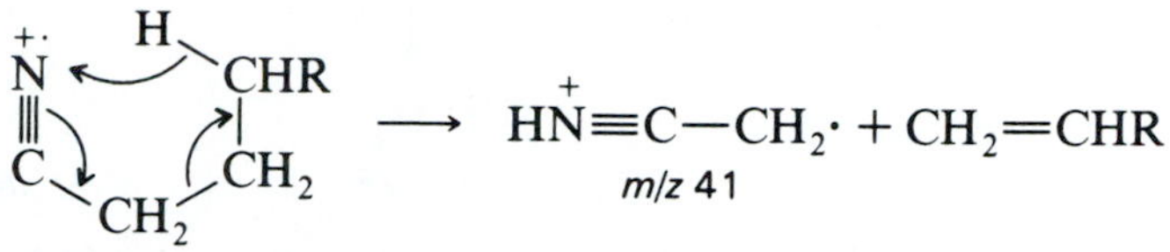

$$HN{\equiv}C-CH_2{\cdot} + CH_2{=}CHR$$
$$m/z\ 41$$

7.8.12 Nitro Compounds

The molecular ion peak of aliphatic nitro compounds is seldom observed. The spectra of such compounds are composed mainly of hydrocarbon ions, with observable peaks containing nitrogen occurring at m/z 30 (NO) and 46 (NO_2). On the other hand, nitrobenzene exhibits an intense molecular ion. A rearrangement reaction also occurs to give $C_6H_5O^+$.

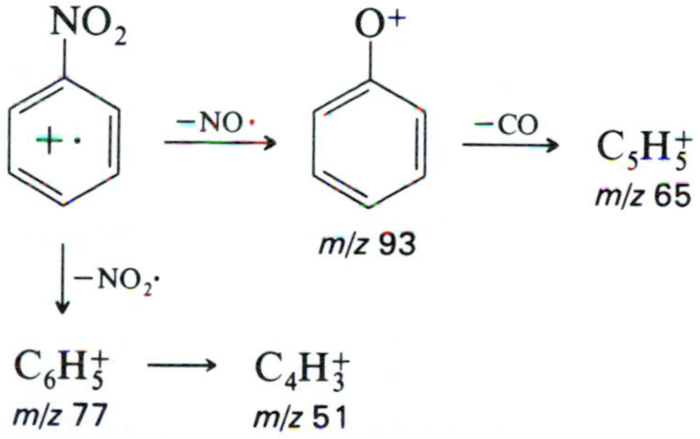

$$\xrightarrow{-NO\cdot}\quad m/z\ 93 \quad\xrightarrow{-CO}\quad C_5H_5^+$$
$$m/z\ 65$$

$$\downarrow{-NO_2{\cdot}}$$

$$C_6H_5^+ \longrightarrow C_4H_3^+$$
$$m/z\ 77 \qquad m/z\ 51$$

7.8.13 Sulfur Compounds

The fragmentation of thiols and sulfides parallels those of alcohols and ethers. In each case, the molecular ion peak of the sulfur compounds is more intense. Cyclic ions of the type

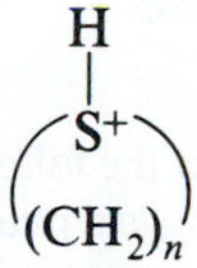

have been postulated in a number of cases.

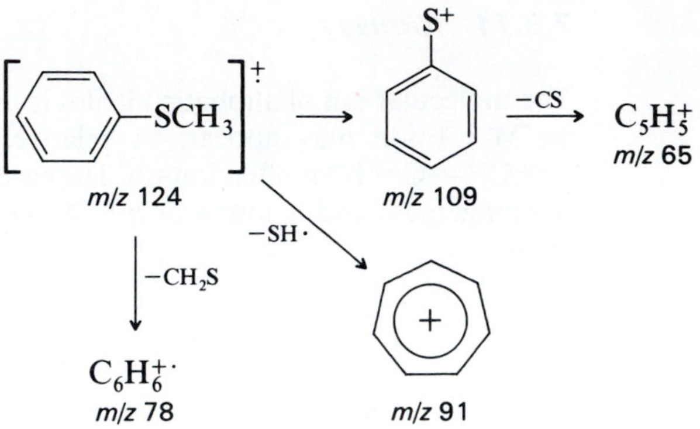

$$\left[\underset{m/z\ 124}{\underline{}} - SCH_3 \right]^{\ddagger} \longrightarrow \underset{m/z\ 109}{\overset{S^+}{\underline{}}} \xrightarrow{-CS} \underset{m/z\ 65}{C_5H_5^+}$$

$-SH\cdot$

$-CH_2S$

$$\underset{m/z\ 78}{C_6H_6^{+\cdot}}$$

$$\underset{m/z\ 91}{(+)}$$

The principal fragments from disulfides are produced by elimination of alkenes.

$$\underset{m/z\ 122}{[CH_3CH_2-S-S-CH_2CH_3]^{\ddagger}} \xrightarrow{-CH_2=CH_2} \underset{m/z\ 94}{[CH_3CH_2SSH]^+}$$

$$\Big\downarrow -CH_2=CH_2$$

$$\underset{m/z\ 66}{[HSSH]^+}$$

The principal fragmentation of a typical aliphatic sulfoxide is shown below.

$$\underset{m/z}{\overset{O}{\underset{\|}{[CH_3CH_2CH_2SCH_2CH_2CH_3]^+}}} \xrightarrow{-C_3H_6} [CH_3CH_2CH_2SOH]^{\ddagger}$$

$\Big\downarrow -OH\cdot$ $\Big\downarrow -CH_3CH_2\cdot$

$$\underset{m/z\ 117}{CH_3CH_2CH_2-\overset{+}{S}=CHCH_2CH_3}$$ $$\underset{m/z\ 63}{CH_2=\overset{+}{S}-OH}$$

Aromatic sulfoxides and sulfones are interesting in that apparently an electron impact-induced aryl migration from sulfur to oxygen occurs.

7.9 MASS SPECTRAL PROBLEMS

Identify each of the following unknowns on the basis of the mass spectral and other structural data given. Suggest plausible explanations for the mass spectral patterns observed.

1. Compound **A** exhibits an intense infrared band at 1715 cm^{-1}. The important ions appearing in the mass spectrum of **A** are listed below:

m/z	Percent Base Peak	m/z	Percent Base Peak
41	44	85	66
42	8	100	12
43	10	113	6
55	10	142	12
57	100		
58	58	metastable peak at 38.3	
71	6		

Compound **A** reacts with semicarbazide to form a single derivative, mp 89–90°C.

2. The infrared spectrum of compound **B** displays strong bands at 1690 and 826 cm^{-1}. The NMR spectrum consists of a pair of AB doublets near δ 7.60 (4) and a singlet at δ 2.45 (3). The peaks at highest m/z in the mass spectrum of **B** are found at 198 (26), 199 (4), 200 (25), 201 (3), and the base peak is found at 183 (100).

3. Extraction of a bacterial culture with dichloromethane gave a compound that readily lost carbon dioxide upon heating. The compound obtained after loss of carbon dioxide gave, upon treatment with iodine and sodium hydroxide followed by neutralization, hexadecanoic acid and a yellow solid. Treatment of the original compound with ethanol and dry HCl at room temperature for several days resulted in compound **C**. The composition of selected ions from HRMS data on compound **C** is shown below.

Compound **C**

m/z	Intensity	Composition* CH	Composition* CHO	Composition* CHO$_2$	Composition* CHO$_3$
327	2				20/39
326	2				20/38
239	15		16/31		
196	2	14/28			
143	23				7/11
131	25				6/11
130	100				6/10
115	6				5/7
102	5				4/6
97	11	7/13	6/9	5/5	
88	20			4/9	
71	17	5/11	4/7		

* Data in each column give the C/H count; for example, 20/39 refers to an ion of composition $C_{20}H_{39}O_3$.

4. Compound **D**, mp 115–116°C, was found to be insoluble in water but soluble in dilute sodium hydroxide. The mass spectrum of **D** is reproduced in Figure 7.23.

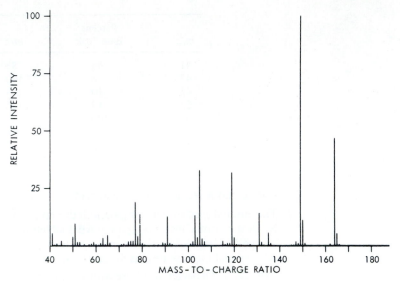

Figure 7.23   Mass spectrum of unknown **D**, problem 4.

5. Compound **E** exhibited two singlets only in its ^{1}H NMR. The EI mass spectrum of **E** is shown in Figure 7.24. HRMS (CI) revealed an M + 1 ion at 115.0758.

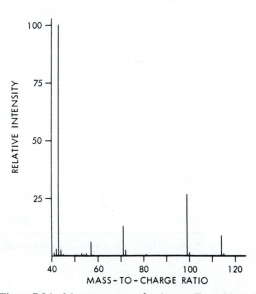

Figure 7.24   Mass spectrum of unknown **E**, problem 5.

6. The infrared spectrum of compound **F** is devoid of absorption in the 1700 cm^{-1} region; mass spectrum (70e) m/z (rel. intensity) 27 (26), 29 (39), 31 (18), 45 (100), 47 (16), 61 (6), 73 (44), 75 (8), 89 (3), 103 (15), 117 (2).

Anal. Found: C, 61.10; H, 11.9; N, 0.00.

Compound **F**, bp 102°C, upon treatment with 2,4-dinitrophenylhydrazine reagent, forms a yellow precipitate, mp 166–168°C.

7. The infrared spectrum of compound **G** contains a carbonyl band at 1830 cm^{-1}. The NMR spectrum consists of a singlet at δ 7.34. The ^{13}C NMR consists of four lines all in the δ 100–200 region. Important ions in the mass spectrum of **G** are indicated below.

m/z	Percent Base Peak
63	48
64	100
92	56.5
136	75.5
137	6.1
138	0.8

Metastable peaks are found at 44.5 and 62.6.

8. Extraction of the leaves of *Nepeta cataria Linneé,* otherwise known as catnip, with methylene chloride followed by chromatography of the crude extract over alumina, employing carbon tetrachloride-acetone as the eluent, resulted in the isolation of an oil, compound **H**, λ_{max}^{EtOH} 240 nm.

Mass Spectral Data for Compound **H**

m/z	Percent Base Peak	m/z	Percent Base Peak
27	43	53	12
28	7	55	100
29	46	70.3 (m*)	—
36.8 (m*)	—	83	96
39	40	84	5
41	12	98	51
43	92	99	3.5
51.1 (m*)	—	100	0.2

9. Compound **J** exhibited an infrared band at 2253 cm^{-1} and gave the mass spectrum shown in Figure 7.25. The ^{1}H NMR consisted of two triplets.

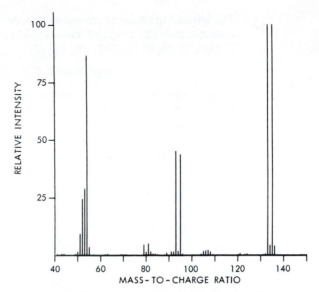

Figure 7.25 Mass spectrum of unknown **J**, problem 9.

10. The ^{13}C NMR of compound **K** revealed four signals. Compound **K** formed an insoluble derivative upon treatment with benzenesulfonyl chloride/NaOH. The EIMS of compound **K** is shown in Figure 7.26.

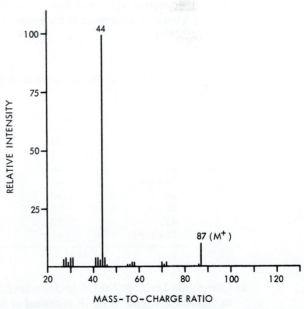

Figure 7.26 Mass spectrum of unknown **K**, problem 10.

7.10 GENERAL REFERENCES ON MASS SPECTROMETRY

1. Beynon, J. H., and A. E. Williams, *Mass and Abundance Tables for Use in Mass Spectrometry.* Amsterdam: Elsevier, 1963.

2. Budzikiewicz, H., C. Djerassi, and D. Williams, *Mass Spectrometry of Organic Compounds.* San Francisco: Holden-Day, 1967.

3. Chapman, J. R., *Practical Organic Mass Spectrometry.* New York: Wiley Interscience, 1985.

4. Harrison, A. G., *Chemical Ionization Mass Spectrometry.* Boca Raton, Fla.: CRC Press, 1983.

5. Lyon, P. A., ed., *Desorption Mass Spectrometry* (ACS Symposium Series, #291). Washington, D.C.: American Chemical Society, 1985.

6. Howe, I., D. H. Williams, and R. D. Bowen, *Mass Spectrometry, Principles and Applications,* 2nd ed. London: McGraw-Hill, 1981.

7. McLafferty, F. W., *Interpretation of Mass Spectra,* 3rd ed. Mill Valley, Calif.: University Science Books, 1980.

8. McLafferty, F. W., ed., *Tandem Mass Spectrometry.* New York: Wiley-Interscience, 1983.

9. Rose, M. E., and R. A. W. Johnstone, *Mass Spectrometry for Chemists and Biochemists.* Cambridge University Press, 1982.

8

Characterization of an Unknown Compound

In Chapters 8 and 9 we shall be concerned with the classification of a compound with respect to the functional groups present and the complete identification of the compound. Such a task is most efficiently accomplished through a systematic integration of physical and chemical methods. Just as the various forms of spectroscopy furnish us information in a synergistic manner, the combination of physical data with chemical observations also acts in a synergistic manner. Through the careful and systematic selection of appropriate physical and chemical methods, the task of reaching our goal — the final identification of an organic structure — can be greatly simplified, and the consumption of both time and material can be greatly reduced.

Of the various forms of spectroscopy, the state of the art, the cost and availability of the instruments, and the ease and rapidity of operation place infrared and nuclear magnetic resonance spectroscopy as the primary spectroscopic methods to which the organic chemist turns when faced with a structural problem. Typically, as soon as there is some reasonable assurance of the purity of a compound, IR and NMR spectra should be obtained. Chemists often find it convenient to examine the IR and/or NMR spectrum of a mixture, not only as an aid in deciding on purification methods, but also as a method of checking to see that no artifacts appear during the purification process, and to provide assurance that all components of a mixture are isolated during a separation process.

Even upon cursory examination of the infrared spectrum of a compound, one can often classify the material with respect to functional groups present. Equally important, and often more so, one can readily establish the absence of many functional groupings, thus immediately eliminating many wet tests, which may be laborious and costly in materials. For example, the casual observation of the absence of infrared bands in the 3600 to 3100 cm^{-1} region and of strong bands in the 1850 to 1650 cm^{-1} region readily eliminates from further consideration all N—H compounds (primary and secondary amines and amides, etc.), all O—H compounds (alcohols, phenols, carboxylic acids, oximes, etc.), and all C=O compounds (acids, esters, ketones, aldehydes, etc.). The same information gained from this quick observation (information that should be accessible in not more than 20 minutes from the time the sample is obtained, provided an IR spectrometer is available for use) might also be obtained in a considerable longer time by the indiscriminant use of the commonly used chemical classification tests. The presence or absence of many functional groups can also be inferred from ^{1}H and ^{13}C NMR. As NMR instruments have become commonly available, simplier, and faster to use, many chemists, particularly those in research environments, often turn to NMR as the primary spectroscopic methods for examination of an unknown.

Compounds will be classified according to functional group; the most useful spectroscopic and chemical methods for the identification of the functional group will be outlined. Finally, the crucial physical and chemical methods that allow structure assignment will be discussed, and the procedures will be outlined in some detail.

8.1 STEPS FOR THE IDENTIFICATION OF AN UNKNOWN

Step 1. Gross Examination

A. Physical state. The physical state of the unknown should be indicated. Additional description, such as "amorphous powder," "short needles," and "viscous liquid," often proves useful.

B. Color. Many organic compounds possess a definite color owing to the presence of chromophoric groups. Among the most common of the simple chromophoric groups are nitro, nitroso, diazo, azo, and quinone. Compounds with extensive conjugation are likely to be colored. The color of many samples is due to impurities frequently produced by air oxidation. Freshly purified aromatic amines and phenols are usually colorless; on storage, small amounts of these compounds are oxidized to highly colored quinone impurities. Stable, colorless liquids or white crystalline solids are not likely to contain the usual chromophoric groups or functional groups that are easily oxidized.

C. Odor (optional). Many organic compounds are exceedingly toxic and/or will produce at least temporary discomfort upon inhalation. *The reader is cautioned about the indiscriminate "whiffing" of compounds or mixtures about which little is known.*

To note the odor of an unknown substance, hold the tube pointed away from the face in one hand and gently wave the vapors from the tube toward the nose with the other hand. With a good nose educated by experience, a chemist can often make a tentative identification of a common chemical or an intelligent guess as to functional groupings present. Alcohols, phenols, amines, aldehydes, ketones, and esters all have odors more or less characteristic of the general group. Mercaptans, low molecular weight amines, and isonitriles (particularly toxic) have characteristic odors that one seldom mistakes once experience has been gained. The commonly used solvents have characteristic odors. Exposure to solvents and other volatile chemicals should be minimized by proper use of hoods.

Step 2. Determination of Purity and Physical Constants

One should be assured that the physical constants used in a structure identification are obtained with pure material. The necessity for purification or fractionation can be indicated by melting point and/or boiling point ranges, behavior on thin-layer or gas-liquid chromatography, or any inhomogeneity or discoloration.

Step 3. Classification by Functional Group

A. Determination of acidity or basicity and solubility behavior.   The acidity or basicity of a compound, as determined with indicators or potentiometer or by solubility in acids or alkali, as well as general solubility behavior (see Section 8.4), is useful, not only in the determination of possible chemical classes to which the compound may belong, but also to serve as a guide in the choice of solvents and procedures used for spectroscopic analyses.

B. Classification of functional groupings by spectroscopic and chemical means.   As indicated earlier, in this text we shall use spectroscopy with significant emphasis on infrared as the primary method of determining the major functional groups present. Supplemental chemical tests will be used to confirm or clarify the assignments made by spectroscopic means.

C. Elemental analysis.   Qualitative elemental analysis (see Section 8.2) by the sodium fusion method is not a procedure that the chemist finds necessary to apply to every unknown compound with which he or she is faced. The presence or, more likely, the absence of many heteroatoms can be inferred from the source and/or history of the sample. The presence or absence of many heteroatoms can also be inferred from spectroscopic data. High-resolution mass spectrometry will provide exact elemental compositions. Evidence of the presence or absence of sulfur, halogens, and nitrogen can usually be obtained from low-resolution mass spectrometry. Infrared spectroscopy, by virtue of detection of various functional groups, can establish the presence or absence of certain elements. The presence of an amino, nitrile, or nitro group in the infrared spectrum provides obligatory evidence for the presence of a nitrogen-containing compound. The basicity of a compound may be sufficient to establish the presence of nitrogen. The presence of sulfur can be inferred from mercaptan, sulfone, sulfoxide, and so on, as bands in the infrared spectrum. A simple Beilstein test is usually reliable for establishing the presence or total absence of halogen in a compound. Finally, quantitative microanalytical data for C, H, and N may often be available. Such data, together with the known presence or absence of oxygen as inferred by chemical or spectroscopic means, may also be sufficient to eliminate or establish the presence of other elements within the molecule.

Step 4. Final Identification

At this point, based on the foregoing chemical and physical data, classification of the unknown compound as belonging to a specific functional group class should be possible. A comparison between the physical data obtained for an unknown compound and information about known compounds listed in the literature should be made, and an initial list of possibilities that melt or boil within 5°C of the value observed for the unknown should be prepared. This initial list is usually obtained by consulting tables of compounds arranged in order of melting points or boiling points and according to the functional groups present. The most extensive of the tables — which also gives data on derivatives — is the *Handbook of Tables for Organic Compound Identification,* published by the Chemical Rubber Publishing Co. It should be apparent that no one compendium is available that provides complete coverage of the literature on physical constants useful in identification work. The tables included in Chapter 9 provide only a sampling of the more commonly encountered compounds.

The initial list of possibilities can be immediately condensed to include only those compounds that conform to the data on elemental composition, solubility, chemical behavior, and other physical and spectral properties. In most cases, a tentative list of possibilities will not contain more than three to five compounds. A more extensive literature search is then made to ascertain additional properties of the compounds listed as possibilities. The final proof of structure of an unknown is accomplished through the preparation of selected derivatives and/or the collection, comparison, and evaluation of other physical and chemical data, such as neutralization equivalent, nuclear magnetic resonance, ultraviolet, and mass spectra, optical rotation or optical rotatory dispersion, and dipole moments.

The classical criteria for the conclusive identification of a compound are that (1) the compound has the expected chemical and spectral properties, and the physical constants match those given in the literature; and (2) the appropriate derivatives prepared from the unknown melt within 1 to 2°C of the melting points given in the literature. The actual criteria necessary to establish the identity of the compound beyond reasonable doubt are, of course, subject to considerable variation, depending on the kind and content of the information available to the experimenter and the complexity of the compound in question. In many cases, one derivative may be sufficient, and, in some cases, it may not be necessary to prepare derivatives at all. Indeed, more rigorous and detailed proof of structure can often be obtained from a combination of carefully obtained and interpreted spectral and other physical data (e.g., molecular weights), chemical reactivity (e.g., kinetics, pK$_a$'s, etc.), and microanalytical data (e.g., found percentages of C, H, and other elements agreeing with the theoretical within 0.4%).

The task of the rigorous establishment of the identity of two compounds is considerably simplified if a known sample is available. In such a case, identity can be establishes by one or more of the following methods.

A. Mixture melting point. Two samples having the same melting point are identical if, upon admixture of the two samples, the melting point observed is not depressed relative to that of pure samples (see Section 3.2.1). This procedure, wherever possible, should be applied not only to the original unknown but also to derivatives thereof.

B. Spectral comparison. In cases of molecules of moderate complexities, the superimposability of the infrared spectrum of the unknown with that of the known establishes the identity of the two. However, different molecules of simple structure, such as undecane and dodecane, may have remarkably similar spectra, whereas in exceedingly complicated molecules, such as steroidal glycosides, peptides, and many antibiotics, the resolution obtained in the infrared region is usually not sufficient to allow identification with certainty by means of spectral comparison. The method applies equally well to the use of spectra recorded in the literature; in either event, care must be taken to be sure that the spectra are recorded under identical conditions.

The comparison of spectra obtained by other spectroscopic methods also provides helpful criteria for the establishment of identity. Since NMR spectra run the gamut from the exceedingly simple to the extraordinarily complex, direct comparison may provide anything from very tenuous to very conclusive information. Mass spectral data can be expected to match very closely only when the spectra are obtained under exactly identical conditions (usually run one after the other on the same instrument). It should be noted that the mass spectral fragmentation pattern is often not sensitive to differences in spatial orientation of atoms or groups; the mass spectra of geometrical isomers are often almost indistinguishable. The superimposability of ultraviolet spectra only serves to indicate that compounds have the same chromophore.

C. Chromatographic comparison. Good but not necessarily rigorous criteria of the identity of two compounds can be established by comparative gas-liquid, thin-layer, and/or paper chromatography (see Section 2.6). The standard procedure involves chromatographing the two compounds separately and as a mixture. Safer conclusions can be drawn if the comparisons are made under several conditions and by more than one method.

D. Selection and preparation of derivatives. Ideally, the derivatives should be easily and quickly prepared from readily available reagents and should be easily purified. The best derivatives meet the following standards:

1. The derivative should be crystalline solid melting between 50 and 250°C. Solids melting below 50°C are often difficult to crystallize and recrystallize. Accurate determinations of melting points above 250°C are difficult.

2. The derivative should have a melting point and other physical properties that are considerably different from the original compound.

3. Most importantly, the derivative chosen should be one whose melting point will single out one of the compounds from the list of possibilities. The melting points of the derivatives to be compared should differ from each other by a minimum of 5°C.

In this text, for each functional group class, detailed procedures will be found only for the most generally suitable derivatives for which extensive data are readily available. In addition to the tables of melting points of derivatives in Chapter 9, more extensive compilations of the melting points of the derivatives chosen will be found in *Handbook of Tables for Organic Compound Identification,* cited earlier.

It should be apparent to the reader that it is impossible to set down an exacting set of conditions that must be met each time a compound is said to be conclusively identified. There are numerous methods by which identity can be established; some are more rigorous than others, and some are more appropriate in one case than in another. In any event, there is no substitute for common sense and the careful and thoughtful planning and execution of crucial experiments, and a thorough evaluation of data.

8.2 QUALITATIVE ELEMENTAL ANALYSIS

Before proceeding with chemical and spectral functional group analysis, it is helpful to determine what elements are present in a molecule and in what ratio these elements are present. The former data are derived via qualitative tests that can be readily carried out by the student, whereas the latter data are usually obtained by trained analysts in laboratories specifically set up for such quantitative analysis. In addition to quantitative analysis of the elements present, these laboratories are usually capable of performing quantitative functional group analyses, for example, for methyl, methoxyl, acetoxyl, and deuterium. The following paragraphs of this section outline the handling and interpretation of such analytical data.

The presence of carbon in organic compounds will be assumed, as will the presence of hydrogen, except in perhalogenated compounds; no specific tests for the presence of carbon and hydrogen will be given. (The presence of hydrogen is readily indicated by infrared and hydrogen nuclear magnetic resonance spectroscopy, as discussed in detail in Chapters 5 and 6.) Heteroatoms most commonly encountered in organic compounds include the halogens, oxygen, sulfur, and nitrogen. The detection of oxygen is relatively difficult by qualitative analytical means, and we shall rely on infrared spectral analysis and on solubility data to indicate its presence.

The presence of metallic elements, for example, lithium, sodium, and potassium (usually as the salts of acidic materials), can be indicated by an ignition test. A few milligrams of the substance are placed on a clean stainless steel spatula and carefully burned in a colorless burner flame. The presence of a noncombustible residue after complete ignition indicates the presence of a metallic element. The color of the flame during ignition may give some indication of the type of metal present. The ignition residue is dissolved in two drops of concentrated hydrochloric acid, and a platinum wire is dipped into the solution and placed in a colorless burner flame. The following colors are produced on ignition: blue by lead, green by copper and boron, carmine by lithium, scarlet by strontium, reddish-yellow by calcium, violet by potassium, cesium, and rubidium, and yellow by sodium. Precautions must be taken to avoid contamination of the sample by sodium because the intense yellow flame produced by sodium generally masks all other elements. If only trace amounts of sodium are present, the initial yellow flame will soon disappear, leaving the final colored flame of the elements present in greater amounts.

The qualitative tests for most of the elements are based on reactions involving an anion of the element, for example, the halides, sulfide, and cyanide; hence, a reductive decomposition of the compound is required. This is generally accomplished by means of fusion with metallic sodium.

$$C, H, O, N, S, X, \xrightarrow[\Delta]{Na} NaX, NaCN, Na_2S, NaSCN$$

8.2.1 Sodium Fusion

To accomplish the decomposition of organic compounds with sodium, the compound is fused with sodium-lead alloy[1] or metallic sodium. Since the sample sizes employed are generally quite small, 1 to 10 mg, and the detection tests are quite sensitive, the student must take certain precautions to avoid contamination of the sample during and after fusion. All glassware must be thoroughly cleaned and rinsed with distilled water, or preferably deionized water, and the solvents and reagents must be of analytical quality. *Safety goggles should be worn during these tests!* The procedure discussed below in (a) is recommended, as it avoids the use of pure metallic sodium.

PROCEDURE: SODIUM FUSION

(a) In a small test tube place about 0.5 g of sodium-lead alloy (dri-Na® from J.T. Baker Chemical Company). Clamp the tube in a vertical position and heat with a flame until the alloy melts and fumes of sodium are seen up the walls of the tube. *Do not heat the alloy to redness.* Add a few drops of liquid sample or 10 mg of solid. During addition be careful not to get any sample on the sides of the tube. Heat gently, if necessary, until the reaction is initiated; remove the flame until the reaction

---

[1] J. A. Vinson and W. T. Grabowski, *J. Chem. Ed.,* **54,** 187 (1977).

subsides, and then heat to redness for 1 or 2 minutes and let cool. Add 3 mL of distilled water and heat gently for a few minutes to react the excess sodium with the water. Filter the solution if necessary. Wash the filter paper or dilute the decanted reaction mixture with about 2 mL of distilled water, and proceed with the elemental analysis. This final solution will be referred to as the fusion solution in discussions that follow.

(b) If the organic sample to be analyzed is quite volatile, a 1- to 10-mg sample of the material is placed in a 4-in. test tube and 30 to 50 mg of sodium, a piece about half the size of a pea, is cautiously added to the test tube. (The mouth of the test tube should be pointed away from the experimenter and other people in the laboratory to prevent their being splashed with material from the tube should a violent reaction ensue on the addition of the sodium.) The test tube is then gently heated until decomposition and charring of the sample occur. When it appears that all the volatile material has been decomposed, the test tube is strongly heated until the residue becomes red. The test tube is allowed to cool to room temperature, and a few drops of methanol are added to decompose the excess sodium. If no gas evolves on the addition of the methanol, an excess of sodium was not present, and there is a distinct possibility of incomplete conversion of the elements to their anions. The fusion should be repeated with a larger quantity of sodium.

The contents from the tube are boiled with 1.5 to 2.0 mL of distilled water, diluted to 10 mL with distilled water, and the mixture is then filtered or centrifuged. The decomposition of the organic material usually leads to the extensive formation of carbon, which may prove very difficult to remove. Occasionally filtration through a filtering aid, for example, Celite, which has been thoroughly washed with distilled water, will remove the finely divided carbon particles. The resulting solution should be clear and nearly colorless. If the solution is highly colored, the entire fusion process should be repeated because the color may interfere with the detection tests. This final solution will be referred to as the fusion solution in the discussions of the following elemental tests.

8.2.2 Detection of Halides

The presence of chlorine, bromine, and iodine can be readily detected by the precipitation of the corresponding silver halides on treatment with silver ion. Fluorine cannot be detected in this test because silver fluoride is soluble.

PROCEDURE: DETECTION OF HALIDES

The presence of nitrogen or sulfur interferes in this test; sulfide and cyanide must be removed before treatment of the solution with silver ion. This is accomplished by acidification of the fusion solution followed by gentle heating to boil off the hydrogen cyanide and hydrogen sulfide formed on acidification. *This procedure must be carried out in a hood.* A great excess of sulfuric acid should be avoided since silver sulfate can precipitate from solutions containing a high concentration of sulfate ions. Take 0.5 mL of the fusion solution and carefully acidify by the dropwise addition of 10% sulfuric acid. (The pH of the solution can be checked readily by immersing a clean stirring rod into the acidified solution and then applying a piece of pHydrion paper to the liquid adhering to the stirring rod.) The solution is gently boiled over a microburner for about 3 to 5 minutes. One or two drops of aqueous silver nitrate solution are added; the formation of a substantial quantity of precipitate indicates the presence of chloride, bromide, or iodide. A white precipitate soluble in ammonium hydroxide indicates the presence of chloride, a pale yellow precipitate slightly soluble in ammonium hydroxide indicates bromide, and a yellow precipitate insoluble in ammonium hydroxide indicates iodide. Fluoride does not form a precipitate with silver ion. The formation of a slight cloudiness in the solution is not indicative of a positive test. It is recommended that a halide analysis on a known compound be carried out for comparison purposes.

The presence of chlorine, bromine, or iodine in organic compounds can be detected by the Beilstein test. The test depends on the production of a volatile copper halide when an organic halide is strongly heated with copper oxide. The test is extremely sensitive, and a positive test should always be confirmed by other methods. The following procedure is normally used.

PROCEDURE: BEILSTEIN TEST

A small loop in the end of a copper wire is heated to redness (an oxide film is formed) in a Bunsen flame until the flame is no longer colored. After the loop has cooled, it is dipped into a little of the compound to be tested and then reheated in the nonluminous Bunsen flame. A blue-green flame produced by volatile copper halides constitutes a positive test for chlorine, bromine, or iodine (copper fluoride is not volatile).

Very volatile compounds may evaporate before proper decomposition occurs, causing the test to fail. Certain compounds such as quinoline, urea, and pyridine derivatives give misleading blue-green flames owing to the formation of volatile copper cyanide.

Mass spectrometric detection of halides. The halogens Cl and Br exist as a mixture of isotopes. The mass spectrum of a Cl- or Br-containing compound will display several peaks in the parent mass region, which will be separated by the difference in mass of the halogen isotopes and will be in an intensity ratio equal to the natural abundance ratio (see Section 7.8.2). Mass spectrometry is also valuable in indicating the number of halogen atoms that occur in a single molecule (see Section 7.4).

8.2.3 Detection of Nitrogen and Sulfur

The detection of nitrogen and sulfur is based on the conversion of these elements present in a given molecule to cyanide and sulfide ions. A sensitive procedure for the detection of these ions utilizes *p*-nitrobenzaldehyde in dimethyl sulfoxide. A more traditional test for cyanide, the cyano complex (Prussian blue) formed from ferrous ammonium sulfate, requires the prior removal of sulfide.

PROCEDURE: DETECTION OF NITROGEN OR SULFUR

Put about 10 drops of the fusion solution in a small test tube and saturate it with solid sodium bicarbonate. Shake to ensure saturation and check to see if excess solid is present. Then add one or two drops of the saturated solution (it is permissible to transfer some of the solid sodium bicarbonate) to a test tube containing about 20 drops of a 1% *p*-nitrobenzaldehyde solution in dimethyl sulfoxide (PNB reagent). This reagent should be discarded if the initial yellow darkens, and it should be stored in a brown bottle. A purple color indicates nitrogen is present. A green color indicates sulfur is present. If both sulfur and nitrogen are present, only a purple color will be visible. Therefore, if a positive test for nitrogen is observed with PNB, a test for sulfur by the lead acetate method should be carried out (see the following procedure). If only a positive test for sulfur is observed by the present method, then nitrogen is definitely absent.

PROCEDURE: DETECTION OF SULFUR

A 1-mL portion of the fusion solution is acidified with acetic acid, and two to three drops of dilute lead acetate solution are added. The formation of a black precipitate indicates the presence of sulfur.

8.3 QUANTITATIVE ELEMENTAL ANALYSIS

Data on the quantitative elemental composition of an unknown are particularly useful in establishing structure. Traditionally, microanalytical data have been accepted as important criteria for proof of structure and purity. In most primary chemical journals, microanalytical data are given for all new compounds reported in a paper; the data are usually presented in the following form:

Anal. Calcd for $C_{10}H_{14}N_2O_4$: C, 53.09; H, 6.24; N, 12.38
Found: C, 53.20; H, 6.35; N, 12.44

(In writing a formula, after C and H, other elements are listed in alphabetical order.) Such quantitative elemental analyses are not generally performed by the individual chemist; they are usually made by trained technicians in laboratories (often outside independent laboratories) specifically designed for this purpose.

The most common, and hence least expensive, analytical data are those for carbon-hydrogen and nitrogen. In these analyses, the sample is quantitatively combusted to yield carbon dioxide, water, and nitrogen. In the classical technique, the water is collected by adsorption on calcium chloride, the carbon dioxide is collected on Ascarite (sodium hydroxide on asbestos), and the nitrogen is determined by volume. In this method, a sample of about 5 mg is required for a carbon-hydrogen determination, and an additional 5 mg for nitrogen and each other determination.

The simultaneous microdetermination of carbon, hydrogen, and nitrogen can be achieved by automatic gas chromatographic methods. With such instruments, a sample of about 1 mg is converted to water, carbon dioxide, and nitrogen, which are recorded as a three-peak chromatogram. The composition can be determined by peak height to an accuracy of about 0.3%.

To derive the empirical formula of a compound that contains n different elements, elemental analyses of $n - 1$ of the elements present must be determined. The percentage of the remaining element can be determined as the difference between the sum of the percentages of the other elements and 100%. (The percentage of oxygen is usually determined by difference.) Certain errors may be introduced by this method. The acceptable experimental error involved in the determination of the percentage of each element is $\pm 0.4\%$. If the analysis of a majority of the elements is high, or low, then the analysis of the element obtained by difference may be anomalously low, or high. With molecules of a high degree of complexity, such errors may result in an inability to distinguish between two or more empirical formulas unless auxiliary information is available. It is therefore recommended that, in cases when complex structures are thought to be involved, the quantitative analysis of each element contained in the compound be obtained. Multiple analyses are also recommended in such cases, using an average value for each element in the calculations of the empirical formula.

For the conversion of microanalytical data to an empirical formula, the first step is to divide the percentage of each element by its atomic weight, and the second is to divide the resulting numbers by the smallest one to determine the atomic ratios.

Element	Percentage		Atomic Weight				Atomic Ratio
C	67.38	÷	12.01	=	5.61		10
H	7.92	÷	1.01	=	7.84	$\times \dfrac{1}{0.56} =$	14
N	15.72	÷	14.01	=	1.12		2
O	8.98	÷	16.00	=	0.56		1

The empirical formula for the above compound is $C_{10}H_{14}N_2O$. Conversion of an empirical formula to a molecular formula requires molecular weight data. Sometimes, it may be known from other data that an unknown contains only one chlorine, two nitrogens, and so on; in such cases, the molecular formula may be determined directly from analytical data.

A compilation of calculated analytical data for a great number of compounds containing C, H, N, O, and S in various combinations has been published in book form and is useful in handling analytical data.[2] Programs are available for personal computers that calculate percent composition from molecular formulas.

It should be obvious from the foregoing paragraphs that great care must be exercised in the preparation of a sample for analysis. The sample must be of very high purity. Recrystallized solids should be heated at low temperatures, below the melting or decomposition points, and under vacuum to remove solvent molecules. Hygroscopic compounds must be protected from moisture. The presence of as little as 1 mol% of water, which is generally a very low weight percent, may be sufficient to produce a bad analysis. The analyst should always be informed about the hygroscopic nature of a compound, and should also be warned of compounds of explosive or very toxic nature.

8.3.1   Interpretation of Empirical Formula Data

Once the chemist has determined the molecular formula of a compound, valuable information can be derived from the molecular formula with respect to gross features of the molecule. In particular, the molecular formula data can be used to calculate the total number of rings and/or double and triple bonds (sites of unsaturation) in the unknown molecule.

The number of sites of unsaturation in a molecule is conveniently calculated by means of Equation (8.1),

$$N = \frac{\sum_i n_i(v_i - 2) + 2}{2} \tag{8.1}$$

where N is the number of sites of unsaturation, n_i is the number of atoms of element i, and v_i is the absolute value of the valence of element i. The following examples will illustrate the use of Equation (8.1).

Example 1

Elemental analysis indicates an empirical formula of C_6H_6. Using v_i of 4 for carbon and 1 for hydrogen, we have

$$N = \frac{6(4 - 2) + 6(1 - 2) + 2}{2} = \frac{8}{2} = 4 \quad \text{sites of unsaturation}$$

Example 2

Elemental analysis indicates an empirical formula of C_7H_6ClNO. Using v_i of 4 for carbon, 1 for hydrogen, 1 for chlorine, 3 for nitrogen, and 2 for oxygen, we have

$$N = \frac{7(4 - 2) + 6(1 - 2) + 1(1 - 2) + 1(3 - 2) + 1(2 - 2) + 2}{2} = 5 \text{ sites of unsaturation}$$

[2] George H. Stout, *Composition Tables for Compounds Containing C, H, N, O, S* (New York: W. A. Benjamin, Inc., 1963).

Example 3

Elemental analysis indicates an empirical formula of $C_{19}H_{18}BrP$. The physical and chemical properties indicate that the bromide is ionic and that we must be dealing with a pentavalent phosphorus. Thus, using v_i of 4 for carbon, 1 for hydrogen and bromine, and 5 for phosphorus, we have

$$N = \frac{19(4-2) + 18(1-2) + 1(1-2) + 1(5-2) + 2}{2} = 12 \text{ sites of unsaturation}$$

In these calculations, no distinction can be made among the various types of sites of unsaturation. The experimentalist must rely on spectral and chemical tests to indicate what types of functional groups, and hence their worth in sites of unsaturation, are present in any given molecule. A single ring, $C=C$, $C=O$, $C=N$, $N=O$, or any other doubly bonded system is considered as a single site of unsaturation; $C\equiv C$ and $C\equiv N$ are considered as two sites of unsaturation. The phosphorus-oxygen and sulfur-oxygen coordinate covalent bonds ($\overset{+}{-P}-O^-$ and $\overset{+}{S}-O^-$) are incorporated as single bonds and thus are *not* sites of unsaturation.

A benzene ring containing a combination of a ring and double bonds represents four sites of unsaturation.

8.4 CLASSIFICATION BY SOLUBILITY

The solubility characteristics of a compound may be very useful in providing structural information. Fairly elaborate solubility and indicator determination schemes have been presented in previous texts on this subject. The authors of this text contend that such elaborate schemes are not required because of the present availability of spectral methods of functional group detection. In addition, the solubility test results obtained with rather complex molecules may not lend themselves to unambiguous interpretation. In view of these comments, a rather limited series of solubility tests are recommended. The solvents recommended for use in these tests are water, 5% hydrochloric acid, 5% sodium hydroxide, and 5% sodium hydrogen carbonate. The effect of structure on the acidity or basicity of organic molecules will not be discussed in this text owing to a necessary limitation of space. For discussions concerning the inductive, resonance, and steric effects on the acidity and basicity of organic compounds, the student is referred to the modern organic chemistry texts.

The quantities of compound and solvent used in the solubility tests are critical if reliable data are to be derived. A compound will be considered "soluble" if the solubility of that compound is greater than 3 parts of compound per 100 parts of solvent at room temperature (25°C). The experimentalist should use as little of the unknown compound as is necessary, generally 10 mg in 0.33 mL of solvent. Employing such small quantities of material requires a fairly accurate determination of the weight of the material to be used; however, it is not necessary to weigh accurately the amount of compound each time. A 10-mg portion of the compound should be accurately weighed out. Subsequent 10-mg portions of the compound can then be estimated visually.

A summary of the solubility characteristics of some of the most important classes of compounds is found in Table 8.1.

8.4.1 Solubility in Water

Water is a highly polar solvent possessing a high dielectric constant and is capable of acting as a hydrogen bond donor or acceptor. As a result, molecules possessing highly polar functional groups capable of entering into hydrogen bonding with water (*hydrophilic* groups) display a greater solubility in water than do molecules without such functional groups. Irrespective of the presence of a highly polar functional group in a molecule, a limiting factor on its solubility in water is the amount of hydrocarbon structure (*lipophilic* portion of the molecule) associated with the functional group. For example, methanol, ethanol, 1-propanol, and 1-butanol all possess solubilities in water at room temperature, which would allow classification as "water-soluble"; however, 1-pentanol would be classified as "insoluble." The solubility within a series

TABLE 8.1   General Solubility Guidelines

Soluble in Water and Ether

Monofunctional alcohols, aldehydes, ketones, acids, esters, amines, amides, and nitriles containing up to five carbons.

Soluble in Water: Insoluble in Ether

Amine salts, acid salts, polyfunctional compounds such as polyhydroxy alcohols, carbohydrates, polybasic acids, amino acids, etc.

Insoluble in Water: Soluble in NaOH and NaHCO₃

High molecular weight acids and negatively substituted phenols.

Insoluble in Water and NaHCO₃: Soluble in NaOH

Phenols, primary and secondary sulfonamides, primary and secondary aliphatic nitro compounds, imides, and thiophenols.

Insoluble in Water: Soluble in Dilute HCl

Amines except diaryl and triaryl amines, hydrazines, and some tertiary amines.

Insoluble in Water, NaOH, and Dilute HCl, but Contain Sulfur or Nitrogen

Tertiary nitro compounds, tertiary sulfonamides, amides, azo compounds, nitriles, nitrates, sulfates, sulfones, sulfides, etc.

Insoluble in Water, NaOH, and HCl; Soluble in H₂SO₄

Alcohols, aldehydes, ketones, esters, ethers (except diaryl ethers), alkenes, alkynes, and polyalkylbenzenes.

Insoluble in Water, NaOH, HCl, and H₂SO₄

Aromatic and aliphatic hydrocarbons and their halogen derivatives, diaryl ethers, perfluoro-alcohols, esters, ketones, etc.

of compounds is also dependent on the extent and position of chain branching, the solubility in water increasing as the branching increases. For example, all of the isomeric pentanols except 3-methyl-1-butanol and 1-pentanol are soluble in water.

Similar solubility trends are observed with aldehydes, amides, amines, carboxylic acids, and ketones, although minor variations in the cutoff solubility point may occur in the various classes of compounds. The upper limit of water solubility for most compounds containing a single hydrophilic group occurs at about five carbon atoms. Most difunctional and polyfunctional derivatives are soluble in water, as are most salts of organic acids and bases.

Distinction between monofunctional and polyfunctional compounds and salts can usually be made by testing their solubilities in diethyl ether, a relatively nonpolar solvent. Glycols, diamines, diacids, or other similar di- and polyfunctional derivatives, as well as salts, are not soluble in diethyl ether.

Further solubility tests of water-soluble compounds in 5% hydrochloric acid, 5% sodium hydroxide, or 5% sodium hydrogen carbonate are meaningless in that their solubilities in these solvents are primarily dependent on the solubility in water and not on the pH of the solvent. It is still possible, however, to detect the presence of acidic and basic functional groups in a water-soluble compound. The pH of the aqueous solution of the compound, derived from the solubility test, is determined by placing a drop of the solution on a piece of pHydrion paper. If the compound contains an acidic functional group, the solution will be acidic (with phenols and enols, the solution will be only very weakly acidic, and a control test should be run on an aqueous solution of a phenol for comparison); for compounds containing a basic functional group, the solution will be basic. It is usually possible to distinguish between a strong and a weak acid by the addition of one drop of dilute sodium hydrogen carbonate solution. Carbon dioxide bubbles will appear if a strong acid is present (see Section 8.4.4).

8.4.2   Solubility in 5% Hydrochloric Acid

Compounds containing basic functional groups, for example, amines (except triarylamines), hydrazines, hydroxylamines, aromatic nitrogen heterocyclics, but *not* amides (although some N,N-dialkyl amides are soluble), are generally soluble in 5% hydrochloric acid. Occasionally an organic base will form an insoluble hydrochloride salt as rapidly as the free base dissolves, thus giving the appearance of an insoluble compound. One should observe the sample carefully as the solubility test is run.

8.4.3   Solubility in 5% Sodium Hydroxide

Compounds containing acidic functional groups with pK_a's of less than approximately 12 will dissolve in 5% sodium hydroxide. Compounds falling into this category include sulfonic acids (and other oxysulfur acids), carboxylic acids, β-dicarbonyl compounds, β-cyanocarbonyl compounds, β-dicyano compounds, nitroalkanes, sulfonamides, enols, phenols, and aromatic thiols. Certain precautions should be observed since some compounds may undergo reaction with sodium hydroxide, for example, reactive esters and acid halides, yielding reaction products that may be soluble or insoluble in the 5% sodium hydroxide solution. Certain active methylene compounds (β-dicarbonyl compounds, etc.) may undergo facile condensation reactions, producing insoluble products. If any reaction appears to occur during the course of the solubility test, it should be noted. Long-chain carboxylic acids (C_{12} and longer) do not form readily soluble sodium salts, but tend to form a "soapy" foam.

8.4.4   Solubility in 5% Sodium Hydrogen Carbonate

The first ionization constant of carbonic acid (H_2CO_3) is approximately 10^{-7}, which is less than the ionization constant of strong acids (carboxylic acids) but is greater than the ionization constant of weak acids (phenols). Therefore, only acids with pK_a's less than 6 will be soluble in 5% sodium hydrogen carbonate. This category includes sulfonic (and other oxysulfur acids) and carboxylic acids and highly electronegatively substituted phenols (for example, 2,4-dinitrophenol and 2,4,6-trinitrophenol).

8.4.5   Additional Solubility Classifications

Two additional solubility tests can be used to further classify compounds, although these tests are not generally necessary in that more valuable and definitive structural information can be derived from the various spectra of the compound.

Compounds containing sulfur and/or nitrogen, but which are not soluble in water, 5% hydrochloric acid, or 5% sodium hydroxide, are almost always sufficiently strong bases to be protonated and dissolved in concentrated sulfuric acid. For such compounds, the sulfuric acid solubility test provides no additional information. Compounds falling into this solubility category include most amides, di- and triarylamines, tertiary nitro, nitroso, azo, azoxy, and related compounds, sulfides, sulfones, disulfides, and so on.

Compounds not soluble in water, 5% hydrochloric acid, or 5% sodium hydroxide and that do not contain nitrogen or sulfur can be further classified based on their solubilities in concentrated sulfuric acid and 85% phosphoric acid. Compounds insoluble in concentrated sulfuric acid include hydrocarbons, unreactive alkenes and aromatics, halides, diaryl ethers, and many perfluoro compounds.

Compounds soluble in concentrated sulfuric acid include compounds that contain very weakly basic functional groups. This solubility class includes alcohols, aldehydes, ketones, esters, aliphatic ethers, reactive alkenes and aromatics, and acetylenes. Dissolution in concentrated sulfuric acid is often accompanied by reaction, as indicated by color changes or decomposition, and should be noted. This solubility class can be subdivided further on the basis of solubility in 85% phosphoric acid. The foregoing compounds that are soluble in sulfuric acid and contain *5 to 9 carbon atoms* generally will be soluble in 85% phosphoric acid.

9

Functional Classification and Characterization

This text gives procedures using semimicro quantities of materials, typically in the 50 to 500 mg range. The experienced and careful chemist will have no trouble manipulating quantities on a much smaller scale. Valuable material can be conserved by running solubility and classification tests in capillary tubes. In most cases, infrared spectra of compounds typical of the functional class under consideration are reproduced in this chapter.

Tables 9.1 to 9.18 provide identification data on representative compounds of various classes. In large part, the compounds listed are readily available from chemical supply houses. For the identification of an unknown, it may be necessary to consult a more extensive compilation of data, for example, *Handbook of Tables for Organic Compound Identification,* CRC Press, Boca Raton, Fla.

Adequate precautions should always be taken to avoid unnecessary exposure to chemicals; see Table 1.1 for chemicals that are known or suspected to be carcinogenic.

9.1 HYDROCARBONS

9.1.1 Alkanes and Cycloalkanes

Classification

The following observations are indicative of saturated hydrocarbons:

1. *An exceptionally simple infrared spectrum.* In the infrared spectra of hydrocarbons, the strongest bands appear near 2900, 1460, and 1370 cm^{-1} (see Figure

9.1). In simple cyclic hydrocarbons containing no methyl groups, the latter peak is absent. Other compounds that typically give rise to rather simple infrared spectra include aliphatic sulfides and disulfides, aliphatic halides, and symmetrically substituted alkenes and alkynes.

2. *Negative iodine charge-transfer test.* When iodine is dissolved in compounds containing π electrons or nonbonded electron pairs, a brown solution results. The brown color is due to the formation of a complex between the iodine and the π or nonbonded electrons. These complexes are called charge-transfer or π complexes. Solutions of iodine in nonparticipating solvents are violet.

$$\cdots I_2, \qquad R_2O \cdots I_2$$

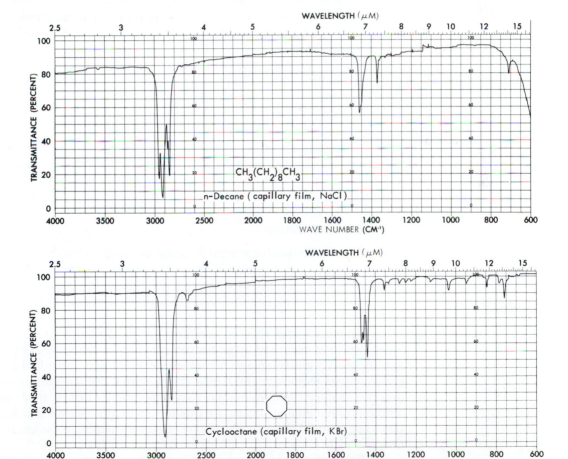

Figure 9.1 Typical infrared spectra of alkanes. Note the absence of a 1370 cm^{-1} band for methyl in the spectrum of cyclooctane.

PROCEDURE: IODINE CHARGE-TRANSFER TEST

On a spot-test plate place a very small crystal of iodine; add one or more drops of the unknown liquid. Saturated hydrocarbons and their fluorinated and chlorinated derivatives, and arenes and halogenated derivatives give violet solutions; all other compounds give brown solutions. The test should be employed only with liquid unknowns. For color comparisons, it is strongly recommended that knowns be run concurrently with the unknowns.

---

3. *Insoluble in cold concentrated sulfuric acid.* Compounds that are unsaturated or possess a functional group containing nitrogen or oxygen are soluble in cold concentrated sulfuric acid. Saturated hydrocarbons, their halogen derivatives, simple aromatic hydrocarbons, and their halogen derivatives are insoluble.

4. *Negative test for halogens.* On the basis of the chemical, physical, and spectroscopic behavior, the most likely compounds to be confused with saturated hydrocarbons are the alkyl halides. If an unknown is suspected of being a hydrocarbon, the absence of halides should be verified by the Beilstein or sodium fusion test. The presence or absence of halogens can also be indicated by quantitative or qualitative measurements of the density of the unknown; hydrocarbons having densities less than one, halides greater than one.

5. *No resonance appears below δ 2 in the NMR spectrum.* In most functionalized compounds, resonance of one or more hydrogens is usually found downfield from δ 2; exceptions include compounds in which the functional group is located on a tertiary carbon and certain alkyl metal and sibyl derivatives.

6. *No absorption above 185 nm in the ultraviolet spectrum.*

Characterization

There are no suitable general chemical methods for the conversion of saturated hydrocarbons to useful derivatives; they are typically inert or undergo nondiscriminate reactions that produce mixtures that are difficult to separate. Saturated hydrocarbons are, therefore, best identified through physical constants and spectroscopic methods.

The most useful physical constants are the boiling point and the refractive index, which is measured by an instrument called a *refractometer* (see Table 9.1). Specific gravity measurements are sometimes employed. Standard pycnometers or gravitometers require careful temperature control and large amounts of material; however, accurate determinations of specific gravity can be made on very small amounts of material by means of a capillary technique (see Section 3.5). Retention times in gas-liquid chromatography are often very helpful in the tentative identification of a hydrocarbon.

Of the spectroscopic methods, mass spectroscopy is perhaps the most useful for the final identification of the saturated hydrocarbons. The utility of the method is

TABLE 9.1 Physical Properties of Alkanes, Alkenes, and Alkynes

Compound	bp (°C)*	D_4^{20}†	n_D^{20}†
Alkanes			
2,2-Dimethylpropane	9.5	0.596	1.3513
2-Methylbutane	28	0.620	1.3580
Pentane	36	0.626	1.3577
Cyclopentane	49.3	0.746	1.4068
2,2-Dimethylbutane	49.7	0.649	1.3689
2,3-Dimethylbutane	58	0.662	1.3750
2-Methylpentane	60.3	0.653	1.3716
3-Methylpentane	63.3	0.664	1.3764
Hexane	68.3	0.659	1.3749
Cyclohexane (mp 6.5°C)	80.7	0.778	1.4264
3,3-Dimethylpentane	86	0.693	1.3911
2-Methylhexane	90	0.679	1.3851
3-Methylhexane	92	0.687	1.3887
3-Ethylpentane	93.5	0.698	1.3934
Heptane	98.4	0.684	1.3877
2,2,4-Trimethylpentane	99.2	0.692	1.3916
Methylcyclohexane	100.9	0.769	1.4231
2,2-Dimethylhexane	106.8	0.695	1.3930
2,5-Dimethylhexane	109.1	0.694	1.3930
2,4-Dimethylhexane	109.4	0.700	1.3953
3,3-Dimethylhexane	112	0.708	1.3992
Cycloheptane	118–20	0.8099	1.4440
Octane	125.7	0.703	1.3975
Ethylcyclohexane	131.7	0.788	1.4332
Nonane	150.8	0.718	1.4056
Cyclooctane (mp 14°C)	150/750 mm	0.8349	1.4586
Cyclononane (mp 9.7°C)	170–2	0.8534	1.4328[16]
Decane	174	0.730	1.4114
trans-Decahydronaphthalene	187.2	0.870	1.4695
cis-Decahydronaphthalene	195.7	0.896	1.4810
Undecane	196	0.702	1.4190
Dodecane	217	0.749	1.4216
Tridecane	235.5	0.7563	1.4256
Tetradecane	253.5	0.764	1.4289
Pentadecane	270.7	0.769	1.4310
Hexadecane	287	0.773	1.4352
Heptadecane	302.6	0.7767	1.4360[25]
Octadecane	317.4	0.7767	1.4367[28]
Nonadecane	331.6	0.7776	
Eicosane	345.1	0.7777	1.4307[50]
Alkenes			
Methylpropene	−6.9	0.6266[−6.6]	1.3462
1-Butene	−6.3	0.6255[−6.5]	1.3462
cis-2-Butene	3.73	0.6303[1]	
3-Methyl-1-butene	20	0.6320	1.3643
1,4-Pentadiene	26.1	0.6607	1.3887

(table continues)

TABLE 9.1   (Continued)

Compound	bp (°C)*	D_4^{20}†	n_D^{20}†
1-Pentene	30.1	0.6410	1.3710
2-Methyl-1-butene	31	0.6504	1.3778
trans-2-Pentene	35.85	0.6486	1.3790
cis-2-Pentene	37.0	0.6562	1.3822
2-Methyl-2-butene	38.5	0.6620	1.3878
3,3-Dimethyl-1-butene	41	0.652	1.3766
Cyclopentene	44.2	0.7736	1.4225
1,5-Hexadiene	60	0.692	1.4045
1-Hexene	64	0.673	1.3875
trans-3-Hexene	67	0.677	1.3940
trans-2-Hexene	69	0.699	1.3929
Cyclohexene	83	0.8088	1.4465
1-Heptene	93.6	0.6971	1.3998
trans-2-Heptene	98.5	0.701	1.4035
2,4,4-Trimethyl-1-pentene	101.2	0.7151	1.4082
Cycloheptene	112	0.824	1.4585
2-Ethyl-1-hexene	120	0.7270	1.4157
1-Octene	121.3	0.7160	1.4088
Styrene	145.2	0.9056	1.5470
1-Nonene	147	0.7292	1.4154
Allylbenzene	157	0.8912	1.5042[25]
α-Methylstyrene	165.4	0.9106	1.5386
1-Decene	170.5	0.7408	1.4220
Indene	181	0.9915	1.5764
trans-Stilbene (mp 180.5°C)	306	0.971	

Alkynes

Compound	bp (°C)*	D_4^{20}†	n_D^{20}†
1-Pentyne	39.7	0.6945	1.3847
2-Pentyne	55	0.710	1.4045
1-Hexyne	71	0.719	1.3990
3-Hexyne	81	0.724	1.4115
1-Heptyne	100	0.7338	1.4084
1-Octyne	131.6	0.7470	1.4172
Ethynylbenzene	141.7	0.9281	1.5485

* bp at 760 mm Hg pressure unless otherwise noted.
† Superscript numbers indicate the temperature (°C) at which the value was determined.

greatly enhanced by the large number of reference spectra available. For the larger saturated hydrocarbons, NMR is of limited utility because of the small differences in chemical shift of the various types of hydrogens. In some cases, integration of the NMR spectrum may provide information on the ratio of methyl to other types of hydrogens in the molecule. In those cases in which the ring bears hydrogens, NMR is the best way to detect the presence of cyclopropanes because of their unique high field resonance. Careful examination of the infrared spectrum can provide information about certain structural features of an unknown hydrocarbon, for example, geminal methyl groups, but typically cannot conclusively lead one to the assignment of structure. There is even some danger in assigning structures to hydrocarbons based upon comparison of infrared spectra because of the small differences in the relatively simple spectra of compounds with closely related structures, for example, 4-methyl- and 5-methylnonane. Comparative gas-liquid chromatography is used in distinguishing hydrocarbons once some idea of the structure or molecular weight is available.

9.1.2 Alkenes

Classification

The infrared spectra of alkenes are generally considerably more complex than those of the saturated hydrocarbons (see Figures 9.1 and 9.2). The $C = C$ absorption appears in the 1680 to 1620 cm^{-1} region; unfortunately, this band varies from very intense to entirely absent in completely symmetrical alkenes. However, unsaturation can also be detected by the vinyl hydrogen band just above 3000 cm^{-1} (if they are not hidden by the stronger, main band below 3000 cm^{-1}) or by the C—H out-of-plane bending bands in the 1000 to 700 cm^{-1} region. In many cases, the latter may be intense and very diagnostic. The presence of vinyl and allylic hydrogens can often be inferred from NMR spectra by the characteristic bands near δ 5 to δ 7 and near δ 2, respectively. Unconjugated alkenes show only simple end absorption in the ultraviolet spectra recorded to 200 nm. The characteristic maxima and extinction coefficients at higher wavelengths for conjugated systems make ultraviolet spectroscopy quite suitable for dealing with conjugated systems.

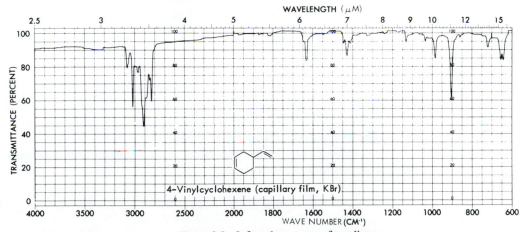

Figure 9.2 Infrared spectrum of an alkene.

As noted in this section, alkenes can be distinguished from saturated hydrocarbons by the positive iodine charge-transfer test and by their solubility in concentrated sulfuric acid.

Compounds suspected of being alkenes or alkynes should be tested with *bromine in carbon tetrachloride.* The majority of alkenes and alkynes add bromine quite rapidly; ethylenes or acetylenes substituted with electronegative groups add bromine slowly or not at all.

Addition:

$$\ce{\underset{/}{\overset{\backslash}{C}}=\underset{\backslash}{\overset{/}{C}} + Br_2 \longrightarrow -\underset{|}{\overset{|}{C}}-\underset{|}{\overset{|}{C}}-}$$

with Br on each carbon.

Substitution: $\ce{-COCH_3 + Br_2 \longrightarrow -COCH_2Br + HBr}$ (gas)

Substitution reactions accompanied by the liberation of hydrogen bromide gas occur with phenols, amines, enols, aldehydes, ketones, and other compounds containing active methylene groups. With amines, the first mole of hydrogen bromide is not evolved but reacts with the amine to form the amine hydrobromide.

PROCEDURE: BROMINE IN CARBON TETRACHLORIDE

Dissolve 50 mg or two drops of the unknown in 1 mL of carbon tetrachloride, and add dropwise a 2% solution of bromine in the solvent until the bromine color persists. If more than two drops of the bromine solution are required to cause the color to remain for 1 minute, an addition or substitution reaction has occurred. Substitution reactions are indicated by the evolution of hydrogen bromide, which is insoluble in carbon tetrachloride. The evolved hydrogen bromide can be detected by blowing across the top of the tube and noting the fog that is produced, or better, by denoting the acidic reaction on dampened pH paper held across the mouth of the tube.

A *permanganate test* can also be used to detect unsaturation. The positive test given by alkenes and alkynes is indicated by the disappearance of the purple permanganate ion and the appearance of the sparingly soluble brown manganese oxide. Positive tests are also given by phenols, aryl amines, aldehydes, primary and secondary alcohols (reaction often slow), and all organic sulfur compounds in which the sulfur is in a reduced state.

$$3 \ce{\underset{/}{\overset{\backslash}{C}}=\underset{\backslash}{\overset{/}{C}}} + 2\ \ce{KMnO_4} + 4\ \ce{H_2O} \longrightarrow 3\ \ce{-\underset{HO}{\overset{|}{C}}-\underset{OH}{\overset{|}{C}}-} + 2\ \ce{MnO_2} + 2\ \ce{KOH}$$

PROCEDURE: PERMANGANATE TEST

Dissolve 25 mg or one drop of the unknown in 2 mL of water or reagent grade acetone in a small test tube. Add dropwise with vigorous shaking a 1% aqueous solution of potassium permanganate. If more than one drop of the permanganate is reduced, the test is positive.

Characterization

Although alkenes undergo numerous reactions, mainly involving addition to or cleavage of the carbon-carbon double bond, relatively few of these reactions have wide general applicability for the preparation of suitable derivatives. Like saturated hydrocarbons, many of the simple alkenes can be identified solely by reference to physical and/or spectral properties (see Table 9.1).

Upon oxidation, alkenes yield aldehydes, ketones, and/or acids. The product obtained depends upon the reagent and the conditions as well as on the degree of substitution about the double bond. *Ozonization* is the classical method for determining positions of double bonds in alkenes. The ozonides that are formed in these reactions are usually not isolated since they are often unstable and may decompose with explosive violence. Hydrolysis of ozonides with water and reduction with zinc yield aldehydes and ketones. When the ozonides are worked up in the presence of acidic hydrogen peroxide, the aldehydic components are oxidized to carboxylic acids. Ozonization reactions have the advantage that they are relatively simple to run and work up. An excellent technique for the ozonolysis of double bonds is to add the ozone to a solution of the alkene in dichloromethane: pyridine (2:1) at $-70°C$. The reaction mixture is then allowed to slowly warm to room temperature, is poured into water, and the aldehydes and/or ketones are isolated by ether extracton. The pyridine acts as a reducing agent. Ozone can be generated from oxygen by a commercially available apparatus or can be purchased diluted with Freon-13 in stainless steel cylinders.

$$R_2C{=}CHR' \xrightarrow{O_3} R_2C\underset{O}{\overset{O-O}{\diagdown \diagup}}CHR' \left\{ \begin{array}{l} \xrightarrow{\text{pyridine}} R_2C{=}O + R'CHO \\ \xrightarrow[H^+]{H_2O_2} R_2C{=}O + R'COOH \end{array} \right.$$

The Diels–Alder reaction may occasionally be found to be the most suitable method for preparing a derivative of an alkene or a diene.[1] Maleic anhydride and *N*-phenylmaleimide are among the most suitable of the dienophiles. They are rather reactive and usually form solid derivatives that can be characterized by further

---

[1] M. C. Kwitzel, *Org. Reactions,* **4,** 1 (1948); H. L. Holmes, *Org. Reactions,* **4,** 60 (1948).

chemical conversion if necessary. Diphenylisobenzofuran[2] is recommended as a highly reactive diene that can be used to trap reactive alkenes and alkynes.[3]

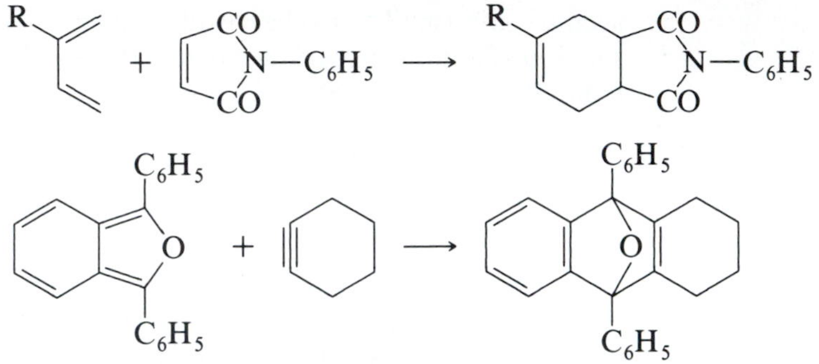

9.1.3 Alkynes and Allenes

Alkynes and allenes, like alkenes, are soluble in concentrated sulfuric acid. In addition they add bromine in carbon tetrachloride and are oxidized by cold aqueous permanganate.

Classification of Alkynes

There are two general types of alkynes—terminal and nonterminal. The terminal alkynes are easily detected by their very characteristic infrared spectra; terminal alkynes show a sharp characteristic $\equiv$C—H band at 3310 to 3200 cm^{-1} and a triple-bond stretch of medium intensity at 2140 to 2100 cm^{-1}. In the NMR, the $\equiv$C—H occurs as a singlet near δ 2.8 to 3.0. The nonterminal alkynes display triple-bond absorption at 2250 to 2150 cm^{-1}, which may be absent in symmetrically substituted cases.

Characterization of Alkynes

Like other hydrocarbons, the characterization of alkynes is heavily dependent on physical properties (Table 9.1). Only a few general methods have been developed for the preparation of solid derivatives.

Alkynes can be *hydrated to ketones* by the action of sulfuric acid and mercuric sulfate in dilute alcohol solution. Terminal alkynes give methyl ketones; nonterminal, unsymmetrically substituted alkynes often give mixtures of ketones.

$$RC\equiv CR + H_2O \xrightarrow[H_2SO_4]{HgSO_4} RCOCH_2R$$

---

[2] M. S. Newman, *J. Org. Chem.*, **26**, 2630 (1961).

[3] G. Wittig and R. Pohlke, *Chem. Ber.*, **94**, 3276 (1961).

PROCEDURE: HYDRATION OF ALKYNES

A solution of 0.2 g of mercuric sulfate and three to four drops of concentrated sulfuric acid in 5 mL of 70% aqueous methanol is warmed to 60°C. The alkyne (0.5 g) is added dropwise, and the solution is maintained at 60°C with stirring for 1 to 2 hours. The methanol is distilled off, and the residue is saturated with salt and extracted with ether. Appropriate spectral and physical properties of the ketone are determined, and a solid derivative is made.

Allenes

Allenes are usually identified by spectral and physical characteristics. The best method of detection is by the presence of a strong stretching vibration, often observed as a doublet, in the 2200 to 1950 cm^{-1} region. Allenes absorb only below 200 nm in the ultraviolet. In the NMR, terminal hydrogens are near δ 4.4 and allenic hydrogens near δ 4.8 with ^4J values of up to 7 Hz. Allenes are capable of exhibiting optical isomerism.

9.1.4 *Aromatic Hydrocarbons*

Classification

Benzene and a large number of its alkylated derivatives are liquids. Most compounds containing more than one aromatic ring, either fused or nonfused, are solids. Aromatic hydrocarbons burn with a characteristic sooty flame and give negative tests with bromine in carbon tetrachloride and alkaline permanganate (see Section 9.1.2). The simple aromatic hydrocarbons are insoluble in sulfuric acid but soluble in fuming sulfuric acid. With two or more alkyl substituents, the nucleus is sufficiently reactive to sulfonate, and such compounds dissolve in concentrated sulfuric acid.

The infrared spectra of aromatic hydrocarbons exhibit the expected aromatic (and, in the case of alkylated nuclei, aliphatic) C—H stretching and bending deformations, plus characteristic ring absorption bands (see Section 5.2.1). The spectra show only bands of moderate intensity above 1000 cm^{-1} and generally only moderate to weak bands between 1000 and 800 cm^{-1} (see Figure 9.3).

In the NMR spectra, the presence, position, and, especially, intensity (integration) of the aryl hydrogens, plus the chemical shift and coupling pattern of any aliphatic protons, provide important and often definitive structural information.

In all alkyl benzenes, the substituted tropylium ion is an important and diagnostic fragment in the mass spectra.

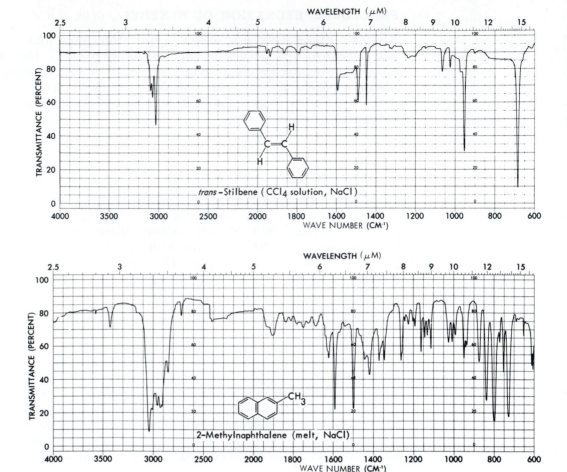

Figure 9.3 Typical infrared spectra of aromatic hydrocarbons. Note the absence of a $v_{C=C}$ in the spectrum of *trans*-stilbene.

Characterization

The *nitro* and *polynitro derivatives* are generally useful in identification of aromatic hydrocarbons. In certain cases, it is advantageous to reduce mononitro compounds to primary amines from which solid derivatives are made.

The nitration of aromatic compounds should always be done with great caution and on a small scale, especially when dealing with unknowns. A number of different procedures are used for nitration. The conditions that should be employed depend on the reactivity of the substrate and on the degree of nitration desired. It is very helpful to know something about the structure and, hence, reactivity of the com-

pound before nitration is attempted. The methods in the following nitration procedures are listed in order of increasingly vigorous conditions. Method (b) usually
leads to mononitro compounds, except for highly reactive substrates such as phenol
and acetanilide. Method (c) usually gives di- or trinitro derivatives.

PROCEDURE: NITRATION

(a) Reflux a mixture of 0.5 g of the compound, 2 mL of glacial acetic acid, and
0.5 mL of fuming nitric acid for 10 minutes. Pour into ice water. Filter the precipitate, wash with cold water, and recrystallize from aqueous ethanol.

(b) Add 0.5 to 1 g of the compound to 2 to 4 mL of concentrated sulfuric acid.
An equal volume of concentrated nitric acid is added drop by drop, with shaking after
each addition. Heat on a water bath at 60°C for 5 to 10 minutes. Remove the tube
from the bath and shake every few minutes. Cool and pour into ice water. Filter off
the precipitate, wash with water, and recrystallize from aqueous ethanol.

(c) Follow the foregoing procedure, substituting fuming nitric acid for concentrated nitric acid. Heat the mixture for 10 minutes on a steam bath.

Aromatic hydrocarbons (and halogen derivatives) react with phthalic anhydride under Friedel–Crafts conditions to yield aroylbenzoic acids, which can be
characterized by their melting points and neutralization equivalents.

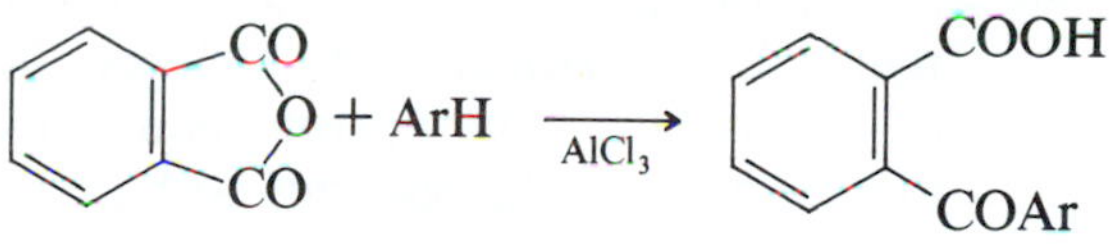

PROCEDURE: AROYLBENZOIC ACIDS

⟿   C A U T I O N :   Carbon disulfide is highly inflammable.

In a small apparatus equipped with a reflux condenser and a hydrogen chloride trap,
place 0.4 g of phthalic anhydride, 10 mL of carbon disulfide, 0.8 g of anhydrous
aluminum chloride, and 0.4 g of the aromatic hydrocarbon. Heat the mixture on a
water bath until no more hydrogen chloride is evolved, or about 30 minutes. Cool
under tap water. Decant the carbon disulfide layer slowly. Add 5 mL of concentrated hydrochloric acid with cooling and then 5 mL of water to the residue. Shake or
stir the mixture thoroughly. Cool if necessary to induce crystallization. Collect the
solid, wash with cold water, and recrystalize from aqueous ethanol. If the product
fails to crystallize, take the oil up in dilute ammonium hydroxide, treat with activated
carbon, filter, cool, and neutralize with concentrated hydrochloric acid.

With chromic acid or alkaline permanganate, *alkyl benzenes are oxidized to aryl carboxylic acids.* This procedure is recommended when there is one side chain, in which case benzoic or a substituted benzoic acid is obtained, or when there are two *ortho* side chains, in which case *o*-phthalic acids are obtained (the melting points of a number of substituted *o*-phthalic acids and anhydrides have been recorded). Permanganate oxidation is usually the preferred procedure.

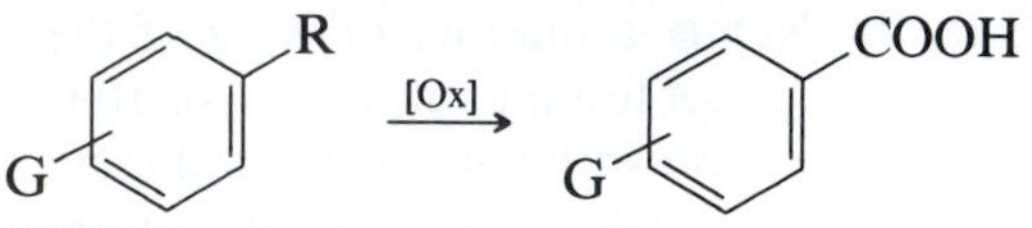

PROCEDURE: PERMANGANATE OXIDATION

In a small flask equipped for reflux, place 1.5 g of potassium permanganate, 25 mL of water, 6.5 mL of 6 N potassium hydroxide, two boiling chips, and 0.4 to 0.5 g of the alkyl benzene. Reflux gently for 1 hour, or until the purple color of the permanganate has disappeared. Cool the reaction mixture and acidify with dilute sulfuric acid. Heat to boiling; add a pinch of sodium hydrogen sulfite if necessary to destroy any brown manganese dioxide. Cool and filter the acid. Purify by recrystallization from water or aqueous ethanol, or by sublimation.

Polynitro aromatic compounds from stable charge-transfer complexes[4] with many electron-rich aromatic systems. The *complexes of 2,4,7-trinitro-9-fluorenone (TNF)* and aromatic hydrocarbons, or derivatives, are of considerable utility in identification work, especially with polynuclear systems.[5]

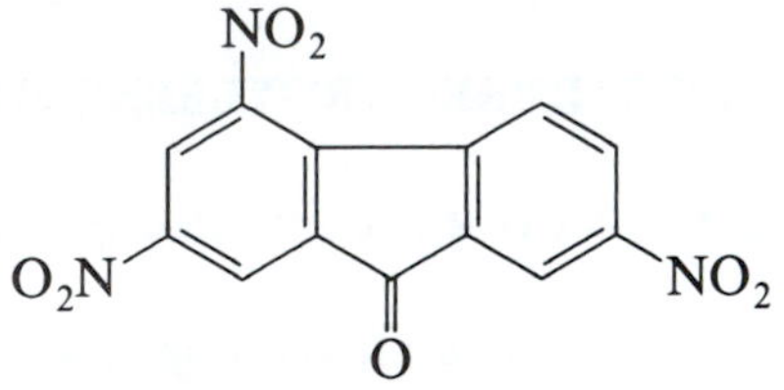

2,4,7-trinitro-9-fluorenone (TNF)

Table 9.2 lists the physical properties and derivatives of selected aromatic hydrocarbons.

---

[4] L. N. Ferguson, *The Modern Structural Theory of Organic Chemistry* (Englewood Cliffs, N.J.: Prentice-Hall, 1963).

[5] M. Orchin and E. O. Woolfolk, *J. Am. Chem. Soc.,* **68,** 1727 (1946); M. Orchin, L. Reggel, and E. O. Woolfolk, *J. Am. Chem. Soc.,* **69,** 1225 (1947); D. E. Laskowski and W. C. McCrone, *Anal. Chem.,* **30,** 1947 (1958).

TABLE 9.2   Physical Properties of Aromatic Hydrocarbons and Derivatives

| | Liquids | | | mp (°C) of Derivatives | | |
| | | | | Nitro | | Aroylbenzoic Acid |
Compound	bp (°C)	D_4^{20}	n_D^{20}	Position	mp (°C)	mp (°C)
Benzene	80.1	0.8790	1.5011	1,3	89	127
Methylbenzene	110.6	0.8670	1.4969	2,4	70	137
Ethylbenzene	136.2	0.8670	1.4959	2,4,6	37	122
1,4-Dimethylbenzene (*p*-Xylene)	138.3	0.8611	1.4958	2,3,5	139	132
1,3-Dimethylbenzene (*m*-Xylene)	139.1	0.8642	1.4972	2,4,6	183	126
1,2-Dimethylbenzene (*o*-Xylene)	144.4	0.8802	1.5054	4,5	118	178
Isopropylbenzene (Cumene)	152.4	0.8618	1.4915	2,4,6	109	133
Propylbenzene	159.2	0.8620	1.4920			125
1,3,5-Trimethylbenzene	164.7	0.8652	1.4994	2,4	86	211
tert-Butylbenzene	169.1	0.8665	1.4926	2,4	62	
1,2,4-Trimethylbenzene	169.2	0.8758	1.5049	3,5,6	185	
Isobutylbenzene	172.8	0.8532	1.4865			
sec-Butylbenzene	173.3	0.8621	1.4901			
1,2,3-Trimethylbenzene	176.2	0.8944	1.5139			
1-Isopropyl-4-methylbenzene	177.1	0.8573	1.4909	2,6	54	123
Indene	182	0.857				
Butylbenzene	183.3	0.8601	1.4898			
1-Methyl-4-propylbenzene	183.5	0.859	1.493			
1,2,3,4-Tetramethylbenzene	197.9	0.8899	1.5125	4,6	181	213
Pentylbenzene	205.4	0.8585	1.4878			
Tetrahydronaphthalene	206	0.971		5,7	95	153
p-Diisopropylbenzene	210.4	0.8568	1.4898			
1-Methylnaphthalene	244.8	1.0200	1.6174	4	71	68
2-Ethylnaphthalene	257.9	0.9922	1.5995			
1-Ethylnaphthalene	258.3	1.0076	1.6052			
1,6-Dimethylnaphthalene	262	1.003	1.607			
1,7-Dimethylnaphthalene	263	1.012	1.607			
1,3-Dimethylnaphthalene	263	1.002	1.6078			
1,4-Dimethylnaphthalene	263	1.008	1.6127			
1,1-Diphenylethane	270	1.003	1.5761			

| | Solids | | mp (°C) of Derivatives | | |
| | | | Nitro | | Aroylbenzoic Acid |
Compound	mp (°C)	bp (°C)	Position	mp (°C)	mp (°C)
Diphenylmethane	25.3		2,2′,4,4′	172	
2-Methylnaphthalene	34.4	241	1	81	
Pentamethylbenzene	51		6	154	
1,2-Diphenylethane	53		4,4′	180	
			2,2′,4,4′	169	
Biphenyl	69.2		4,4′	237	224
			2,2′,4,4′	150	
Naphthalene	80.2	218	1	61	172

(table continues)

TABLE 9.2   (Continued)

Compound	Solids mp (°C)	bp (°C)	mp (°C) of Derivatives Nitro Position	Nitro mp (°C)	Aroylbenzoic Acid mp (°C)
Triphenylmethane	92		4,4′,4″	206	
Acenaphthene	96.2		5	101	198
Phenanthrene	96.3				
2,3-Dimethylnaphthalene	104				
2,6-Dimethylnaphthalene	111				
Fluorene	113.5	295	2	156	227
Pyrene	148	385			
Hexamethylbenzene	165				
Anthracene	216				

9.2   HALIDES

General Classification

Halogen substituents can be found in combination with all other functional groupings. When another functional group is present, chemical transformations for the purpose of making derivatives are usually carried out on that functional group rather than the halogen, for example, esters are made from chloroacids, carbonyl derivatives are made from chloroketones, and urethanes are made from bromoalcohols.

The presence or absence of halogen in an unknown can seldom be inferred from its infrared spectrum (see Figure 9.4). The carbon fluorine stretching absorptions occur in the 1350 to 960 cm^{-1} region, those for chlorine occur in the 850 to 500 cm^{-1} region (often obscured by chlorinated solvents or aromatic substitution bands), while those for bromine and iodine occur below 667 cm^{-1} and are not observed in the normal infrared region.

The presence or absence of halogen can be determined by the Beilstein test, analysis of the filtrate from a sodium fusion, or by application of the silver nitrate or sodium iodide tests (see below). The very characteristic cracking patterns and/or isotope peaks observed in the mass spectra of halogenated compounds provide definitive evidence for the presence and types of halogen(s). The presence of halogen is sometimes suggested by the characteristic chemical shift in the NMR of hydrogens on the carbon atom bearing the halogen. The presence of fluorine can be demonstrated by NMR spectroscopy. In hydrogen spectra, hydrogen-fluorine coupling can be observed; in fluorine spectra, the relative number and kinds of fluorines present can be determined through integration and measurement of characteristic fluorine chemical shifts.

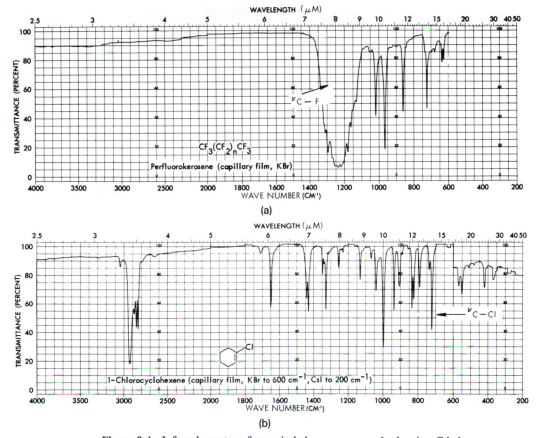

Figure 9.4 Infrared spectra of organic halogen compounds showing C-halogen stretching bands.

In addition to alkyl and aryl halides, halogen occurs in the following types of compounds:

1. Salts (amine hydrochlorides, ammonium, phosphonium and sulfonium salts, etc.).
2. Acid halides (acyl halides, sulfonyl halides, etc.).
3. Haloamines and haloamides (*N*-bromosuccimide, Chloramine-T, etc.).
4. Miscellaneous compounds (iodobenzene dichlorides, alkyl hypochlorites, halosilanes, alkyl mercuric halides, and other organometallic halides, etc.).

Alkyl and aryl halides (along with hydrocarbons, diaryl ethers, and most perfluoro compounds) are insoluble in concentrated sulfuric acid.

The following tests are useful in classifying the various types of alkyl and aryl halogen compounds. It is advisable to test the reactivity of any unknown halide toward each reagent.

Alcoholic Silver Nitrate

$$AgNO_3 + X^- \longrightarrow AgX + NO_3^-$$

The test depends on the rapid and quantitative reaction of silver nitrate with halide ion to produce insoluble silver halide (excepting the fluoride).

Silver nitrate in aqueous or ethanolic solution gives an immediate precipitate with compounds that contain ionic halide, or with compounds such as acid chlorides, which react immediately with water or ethanol to produce halide ion. Many other halogen-containing substances react with silver nitrate to produce insoluble silver halide. The rate of such reactions is a measure of the reactivity of the substrate in S_N1 reactions.

$$RX \longrightarrow R^+ + X^-$$

Silver and other heavy metal salts catalyze S_N1 reactions of alkyl halides by complexation with the unshared electrons of the halide, making the leaving group a metal halide rather than halide ion.

$$R-\ddot{X}: \xrightarrow{Ag^+} [R-\overset{\delta+}{\ddot{X}}:\cdots\overset{\delta+}{Ag}] \xrightarrow{slow} R^+ + AgX \downarrow$$

$$R^+ \longrightarrow \text{alcohols, ethers, or alkenes}$$

The rate of precipitation of silver halide depends on the leaving group, $I >$ Br $>$ Cl (silver fluoride is soluble), and upon the structure of the alkyl group. Any structural factors that stabilize the electron-deficient carbocation (R^+) will accelerate the reaction. We thus find the expected order of reactivity

benzyl ≈ allyl > tertiary > secondary > primary > methyl ≫ vinyl ≈ aryl

The following equations illustrate other structural classes that are unusually reactive in S_N1 reactions.

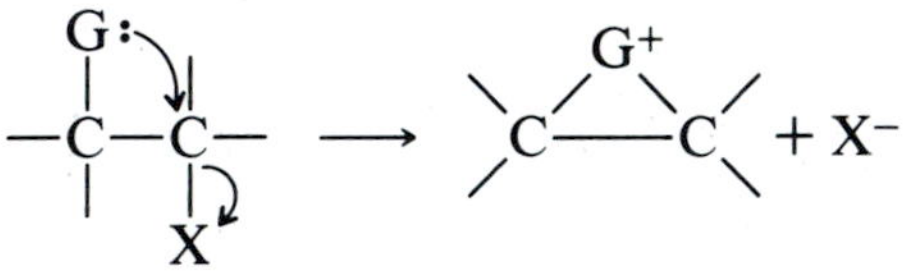

G = amino, sulfide, iodide, aryl, etc.

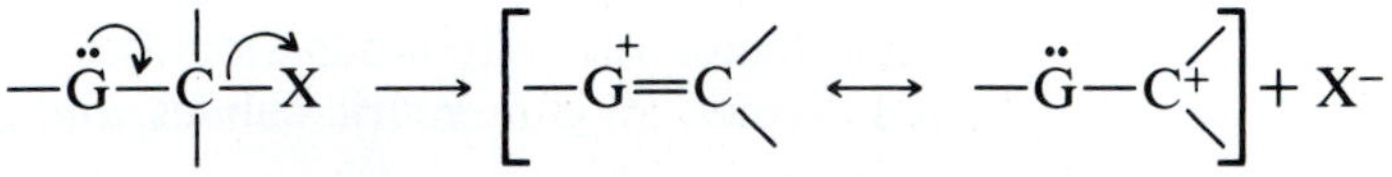

G = amino, alkoxy, thioalkoxy, etc.

Factors that tend to destabilize the incipient carbonium ion decelerate the rate of S_N1 reactions. The following compounds are not reactive under S_N1 conditions:

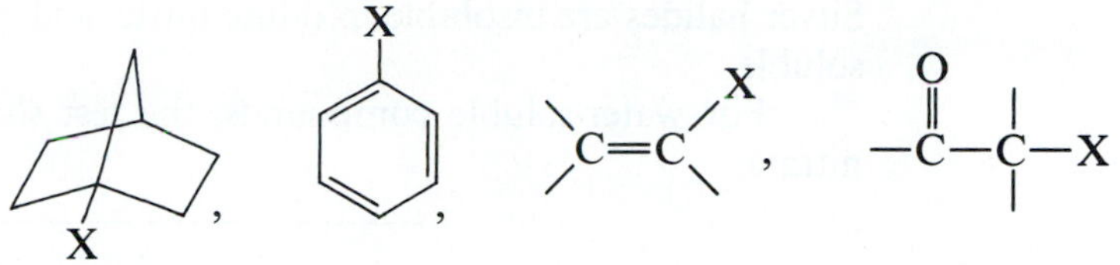

A summary of types of compounds in order of decreasing reactivity of silver nitrate with various organic compounds containing halogen is as follows (X=I, Br, or Cl):

1. Water-soluble compounds that give immediate precipitates with aqueous silver nitrate.
   (a) salts
   (b) halogen compounds that are hydrolyzed immediately with water, such as low molecular weight acyl or sulfonyl halides
2. Reactivity of water-insoluble compounds with alcoholic silver nitrate.
   (a) immediate precipitation

$RCOCl$ RSO_2Cl $RO{-}\overset{|}{\underset{|}{C}}{-}X$

$R{-}CH{=}CH{-}CH_2{-}X$ $R{-}CHBrCH_2Br$

$ArCH_2X$ R_3CX RI

   (b) silver halide precipitates slowly or not at all at room temperature, but readily when warmed on the steam bath

RCH_2Cl $R_2CHClBr$ $RCHBr_2$

   (c) inert toward hot silver nitrate solutions

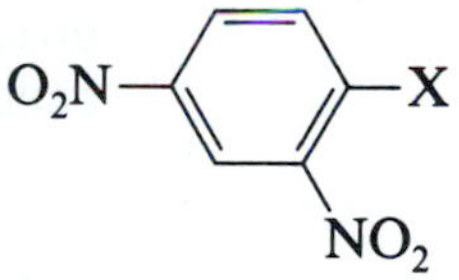

ArX $RCH{=}CHX$ $HCCl_3$ $-\overset{O}{\overset{\|}{C}}-CH_2X$

PROCEDURE: SILVER NITRATE TEST

Add one drop (or several drops of an ethanol solution) of the halogen compound to 2 mL of 2% ethanolic silver nitrate. If no reaction is observed after 5 minutes, heat the solution to boiling for several minutes. Note the color of any precipitate formed. Add two drops of 5% nitric acid. Certain organic acids give insoluble silver salts.

Silver halides are insoluble in dilute nitric acid; the silver salts of organic acids are soluble.

For water-soluble compounds, the test should be run using aqueous silver nitrate.

Sodium Iodide in Acetone

$$NaI + RX \xrightarrow[\text{acetone}]{} RI + NaX$$
$$X = Br \text{ or } Cl$$

The rate of the reaction of sodium iodide in acetone with a compound containing covalent bromine or chlorine is a measure of the reactivity of the compound toward bimolecular nucleophilic substitution (SN2 reaction). The test depends on the fact that sodium iodine is readily soluble in acetone, whereas sodium chloride and bromide are only slightly soluble. (Sodium fluoride is soluble in acetone; thus no precipitate formation is observed.)

SN2 reactions are concerted and involve backside nucleophilic displacements that result in inversion of the configuration of the carbon atom at the reaction site.

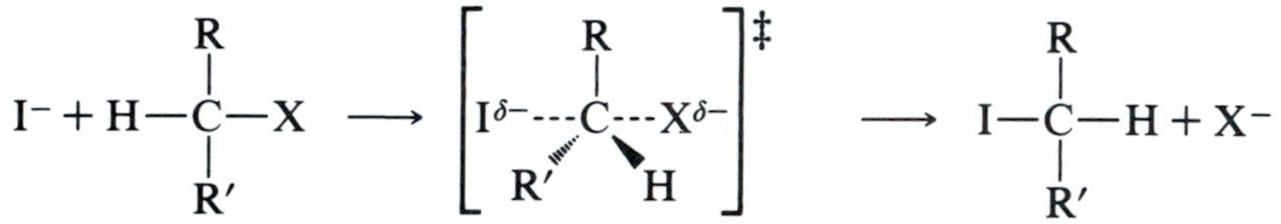

$$I^- + H - \underset{R'}{\overset{R}{\underset{|}{\overset{|}{C}}}} - X \longrightarrow \left[I^{\delta-} \cdots \underset{R' \quad H}{\overset{R}{\underset{|}{C}}} \cdots X^{\delta-} \right]^{\ddagger} \longrightarrow I - \underset{R'}{\overset{R}{\underset{|}{\overset{|}{C}}}} - H + X^-$$

As might be expected from the consideration of the transition state, such reactions are relatively insensitive to electronic effects of the R groups, but steric effects are of primary importance. In contrast to SN1 reactions, the rates of SN2 reactions of alkyl halides follow the order

$$methyl > primary > secondary > tertiary$$

It should also be noted that α-halocarbonyl compounds and α-halonitriles are highly reactive under SN2 conditions. This has been attributed to stabilization of the transition state by a partial distribution of the charge from the entering and leaving groups onto the carbonyl oxygen. Also important may be the fact that the stereochemistry of the carbonyl and nitrile groups provides a favorable environment for nucleophilic attack.

A summary of the results to be expected in the sodium iodide in acetone test follows:

1. Precipitate within 3 minutes at room temperature.
   (a) primary bromides
   (b) acyl halides

     (c) allyl halides

     (d) α-haloketones, -esters, -amides, and -nitriles

     (e) carbon tetrabromide

2. Precipitate only when heated up to 6 minutes at 50°C.

     (a) primary and secondary chlorides

     (b) secondary and tertiary bromides

     (c) geminal polybromo compounds (bromoform)

3. Do not react in 6 minutes at 50°C.

     (a) vinyl halides

     (b) aryl halides

     (c) geminal polychloro compounds (chloroform, carbon tetrachloride, trichloroacetic acid)

4. React to give precipitate and also liberate iodine.

     (a) vicinal halides

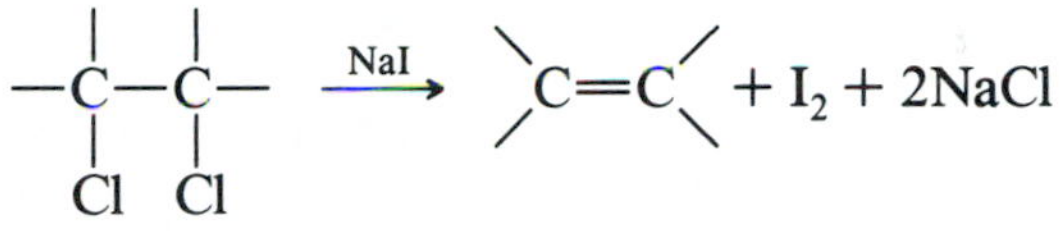

$$-\underset{\underset{Cl}{|}}{\overset{|}{C}}-\underset{\underset{Cl}{|}}{\overset{|}{C}}- \xrightarrow{\text{NaI}} ^{\backslash}\!/ C\!=\!C^{\diagup}_{\backslash} + I_2 + 2NaCl$$

     (b) sulfonyl halides

$$ArSO_2Cl + NaI \longrightarrow ArSO_2I + NaCl \quad \text{(immediate)}$$

$$\text{NaI} \downarrow$$

$$ArSO_2Na + I_2$$

     (c) triphenylmethyl halides

$$(C_6H_5)_3CX + NaI \longrightarrow NaX + (C_6H_5)_3CI \xrightarrow{\text{NaI}} \text{Dimer} + I_2$$

PROCEDURE: SODIUM IODIDE IN ACETONE

Reagent: Dissolve 15 g of sodium iodide in 100 mL of pure acetone. The reagent should be stored in a dark bottle. On standing, it slowly discolors. It should be discarded when a definite red-brown color develops.

    To 1 to 2 mL of the reagent in a small test tube, add two drops (or 0.1 g in a minute volume of acetone) of the compound. Mix well and allow the solution to stand at room temperature for 3 to 4 minutes. Note whether a precipitate forms and whether the solution acquires a red-brown color. If no change occurs at room temperature, warm the tube in a water bath at 50°C for 6 minutes. Cool to room temperature and record any changes.

9.2.2 Alkyl Halides

Characterization

A number of the commonly encountered alkyl halides can be identified by reference to physical and spectral properties or conversion to suitable derivatives. A good number of alkyl halides are readily available in most organic laboratories. The utility of comparative glc and infrared spectroscopy should not be overlooked. Useful methods for preparing solid derivatives of polyhalogenated nonbenzenoid hydrocarbons are not available. Identification in such cases rests solely on spectral and physical data and chromatographic comparisons.

The most readily formed solid derivatives of primary and secondary alkyl halides are the *S-alkylthiuronium picrates*. The initial reaction involves the direct displacement of halide ion by the strongly nucleophilic thiourea; tertiary halides do not react.

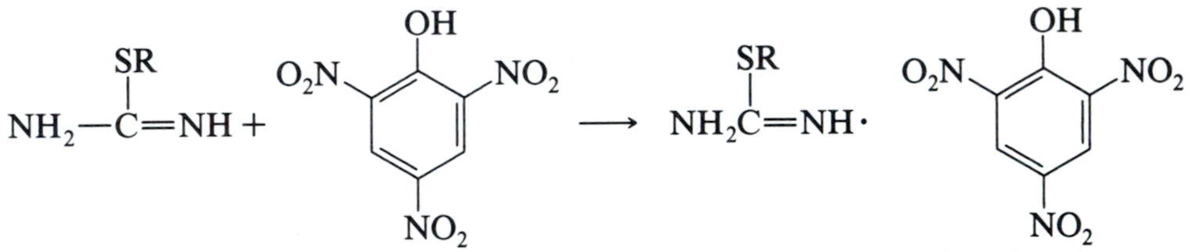

S-alkylthiuronium picrate

PROCEDURE: S-ALKYLTHIURONIUM PICRATES

C A U T I O N : Thiourea may be a carcinogen. Picric acid and picratis are explosive.

The alkyl halide (1 mmol) and thiourea (2 mmol) are dissolved in 5 mL of ethylene glycol contained in a small tube or flask fitted with a condenser. The mixture is heated in an oil bath for 30 minutes. Primary alkyl iodides react at 65°C, others require temperatures near 120°C. Add 1 mL of a saturated solution of picric acid in ethanol and continue to heat for an additional 15 minutes. Cool the reaction mixture and add 5 mL of cold water. Allow the mixture to stand in an ice bath. Collect the precipitate and recrystallize from methanol.

Alkyl 3,5-dinitrobenzoates can be prepared by reaction of alkyl iodides with the silver salt of 3,5-dinitrobenzoic acid.

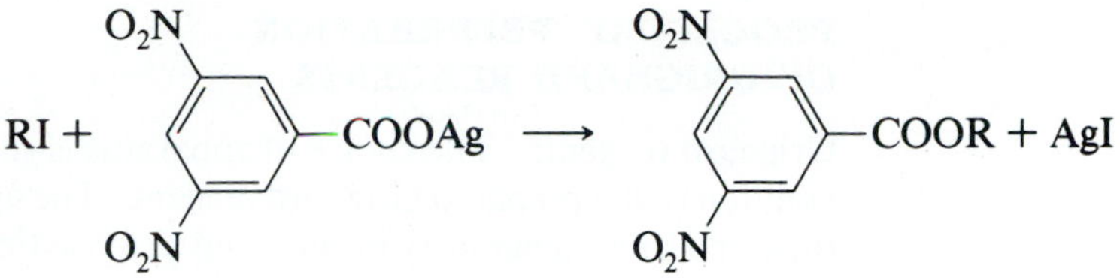

$$RI + \underset{O_2N}{\overset{O_2N}{}}\!\!-\!\!COOAg \longrightarrow \underset{O_2N}{\overset{O_2N}{}}\!\!-\!\!COOR + AgI$$

PROCEDURE: 3,5-DINITROBENZOATES

The alkyl iodide and a slight excess of finely powered silver 3,5-dinitrobenzoate are refluxed in a small volume of alcohol until conversion to silver iodide appears complete. Evaporate the mixture to dryness, and extract the ester with ether. Recrystallize from ethanol.

Derivative Prepared from Grignard Reagents

The alkyl halide can be converted to a Grignard reagent, which, in turn, can be treated with an isocyanate to form a substituted amide, with mercuric halide to form an alkylmercuric halide, or with carbon dioxide to form an acid. The latter should be attempted only in cases where solid acids are obtained.

$$RX \xrightarrow[\text{Et}_2\text{O}]{\text{Mg}} RMgX \begin{cases} \xrightarrow[\text{(2) H}_2\text{O}]{\text{(1) ArNCO}} RCONHAr \\[4pt] \xrightarrow{\text{HgX}_2} RHgX + MgX_2 \\[4pt] \xrightarrow[\text{(2) H}_3\text{O}^+]{\text{(1) CO}_2} RCOOH \end{cases}$$

Grignard reagents of the allyl or benzyl type are likely to give rearranged and/or dimeric products.

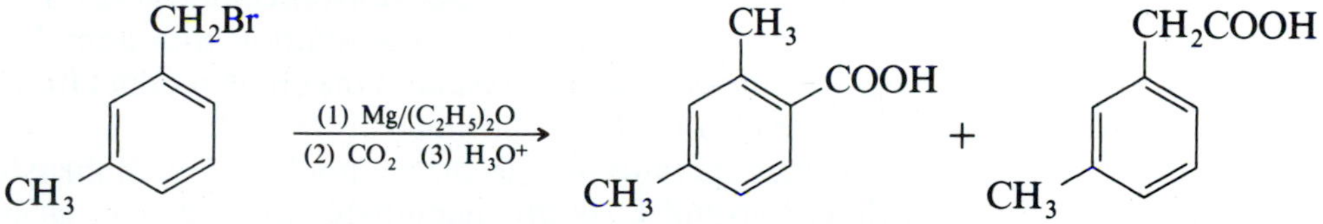

$$CH_3CH=CH-CH_2Br \xrightarrow[\text{(2) CO}_2 \quad \text{(3) H}_3\text{O}^+]{\text{(1) Mg/(C}_2\text{H}_5)_2\text{O}} \underset{\underset{\text{(in part)}}{COOH}}{CH_3\overset{|}{C}HCH=CH_2}$$

$$CH_2=CH-CH_2Br \xrightarrow[(C_2H_5)_2O]{Mg} \underset{\text{(in part)}}{CH_2=CH-CH_2-CH_2-CH=CH_2}$$

PROCEDURE: PREPARATION OF GRIGNARD REAGENTS

Grignard reagent: The choice of apparatus is governed by the steps involved in the reaction of the prepared Grignard reagent. The apparatus should consist of a flask, a tube or a small separatory funnel used as a reaction vessel, a condenser, and a drying tube. Into the carefully dried apparatus, place about 0.5 mmol (120 mg) of finely cut magnesium turnings, 0.55 mmol of the halide, 5 mL of absolute ether, and a small crystal of iodine. If the reaction is low in starting, warm it with a beaker of warm water. The reaction is normally complete in 5 to 10 minutes.

PROCEDURE: ALKYLMERCURIC HALIDES

 C A U T I O N : These materials are highly toxic and should be handled with care.

The Grignard reaction mixture (equivalent to 0.5 mmol Grignard) is filtered through a microporous disk or a little glass wool into a test tube containing 1 to 2 g of mercuric chloride, bromide, or iodide (corresponding to the halogen of the alkyl halide). Stopper the tube and shake the mixture vigorously. Remove the stopper, warm the tube on the steam bath for 1 or 2 minutes, and shake again. Evaporate to dryness, add 8 to 10 mL of ethanol, and heat on the steam bath until the alcohol boils. Filter and dilute with $\frac{1}{2}$ volume of water. Cool in an ice bath; when crystallization is complete collect the product and recrystallize from dilute ethanol.

PROCEDURE: *N*-ARYL AMIDES

Prepare the Grignard solution as above. Immerse the reaction vessel in cool water. Dissolve 0.5 mmol of α-naphthyl, phenyl, or p-tolyl isocyanate in 10 mL of absolute ether. By means of a dropper, add the isocyanate solution in small portions through the condenser into the Grignard solution. Shake the mixture and allow it to stand for 10 minutes. Pour the reaction mixture into a separatory funnel containing 20 mL of ice water and 1 mL of concentrated hydrochloric acid. Shake well and discard the lower aqueous layer. Dry the ether solution with anhydrous magnesium sulfate. Evaporate the ether and recrystallize the crude product from alcohol, aqueous alcohol, or petroleum ether.

If the apparatus or reagents are not kept dry, diphenylurea (mp 241°C), di-*p*-tolylurea (mp 268°C) or di-1-naphthylurea (mp 297°C) is obtained. These isocyanates are liquids and should not be used without purification if crystals (substituted urea formed by hydrolysis) are present in the reagent bottle.

PROCEDURE: CARBOXYLIC ACIDS

(a) Filter the Grignard reagent into a mixture of finely crushed Dry Ice and ether. Pour the reaction mixture into 20 mL of water and 1 mL of concentrated hydrochloric acid contained in a separatory funnel. After the reaction subsides, separate the ether layer.

(b) In areas of high humidity, large amounts of water may condense in the Dry Ice/ether mixture described above. In such cases it is advantageous to bubble carbon dioxide into, or over, the Grignard solution with stirring. The carbon dioxide can be obtained from a tank or from Dry Ice.

9.2.3 Alkyl Fluorides and Fluorocarbons

Characterization

Simple alkyl and cycloalkyl fluorides are identified by reference to physical constants and spectral characteristics.

In recent years, the chemistry of highly fluorinated compounds ("fluorocarbons") has attracted considerable attention. Perfluoro compounds (all hydrogen atoms, except those in the functional group, are replaced by fluorine) and compounds contaning the trifluoromethyl group are commonplace. The chemistry and physical properties of these materials differ considerably from their nonfluorinated analogues. The boiling points are usually lower than those of the hydrogen analogues, for example, perfluoropentane (29°C) and pentane (36°C), acetophenone (202°C) and trifluoromethyl phenyl ketone (152°C).

Highly fluorinated olefins do not add bromine under the usual conditions. They do, however, react with potassium permanganate. Most perhalogenated compounds fail to burn.

Fluorination increases the acidity of acids and alcohols and reduces the basicity of amines. Perfluoroaldehydes and ketones exhibit abnormally high carbonyl frequencies in the infrared. They are readily cleaved into acids and highly volatile monohydrogen perfluoroalkanes by the action of hydroxide (haloform reaction). The usual carbonyl derivatives of these compounds can be made by the appropriate carbonyl reagents (semicarbazide hydrochloride, etc.) in alcohol, but require long reaction times.

9.2.4 Vinyl Halides

Characterization

Vinyl halides, like aryl halides, are much less reactive than saturated halides. Identification can usually be made by reference to spectral and physical data. The nature and stereochemistry of substitution at the double bond can be inferred from the $C{=}C$ stretching and $C{-}H$ bending bands in the infrared spectrum (see Section 5.2.1) and from the coupling constants in the NMR (see Section 6.5). For vinyl fluorides, J_{FH} may be particularly useful.

Grignard reagents can be made from vinyl chlorides, iodides, and bromides, provided that dry tetrahydrofuran is substituted for diethyl ether as solvent (see Section 9.2.2).

9.2.5 Aryl Halides

Characterization

As in the case of many other types of aromatic compounds, the best procedures for making solid derivatives of aryl halides involve additional substitution of the aromatic nucleus. Nitration by the procedures outlined for aromatic hydrocarbons (see Section 9.1.4) is generally useful.

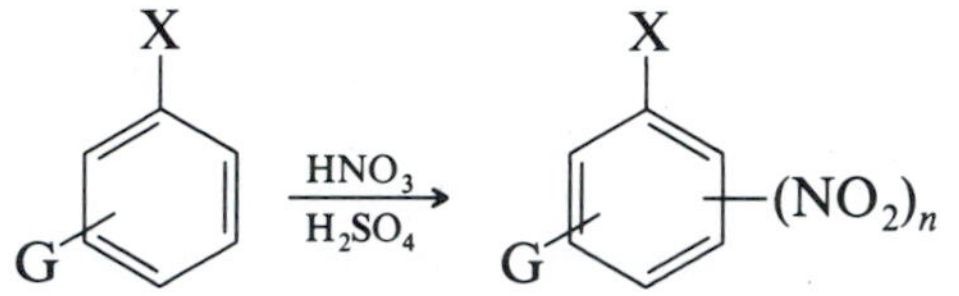

$$\text{(structure with X and G)} \quad \xrightarrow[\text{H}_2\text{SO}_4]{\text{HNO}_3} \quad \text{(structure with X, G)} - (\text{NO}_2)_n$$

The aryl halides react readily with the chlorosulfonic acid to produce sulfonyl chlorides, which in turn can be treated with ammonia to yield sulfonamides.

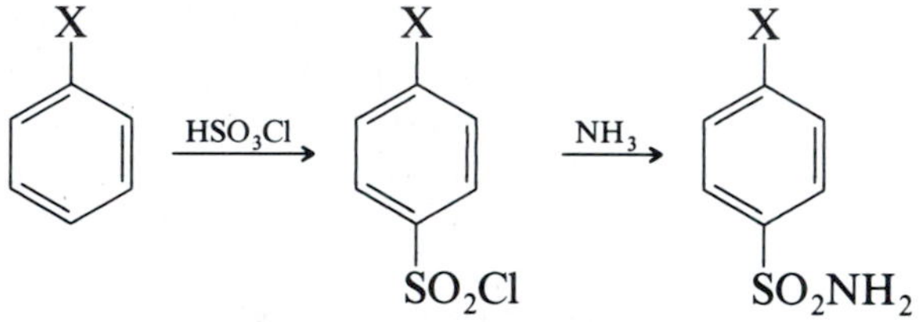

$$\text{(X-benzene)} \quad \xrightarrow{\text{HSO}_3\text{Cl}} \quad \text{(X-benzene-SO}_2\text{Cl)} \quad \xrightarrow{\text{NH}_3} \quad \text{(X-benzene-SO}_2\text{NH}_2)$$

The procedure outlined under aromatic ethers (see Section 9.4) can be employed. Sometimes the intermediate sulfonyl chloride will serve as a suitable derivative. With polyhalo aromatics, more vigorous conditions may be necessary to produce the sulfonyl chloride; the chloroform solution can be warmed or the reactants warmed neat to 100°C. Sometimes side reactions become predominant, products resulting from nuclear chlorination are obtained, or the reaction can provide diaryl sulfones.

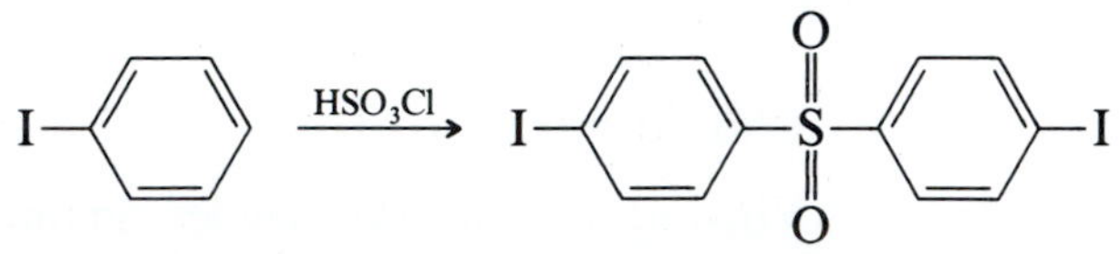

$$\text{I}-\text{(benzene)} \quad \xrightarrow{\text{HSO}_3\text{Cl}} \quad \text{I}-\text{(benzene)}-\overset{\overset{\text{O}}{\|}}{\underset{\underset{\text{O}}{\|}}{\text{S}}}-\text{(benzene)}-\text{I}$$

Aryl bromides and chlorides having aliphatic side chains can be oxidized to the corresponding acids. Aryl iodides and bromides can be converted to Grignard reagents and subsequently carbonated to yield acids or reacted with isocyanates to form amides (see Secton 9.2.2). Aryl chlorides form Grignard reagents only in tetrahydrofuran; in ether, bromochlorobenzenes form the Grignard reagent exclusively at the bromo position.

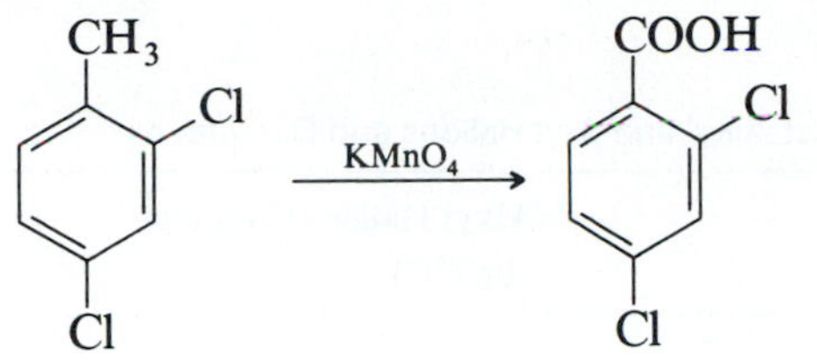

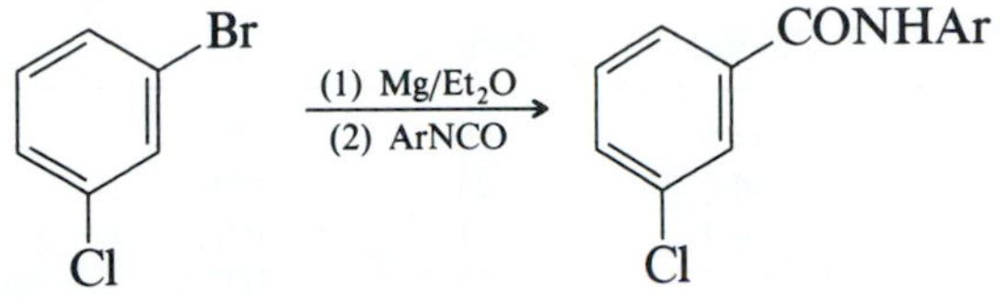

Occasionally aromatic polyvalent iodine compounds are encountered; representative examples are as follows:

$$ArICl_2 \qquad\qquad ArI(OAc)_2$$

iodoarene dichloride   iodoarene diacetate

$$Ar_2I^+X^- \qquad\qquad ArIO \qquad\qquad ArIO_2$$

diaryliodonium salts   iodosoarene   iodoxyarene

Iodoarene dichlorides act as a ready source of chlorine. They will add chlorine to alkenes. The oxygen derivatives all act as oxidants. With the exception of the diaryliodonium salts, all will liberate iodine from acidified starch iodide paper. The diaryliodonium salts are arylating agents; they can be pyrolyzed to yield ArI and ArX.

Table 9.3 lists the physical properties of organic halides and their derivatives.

9.3 ALCOHOLS AND PHENOLS

In the absence of carbonyl absorption, the appearance of a medium-to-strong band in the infrared spectrum in the 3600 to 3400 cm^{-1} region indicates an alcohol, a phenol, or a primary or secondary amine, or possibly a wet sample (see Figure 9.5 for the infrared spectrum of water). Amines can be distinguished from alcohols and phenols by taking advantage of their basic characters. Water-insoluble amines are soluble in dilute hydrochloric acid. The water-soluble amines have a characteristic ammoniacal odor, and their aqueous solutions are basic to litmus. Phenols are considerably more acidic than alcohols and can be differentiated from the latter by their solubility in 5% sodium hydroxide solution and by the colorations usually produced when treated with ferric chloride solution. For additional details, the classification sections on alcohols (9.3.1), phenols (9.3.2), and amines (9.10) should be consulted.

TABLE 9.3   Physical Properties of Alkyl and Aryl Halides and Derivatives

	Alkyl Halides (Liquids) bp (°C)			mp (°C) of Derivatives		
	Chloride	Bromide	Iodide	Anilide	S-Alkyl-thiuronium Picrate	Alkyl-mercuric Chloride
Isopropyl	36.5	60	89.8	103	196	
Allyl	44.6	71	103	114	155	
Propyl	46.4	71	102.5	92	181	147
t-Butyl	50.7	73.2	103.3	128	161	123
sec-Butyl	68	91.2	120	108	190	30.5
Isobutyl	68.9	91	120.4	109	174	
Butyl	77.8	101.6	130	63	180	128
Neopentyl	85	109		131		118
t-Pentyl	86	108	128	92		
Pentyl	106	129	155	96	154	110
Hexyl	133	157	179	69	157	125
Cyclohexyl	143	165		146		
Heptyl	159	180	204	57	142	119.5
Benzyl	179.4	198	(mp 24)	117	188	104
Octyl	184	201.5		57	134	115
2-Phenylethyl	190	218		97		
1-Phenylethyl	195	205				

Aryl Halides (Liquids)					
	bp (°C)	D_4^{20}	n_D^{20}	Nitro derivative Position	mp (°C)
Fluorobenzene	87.4	1.024	1.466		
Chlorobenzene	131.8	1.107	1.525	2,4	52
Bromobenzene	156.2	1.494	1.560	2,4	72
o-Chlorotoluene	159.3	1.082	1.524	3,5	63
m-Chlorotoluene	162.3	1.072	1.521	4,6	91
p-Chlorotoluene	162.4	1.071	1.521	2	38
p-Dichlorobenzene (mp 53°C)	173				
m-Dichlorobenzene	173	1.288	1.546	4,6	103
o-Dichlorobenzene	179	1.305	1.552	4,5	110
o-Bromotoluene	181.8	1.425		3,5	82
p-Bromotoluene (mp 28.5°C)	184.5	1.390		2	47
Iodobenzene	188.6	1.831	1.620	4	171
o-Chloroanisole	195	1.248	1.5480		
p-Chloroanisole	200	1.1851			
2,6-Dichlorotoluene	199				
2,4-Dichlorotoluene	200	1.249	1.549	3,5	104
o-Dibromobenzene	219.3	1.956	1.609	4,5	114
1-Chloronaphthalene	259.3	1.191	1.633	4,5	180
1-Bromonaphthalene	281.2	1.484	1.658	4	85

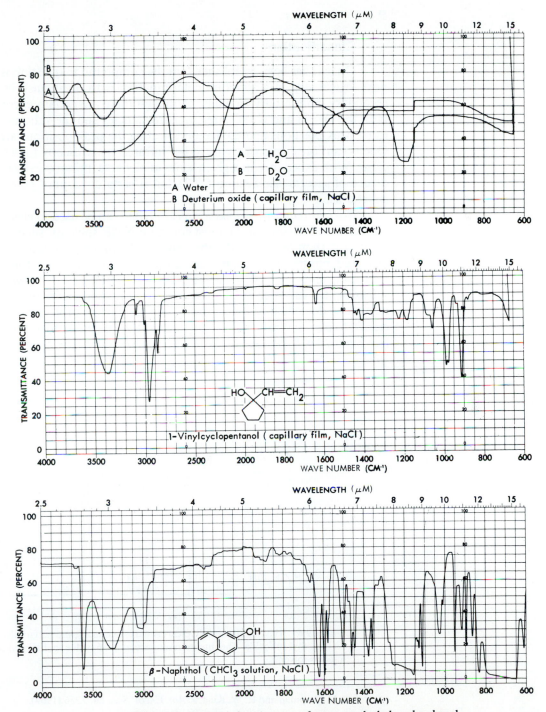

Figure 9.5 Typical infrared spectra of water, an alcohol, and a phenol.

9.3.1 Alcohols

Classification

Once it has been established that the compound in question is an alcohol, the next logical step in a structure determination is to distinguish between primary, secondary, and tertiary alcohols. Although there are certain correlations between both the O—H and C—O stretching frequencies and the subclass of alcohols, the exact peak positions and the factors affecting the peak positions are quite variable. One is cautioned against trying to use infrared spectroscopy to determine the subclass. With experience in a given series of alcohols, infrared can be used with some degree of success (see Section 5.2.2), but it certainly cannot be recommended for general use.

Nuclear magnetic resonance spectroscopy can be used to classify alcohols with respect to subclasses. In the most common NMR solvents, such as deuterochloroform or carbon tetrachloride, the hydroxyl hydrogen resonance is not only often obscured, but traces of acid present in the solvents catalyze proton exchange so that spin-spin splitting of the hydroxyl peaks is rarely observed. However, in dimethyl sulfoxide, the strong hydrogen bonding to the solvent shifts the hydroxyl hydrogen resonance downfield to δ 4 or lower, and the rate of proton exchange slows sufficiently to permit observation of the carbinol hydrogen-hydroxyl hydrogen coupling (see Figure 9.6).[6] Because of the low-field resonance of the hydroxyl hydrogens, it is

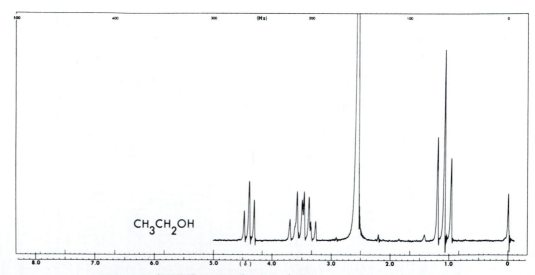

Figure 9.6 ¹H NMR spectra of typical primary, secondary, and tertiary alcohols run in dimethyl sulfoxide. The intense resonance at δ 2.6 is due to dimethyl sulfoxide used as solvent.

---

[6] O. L. Chapman and R. W. King, *J. Am. Chem. Soc.,* **86,** 1256 (1964).

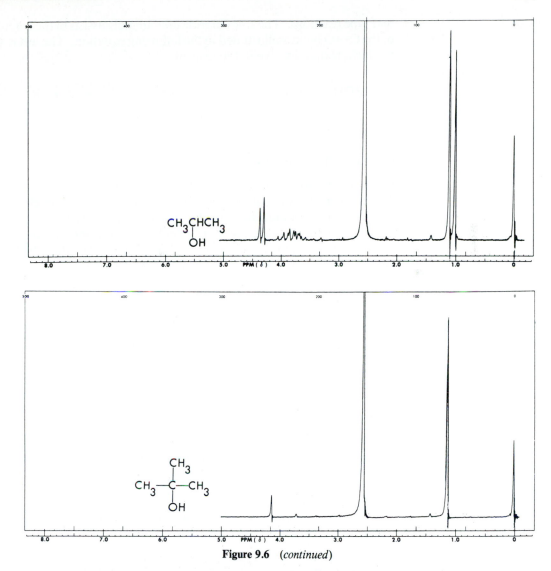

Figure 9.6 (*continued*)

possible to use dimethyl sulfoxide rather than its deuterated analog. The hydroxyl hydrogen in methanol gives a quartet; primary, secondary, and tertiary alcohols give clearly resolved triplets, doublets, and singlets, respectively. Polyhydroxy compounds often give separate peaks for each hydroxyl hydrogen. This method is not reliable when acidic or basic groups are present, for example, amino alcohols, hydroxy acids, or phenols.

A solution of zinc chloride in concentrated hydrochloric acid (Lucas Reagent) has been widely used as a classical reagent to differentiate between the lower primary, secondary, and tertiary alcohols. With this reagent, the order to reactivity is typical of

S_N1 type reactions. The zinc chloride undoubtedly assists in the heterolytic breaking of the C—O bond as illustrated in the following equation. The test is applicable only to alcohols that dissolve in the reagent.

$$ROH + ZnCl_2 \rightleftharpoons R\overset{\delta+}{-}\overset{\delta-}{\underset{H}{O}}\text{---}ZnCl_2 \xrightarrow{-[HOZnCl_2]^-} R^+ \xrightarrow{Cl^-} RCl$$

PROCEDURE: LUCAS TEST

Reagent: Dissolve 16 g of anhydrous zinc chloride in 10 mL of concentrated hydrochloric acid with cooling to avoid a loss of hydrogen chloride.

Add three to four drops of the alcohol to 2 mL of the reagent in a small test tube. Shake the test tube vigorously, and then allow the mixture to stand at room temperature. Primary alcohols lower than hexyl dissolve; those higher than hexyl do not dissolve appreciably, and the aqueous phase remains clear. Secondary alcohols react to produce a cloudy solution of insoluble alkyl chloride after 2 to 5 minutes. With tertiary, allyl, and benzyl alcohols, there is almost immediate separation of two phases owing to the formation of the insoluble alkyl chloride. If any question remains about whether the alcohol is secondary or tertiary, the test can be repeated employing concentrated hydrochloric acid. With this reagent, tertiary alcohols react immediately to form the insoluble alkyl chloride, whereas secondary alcohols do not react.

A second method to differentiate tertiary alcohols from primary and secondary alcohols takes advantage of the inertness of tertiary alcohols toward oxidation with *chromic acid.*

$$3\text{ H}-\underset{|}{\overset{|}{C}}-OH + 2\text{ CrO}_3 + 6\text{ H}^+ \longrightarrow 3\text{ }\diagdown\text{C}=O + 2\text{ Cr}^{+3} + 6\text{ H}_2O$$

PROCEDURE: CHROMIC ACID TEST

Reagent: Dissolve 1 g of chromic oxide in 1 mL of concentrated sulfuric acid, and carefully dilute with 3 mL of water.

Dissolve 20 mg or one drop of the alcohol in 1 mL of reagent grade acetone, and add one drop of the reagent. Shake the mixture. Primary and secondary alcohols react within 10 seconds to give an opaque blue-green suspension. Tertiary alcohols do not react with the reagent. Other easily oxidized substances such as aldehydes, phenols, and enols also react with this reagent.

Characterization

Phenyl and α-naphthylurethanes

$$ROH + ArN{=}C{=}O \longrightarrow ArNH\overset{\overset{\displaystyle O}{\|}}{C}{-}OR$$

isocyanate   urethane

The most generally applicable derivatives of primary and secondary alcohols are the urethanes, prepared by reaction of the alcohol with the appropriate isocyanate. *For the preparation of the urethanes, the alcohols must be anhydrous;* water hydrolyzes the isocyanates to give aryl amines, which combine with the reagent to produce disubstituted ureas.

$$ArNCO + H_2O \longrightarrow ArNH_2 + CO_2$$

$$ArNH_2 + ArNCO \longrightarrow ArNHCONHAr$$

These ureas, which are high melting and less soluble than the urethanes, make isolation and purification of these derivatives very difficult. It is not advisable to attempt the preparation of these urethanes from tertiary alcohols. At low temperatures the reaction is too slow; at higher temperatures the reagents cause dehydration to occur with the formation of an alkene and water, which, in turn, react to produce diaryl ureas.

PROCEDURE: URETHANES

Thoroughly dry a small test tube over a flame or in an oven; cork it and allow to cool. By means of a pipette, place 0.2 mL of anhydrous alcohol and 0.2 mL of α-naphthyl or phenyl isocyanate into the tube (if the reactant is a phenol, add one drop of pyridine), and immediately replace the cork. If a spontaneous reaction does not take place, the solution should be warmed in a water bath at 60 to 70°C for 5 minutes. Cool in an ice bath and scratch the sides of the tube with a glass rod to induce crystallization. The urethane is purified by recrystallization from petroleum ether or carbon tetrachloride. Filter the hot solution to remove the less soluble urea, which may form owing to traces of moisture. Cool the filtrate in an ice bath and collect the crystals. Diphenylurea, di-*p*-tolylurea, and di-1-naphthylurea have melting points of 241, 268, and 297°C, respectively.

Arenesulfonylurethanes. Benzenesulfonyl and *p*-toluenesulfonylisocyanate, which are considerably more reactive than phenyl and related isocyanates, may be used to advantage in the preparation of urethanes of tertiary and other highly hindered alcohols.[7] Extensive compilations of data are not available.

$$R-\!\!\!\underset{}{\bigcirc}\!\!\!-SO_2N\!=\!C\!=\!O + R_3'COH \longrightarrow R-\!\!\!\underset{}{\bigcirc}\!\!\!-SO_2NH\overset{\overset{\displaystyle O}{\|}}{C}OCR_3'$$

R = H or CH$_3$

Esters.   Numerous esters have been used to aid in the characterization of alcohols. Among these are a wide variety of phthalic acid esters, xanthates, benzoates, and acetates. The latter two are especially useful for the characterization of glycols and polyhydroxy compounds. Among the most generally useful esters are the *3,5-dinitro* and *p-nitrobenzoates*. These esters can be prepared from the corresponding benzoyl chlorides by either of the two procedures given below. They are to be especially recommended for water-soluble alcohols that are likely to contain small amounts of moisture and hence produce trouble in the formation of the urethane derivatives. The method below employing pyridine as a solvent is one of the most useful methods for making derivatives of tertiary alcohols.

The acyl halides tend to hydrolyze on storage; it is advisable to check the melting point of the reagent prior to use [3,5-dinitrobenzoyl chloride, mp 74°C (acid, 202°C) *p*-nitrobenzoyl chloride, mp 75°C (acid 241°C)].

PROCEDURE: 3,5-DINITRO- AND *p*-NITROBENZOATES

(a) In a small test tube place 200 mg of pure *3,5-dinitrobenzoyl* or *p-nitrobenzoyl chloride* and 0.1 mL (two drops) of alcohol. Heat the tube for 5 minutes (10 minutes if the alcohol boils above 160°C), employing a microburner so that the melt at the bottom of the tube is maintained in the liquid state. Do not overheat. Allow the melt to cool and solidify. Pulverize the crystalline mass with a glass rod or spatula. Add 3 or 4 mL of a 2% sodium carbonate solution and thoroughly mix and grind the mixture. Warm the mixture gradually to 50 or 60°C and stir thoroughly. Collect the remaining precipitate and wash several times with small portions of water. Recrystallize from ethanol or aqueous ethanol.

(b) In a small test tube or round-bottomed flask, place 100 mg of alcohol or phenol, 100 mg of *p*-nitrobenzoyl or 3,5-dinitrobenzoyl chloride, 2 mL of pyridine, and a boiling chip. Put a condenser in place and reflux gently for 1 hour or allow to stand overnight. Cool the reaction mixture and add 5 mL of water and two or three drops of sulfuric acid. Shake or stir well and collect the crystals. Suspend the crystals

---

[7] J. W. McFarland and J. B. Howard. *J. Org. Chem.*, **30**, 957 (1965); J. W. McFarland, D. E. Lenz, and D. J. Grosse, *J. Org. Chem.*, **31**, 3798 (1966).

in 5 mL of 2% sodium hydroxide, and shake well to remove the nitrobenzoic acid. Filter, wash several times with cold water, and recrystallize the derivative from ethanol or aqueous ethanol.

Acetates, hydrogen phthalates, and other esters of highly hindered alcohols (as well as phenols) can be prepared in excellent yield by use of the "hypernucleophilic" acylation catalyst, 4-dimethylaminopyridine.[8]

Oxidation to carbonyl compounds. The oxidation of primary and secondary alcohols to carbonyl compounds often proves to be an exceedingly useful method of identification. From the carbonyl compounds, easily prepared derivatives can be made on a small scale, and, equally important, the spectroscopic features of the carbonyl group can provide much valuable structural information.

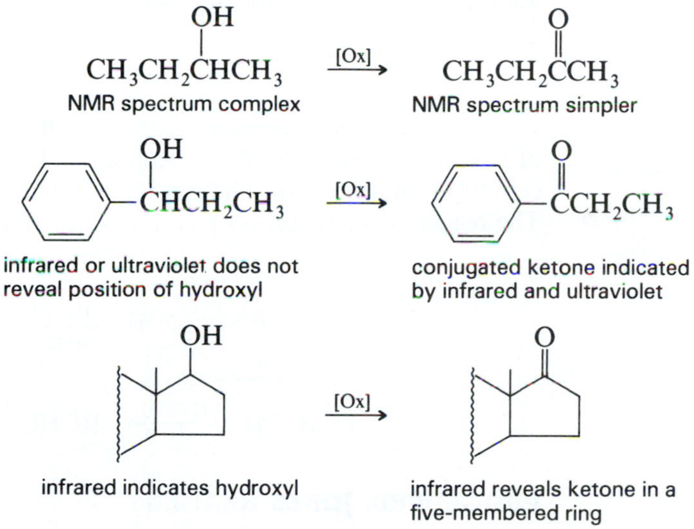

$$CH_3CH_2\overset{\overset{\displaystyle OH}{|}}{C}HCH_3 \xrightarrow{[Ox]} CH_3CH_2\overset{\overset{\displaystyle O}{\|}}{C}CH_3$$

NMR spectrum complex   NMR spectrum simpler

infrared or ultraviolet does not reveal position of hydroxyl

conjugated ketone indicated by infrared and ultraviolet

infrared indicates hydroxyl

infrared reveals ketone in a five-membered ring

Numerous methods are available for the oxidation of alcohols to carbonyl compounds. Only a few are included here; they are chosen for their mild conditions and suitability for small-scale operation.

The chromic anhydride-pyridine complex (Sarett reagent)[9] is particularly useful for the oxidation of substances containing acid-sensitive groups. Alcohols, including allylic and benzylic, can be oxidized to the corresponding aldehydes or ketones.

---

[8] G. Hofle and W. Steglick, *Synthesis,* 619 (1972).

[9] G. I. Poos, G. E. Arth, R. E. Beyler, and L. H. Sarett, *J. Am. Chem. Soc.,* **75,** 422 (1963).

PROCEDURE: SARETT REAGENT

Two mL of pyridine in a small round-bottomed flask equipped with a magnetic stirrer is cooled to 15 to 20°C. To the pyridine is added, in portions, 200 mg of chromium trioxide at such a rate that the temperature does not rise above 30°C.

C A U T I O N : *If the pyridine is added to the chromium trioxide, the mixture will ignite.* A slurry of the yellow complex in pyridine remains after the last addition.

To the slurry is added 100 mg of the alcohol in 1 mL of pyridine. The flask is stoppered and allowed to stir overnight. The reaction mixture is poured into 20 mL of ether; the precipitated chromium salts are removed by filtration. The filtrate is placed in a separatory funnel and washed at least three times with water to remove the pyridine. The ether layer is dried over sodium sulfate and concentrated to provide the aldehyde or ketone. Low molecular weight, water-soluble ketones can be isolated by pouring the reaction mixture directly into water, neutralizing the pyridine by addition of acid and extraction with an appropriate solvent.

Chromic acid in acetone (Jones reagent)[10] is very convenient for the oxidation of acetone-soluble secondary alcohols to ketones. The reagent does not attack centers of unsaturation. The method is applicable on scales from 1 mmol to 1 mol. The reagent oxidizes primary alcohols and aldehydes to acids.

$$RCH_2OH \xrightarrow{H_2CrO_4} RCHO \xrightarrow{H_2CrO_4} RCOOH$$

PROCEDURE: JONES REAGENT

Reagent: Dissolve 6.7 g of chromium trioxide (CrO_3) in 6 mL of concentrated sulfuric acid and carefully dilute with distilled water to 50 mL. One mL of this reagent is sufficient to oxidize 2 mmol of a secondary alcohol to a ketone, 2 mmol of aldehyde to an acid, or 1 mmol of primary alcohol to an acid.

The oxidation is carried out by the addition of the reagent from an addition funnel to a stirred acetone (reagent grade) solution of the alcohol maintained at 15 to 20°C. The reaction is nearly instantaneous; the mixture separates into a lower green layer of chromous salts and an upper layer that is an acetone solution of the oxidation

---

[10] K. Bowden, I. M. Heilbron, E. R. H. Jones, and B. C. L. Weedon, *J. Chem. Soc.,* 39 (1946); A. Bowers, T. G. Halsall, E. R. H. Jones, and A. J. Lemin, *J. Chem. Soc.,* 2555 (1943).

product. The reaction mixture can be worked up by the addition of water or by other means, depending on the properties of the oxidation product. Any brown coloration, caused by an excess of the oxidizing agent remaining in the mixture, in the upper layer can be removed by a small pinch of sodium hydrogen sulfite or a few drops of methanol to destroy the excess reagent.

Pyridinium chlorochromate (PCC) ($C_5H_6N^+$ $ClCrO_3^-$), a stable crystalline reagent that can be readily synthesized or purchased, oxidizes alcohols to the corresponding aldehydes or ketones in high yield under mild conditions.

PROCEDURE: PYRIDINIUM CHLOROCHROMATE OXIDATION

Reagent: To 18.4 mL of $6N$ hydrochloric acid (0.11 mol) add 10 g (0.1 mol) of chromium trioxide rapidly with stirring. After 5 minutes, cool the homogeneous solution to 0°C and add dropwise, over a period of 10 minutes, 7.9 g (0.1 mol) of pyridine. Recool to 0°C and collect the yellow-orange pyridinium chlorochromate in a sintered glass funnel. After drying for 1 hour in vacuo, the reagent can be stored.

Pyridinium chlorochromate (0.32 g, 1.5 mmol) is suspended in 2 to 3 mL of dichloromethane in a 15-cm test tube, and the alcohol (1 mmol) in 1 mL of dichloromethane is added. After 1 to 2 hours, the oxidation is complete and black reduced reagent has precipitated. The reaction mixture is diluted with 10 to 12 mL of diethyl ether, the solids are removed by filtration, and the product is obtained by evaporation of the solvent. If the alcohol contains acid-labile functions (e.g., esters, etc.), sodium acetate (25 mg, 0.3 mmol) should be added to the reagent suspension as a buffer prior to the addition of the alcohol.

The *trimethylsilylation* reaction proceeds smoothly, quickly, and often quantitatively to provide derivatives of mono- and polyhydric alcohols, which show high volatility and often are easily separated from closely related compounds by glc.

$$2\ ROH + (CH_3)_3SiNHSi(CH_3)_3 \longrightarrow 2\ ROSi(Ch_3)_3 + NH_3$$

PROCEDURE: TRIMETHYLSILYL DERIVATIVES

Reagent: Mix anhydrous pyridine (10 mL), hexamethyldisilazane (2 mL), and trimethylchlorosilane (1 mL). (A reagent of similar composition is marketed by the Pierce Chemical Company under the name TRI-SIL.)

Place 10 mg or less of the sample in a small vial. Add 1 mL of the reagent, place the top on the vial, and shake vigorously for 30 seconds. Warm if necessary to effect

TABLE 9.4 Physical Properties of Alcohols and Derivatives

| | Liquids | | | | |
| | | mp (°C) of Derivatives | | | |
Compound	bp (°C)*	α-Naphthyl-urethan	Phenyl-urethan	3,5-Dinitro-benzoate	p-Nitro-benzoate
Methanol	64.65	124	47	108	96
Ethanol	78.32	79	52	93	57
2-Propanol	82.4	106	75	123	110
2-Methyl-2-propanol (mp 25.5°C)	82.5				44
3-Buten-2-ol	94.6				
2-Propen-1-ol (Allyl alcohol)	97.1	108	70	49	28
1-Propanol	97.15	80	57	74	35
2-Butanol	99.5	97	64.5	76	26
2-Methyl-2-butanol	102.35	72	42	116	85
2-Methyl-1-propanol	108.1	104	86	87	69
3-Buten-1-ol	112.5				
3-Methyl-2-butanol	114	109	68		
3-Pentanol	116.1	95	49	101	17
1-Butanol	117.7	71	61	64	36
2,3-Dimethyl-2-butanol	120.5	101	66	111	
2-Methyl-2-pentanol	121		239 *86*	72	
3-Methyl-3-pentanol	123	83.5	43.5	96.5	
2-Methyl-1-butanol	128.9	82	31	70	
4-Methyl-2-pentanol	132	88	143	65	26
3-Methyl-2-pentanol	134/749 mm	72		43.5	
3-Hexanol	136			77	
2,2-Dimethyl-1-butanol	136.7	81	66	51	
1-Pentanol	138	68	46	46.4	
2-Hexanol	138–9/745 mm	60.5		38.5	
Cyclopentanol	140.8	118	132		
3-Ethyl-3-pentanol	142				
2,3-Dimethyl-1-butanol	145		29	51.1	
2-Methyl-1-pentanol	148	76		50.5	
3-Methyl-1-pentanol (mp 8°C)	151–2	58		38	
3-Heptanol	156				
4-Heptanol	156	80		64	35
2-Heptanol	158.7	54		49	
1-Hexanol	157.5	59	42	58.4	
Cyclohexanol	161.1				
Furfuryl alcohol	172	130	45	81	76
1-Heptanol	176.8	62	60	47	10
Tetrahydrofurfuryl alcohol	178/743 mm		61	83–4	48
2-Octanol	179	64		32	28
Cyclohexylmethanol	182				
2,3-Butanediol	182.5	201 (bis)			
1-Phenylethanol (mp 20°C)	202	106	92	95	43
Benzyl alcohol	205.5	134	77	113	85
1-Nonanol	215	65.5	60	52	10
1-Phenylpropanol	219	102			60
1,4-Butanediol (mp 19.5°C)	230	199 (bis)	183.5 (bis)		175 (bis)

TABLE 9.4 (Continued)

| | | Liquids | | | |
| | | mp (°C) of Derivatives | | | |
Compound	bp (°C)*	α-Naphthyl-urethan	Phenyl-urethan	3,5-Dinitro-benzoate	p-Nitro-benzoate
3-Phenylpropanol	237.4		45	92	47
1-Undecanol (mp 15.85°C)	243		62		
Cinnamyl alcohol (mp 33°C)	257	114	91.5	121	78
2,3-Dimethyl-2,3-butanediol (mp 43°C)	173				
L-Menthol (mp 44°C)	216	119 (126)	112	153	62
1,2-Diphenylethanol	167/10 mm				
Diphenylmethanol (mp 68°C)	180/20 mm	139	140	141	132
o-Nitrobenzyl alcohol (mp 74°C)	270				
2-Hydroxy-1-phenyl-1-ethanone (mp 86°C)	120/11 mm				128.6
Cholesterol (mp 148.5°C)	360d†	176	168		185
Triphenylmethanol (mp 162°C)					

* bp at 760 mm Hg pressure unless otherwise noted.
† d denotes decomposition occurs.

solution. Allow the mixture to stand for 5 minutes. The mixture can be injected directly into the gas chromatograph.

Table 9.4 lists the physical properties of alcohols and their derivatives.

9.3.2 Phenols

Classification

Phenols are compounds of acidity intermediate between that of carboxylic acid and alcohols. Alcohols do not show acid properties in aqueous systems, whereas acids and phenols react with and are soluble in 5% aqueous sodium hydroxide solution (exceptions are the highly hindered phenols, such as 2,6-di-t-butylphenol, which are insoluble in alkali). Acids and phenols can be differentiated on the basis of the insolubility of phenols in 5% aqueous sodium hydrogen carbonate solution. Phenols that contain highly electronegative substituents, such as 2,4-dinitrophenol and 2,4,6-tribromophenol, show increased acidity and are soluble in sodium hydrogen carbonate solution.

Most phenols yield intense red, blue, purple, or green colorations in the *ferric chloride test*. All phenols do not produce color with this reagent; a negative test must not be taken as confirming the absence of a phenol grouping without additional supporting evidence. Other functional groups also produce color changes with ferric chloride: aliphatic acids give a yellow solution; aromatic acids may produce a tan precipitate; enols give a red-tan to red-violet color; oximes, hydroxamic acids, and sulfinic acids give red to red-violet colorations.

PROCEDURE: FERRIC CHLORIDE TEST

To 1 mL of a dilute aqueous solution (0.1 to 0.3%) of the compound in question, add several drops of a 2.5% aqueous solution of ferric chloride. Compare the color produced with that of pure water containing an equivalent amount of the ferric chloride solution. The color produced may not be permanent, thus the observation should be made at the time of addition.

Certain phenols do not produce coloration with the foregoing procedure. As an alternative procedure, dissolve or suspend 20 mg of a solid or one drop of a liquid in 1 mL of dichloromethane, and add two drops of a saturated solution of anhydrous ferric chloride in dichloromethane. To the test solution add one drop of pyridine and observe the color change.

Phenols rapidly react with *bromine water* to produce insoluble substitution products; all available positions ortho and para to the hydroxyl are brominated. Advantage can be taken of this reaction both as a qualitative test for the presence of the phenolic grouping and for the preparation of solid derivatives. The tests should be applied with some discrimination since aniline and its substituted derivatives also react rapidly with bromine water to produce insoluble precipitates.

PROCEDURE: BROMINE WATER TEST

To a 1% aqueous solution of the suspected phenol, add a saturated solution of bromine water, drop by drop, until bromine color is no longer discharged. A positive test is indicated by the precipitation of the sparingly soluble bromine substitution product and the production of a very strongly acidic reaction mixture. In the case of phenol, the product is 2,4,6-tribromophenol.

In the ultraviolet region, ionization of a phenol by a base increases both the wavelengths and the intensities of the absorption bands.

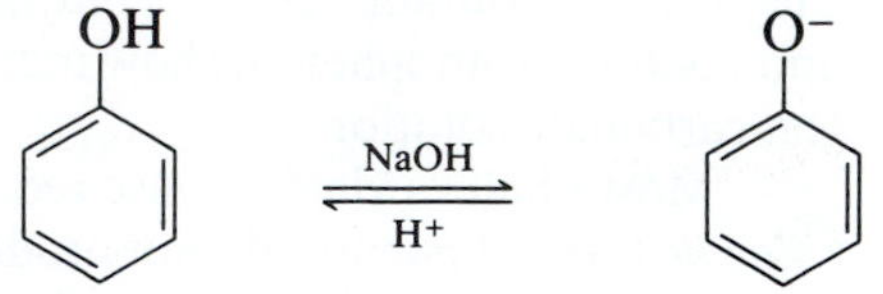

$\lambda_{max}^{H_2O}$ 211,270 nm
(6200, 1450)

$\lambda_{max}^{0.1N\ NaOH}$ 235,287 nm
(9400, 2600)

This shift can be visually observed in the case of *p*-nitrophenol; *p*-nitrophenol is yellow, whereas sodium *p*-nitrophenolate is red.

Characterization

Phenols, like alcohols, react with isocyanates to produce urethanes. The α-naphthylurethanes are the derivatives generally used for the identification of phenols. The procedure employed is that given for alcohols (see Section 9.3.1). For highly hindered phenols, benzenesulfonylisocyanate is the reagent of choice.

The melting points of a large number of 3,5-dinitrobenzoates of phenols have also been recorded. These derivatives can be prepared using the pyridine method discussed under alcohols (see Section 9.3.1).

The preparation of *brominated phenols* is an exceedingly simple procedure, and the bromo-substituted phenols are very useful derivatives. Although saturated bromine-water can be used satisfactorily for this bromination, the following procedure usually yields better results on a preparative scale.

PROCEDURE: BROMINATION

In a test tube or an Erlenmeyer flask, dissolve 1 g of potassium bromide in 6 mL of water. Carefully add 0.6 g of bromine. In a second test tube, place 100 mg of the phenol, 1 mL of methanol, and 1 mL of water. Add 1 mL of the prepared bromine solution and shake. Continue the addition of bromine solution in small portions until the mixture retains a yellow color after shaking. Add 3 to 4 mL of cold water and shake vigorously. Filter the bromophenol and wash the precipitate well with water. Dissolve the crystals in hot methanol, filter the solution, and add water dropwise to the filtrate until a permanent cloudiness results. Allow the mixture to cool to complete crystallization.

The phenoxide ions, produced by dissolution of phenols in aqueous alkali, react readily with chloracetic acid to give *aryloxyacetic acids*. These derivatives are very useful; they crystallize well from water, have well-defined melting points, and can be characterized by reference to their neutralization equivalents.

$$G{-}C_6H_4{-}OH + Cl{-}CH_2COOH \xrightarrow[\text{(2) } H_3O^+]{\text{(1) NaOH}} G{-}C_6H_4{-}O{-}CH_2COOH$$

PROCEDURE: ARYLOXYACETIC ACIDS

Approximately 200 mg of the phenol is dissolved in 1 mL of 6 N sodium hydroxide in a small test tube. Additional water should be added if necessary to completely dissolve the sodium phenoxide. To the solution, 0.5 mL of a 50% aqueous solution of chloroacetic acid is added. The tube is provided with a microcondenser, and the

TABLE 9.5 Physical Properties of Phenols and Derivatives

Compound	mp (°C)	bp (°C)	α-Naphthyl-urethan	3,5-Dinitro-benzoate	N-Phenyl-urethan	Bromo Derivative	Aryloxy-acetic Acid
				mp (°C) of Derivatives			
o-Chlorophenol	7	175.6	120	143	121	49 (mono) 76 (di)	145
Phenol	41.8	182					
o-Cresol	31	191–2					
o-Bromophenol	5	195	129			95	143
Salicylaldehyde		197			133		
p-Cresol	36	202					
m-Cresol	12	203	128	165.4	125	84 (tri)	103
o-Ethylphenol		207		108	143.5		141
2,4-Dimethylphenol	27	211.5					
o-Hydroxyacetophenone	28	215					
m-Ethylphenol		217			137		77
p-Propylphenol	22	232			129		
p-Isobutylphenol		236					
m-Methoxyphenol		243	129			104 (tri)	118
p-Butylphenol	22	248			115		81
2-Methoxy-4-(2-propenyl)phenol	19	254.8	122	130.8	95.5	118 (tetra)	81(100)
2-Methoxy-4-(1-propenyl)phenol		267.5	150	158.4	118 (cis) 152 (trans)		94(116)
2,4-Dibromophenol	36	238–9				95	153
p-Chlorophenol	45	217	166	186	148.5		156
2,4-Dichlorophenol	43	209				68	141(135)
o-Nitrophenol	45		113	155		177 (di)	158
p-Ethylphenol	47	219	128	133	120		97
2,6-Dimethylphenol	49	203	176.5	158.8	133	79	139.5
2-Isopropyl-5-methylphenol	49.7	233	160	103.2	107	55	149
o-Phenylphenol	57.5	275					
3,4-Dimethylphenol	62.5	225	141–2	181.6	120	171 (di)	162.5
p-Bromophenol	64		169	191		95 (di)	160
2,4,6-Trichlorophenol	68	245					
3,5-Dimethylphenol	63	220	195.4	148		166 (tri)	111
2,5-Dimethylphenol	74.5	212	173	137.2	161(166)	178 (di)	118
2,3-Dimethylphenol	75						187
4-Hydroxy-3-methoxy-benzaldehyde	81						187
4-Chloro-2-nitrophenol	86						
1-Naphthol	94	280	152	217.4	178	105(2,4-di)	193.5
p-Iodophenol	94				148		156
2,4,6-Tribromophenol	95		153	174			
p-tert-Butylphenol	100	237	110		148.5	50	86
m-Hydroxybenzaldehyde	102						148
1,3-Benzenediol	110	280.8		201	164	112(4,6-di)	195
p-Nitrophenol	114		151	186		142(2,6-di)	187
p-Hydroxybenzaldehyde	116						198
2,4,6-Trinitrophenol	122						
2-Naphthol	123	286	157	210.2	156	84	154
1,2,3-Benzenetriol	133			205 (tri)	173 (tri)	158 (di)	198
o-Aminophenol	174						
p-Aminophenol	184						

reaction is heated on a water bath at 90 to 100°C for 1 hour. The solution is cooled, 2 mL of water is added, and the solution is acidified with dilute hydrochloric acid. The mixture is extracted several times with small portions of ether. The ether extract is washed with 2 ml of water and then extracted with 5% sodium carbonate solution. The sodium carbonate extract is acidified with dilute HCl to precipitate the aryloxya-cetic acid. The derivatives should be recrystallized from hot water. The melting point, along with the neutralization equivalent, if necessary, is determined.

Table 9.5 lists the physical properties of phenols and their derivatives.

9.4 ETHERS

Classification

The identification of the ether grouping in the infrared is complicated by the fact that other functional groups contain the C—O bond and, consequently, have bands in the same region. As a general guide, a relatively strong band in the 1250 to 1100 cm^{-1} region and the absence of C=O and O—H bands are good indications of an ether (see Figure 9.7). The aliphatic ethers absorb at the lower end of the range; conjugation raises the frequency.

Like hydrocarbons, ethers are quite unreactive, but they can be chemically distinguished from saturated hydrocarbons by the iodine charge-transfer test (see Section 9.1.1) and by their solubility in sulfuric acid (except diaryl ethers). Dialkyl ethers are also soluble in concentrated hydrochloric acid, whereas others are not.

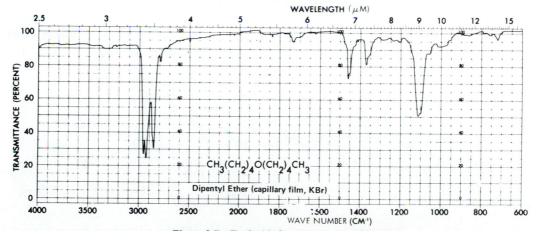

Figure 9.7 Typical infrared spectrum of an ether.

In the absence of hydroxyl and carbonyl absorption in the infrared, a methyl (especially with characteristic ringing) in NMR spectra near δ 4 or a methine or methylene at slightly lower field should alert the investigator to the possibility of the presence of an alkyl ether.

C A U T I O N : Ethers tend to form highly explosive peroxides on standing, particularly when exposed to air and light. If peroxides are present, they accumulate in the pot during distillation and may result in a violent explosion.

The presence of peroxides can be detected by the starch-iodide test under acidic conditions (see Section 1.15.8). A positive test is indicated by the blue starch-iodine color. Peroxides (as well as water and alcohols) can be removed from ethers by filtering the ether through a short column of highly active alumina (Woelm basic alumina, activity grade 1, is recommended). Hydroperoxides (but not water-insoluble dialkyl or diacyl peroxides) can be removed from organic materials by treatment with ferrous sulfate; wash 10 mL of peroxide-containing ether with 3 to 5 mL of 1% ferrous sulfate solution acidified with one drop of sulfuric acid.

Characterization

Aliphatic ethers.   Although ethers are less reactive than most other functions, cleavage and conversion to other more easily characterized functions can be achieved under relatively mild conditions.

Dialkyl and alkyl aryl ethers are cleaved by hydriodic acid.

$$R_2O + 2\ HI \longrightarrow 2\ RI + H_2O$$

$$ArOR + HI \longrightarrow ArOH + RI$$

To obtain sufficient product, especially for characterization of the alkyl iodides, it is necessary to use a sample of 4 to 5 g. The alkyl iodides and phenols can be transformed into suitable derivatives.

Aromatic ethers.   The derivatives of aromatic ethers most frequently employed are those obtained by electrophilic aromatic substitution.

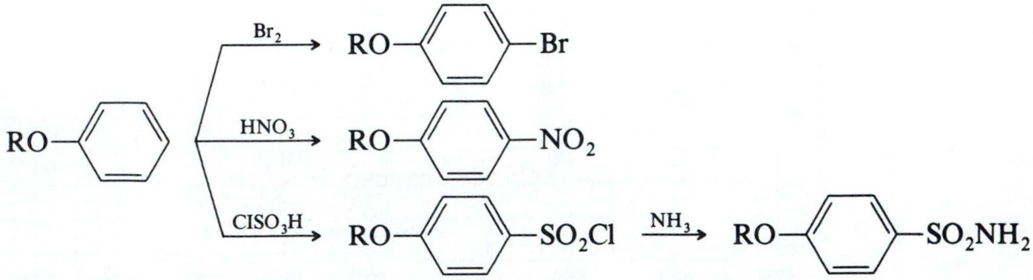

The extent of *bromination* of aromatic ethers depends on the groups already present.

PROCEDURE: BROMINATION

The aryl ether (5 to 100 mg) is dissolved in glacial acetic acid and placed in an ice bath; a slight excess of bromine is added with cooling (liquid bromine can be used, although for small-scale operations, a solution of bromine in glacial acetic acid is recommended). The reaction mixture is allowed to stand for a short while in the ice bath. It is then removed from the ice bath and allowed to stand at room temperature for 10 to 15 minutes. The bromo compound is separated by the addition of water. The crude product is recrystallized from dilute ethanol or petroleum ether.

Aromatic ethers are readily chlorosulfonated with chlorosulfonic acid. The intermediate sulfonyl chlorides can be isolated, but it is usually more convenient to convert them directly to *sulfonamides.*

PROCEDURE: SULFONAMIDES

A solution of 0.25 g, or 0.25 mL of aromatic ether in 2 mL of chloroform, in a test tube is cooled in an ice bath. About 1 g of chlorosulfonic acid is added, drop by drop, over a 5 minute period. The tube is removed from the ice bath and allowed to stand for 30 minutes. The reaction mixture is then poured into a small separatory funnel containing 5 mL of ice water. The chloroform layer is separated and washed with water. This layer is then added, with stirring, to 3 mL of concentrated ammonia solution (or ammonia gas is bubbled into the chloroform solution). The solution is stirred for 10 minutes and then the chloroform is evaporated off. The residue is dissolved in 3 mL of 5% sodium hydroxide, and the solution is filtered to remove insoluble material. The filtrate is acidified with dilute hydrochloric acid and cooled in an ice bath. The sulfonamide is then collected and recrystallized from dilute ethanol.

Nitration can be accomplished by any of the procedures outlined for aromatic hydrocarbons (see Section 9.1.4). Again the extent of nitration depends on the conditions employed and the groups already present.

9.5 ALDEHYDES AND KETONES

9.5.1 General Characterization

Classification

The presence of an intense band in the 1850 to 1650 cm^{-1} region of the infrared indicates a carbonyl compound. The lack of absorption in the 3600 to 3300 cm^{-1} region eliminates acids and primary and secondary amides as possibilities. The

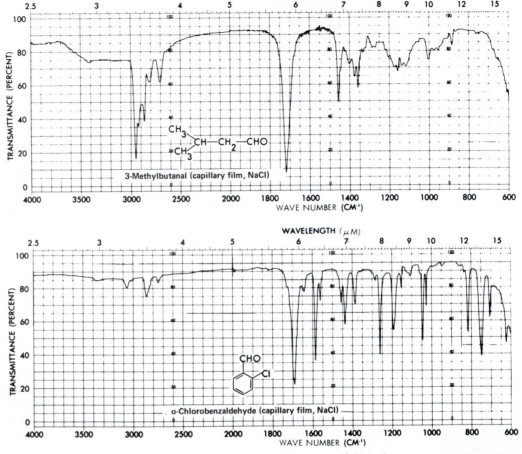

Figure 9.8   Typical infrared spectra of aldehydes.

differentiation between aldehydes and ketones and the other remaining carbonyl compounds (esters, anhydrides, tertiary amides, etc.) can be accomplished by the 2,4-dinitrophenylhydrazine test. From a cursory inspection of infrared spectra, the most likely compounds to be confused with aldehydes and ketones are the esters. Compare the spectra of aldehydes (see Figure 9.8) and ketones (see Figure 9.9) with those of esters (see Figure 9.12 on page 349). Note the two very intense C—O bands in the 1250 to 1050 cm^{-1} region in the ester spectrum, which are absent in the aldehydes and ketones. With the latter, the carbonyl peak is usually the most intense in the spectrum.

2,4-Dinitrophenylhydrazine Test

The utility of 2,4-dinitrophenylhydrazine lies in the fact that almost all aldehydes and ketones readily yield insoluble, solid 2,4-dinitrophenylhydrazones. There are a few exceptions among the long chain aliphatic ketones that yield oils.

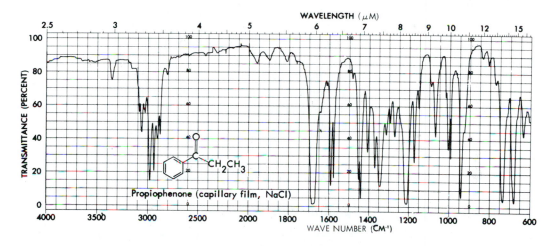

R and R′ = alkyl, aryl, or hydrogen

It is wise to apply the test with some discrimination, for example, only to compounds that give infrared evidence of being an aldehyde or ketone. The reagent can react with anhydrides or highly reactive esters; it is capable of oxidizing certain allylic and benzylic alcohols to aldehydes and ketones, which then give positive tests. The reagent can also give insoluble charge transfer complexes with amines and phenols. These, however, are usually considerably more soluble than the phenylhydrazones and can be distinguished on this basis. Indiscriminate use of the reagent

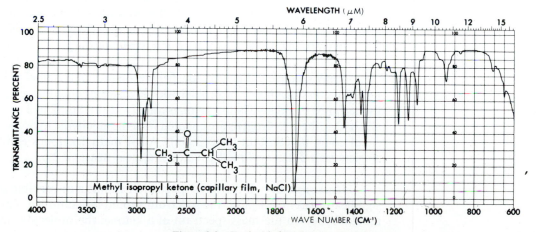

Figure 9.9 Typical infrared spectra of ketones.

may provide misleading positive tests owing to the hydrolysis of compounds such as ketals or trace impurities of carbonyl compounds.

The color of a 2,4-dinitrophenylhydrazone can provide a qualitative indication about conjugation, or lack thereof, in the starting carbonyl compound. Dinitrophenylhydrazones of saturated aldehydes and ketones are typically yellow. Conjugation with a carbon-carbon double bond or with an aromatic ring changes the color from yellow to orange-red. It should be noted that 2,4-dinitrophenylhydrazone itself is orange-red, and sometimes under the reaction conditions this material precipitates or contaminates the dinitrophenylhydrazone. Any orange-red precipitate that melts near 2,4-dinitrophenylhydrazine (198°C with decomposition) should be looked at with due suspicion.

PROCEDURE: 2,4-DINITROPHENYLHYDRAZINE TEST

Reagent: Dissolve 3 g of 2,4-dinitrophenylhydrazine in 15 mL of concentrated sulfuric acid. Add this solution with stirring to 20 mL of water and 70 mL of 95% ethanol. Mix this solution thoroughly and filter.

Place 1 mL of the 2,4-dinitrophenylhydrazine reagent in a small test tube. Add one drop of the liquid carbonyl compound to be tested or an equivalent amount of solid compound dissolved in a minimum amount of ethanol. Shake the tube vigorously; if no precipitate forms immediately, allow the solution to stand for 15 minutes. The formation of a yellow to orange-red precipitate is considered a positive test.

9.5.2 *Differentation of Aldehydes and Ketones*

Spectral Methods

Infrared spectra. The infrared spectra of ketones exhibit carbonyl bands in the 1780 to 1665 cm^{-1} region. Aldehyde carbonyl bands appear in the more restricted 1725 to 1685 cm^{-1} region. In the aldehyde spectra, additional bands that are not present in the spectra of ketones appear near 2820 and 2700 cm^{-1} (see Figure 9.8). Regrettably, these bands are often weak and/or poorly resolved.

NMR spectra. The very low field resonance of the aldehydic hydrogen (δ 9.5 to 10.1) is very diagnostic of an aldehyde. Hydrogens on saturated carbon adjacent to the carbonyl of aldehydes and ketones usually fall in the δ 2 to 3 region. In the case of aldehydes, these hydrogens are coupled with the aldehydic hydrogen with a relatively small coupling constant (1 to 3 Hz). Alkyl methyl ketones exhibit sharp methyl singlets near δ 2.2; aryl methyl ketones appear nearer δ 2.6.

Mass spectra. If the mass spectrum exhibits a discernible parent peak and a significant parent − 1 peak, the spectrum should be examined for other evidence of an aldehyde. In the mass spectrum of a ketone, the molecular ion is usually pronounced. Major fragmentation peaks for acyclic ketones result from alpha bond cleavage to give acylium ions. In the alkyl aryl ketones, the $ArCO^+$ fragment is usually the base peak.

Chemical Methods

The lability of aldehydes toward oxidation to acids forms the basis for a number of tests to differentiate between aldehydes and ketones.

Most aldehydes reduce silver ion from *ammoniacal silver nitrate solution (Tollen's reagent)* to metallic silver.

$$RCHO + 2\ Ag(NH_3)_2OH \longrightarrow 2\ Ag \downarrow + RCOO^-NH_4^+ + H_2O + NH_3$$

Positive tests are also given by hydroxyketones, reducing sugars, and certain amino compounds such as hydroxylamines. Normal ketones do not react. The test should be run only after it has been established that the unknown in question is an aldehyde or ketone.

PROCEDURE: TOLLEN'S TEST

Reagent: Solution A: Dissolve 3 g of $AgNO_3$ in 30 mL of water. **Solution B:**   10% sodium hydroxide. When the reagent is required, mix 1 mL of solution **A** with 1 mL of solution **B** in a *clean* test tube, and add dilute ammonia solution dropwise until the silver oxide is just dissolved.

C A U T I O N : *Prepare the reagent only immediately prior to use.* Do not heat the reagent during its preparation or allow the prepared reagent to stand, since the *very explosive* silver fulminate may be formed. Wash any residue down the sink with a large quantity of water. Rinse the test tube with dilute HNO_3 when the test is completed.

Add a few drops of a dilute solution of the compound to 2 to 3 mL of the prepared reagent. In a positive test, a silver mirror is deposited on the walls of the tube either in the cold or upon warming in a beaker of hot water.

Chromic acid in acetone rapidly oxidizes aldehydes to acids (the reagent also attacks primary and secondary alcohols, see Section 9.3.1), whereas ketones are attacked slowly or not at all. Aliphatic aldehydes are oxidized somewhat faster than aromatic aldehydes; this rate difference has been used to distinguish between aliphatic and aromatic aldehydes.[11]

PROCEDURE: CHROMIC ACID TEST

Reagent: Dissolve 1 g of chromium trioxide in 1 mL of concentrated sulfuric acid, and dilute with 3 mL of water.

---

[11] J. D. Morrison, *J. Chem. Ed.,* **42,** 554 (1965).

Dissolve one drop or 10 mg of the compound in 1 mL of reagent grade acetone (or better, in acetone that has been distilled from permanganate). Add several drops of chromic acid reagent. A positive test is indicated by the formation of a green precipitate of chromous salts. With aliphatic aldehydes, the solution turns cloudy within 5 seconds, and a precipitate appears within 30 seconds. Aromatic aldehydes generally require 30 to 90 seconds for the formation of a precipitate.

Aldehydes react with methone to give dimethone derivatives. Ketones do not react, and the method can serve to differentiate between the two (see next section).

9.5.3 *Aldehydes*

Characterization

The most frequently employed derivatives for aldehydes include the 2,4-dinitro-, the *p*-nitro-, and the unsubstituted phenylhydrazones, the semicarbazones, and the oximes. Discussion and procedures for these derivatives, as well as a method for reduction to a primary alcohol, will be presented in Section 9.5.5.

The *oxidation of an aldehyde to the corresponding acid* is a particularly useful method of identification, especially in the case of aromatic aldehydes that yield solid acids (characterized by their melting point and neutralization equivalent).

PROCEDURE: OXIDATION TO ACIDS

(a) The oxidation can be carried out by the chromic acid-in-acetone method outlined under alcohols (see Section 9.3.1).

(b) In a test tube place 0.5 g of aldehyde, 5 to 10 mL of water, and several drops of 10% sodium hydroxide. Add saturated aqueous potassium permanganate several drops at a time. Shake the mixture vigorously. Add additional permanganate until a definite purple color remains. Acidify the mixture with dilute sulfuric acid, and add sodium bisulfite solution until the excess permanganate and manganese dioxide have been converted to the colorless soluble manganese sulfate. Remove the acid by filtration or extraction with dichloromethane or ether. The acid can be recrystallized from water or aqueous ethanol, or it can be purified by sublimation.

The equations following this paragraph illustrate the reaction between 5,5-dimethylcyclohexane-1,3-dione (methone, dimethyldihydroresorcinol) and an aldehyde. One mole of the aldehyde condenses with 2 mol of the reagent to give the *dimethone,* sometimes called a *dimedone derivative.* As ketones do not yield derivatives, the reaction can serve as another method of differentiating between aldehydes and ketones. Aldehydes can be made to react in the presence of ketones. Also, the reaction can serve as a method for the quantitative determination of formaldehyde. The dimethones are generally more suitable derivatives for low molecular weight and aldehydes than are 2,4-dinitrophenylhydrazones.

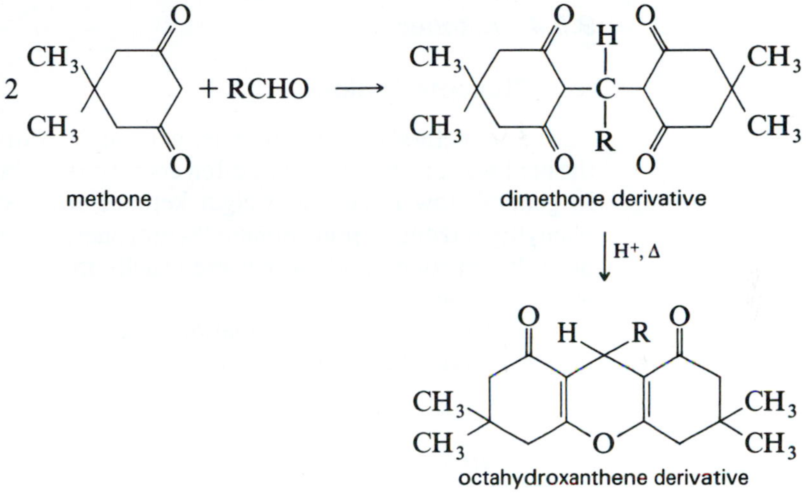

methone   dimethone derivative

octahydroxanthene derivative

PROCEDURE: DIMETHONE DERIVATIVES

To a solution of 100 mg of the aldehyde in 4 mL of 50% ethanol, add 0.3 g of methone and one drop of piperidine. Boil the mixture gently for 5 minutes. If the solution is clear at this point, add water dropwise to the cloud point. Cool the mixture in an ice bath until crystallization is complete. Crystallization is often slow, and one should allow, if necessary, 3 to 4 hours. Collect the crystals by vacuum filtration, and wash with a minimum amount of cold 50% ethanol. If necessary, the derivative can be recrystallized from mixtures of alcohol and water.

If the dimedone derivatives are heated with acetic anhydride or with alcohol containing a small amount of hydrochloric acid, cyclization to the *octahydroxanthene (methone anhydride)* occurs. The cyclization is rapid and quantitative, and the xanthenes can serve as a second derivative.

PROCEDURE: OCTAHYDROXANTHENES

The foregoing dimedone derivative can be converted to the octahydroxanthenes by boiling a solution of 100 mg of the derivative in a small volume of 80% ethanol to which one drop of concentrated hydrochloric acid has been added. The cyclization is complete in 5 minutes; water is added to the cloud point, and the solution is cooled to induce crystallization. The product is usually pure, and further recrystallization is not necessary.

9.5.4 Ketones

Characterization

The methods described under general derivatives of aldehydes and ketones in the next section are those most often employed to obtain solid derivatives of ketones. In general, low molecular weight ketones are best characterized by 2,4-dinitrophenylhydrazones, p-nitrophenylhydrazones, or semicarbazones. Hydrazones, phenylhydrazones, and oximes are usually more suitable for the higher molecular weight ketones.

Active hydrogen determination by deuterium exchange is sometimes useful in the structural determination of ketones.

$$RCH_2-\overset{\overset{\text{O}}{\|}}{C}-CH_3 \xrightarrow[Na_2CO_3]{D_2O} RCD_2-\overset{\overset{\text{O}}{\|}}{C}-CD_3 \quad \begin{cases} \text{mass spectrum} \\ \text{NMR} \\ \text{microanalysis} \end{cases}$$

The *haloform reaction* is useful in classification and derivatization of methyl ketones. The halogenation of aldehydes and ketones is catalyzed by acids and bases. With acid catalysis, the rate-determining step is enol formation, and with base catalysis, it is the formation of the enolate. Hence the reaction is independent of the concentration or kind of halogen. With base catalysis, the initial reaction occurs at the least substituted carbon.

$$R-\overset{\overset{\text{}}{|}}{\underset{\underset{\text{O}}{\|}}{C}}-CH_3 \xrightarrow{B^-} R-\overset{\overset{\text{}}{|}}{\underset{\underset{\text{O}^-}{|}}{C}}=CH_2 \xrightarrow{X_2} R-\overset{\overset{\text{}}{|}}{\underset{\underset{\text{O}}{\|}}{C}}-CH_2X$$

Since halogen is highly electronegative, successive hydrogens are replaced more readily, and unsymmetrical polysubstitution occurs.

$$R-\overset{\overset{\text{}}{|}}{\underset{\underset{\text{O}}{\|}}{C}}-CH_2X \xrightarrow[X_2]{B^-} R-\overset{\overset{\text{}}{|}}{\underset{\underset{\text{O}}{\|}}{C}}-CX_3$$

With methyl ketones or acetaldehyde, a trihalogenated product is formed. These trihalomethylcarbonyl compounds are readily cleaved by base to give a haloform and a salt of a carboxylic acid.

$$R\overset{\overset{\text{O}}{\|}}{C}-CX_3 \xrightarrow{OH^-} R\overset{\overset{\text{}}{|}}{\underset{\underset{\text{OH}}{|}}{C}}-CX_3 \longrightarrow RCOH + CX_3^- \longrightarrow R\overset{\overset{\text{O}}{\|}}{C}O^- + HCX_3$$

The haloform reaction will proceed with aldehydes or ketones that contain the groupings CH_3CO-, CH_2XCO-, CHX_2-CO, or CX_3CO-. The reaction will also proceed with compounds that will react with the reagent to give a derivative containing one of the necessary groupings, for example,

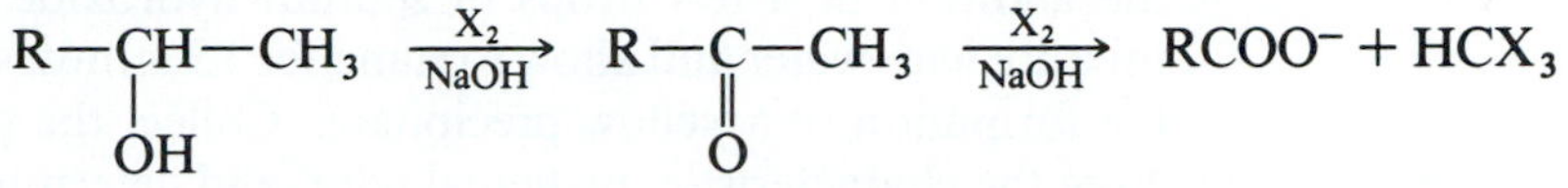

$$R-\underset{\underset{OH}{|}}{CH}-CH_3 \xrightarrow[NaOH]{X_2} R-\underset{\underset{O}{\|}}{C}-CH_3 \xrightarrow[NaOH]{X_2} RCOO^- + HCX_3$$

Acetaldehyde is the only aldehyde, and ethyl alcohol is the only primary alcohol, that produces haloform. The principal types of compounds that give the haloform reaction are methyl ketones and alkylmethyl carbinols

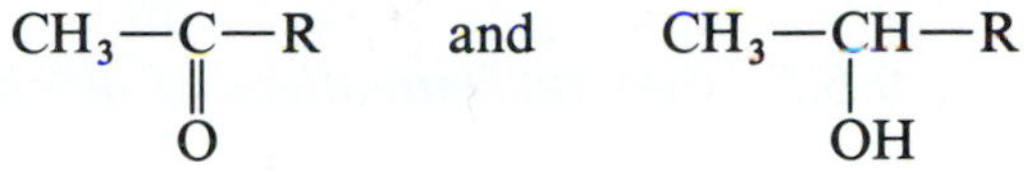

$$CH_3-\underset{\underset{O}{\|}}{C}-R \qquad and \qquad CH_3-\underset{\underset{OH}{|}}{CH}-R$$

R = alkyl, vinyl, or aryl

and compounds that have the structures

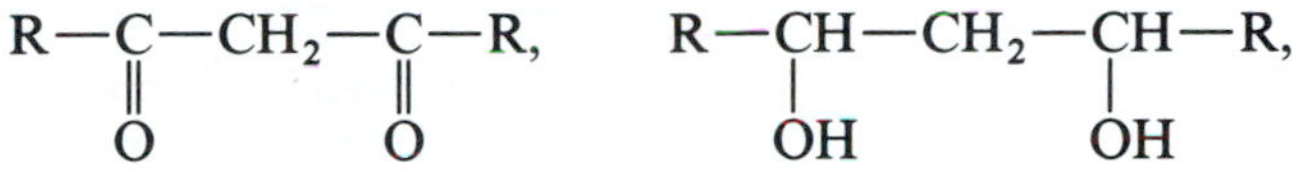

$$R-\underset{\underset{O}{\|}}{C}-CH_2-\underset{\underset{O}{\|}}{C}-R, \qquad R-\underset{\underset{OH}{|}}{CH}-CH_2-\underset{\underset{OH}{|}}{CH}-R, \qquad etc.$$

which are, or which upon oxidation yield, ketoalcohols or diketones that are cleaved by alkali to methyl ketones.

$$R-\underset{\underset{O}{\|}}{C}-CH_2-\underset{\underset{OH}{|}}{CH}-R' \xrightarrow{OH^-} R-\underset{\underset{O}{\|}}{C}-CH_3 + R'CHO$$

Acetic acid and its esters, and compounds such as

$$CH_3-\underset{\underset{O}{\|}}{C}-CH_2-COOR, \qquad CH_3-\underset{\underset{O}{\|}}{C}-CH_2CN$$

which substitute on the methylene rather than the methyl, do not produce haloforms.

As a method of preparation of acids from methyl ketones and related substances, sodium hypochlorite (bleaching powder or bleaching solution) is often used. As a qualitative test, the reaction is usually run with alkali and iodine, since iodoform is a water-insoluble crystalline solid readily identified by its melting point (*iodoform test*).

PROCEDURE: IODOFORM TEST

Reagent: Dissolve 20 g of potassium iodide and 10 g of iodine in 100 mL of water.

Dissolve 0.1 g or four to five drops of compound in 2 mL of water; if necessary, add 1,2-dimethoxyethane to produce a homogeneous solution. Add 1 mL of 10% sodium hydroxide and the potassium iodide-iodine reagent dropwise, with shaking, until a definite dark color of iodine persists. Allow the mixture to stand for several minutes; if no iodoform separates at room temperature, warm the tube in a 60°C water bath. If the color disappears, continue to add the iodine reagent dropwise until its color does not disappear after 2 minutes of heating. Remove the excess iodine by the addition of a few drops of sodium hydroxide solution. Dilute the reaction mixture with water and allow to stand for 15 minutes. A positive test is indicated by the formation of a yellow precipitate. Collect the precipitate; dry on filter paper. Note the characteristic medicinal odor, and determine the melting point; iodoform melts at 119 to 121°C.

9.5.5 General Derivatives of Aldehydes and Ketones

2,4-Dinitrophenylhydrazones and p-Nitrophenylhydrazones

$$ArNHNH_2 + RCOR' \longrightarrow ArNHN{=}CRR'$$

Ar = 2,4-dinitrophenyl, or p-nitrophenyl
R and R' = alkyl, aryl, or hydrogen

The 2,4-dinitrophenylhydrazones are excellent derivatives for small-scale operations. It is often possible to obtain sufficient amounts of the derivative for characterization from the precipitate formed in the classification test described earlier (see Section 9.5.1, page 319). There are certain limitations to the use of nitrophenylhydrazones as derivatives. From derivative tables one will observe that, in many cases, the melting points of the nitrophenylhydrazones are too high for practical utility; hence, the use of other derivatives such as phenylhydrazones, semicarbazones, or oximes is recommended. In general, mixtures of (E)- and (Z)-isomers about the C=N are obtained. This aspect was not recognized by earlier chemists, and the melting points of derivatives are often those of mixtures. A single recrystallization does not result in the separation of the isomers, and the melting point thus obtained generally corresponds closely with the literature value. Repeated careful fractional recrystallization often results in the separation and isolation of a single stereoisomer whose melting point does not agree with that reported. (In Tables 9.6 and 9.7, more than a single value for the melting point of a derivative is often given due to these complications.) Determination of the configuration of oximes can be accomplished by ^{13}C NMR.[12] In addition to E-,Z-isomerism about the C=N, the 2,4-dinitrophenylhydrazones tend to form several crystalline modifications with different melting points. One may find it useful to redetermine the melting points after allowing the melt to resolidify.

[12] G. E. Hawkes, K. Herwig, and J. D. Roberts, *J. Org. Chem.*, **39**, 1017 (1974); G. C. Levy and G. L. Nelson, *J. Am. Chem. Soc.*, **94**, 4897 (1972).

TABLE 9.6 Physical Properties of Aldehydes and Derivatives

Compound	bp (°C)†	Semi-carbazone	2,4-Dinitro-phenyl-hydrazone	p-Nitro-phenyl-hydrazone	Phenyl-hydra-zone	Oxime	Methone
Liquids			mp (°C) of Derivatives*				
Acetaldehyde (Ethanal)	20.2	162	148(168)	128.5	57(63,99)	47	140
Propionaldehyde (Propanal)	49.0	89	155	125	(oil)	40	155
Glyoxal (mp 15°C)	50	270	328		180	178	228
Acrolein	52.4	171	165	151	52 (pyrazonline)		192
2-Methylpropanal	64	125–6	187	131	(oil)		154
Butanal	74.7	106(96)	123	95(87)			135(142)
2,2-Dimethylpropanal	75	190	209	119		41	
3-Methylbutanal	92.5	107	123	111	(oil)	48.5	154
2-Methylbutanal	92–3	103	120				
Chloral	98	90d	131			56(40)	
Pentanal	103		107(98)			52	104.5
Hexanal	131	106	104(107)			51	108.5
Tetrahydrofurfural	144/740 mm	166					123
Heptanal	153	109	108	73		57	135(103)
Furfural	161.7 (90/65 mm)	202	230(214)	154(127)	97	92(76)	160
Cyclohexanecarboxyaldehyde	162	173				91	
Octanal	171 (81/32 mm)	101(98)	106	80		60	90
Benzaldehyde	179	222	237	192	158(155)	35	193
Nonanal	185	100(84)	100			64(69)	86
Salicylaldehyde	197	231	252	227	142	57(63)	206
m-Tolualdehyde	199	204(224)		157	91(84)	60	
o-Tolualdehyde	200	218(212)	194	222	101(106)	49	
p-Tolualdehyde	204	234(215)	234	200.5	114(121)	80	
n-Decanal	209	102	104			69	91.7
p-Methoxybenzaldehyde	248	210	253-4d	160–1	120–1	133	145
Cinnamaldehyde (trans)	252	215(208)	255d	195	168	138	213
Solids mp (°C)							
1-Naphthaldehyde (bp 292°C)	34	221		234	80	90	
Phenylacetaldehyde (bp 195°C)	34	156	121	151	63(58,102)	99(103)	165
o-Nitrobenzaldehyde	44	256	250(265)	263	156	102(154)	
p-Chlorobenzaldehyde (bp 214.5–216.5°C)	47	230	265	237(220)	127	110(146)	
Phthalaldehyde	56				191(di)		
2-Naphthaldehyde	60	245	270	230	206(218)	156	
p-Aminobenzaldehyde	72	153			156	124	
p-Dimethylaminobenzaldehyde	74	222	325	182	148	185	
m-Hydroxybenzaldehyde	102	198	260d	221–2 147	131	90	
p-Nitrobenzaldehyde	106	221	320	249	159	133	

* d denotes decomposition.

† bp at 760 mm Hg pressure unless otherwise noted.

TABLE 9.7 Physical Properties of Ketones and Derivatives

Compound	bp (°C)	Liquids mp (°C) of Derivatives				
		Semi-carbazone	2,4-Dinitro-phenyl-hydrazone	p-Nitro-phenyl-hydrazone	Phenyl-hydrazone	Oxime
2-Propanone (Acetone)	56.11	190	126	149	42	59
2-Butanone	80	135	117	129		(bp 152°C)
3-Buten-2-one	81	141				
Butane-2,3-dione	88	279(di)	315(di)	230(mono)	243(di)	76(mono) 245–6(di)
2-Methyl-3-butanone	94.3	114	120	109	(oil)	(oil)
3-Pentanone	102	138–9	156	144(139)	(oil)	(bp 165°C)
2-Pentanone	102.3	112	144	117	(oil)	(bp 167°C)
3,3-Dimethyl-2-butanone	106	158	125		(oil)	75(79)
4-Methyl-2-pentanone	116.8	132(134)	95			(bp 176°C)
2,4-Dimethyl-3-pentanone	124	160	88(98)			
2-Hexanone	128	125(122)	110	88	(oil)	49
4-Methylpent-3-en-2-one	130	164(133)	200	134	142	49
Cylopentanone	130.7	210	146	154	55(50)	56.5
2,4-Pentanedione	139	122(mono) 209(di)				149(di)
4-Heptanone	144	132	75			(bp 193°C)
3-Hydroxy-2-butanone	145	185(202)	318			
1-Hydroxy-2-propanone	146	196	128.5	173	103	71
2-Heptanone	151.2	123	89		207	
Cyclohexanone	156	167	162	147	81	91
3,5-Dimethyl-4-heptanone	162(173)	84				
2-Methylcyclohexanone	165	197d	137	132		43
2,6-Dimethyl-4-heptanone	168	122(126)	66(92)			
3-Methylcyclohexanone	170	179	155	119	94	
4-Methylcyclohexanone	171	203		128	110	39
2-Octanone	173	122	58	93		
Cycloheptanone	181	163	148	137		23
Ethyl acetoacetate	181	133(129d)	93			
5-Nonanone	187	90				
2,5-Hexanedione	194	185d(mono) 224(di)	257(di)	212(di)	120(di)	137(di)
Acetophenone (mp 20°C)	202	198–9(203)	240(250)	184–5	106	60
1-(2-Methylphenyl)ethanone	214	203	159			
1-(4-Methylphenyl)ethanone	214	203–5	159			
Isophorone	215	199.5d			68	79.5
1-Phenyl-2-propanone (mp 27°C)	216					
Propiophenone (mp 20°C)	218	182(174)	191			54
n-Butyrophenone (mp 13°C)	221	188				50
Isobutyrophenone	222	181	163		73	94
Isovalerophenone	226	210				72(64.5)
1-Phenyl-2-butanone	226	135(146)				
1-Phenyl-1-butanone	230	191	190			50

TABLE 9.7 (Continued)

		Solids				
		mp (°C) of Derivatives				
Compound	mp (°C)	Semi-carbazone	2,4-Dinitro-phenyl-hydrazone	p-Nitro-phenyl-hydrazone	Phenyl-hydrazone	Oxime
1,3-Diphenyl-2-propanone (bp 330°C)	34	146(126)	100		121(129)	125
α-Naphthayl methyl ketone (bp 302°C)	34	229			146	139
4-Phenyl-3-buten-2-one (bp 262°C)	42	188	227	165–7	159	117
1-Indanone	42	233	258	234–5	130–1	146
Benzophenone (bp 360°C)	48	167	238–9	154–5	137–8	144
Phenacyl bromide	50	146				89.5(97.0)
Benzalacetophenone	58	168	245		120	140(116)
4-Methylbenzophenone	60	122	202.4		109	154(115)
Desoxybenzoin (bp 320°C)	60	148	204	163	116	98
1-Phenyl-1,3-butanedione	61					
4-Methoxybenzophenone	62		180	199	132(90)	147(138) 115–6
1,3-Diphenyl-1,3-propanedione	78	205				
Fluorenone	83		284	269	152	196
Benzil	95	175(182) (mono) 243–4(di)	189	290	134(mono) 255(di)	138(mono) 237(di)
Benzoquinone	116	243(di) 166(mono) 178(mono)	231(di) 185.6(mono)		152	240
1,4-Naphthoquinone	125	247(mono)	278(mono)	279(mono)	206(mono)	198(mono)
Benzoin	137	206	245(234)		159(syn) 106(anti)	99(syn) 152(anti)
Anthraquinone	286				183	

PROCEDURE: 2,4-DINTRO- AND *p*-NITROPHENYLHYDRAZONES

Place 100 mg of 2,4-dinitro- or *p*-nitrophenylhydrazine in a test tube or Erlenmeyer flask containing 10 mL of methanol. Cautiously add five drops of concentrated hydrochloric acid and warm the solution on the steam bath, if necessary, to complete solution. Dissolve approximately 1 mmol of the carbonyl compound in 1 mL of methanol, and add this to the reagent. Warm the mixture on a water or steam bath for 1 to 2 minutes and allow to stand for 15 to 30 minutes. Most of the derivatives precipitate out on cooling, but for complete precipitation, it is advisable to add water to cloudiness. The derivative can be purified by crystallization from alcohol-water mixtures. In the case of less soluble materials, ethyl acetate is often found to be a more suitable solvent. This procedure is generally suitable for the preparation of both dinitrophenylhydrazones and nitrophenylhydrazones.

Semicarbazones

Semicarbazones prepared from carbonyl compounds and semicarbazide hydrochloride are excellent derivatives for ketones and aldehydes above five carbon atoms because of their ease of formation, highly crystalline properties, sharp melting points, and ease of recrystallization. The lower aldehydes react slowly and/or give soluble derivatives. As in the case of most carbonyl derivatives, semicarbazones from unsymmetrical carbonyl compounds are capable of existing in two isomeric forms. A number of substituted semicarbazones and thiosemicarbazones are occasionally used.

$$\begin{array}{c} R \\ \diagdown \\ \diagup \\ R' \end{array} C{=}O + NH_2NHCONH_2 \longrightarrow \begin{array}{c} R \\ \diagdown \\ \diagup \\ R' \end{array} C{=}NNHCONH_2 + H_2O$$

semicarbazide   semicarbazone

PROCEDURE: SEMICARBAZONES

In an 8-in. test tube, place 100 mg of semicarbazide hydrochloride, 150 mg of sodium acetate, 1 mL of water, and 1 mL of alcohol. Add 100 mg (0.1 mL) of the aldehyde or ketone. If the mixture is turbid, add alcohol until a clear solution is obtained. Shake the mixture for a few minutes and allow to stand. The semicarbazone typically will crystallize from the cold solution on standing, the time varying from a few minutes to several hours. In the case of less reactive carbonyl compounds, the reaction can be accelerated by warming the mixture on a water bath for 10 minutes and then cooling in ice water. The crystals are filtered off and washed with a little cold water; they can usually be recrystallized from methyl or ethyl alcohol alone or mixed with water.

Oximes

Oximes do not have as wide a utility as other carbonyl derivatives for the identification of aldehydes and ketones. They are somewhat more difficult to obtain crystalline and are more likely to exist as mixtures of geometrical isomers. However, they are occasionally useful and should be considered.

$$\begin{array}{c} R \\ \diagdown \\ \diagup \\ R' \end{array} C{=}O + NH_2OH \longrightarrow \begin{array}{c} R \\ \diagdown \\ \diagup \\ R' \end{array} C{=}N \diagdown_{OH} \quad \text{and/or} \quad \begin{array}{c} R \\ \diagdown \\ \diagup \\ R' \end{array} C{=}N \diagup^{OH} + H_2O$$

(E)- and *(Z)-*isomeric oximes

PROCEDURE: OXIMES

(a) The method described for preparation of semicarbazones can be employed. One substitutes hydroxylamine hydrochloride for the semicarbazide hydrochloride. The heating period is usually necessary.

(b) Reflux a mixture of 0.1 g of the aldehyde or ketone, 0.1 g of hydroxylamine hydrochloride, 2 mL of ethanol, and 0.5 mL of pyridine on a water bath for 15 to 60 minutes. Remove the solvent at reduced pressure or by evaporation with a current of air in a hood. Add several milliliters of cold water, and triturate thoroughly. Collect the oxime and recrystallize from ethanol, aqueous ethanol, or toluene.

Phenylhydrazones

Phenylhydrazones are recommended particularly for aryl carbonyl compounds whenever the dinitrophenylhydrazone is not suitable or adequate for identification. The manipulation of these derivatives should be as rapid as possible since they undergo slow decomposition in air.

PROCEDURE: PHENYLHYDRAZONES

C A U T I O N : Phenylhydrazine is a suspected carcinogen.

In a test tube place 100 mg of aldehyde or ketone, 4 mL of methanol, and four drops of phenylhydrazine. Boil the mixture for 1 minute, add one drop of glacial acetic acid, and boil gently for 3 minutes. Add cold water dropwise until a permanent cloudiness results, cool, collect the crystals, and wash with 1 mL of water containing one drop of acetic acid. Recrystallize the product immediately by dissolving in hot methanol, add water to the solution until cloudiness appears, cool, and scratch the sides of the tube if necessary to crystallize. Collect the crystals, wash with a few drops of dilute methanol, and determine the melting point as soon as possible. Dry the crystals by pressing between filter paper.

Reduction to Alcohols

$$\text{NaBH}_4 + 4\,\text{RR}'\text{C}=\text{O} \xrightarrow[\text{(2) NaOH,H}_2\text{O}]{} 4\,\text{RR}'\text{CHOH} + \text{NaBO}_2$$

Alcohols are likely to be liquids, oils, or low-melting solids; however, there are occasions when the alcohols formed by reduction of aldehydes or ketones are well-defined crystalline solids suitable for identification purposes. Metal hydride reductions of carbonyl compounds to the corresponding alcohols generally proceed smoothly and give a high yield of the desired product. Sodium borohydride is less reactive than other hydrides such as lithium aluminum hydride, but it has the advantage of being stable in the presence of air and moisture. Sodium borohydride can be dissolved in cold water without extensive decomposition, whereas lithium aluminum hydride decomposes explosively on contact with water, and one must use an ether solvent for the reduction medium.

A commonly used method of reduction is to dissolve the organic compound in ethanol and to add an aqueous solution of sodium borohydride. Sodium borohydride, although stable in water, reacts at an appreciable rate with ethanol. It is therefore necessary to use an excess of the borohydride in order to offset losses sustained through solvolysis. Excess borohydride remaining in the reaction is destroyed by boiling the reaction mixture for a few minutes. The borate ester formed in the reduction is hydrolyzed during the boiling step by the addition of aqueous base.

PROCEDURE: SODIUM BOROHYDRIDE REDUCTION

In a small test tube, dissolve 2 mmol of the carbonyl compounds in 3 to 5 mL of ethanol. In a second test tube, place 50 mg of sodium borohydride and dissolve in five drops of distilled water. Add the ethanol solution of the carbonyl compound to the borohydride solution. Shake the reaction mixture several times during a 15-minute period. Add 0.5 mL of 6N aqueous sodium hydroxide solution and a boiling chip to the reaction mixture. Boil it gently on the steam bath for 5 minutes. Pour the reaction mixture onto a small amount of ice, and collect the resulting precipitate by vacuum filtration. If the alcohol does not crystallize, it can be recovered by extraction with ether and then purified.

9.5.6 Special Method of Separation and Purification of Aldehydes and Ketones

Most aldehydes, alkyl methyl ketones, and unhindered cyclic ketones react with saturated sodium hydrogen sulfite (bisulfite) solution to form crystalline *bisulfite addition compounds.* Other aliphatic and aromatic ketones do not yield addition compounds.

$$\diagup\hspace{-0.5em}\overset{\displaystyle\diagdown}{}C{=}O \underset{H^+ \text{ or } OH^-}{\overset{NaHSO_3}{\rightleftharpoons}} \overset{\displaystyle\diagdown}{\underset{\diagup}{}}C\overset{SO_3Na}{\underset{OH}{}}$$

The reaction is reversible, and the aldehyde or ketone can be recovered by reaction of the addition compound with aqueous sodium carbonate or dilute hydrochloric acid. The sequence of formation and decomposition of the addition compounds can be used for the purification and/or separation of aldehydes and certain ketones from other materials.

PROCEDURE: BISULFITE ADDITION COMPOUNDS

Shake or stir thoroughly the aldehyde- or ketone-containing mixture with a saturated solution of sodium hydrogen sulfite. The temperature will rise as the exothermic addition reaction takes place. Extract the residual organic material into ether and recover. Collect the crystalline precipitate, wash it with a little ethanol followed by ether, and allow it to dry.

To decompose the addition compound, use 10% sodium carbonate solution or dilute hydrochloric acid.

9.6 CARBOHYDRATES

Classification

Mono- and disaccharides are colorless solids or syrupy liquids. They are readily soluble in water but are almost completely insoluble in most organic solvents. The solids typically melt at high temperatures (over 200°C) with decomposition (browning and producing a characteristic caramel-like odor), and they char when treated with concentrated sulfuric acid. The polysaccharides possess similar properties but are generally insoluble or only slightly soluble in water.

The infrared spectra of carbohydrates obtained as mulls or potassium bromide pellets exhibit prominent hydroxyl absorption with correspondingly strong absorption in the C—O stretching region. Almost all sugars of four or more carbons exist in cyclic hemiacetal, acetal, hemiketal, or ketal forms, and hence exhibit no carbonyl band in the infrared (see Figure 9.10).

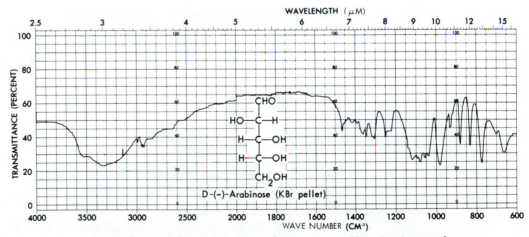

Figure 9.10 Typical carbohydrate infrared spectrum. Note the absence of a carbonyl band.

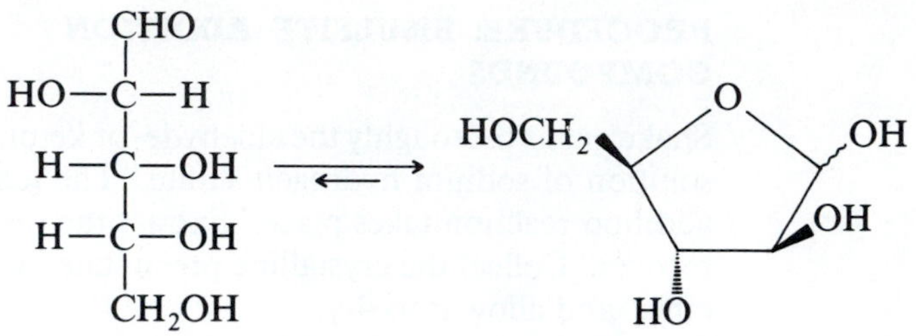

$$
\begin{array}{c}
\text{CHO} \\
| \\
\text{HO—C—H} \\
| \\
\text{H—C—OH} \\
| \\
\text{H—C—OH} \\
| \\
\text{CH}_2\text{OH}
\end{array}
\quad \underset{\longleftarrow}{\longrightarrow} \quad
$$

Sugars containing an aldehyde or α-hydroxy ketone grouping present in the free or hemiacetal (but not acetal) form are oxidized by Benedict's and Tollen's reagents (see Section 9.5.2 for Tollen's test). Such sugars are referred to as "reducing sugars."

Pentoses and hexoses are dehydrated by concentrated sulfuric acid to form furfural or hydroxymethylfurfural, respectively. In the *Molisch test,* these furfurals condense with 1-naphthol to give colored products.

PROCEDURE: MOLISCH TEST

Place 5 mg of the substance in a test tube containing 0.5 mL of water. Add two drops of a 10% solution of 1-naphthol in ethanol. By means of a dropper, allow 1 mL of concentrated sulfuric acid to flow down the side of the inclined tube so that the heavier acid forms a bottom layer. If a carbohydrate is present, a red ring appears at the interface of the two liquids. A violet solution is formed on mixing. Allow the mixed solution to stand for 2 min; then dilute with 5 mL of water. A dull-violet precipitate will appear.

Characterization

Since sugars decompose on heating, they do not have well-defined characteristic melting points. The same is true of some of their derivatives, for example, osazones. Fortunately, the number of sugars normally encountered in identification work is relatively small. Authentic samples of most are readily available for comparative purposes.

Thin-layer and paper chromatographic comparison[13] of an unknown with authentic samples provides an excellent means for tentative identification. For simple free sugars, paper chromatography is generally superior to thin-layer chromatography. Recommended solvent systems for paper chromatography of sugars include:

1. 1-Butanol, ethanol, water (40, 11, 19 parts by volume).
2. 1-Butanol, pyridine, water (9, 5, 4).

---

[13] E. Heftmann, *Chromatography,* 2nd ed. (New York: Reinhold, 1967); R. J. Block, E. L. Durram, and G. Zweig, *Paper Chromatography and Paper Electrophoresis* (New York: Academic Press, 1958); E. Lederer and M. Lederer, *Chromatography,* 2nd ed. (Amsterdam: Elsevier, 1957).

3. 2-Propanol, pyridine, water (7, 7, 5).
4. 1-Butanol, acetic acid, water (2, 1, 1).

For the *detection of reducing sugars* on paper chromatograms, the authors suggest the following spray.

PROCEDURE: SPRAY FOR REDUCING SUGARS

The dried chromatogram is sprayed with a 3% solution of *p*-anisidine hydrochloride in 1-butanol and heated to 100°C for 3 to 10 minutes. Aldo- and ketohexoses, as well as other sugars, give different colored spots with this reagent.

The structure and stereochemistry of a number of sugars and sugar derivatives have been examined by NMR and mass spectrometry. The primary literature should be consulted for details. The relationship between dihedral angle and coupling constants between the hydrogens has been of significant utility in determining the stereochemical relationships of the various ring hydrogens in cyclic sugar derivatives.

The specific rotation of sugars and derivatives is a useful means of identification. Rotations must be measured under specified conditions of concentration, solvent, and temperature, employing pure samples.

Reaction of Sugars with Phenylhydrazine

Sugars containing an aldehyde or keto group (as the hemiacetal or hemiketal) react with an equivalent of phenylhydrazine in the cold to produce the corresponding phenylhydrazones. These derivatives are water-soluble and do not precipitate. Heating these sugars with excess (3 to 4 equivalents) phenylhydrazine produces *osazones* and *polyzones*.[14]

$$
\begin{array}{c}
H \\
| \\
C=O \\
| \\
H-C-OH \\
| \\
R
\end{array}
+ 3\ C_6H_5NHNH_2 \longrightarrow
\begin{array}{c}
H \\
| \\
C=N-NHC_6H_5 \\
| \\
C=N-NHC_6H_5 \\
| \\
R
\end{array}
+ C_6H_5NH_2 + 2\ H_2O + NH_3
$$

an osazone

It should be noted that, in the formation of osazones, one carbinol grouping is oxidized, and hence a number of isomeric sugars give the same osazone. The osazones exist in chelated structures in equilibrium with a nonchelated isomer.[14]

---

[14] O. L. Chapman, R. W. King, W. J. Welstead, Jr., and T. J. Murphy, *J. Am. Chem. Soc.*, **86**, 4968 (1964); O. L. Chapman, *Tetrahedron Letters*, 2599 (1966).

PROCEDURE: OSAZONES

➤ **C A U T I O N :** Phenylhydrazine is a suspected carcinogen.

Place a 0.1-g sample of the unknown sugar and a 0.1-g sample of a known sugar (suspected to be the unknown) in separate test tubes. To each sample add 0.2 g of phenylhydrazine hydrochloride, 0.3 g of sodium acetate, and 2 mL of distilled water. Stopper the test tubes with vented corks, and place them together in a beaker of boiling water. Note the time of immersion and the time of precipitation of each osazone. Shake the tubes occasionally. The time required for the formation of the osazone can be used as evidence for the identification of the unknown sugar.

Under these conditions, fructose osazone precipitates in 2 minutes, glucose in 4 to 5 minutes, xylose in 7 minutes, arabinose in 10 minutes, and galactose in 15 to 19 minutes. Lactose and maltose osazones are soluble in hot water.

After 30 minutes, remove the tubes from the hot water and allow them to cool. Carefully collect the crystals, and compare the unknown with the known crystals under a low-power microscope. The melting points (decomposition) of osazones depend on the rate of heating and lie too close together to be of value.

Osazones can be converted to osotriazoles, which have sharp melting points.[15]

$$
\begin{array}{ccc}
\text{H}-\text{C}=\text{NNHC}_6\text{H}_5 & & \text{H}-\text{C}=\text{N} \\
| & \xrightarrow{\text{Cu}^{++}} & | \qquad\quad \diagdown \\
\text{C}=\text{NNHC}_6\text{H}_5 & & \quad\qquad \text{N}-\text{C}_6\text{H}_5 \\
| & & \text{C}=\text{N} \diagup \\
\text{R} & & \text{R}
\end{array}
$$

an osotriazole

Other derivatives of value in identification of sugars include acetates, and acetates of thioacetals, benzoates, acetonides, and benzylidene derivatives. The trimethylsilyl ethers (see Section 9.3.2) are useful for mass spectral and glc analysis.

Table 9.8 lists the physical properties of carbohydrates and their derivatives.

9.7 CARBOXYLIC ACIDS

Classification

Acidic compounds containing only carbon, hydrogen, and oxygen are either carboxylic acids, phenols, or possibly enols. An indication of whether a water-insoluble compound is an acid or a phenol can be obtained from simple solubility tests.

---

[15] W. T. Haskins, R. M. Hahn, and C. S. Hudson, *J. Am. Chem. Soc.,* **69,** 1461 (1947) and previous papers.

TABLE 9.8   Physical Properties of Carbohydrates and their Phenylosazones

Compound		mp (°C)	$[\alpha]_D^{20}$ (water)*	Phenylosazone mp (°C)
D-Erythrose		Syrup	$+1 \rightarrow -14.5°(c = 11)$	164
L-Erythrulose		Syrup	$+11.4°(c = 2.4)$	164
α-L-Rhamnose (H$_2$O)		82–92	$-7.7 \rightarrow +8.9°$	222
D-Ribose		87–95	$-19.7°(c = 4)$	164
2-Deoxy-D-ribose		90	$+2.88 \rightarrow -56.2°$	
β-Maltose(H$_2$O)		102–3	$+111.7 \rightarrow +130.4°(c = 4)$	206
D-Fructose		102–4	$-132.2 \rightarrow -92°(c = 2)$	210
D-Altrose		105	$+32.6(c = 7.6)$	178
α-D-Lyxose		106–7	$+5.5 \rightarrow -14.0°(c = 0.8)$	164
D-Threose		126–32	$? \rightarrow -12.3°(c = 4)$	164
β-D-Allose		128	$+0.58 \rightarrow +14.4°(c = 5)$	178
α-D-Mannose		133	$+29.3 \rightarrow +14.2°(c = 4)$	210
β-D-Mannose		132	$-17.0 \rightarrow +14.2°(c = 4)$	210
D-Xylose		144	$+92 \rightarrow +18.6°(c = 10)$	164
α-D-Glucose		146(anh.)	$+112.2 \rightarrow +52.7°(c = 10)$	210
β-D-Glucose		148–50	$+18.7 \rightarrow +52.7°(c = 10)$	210
β-L-Arabinose		158–9	$+173 \rightarrow +105(c = 3)$	163
L-Sorbose		165	$-43.7°(c = 5)$	156
α-D-Galactose		167	$+150.7 \rightarrow +80.2°$	196
Sucrose		169–70	$+66.5°(c = 2)$	
Lactose	α-form	223(anh.)	$+92.6 \rightarrow +52.3°(c = 4.5)$	200
	β-form	252(anh.)	$+34 \rightarrow 52.3°(c = 4.5)$	200

* When two rotations are shown, the first is the initial rotation and the second is that observed at equilibrium (mutarotation).

Both classes of compounds are soluble in sodium hydroxide, but only carboxylic acids are soluble in 5% sodium hydrogen carbonate with the liberation of carbon dioxide (for exceptions, see Section 8.4.4). Water-soluble acids also liberate carbon dioxide from sodium hydrogen carbonate solution. Other classes of compounds that liberate carbon dioxide from sodium hydrogen carbonate solution include salts of primary and secondary amines (these can be differentiated readily on the basis of the liberation of free amines, their melting point behavior, and their elemental analysis); sulfinic and sulfonic acids (differentiated on the basis of elemental analysis, infrared spectra, and their more acidic character); and a variety of substances such as acid halides and acid anhydrides that are readily hydrolyzed to acidic materials. Phenols and enols give positive color tests with ferric chloride solution. Fortunately, all these categories of acidic compounds can be readily differentiated from carboxylic acids on the basis of the extremely characteristic infrared spectrum of a carboxylic acid. Carboxylic acids give rise to a very broad and characteristic O—H band and a carbonyl band near 1700 cm^{-1} (see Figure 9.11). The chemist with even minimum spectroscopic experience will soon learn to distinguish other compounds from carboxylic acids by infrared spectroscopy.

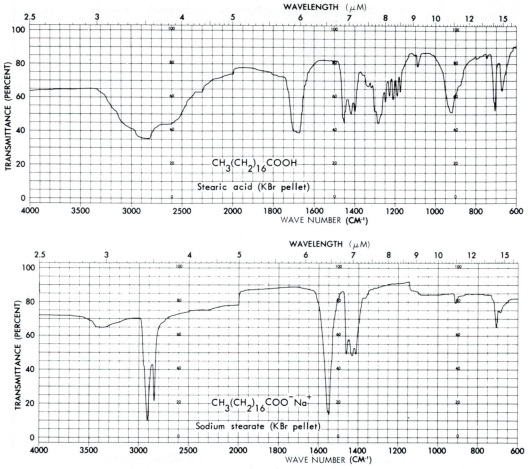

Figure 9.11   Infrared spectra of a typical acid and its sodium salt. Note the shift in the carbonyl frequency.

Characterization

One of the simplest and most informative ways to characterize a carboxylic acid is to determine its *neutralization equivalent.* The neutralization equivalent of an acid is its equivalent weight; the molecular weight can be determined from the neutralization equivalent by multiplying that value by the number of acidic groups in the molecule.

Since the pK$_a$'s of both the organic acid and the indicator are sensitive to solvent changes, one should employ only enough ethanol to dissolve the organic acid. With high concentrations of ethanol, sharp endpoints are not obtained with phenolphthalein. If it is necessary to employ absolute or 95% ethanol, bromthymol blue should be used as the indicator.

For accurate results, a blank should always be run on the solvent, and one should take care that the neutralization equivalent is determined from a substance that is pure and anhydrous. Neutralization equivalents can be obtained with an accuracy of 1% or less.

With good technique, one can obtain an accurate neutralization equivalent with samples as small as 5 mg, employing more dilute standard alkali in burettes designed for greater accuracy with small volumes.

PROCEDURE: NEUTRALIZATION EQUIVALENTS

Approximately 200 mg of the acid is accurately weighed and dissolved in 50 to 100 mL of water or aqueous ethanol. This solution is titrated with standard sodium hydroxide solution (approximately $0.1N$), employing phenolphthalein as the indicator or employing a pH meter.

$$\text{Neutralization equivalent} = \frac{\text{weight of sample in milligrams}}{\text{milliters of base} \cdot \text{normality}}$$

If an acid has been well characterized, it is often sufficient for identificaton purposes to obtain the neutralization equivalent and the melting point of a carefully chosen derivative. More than 70 types of derivatives have been suggested at various times for the identification of carboxylic acids. The majority of these derivatives fall in the categories of amides, esters, and salts of organic bases. Representative and recommended examples from each of these categories will be discussed in the following paragraphs.

Amides and Substituted Amides

Acids or salts of acids can be converted directly to acid chlorides by the action of thionyl chloride. The acid chlorides are converted to the *amides* and *substituted amides* by reaction with excess ammonia or an amine, as illustrated below.

$$RCOOH + SOCl_2 \longrightarrow RCOCl + SO_2 + HCl$$

$$RCOCl + H_2NR' \longrightarrow RCONHR' + HCl$$

PROCEDURE: AMIDES FROM ACIDS

C A U T I O N : Noxious gases may be evolved. Do these reactions in a hood.

(a) *Acid chloride.* In a 25-mL round-bottomed flask fitted with a condenser and a calcium chloride tube or a cotton plug in the top of the condenser to exclude moisture, place 0.5 to 1.0 g of the anhydrous acid or anhydrous sodium salt and add 2.5 to 5 mL of thionyl chloride. Reflux the mixture gently for 30 minutes. Arrange for distillation, and distill off the excess thionyl chloride (bp 78°C). For acids below four carbon atoms, the bp of the acid chloride may be too near that of thionyl chloride to afford practical separation by distillation. In this event, the excess of the reagent can be destroyed by the addition of formic acid.

$$HCOOH + SOCl_2 \longrightarrow CO + SO_2 + 2\ HCl$$

(b) *Primary amides.* For the preparation of primary amides, it is unnecessary to distill off the excess thionyl chloride. The entire reaction mixture can be cautiously poured into 15 mL of ice-cold concentrated ammonia. The precipitated amide is collected by vacuum filtration and purified by recrystallization from water or aqueous ethanol.

As an alternate method, the acid chloride is dissolved in 5 to 10 mL of dry toluene and the excess ammonia gas is passed through the solution. If the amide does not precipitate, it can be recovered by evaporation of the solvent.

(c) *Anilides and substituted anilides.* Dissolve the acid chloride in 5 mL of toluene and add a solution of 2 g of pure aniline, *p*-bromoaniline, or *p*-toluidine in 15 mL of toluene. It may be convenient to run this reaction in a small separatory funnel. Shake the reaction mixture with 5 mL of dilute hydrochloric acid to remove the excess aniline, wash the toluene layer with 5 mL of water, evaporate the solvent, and recrystallize the anilide from water or aqueous ethanol.

Anilides and *p-toluides* can also be prepared directly from the acids or from alkali metal salts of the acids by heating them directly with the aniline or the hydrochloride salt of the aniline.

$$RCOOH + C_6H_5NH_2 \longrightarrow RCOO^{-+}NH_3C_6H_5 \xrightarrow[\Delta]{} RCONHC_6H_5 + H_2O$$

$$RCOONa + C_6H_5NH_3^+\ Cl^- \xrightarrow[\Delta]{} RCONHC_6H_5 + NaCl + H_2O$$

PROCEDURE: ANILIDES AND *p*-TOLUIDES

Place 0.5 g of the acid and 1 g of aniline or *p*-toluidine in a small, dry, round-bottomed flask. Attach a short air condenser and heat the mixture in an oil bath at 140 to 160°C for 2 hours. Caution should be used to avoid heating the mixture too vigorously, thus causing loss of the acid by distillation or sublimation. If the material is available as the sodium salt, use 0.5 g of the salt and 1 g of the amine hydrochloride. If there is evidence that the substance is a diacid, use double the quantity of the amine and increase the reaction temperature to 180 to 200°C. At the end of the reaction time, cool the reaction mixture and triturate it with 20 to 30 mL of 10% hydrochloric acid, or dissolve the residue in an appropriate solvent and wash it with hydrochloric acid, dilute sodium hydroxide, and water. Then evaporate the solvent. The amides can usually be recrystallized from aqueous ethanol.

Esters

Solid esters form a second series of compounds useful in the identification and characterization of acids. The most commonly used are the *p*-nitrobenzyl and *p*-bromophenacyl esters, although the phenacyl, *p*-chlorophenacyl, and *p*-phenylphenacyl are also occasionally used. The halides corresponding to these esters, for example, the *p*-nitrobenzyl halides and the phenacyl halides, all undergo facile SN2 displacement reactions. These esters are prepared in aqueous alcoholic solution by displacement of the corresponding halide with the weakly nucleophilic carboxylate anions, as illustrated in the following equation.

$$\text{RCOO}^-\text{Na}^+ + \text{BrCH}_2\overset{\overset{\displaystyle O}{\|}}{\text{C}}\!\!-\!\!\langle\ \rangle\!\!-\!\!\text{Br} \longrightarrow \text{RCOOCH}_2\text{CO}\!\!-\!\!\langle\ \rangle\!\!-\!\!\text{Br} + \text{NaBr}$$

The method is particularly advantageous because it does not require the acid to be anhydrous, and it can be run with equal facility with the alkali salts of the acids.

Although the phenacyl and benzyl esters are extremely useful for making derivatives of acids that cannot be easily obtained in anhydrous conditions (e.g., from saponification of esters of lower carboxylic acids) or directly from alkali metal salts, there are certain disadvantages that make them undesirable for routine use in identification. *The benzyl and phenacyl halides all have severe lachrymatory and blistering properties.* The formation of the esters is generally slow, and any unreacted halide is often difficult to separate from the ester. For this reason, less than one equivalent of the halide should be used, and the reaction should be continued for $1\frac{1}{2}$ to 2 hours to ensure completeness. The traces of the halides remaining with the esters impart irritating properties to the esters and severely depress the melting point.

PROCEDURE: PHENACYL AND *p*-NITROBENZYL ESTERS

C A U T I O N : Irritants involved. See proceeding paragraph.

In a small test tube or a small, round-bottomed flask equipped with a reflux condenser, place 1 mmol of the acid and 1 mL of water. Add one drop of phenolphthalein, and carefully neutralize by the dropwise addition of 10% sodium hydroxide solution until the color of the solution is just pink. Add one or two drops of dilute hydrochloric acid to discharge the pink color of the indicator. Add a solution of 0.9 mmol of the phenacyl or benzyl halide dissolved in 5 to 8 mL of ethanol. Reflux the solution for $1\frac{1}{2}$ to 2 hours, cool, and add 1 mL of water, and scratch the sides of the tube. When precipitation is complete, collect the ester by filtration; wash with a small amount of 5% sodium carbonate solution and then several times with small quantities of cold water. The esters can usually be recrystallized from aqueous ethanol. A frequently used procedure is to dissolve the crystals in hot alcohol, filter, and add water to the hot filtrate until a cloudiness appears. Rewarm the solution until the

TABLE 9.9 Physical Properties of Carboxylic Acids and Derivatives

Name of Acid	bp(°C)*	mp(°C)	mp (°C) of Derivatives Amide	Anilide	p-Toluide
Acetic	118	16	82	114	147
Acrylic	140	13	85	105	141
Propionic	141		81	106	126
Propynoic	144d	18		87	62
2-Methylpropanoic	154		128	105	107
Methacrylic	161	16	106		
Butyric	162.5		115	96	75
3-Methylbutanoic	176			137	107
Pentanoic	186		106	63	74
Dichloroacetic	194		98	118	153
Hexanoic	205		82	95	75
Heptanoic	223		96	70	81
Octanoic	239		110	57	70
Nonanoic	255		99	57	84
Oleic	216/5 mm		76	41	42
Undecanoic	284	30	103(99)	71	80
4-Oxopentanoic (levulinic)	246	35	108d	102	109
Decanoic	270	31	108(98)	70	78
2,2-Dimethylpropanoic	164	35	178	129	120
Dodecanoic (lauric)	299	44	110(102)	78	87
trans-9,10-Octadec-enoic (elaidic)	234/15 mm	45	94		
(Z)-2-Methyl-2-bute-noic (angelic)	185	46	127–8	126	
3-Phenylpropanoic	280/754 mm	48	105	98	135
4-Phenylbutanoic	290	52	84		
Tetradecanoic	202/16 mm	54	107(103)	84	93
Trichloroacetic	197	58	141	97	113
Hexadecanoic	222/16 mm	63	107	90	98
Chloroacetic	189	63	121	134	162
Phenylacetic	265	76.5	156	118	136
Glutaric	200/20 mm	97	137	224	218
Phenoxyacetic		100	148	99	
Oxalic (dihydrate)		101	419d(di) 210(mono)	254(di) 148–9(mono)	268
o-Toluic		104	143	125	144
m-Toluic		113	97	126	118
D,L-Mandelic		118	132	152	172
D or L-Mandelic		133			
Benzoic		122	110	163	158
Maleic		130	266(di)	187(di)	142(di)
Furancarboxylic		134	143	123.5	107.5
Cinnamic (trans)		133	148	153	168
Malonic		135	50(mono) 170(di)	132(mono) 230(di)	86(mono) 235(di)
o-Nitrobenzoic		146	176	155	
Diphenylacetic		148	168	180	173
1-Naphthoic		162	202	163	

TABLE 9.9 (Continued)

Name of Acid	bp(°C)	mp(°C)	mp (°C) of Derivatives		
			Amide	Anilide	p-Toluide
4-Nitrophthalic		165	200d	192	172(mono)
D-Tartaric		171	196d	180d(mono)	
				264d(di)	
p-Toluic	275	180	160	148	160(165)
Acetylenedicarboxylic		179	294d(di)		
4-Methoxybenzoic		186	167	169–71	186
2-Naphthoic		185.5	195	171	192
Butanedioic (succinic)	235d	186	157(mono)	148(mono)	180(mono)
		185	260(di)	230(di)	255(di)
p-Aminobenzoic		188	193(di)		
Terephthalic		300		237	
Fumaric		302	266d(di)	314	

* bp at 760 mm Hg pressure unless otherwise noted.

cloudiness disappears, and then cool. Scratch the sides of the test tube, if necessary to induce crystallization. *Avoid handling or contact of the crystals with the skin.*

If the original acid is available as an alkali salt, dissolve one equivalent of it in a minimum amount of water, add a drop of phenolphthalein, and adjust the acidity as above.

Methyl esters are also occasionally used as derivatives. In certain cases these are solids, but in the majority of the cases they appear as oils. When the methyl esters are relatively high-melting solids, they are often very respectable derivatives for the purposes of melting point determinations. Frequently one may wish to make the methyl ester for purposes of vapor phase chromatographic comparison of retention times of unknown esters with known methyl esters. The methyl esters have the advantage in that they are often more volatile, more easily separated, and they elute as sharper peaks from the vapor phase chromatograph than do the corresponding acids. Methyl esters can be made quantitatively on a very small scale, employing diazomethane. They can also be made conveniently on a small scale employing 2,2-dimethoxypropane with a small amount of p-toluenesulfonic acid or concentrated hydrochloric acid.

$$RCOOH + CH_2N_2 \longrightarrow RCOOCH_3 + N_2$$

$$RCOOH + CH_3\overset{\overset{\displaystyle OCH_3}{|}}{\underset{\underset{\displaystyle OCH_3}{|}}{C}}CH_3 \xrightarrow{H^+} RCOOCH_3 + CH_3OH + CH_3COCH_3$$

Table 9.9 lists the physical properties of carboxylic acids and their derivatives.

9.7.2 Salts of Carboxylic Acids

Classification

Metallic salts of carboxylic acids—the most commonly encountered are the sodium and potassium salts—will generally be suspected from their solubility, melting point, and ignition-test behavior. They are quite water soluble, providing slightly alkaline solutions from which free acids sometimes precipitate upon neutralization with mineral acid. They are insoluble in most organic solvents. The salts melt or decompose only at very high temperatures. They leave a light-colored alkaline residue of oxide or carbonate upon ignition. If necessary, the identity of the metal can be made by the standard tests outlined in inorganic qualitative analysis texts or by flame photometry.

Occasionally one encounters the carboxylic acid salts of ammonia or amines. The ammonia or amine is liberated from the salt on dissolution in aqueous alkali. Place 20 to 30 mg of the sample in a small test tube containing 0.5 mL of dilute sodium hydroxide. Ammonia or a low molecular weight amine can be detected by placing a piece of dampened litmus paper in the vapor produced by warming the tube with a small flame. For a more sensitive test, dampen filter paper with a 10% solution of copper sulfate and hold the paper in the vapors. A positive test is indicated by the blue color of the ammonia or amine complex of copper sulfate. The amine can be isolated by extraction of a basic solution of the salt and identified as described in Section 9.10.

The infrared spectra of acid salts, usually taken as Nujol mulls, show strong carbonyl bands at 1610 to 1550 cm^{-1} and near 1400 cm^{-1} (Figure 9.11).

Characterization

For characterization of carboxylic acid salts, it is necessary to rely on the chemistry and physical constants of the acid. Addition of sufuric or other strong mineral acids to an aqueous solution of a salt of a carboxylic acid liberates the free acid, which can be isolated by extraction, steam distillation, or filtration. Suitable derivatives are then made from the free acid.

9.8 ACID ANHYDRIDES AND ACID HALIDES

Classification

The combination of high reactivity and unique spectral properties greatly facilitates the classification and identification of acid halides and acid anhydrides.

Acid anhydrides characteristically exhibit two bands in the carbonyl region of the infrared spectrum; in acyclic aliphatic anhydrides, these bands appear near 1820 and 1760 cm^{-1}. They show the usual variation with unsaturation and ring size. The relative intensity of the two bands is variable; the higher wave number band is stronger in acyclic anhydrides, and the lower wave number band is usually stronger in cyclic anhydrides. Diacyl peroxides also exhibit double carbonyl bands in the infrared.

Acid chlorides, which are by far the most common of the acid halides, have a strong infrared band near 1800 cm^{-1}. The band is only slightly shifted to lower wave number on conjugaton; in aroyl halides, a prominent shoulder usually appears on the lower wave number side of the carbonyl band. As would be predicted, acid fluorides are shifted to higher wave numbers, whereas the bromides and iodides are at lower wave numbers. The presence of this strong carbonyl band and the absence of bands in the O—H, N—H, and C—O regions of the spectrum, the presence of reactive halogen (silver nitrate test, see Section 9.2.1), and the high reactivity of the substance with water, alcohols, and amines provide sufficient diagnosis for the acid halide grouping. Table 9.10 lists the physical properties of commonly encountered acid anhydrides and acid chlorides.

TABLE 9.10 Physical Properties of Acid Anhydrides and Acid Chlorides

Acid Anhydrides	bp (°C)*
Trifluoroacetic	39
Acetic	140
Propanoic	167
2-Methylpropanoic	182
Butanoic	198
2-Methyl-2-butenedioic	214
Pentanoic	218
trans-2-Butenoic	248
Hexanoic	254
Benzoic (mp 42°C)	360
Maleic (mp 52°C)	200(82/14 mm)
Glutaric (mp 56°C)	
Phenylacetic (mp 72°C)	

Acid Chlorides	bp (°C)*
Acetyl	51
Oxalyl	64
Propanoyl	80
2-Methylpropanoyl	92
Butanoyl	101
2,2-Dimethylpropanoyl	105
Chloroacetyl	108
Dichloroacetyl	108
3-Methylbutanoyl	115
Pentanoyl	126
Succinoyl	190d
Benzoyl	197
Phenylacetyl	210
Phthaloyl (mp 15°C)	281
3,5-Dinitrobenzoyl (mp 69°C)	
p-Nitrobenzoyl (mp 75°C)	150/15 mm

* bp at 760 mm Hg pressure unless otherwise noted.

Characterization

The most commonly encountered anhydrides are the simple symmetrical acetic and benzoic anhydrides and the cyclic anhydrides of succinic, maleic, and phthalic acids. Mixed anhydrides formed from two carboxylic acids or from a carboxylic acid and a sulfonic acid are known, but are seldom encountered.

Acid halides and anhydrides are most often converted directly to amides for identification purposes. Recall that, in the identification of a carboxylic acid, the acids are usually converted to the acid chlorides in order to prepare amides. The anilides are highly recommended (see Section 9.7.1 on amides and substituted amides).

$$RCOCl + 2\ NH_2C_6H_5 \longrightarrow RCONHC_6H_5 + C_6H_5\overset{+}{N}H_3\ Cl^-$$

$$(RCO)_2O + 2\ NH_2C_6H_5 \longrightarrow RCONHC_6H_5 + C_6H_5\overset{+}{N}H_3\ CH_3COO^-$$

Cyclic anhydrides, on reaction with amines, may give the monamide or the imide, depending upon the conditions of the reaction.

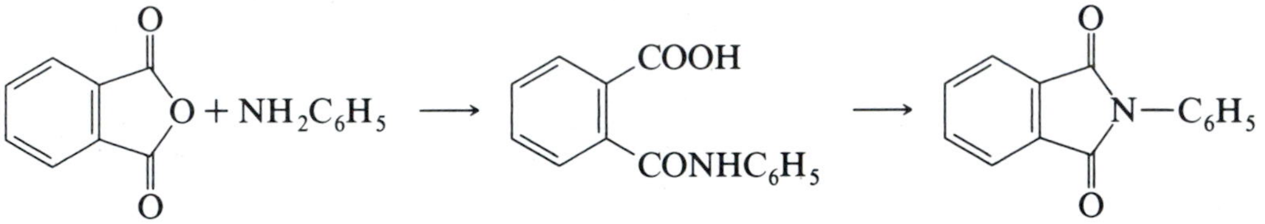

9.9 ESTERS

Classification

The majority of esters are liquids or low-melting solids, many with characteristic flowery or fruity odors. The presence of the ester function can usually be established by infrared spectroscopy. Esters have strong carbonyl bands in the infrared region in the 1780 to 1720 cm^{-1} region accompanied by two strong C—O absorptions in the 1300 to 1050 cm^{-1} region (see Figure 9.12). The higher wave number C—O band is usually stronger and broader than the carbonyl band. As a general rule, in the spectra of aldehydes, ketones, and amides, the carbonyl band is the strongest in the spectrum (see Figures 9.8, 9.9, 9.12, and 9.16).

Carbonyl compounds whose infrared spectra have the foregoing characteristics, which lack NH and OH absorption, and that do not give a positive 2,4-dinitrophenylhydrazine test, are most likely esters. An ester spectrum is most likely to be mistaken for that of a ketone. If the NMR spectrum of the suspected ester is available, evidence for an ester can be provided by the chemical shifts and coupling patterns of aliphatic hydrogens attached to the ethereal oxygen (near δ 4) and hydrogens *alpha* to the carbonyl group (near δ 2). Esters, like other carbonyl compounds, exhibit very characteristic fragmentation patterns in mass spectrometry.

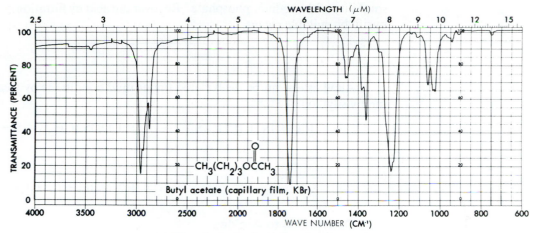

Figure 9.12   Typical infrared spectrum of an ester.

Characterization

The principal procedure for the characterization of an ester involves the identification of the parent acid and alcohol. Because it is often difficult to separate and purify the hydrolysis products of esters on a small scale, it is usually advantageous to prepare derivatives of the acid and alcohol portion by reactions on the original ester. The sample should be apportioned accordingly. Fortunately, a large number of the esters encountered are derivatives of simple acids (e.g., acetates) or alcohols (e.g., methyl or ethyl esters); it is often possible to unequivocally establish the presence of an acetyl or methyl or ethyl grouping by spectroscopic means.

Aromatic esters can sometimes be identified through solid derivatives prepared directly by aromatic substitution (nitration, halogenation, etc.), thus eliminating the necessity of preparing separate derivatives of both the alcohol and acid portion.

9.9.2   Derivatives of the Acyl Moiety

Hydrolysis to Acids

If the acid obtained by hydrolysis of the ester is a solid, it will serve as an excellent derivative (mp and neutralization equivalent).

PROCEDURE: HYDROLYSIS OF ESTERS

Into a small, round-bottomed flask place 0.2 to 1 g of the ester; add 2 to 10 mL of 25% aqueous sodium hydroxide and a boiling chip. Attach a condenser and reflux for 30 minutes for esters boiling below 110°C, 1 to 2 hours for esters boiling between 110 to 200°C, or until the oily layer and/or the characteristic ester odor disappears. Cool the flask and acidify with dilute acid. Phosphoric is recommended because of the high solubility of the sodium phosphate. Recover the acid by filtration or extraction, and purify by suitable means.

Alkyl esters can be dealkylated under mild conditions by treatment with trimethylsilyl iodide.[16]

$$R'COOR + (CH_3)_3SiI \longrightarrow R'COOSi(CH_3)_3 + R-I$$

$$\downarrow H_2O$$

$$R'COOH + (CH_3)_3SiOH$$

Conversion to Amides

Esters can be heated with aqueous or alcoholic ammonia to produce amides. Some simple esters react on standing or at reflux; however, most must be heated under pressure.

$$RCOOCH_3 + NH_3 \longrightarrow RCONH_2 + CH_3OH$$

Many esters can be converted to crystalline *N-benzylamides* by refluxing with benzylamine in the presence of an acid catalyst. This is often the most preferred method of making a derivative of the acyl moiety.

$$RCOOR' + CH_3OH \xrightarrow[\text{or HCl}]{CH_3O^-} RCOOCH_3 + R'OH$$

$$RCOOCH_3 + C_6H_5CH_2NH_2 \xrightarrow{NH_4Cl} RCONHCH_2C_6H_5 + CH_3OH$$

PROCEDURE: *N*-BENZYLAMIDES FROM ESTERS

Into a test tube or a 10-mL round-bottomed flask equipped with a reflux condenser, place 100 mg of ammonium chloride, 1 mL or 1 g of the ester, 3 mL of benzylamine, and a boiling stone. Reflux gently for 1 hour. Cool and wash the reaction mixture with water to remove the excess benzylamine and soluble salts. If crystallization does not occur, add one or two drops of dilute hydrochloric acid. If crystallization still does not occur, it may be due to excess ester. It is often possible to remove the ester by boiling the reaction mixture for a few minutes in an evaporating dish. Collect the solid *N*-benzylamide by filtration, wash it with a little hexane, and recrystallize from aqueous ethanol or acetone.

The aminolysis of esters higher than ethyl is usually quite slow. In these cases, it is best to convert the ester to the methyl ester before attempting to make the benzylamide. This can be done by refluxing 1 g of ester with 5 mL of absolute methanol in which about 0.1 g of sodium has been dissolved, or by refluxing in 3% methanolic

---

[16] M. E. Jung and M. A. Lyster, *J. Am. Chem. Soc.*, **99**, 968 (1977).

hydrogen chloride. Remove the methanol by distillation, and treat the residue as described above.

9.9.3 Derivatives of the Alcohol Moiety

Occasionally the alcohol derived from the ester by hydrolysis is a solid, in which case it and subsequent derivatives can serve to identify that portion of the ester. In general, alcohols above four carbons can be recovered from aqueous hydrolysates by extraction procedures. In many cases, it is much more convenient to identify alcohol moiety by a transesterification reaction with 3,5-dinitrobenzoic acid to produce the *alkyl 3,5-dinitrobenzoate.*

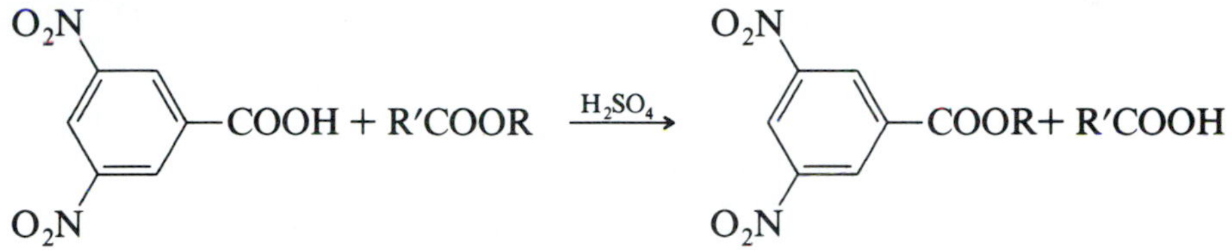

PROCEDURE: 3,5-DINITROBENZOATES FROM ESTERS

Mix about 0.5 g of the ester with 0.4 g of 3,5-dinitrobenzoic acid, and add one drop of concentrated sulfuric acid. If the ester boils below 150°C, heat to reflux with the condenser in place. Otherwise run the reaction in an open test tube placed in an oil bath at 150°C. Stir the mixture occasionally. If the acid dissolves within 15 minutes, heat the mixture for 30 minutes; otherwise continue the heating for 1 hour. Cool the reaction mixture, dissolve it in 20 mL of ether, and extract thoroughly with 10 mL of 5% sodium carbonate solution. Wash with water and remove the ether. Dissolve the crystalline or oily residue in 1 to 2 mL of hot ethanol and add water slowly until the 3,5-dinitrobenzoate begins to crystallize. Cool, collect, and recrystallize from aqueous ethanol if necessary.

Table 9.11 lists physical properties of esters.

9.10 AMINES

General Classification

Most simple amines are readily recognized by their solubility in dilute mineral acids. However, many substituted aromatic amines (e.g., diphenylamine) and aromatic nitrogen heterocycles, which might formally be classified as amines, fail to

TABLE 9.11   Physical Properties of Esters

Name of Compound	mp (°C)	bp (°C)	D_4^{20}	n_D^{20}
Ethyl formate	−79.4	54.2	0.92247	1.35975
Ethyl acetate	−83.6	77.15	0.90055	1.372
Methyl propanoate	−87.5	79.9	0.9151	1.3779
Methyl acrylate		80.3	0.961	1.3984
Isopropyl acetate	−73.4	91	0.873	1.377
Methyl 2-methylpropanoate	−84.7	92.6	0.8906	1.3840
t-Butyl acetate		97.8	0.867	1.386
Ethyl propanoate	−73.9	99.1	0.8889	1.3853
Ethyl acrylate		101	0.9136	1.4059
Methyl methacrylate	−50	101	0.936	1.413
Propyl acetate	−95	101	0.8834	1.38468
Allyl acetate		104	0.9276	1.40488
sec-Butyl acetate		112	0.872	1.3865
Isobutyl acetate		117.2	0.8747	1.39008
Ethyl butanoate	−100.8	121.6	0.87917	1.40002
Butyl acetate	−73.5	126.1	0.881	1.3947
Isoamyl acetate		142	0.8674	1.40034
Amyl acetate	−70.8	148.8	0.8756	1.4031
Methyl hexanoate	−71.0	151.2	0.88464	1.405
Cyclohexyl acetate		175	0.970	1.442
Butyl chloroacetate		175	1.081	
Ethyl acetoacetate		181	1.025	1.41976
Dimethyl malonate	−62	181.5	1.1539	1.41398
Dimethyl succinate	18.2	196.0	1.1192	1.41965
Phenyl acetate		196.7	1.078	1.503
Methyl benzoate	−12.5	199.5	1.089	1.5164
Ethyl benzoate	−34.2	213.2	1.0465	1.506
Benzyl acetate		217	1.055	1.5200
Diethyl succinate	−21	217.7	1.0398	1.41975
Diethyl fumarate	0.2	218.4	1.052	1.44103
Methyl phenylacetate		220	1.068	1.507
Diethyl maleate	−17	222.7	1.066	1.44156
Allyl benzoate		230	1.067	
Ethyl cinnamate	6.5	271	1.0490	1.55982
Diethyl phthalate		283.8	1.191	1.5138
Benzyl benzoate	21	324	1.1224	1.5681

dissolve in dilute acids. (The parent ring system of the latter can often be determined from their characteristic ultraviolet spectra.) Water-soluble amines can be detected by their basic reaction to litmus or other indicators, or by the *copper ion test*.

PROCEDURE: COPPER ION TEST

Add 10 mg or a small drop of the unknown to 0.5 mL of a 10% solution of copper sulfate. A blue to blue-green coloration or precipitate is indicative of an amine. The test can be run as a spot test on filter paper that has been treated with the copper sulfate solution. Ammonia also gives a positive test.

Infrared spectroscopy can be very useful in the recognition and classification of amines (see Figures 9.13 and 9.14). Primary amines, both aliphatic and aromatic, exhibit a weak but recognizable doublet in the 3500 to 3300 cm^{-1} region (symmetric and asymmetric NH stretch) and strong absorption due to NH bending in the 1640 to 1560 cm^{-1} region (see Figures 5.7 and 9.13). The 3500 to 3000 cm^{-1} bands in aliphatic amines are quite broad due to hydrogen bonding, but are reasonably sharp in aromatic amines. Secondary amines exhibit a single band in the 3450 to 3310 cm^{-1} region; aromatic amines absorb nearer the higher end of the range, and the aliphatics absorb at the lower end. The NH bending band in secondary amines in the 1580 to 1490 cm^{-1} region is weak and generally of no diagnostic value. Tertiary amines have no generally useful characteristic absorptions. (However, the absence of characteristic NH absorptions is a helpful observation in the classification of a compound known to be an amine.) The presence of a tertiary amine grouping can be established by infrared examination of the hydrochloride or other proton-acid salt of the amine; $\overset{\scriptscriptstyle +}{\underset{\diagdown}{\diagup}}$NH absorption occurs in the 2700 to 2250 cm^{-1} region.

Color reactions with sodium nitroprusside have been recommended as classification tests to differentiate between the various classes of amines.[17]

Hinsberg Test

This test is based on the fact that primary and secondary amines react with arenesulfonyl halides to give *N*-substituted sulfonamides. The tertiary amines do not give derivatives. The sulfonamides from primary amines react with alkali to form soluble salts (the sodium salts of certain primary alicyclic amines and some long chain alkyl amines are relatively insoluble and give confusing results).

Primary amines

$$ArSO_2Cl + RNH_2 + 2\,NaOH \longrightarrow \underset{\text{soluble}}{[ArSO_2NR]^- Na^+} + NaCl + 2\,H_2O$$

$$[ArSO_2NR]^- Na^+ + HCl \longrightarrow \underset{\text{insoluble}}{ArSO_2NHR} + NaCl$$

Secondary amines

$$ArSO_2Cl + R_2NH + NaOH \longrightarrow \underset{\text{insoluble}}{ArSO_2NR_2} + H_2O + NaCl$$

---

[17] R. L. Baumgarten, C. M. Dougherty, and O. N. Nercissian, *J. Chem. Ed.*, **54**, 189 (1977).

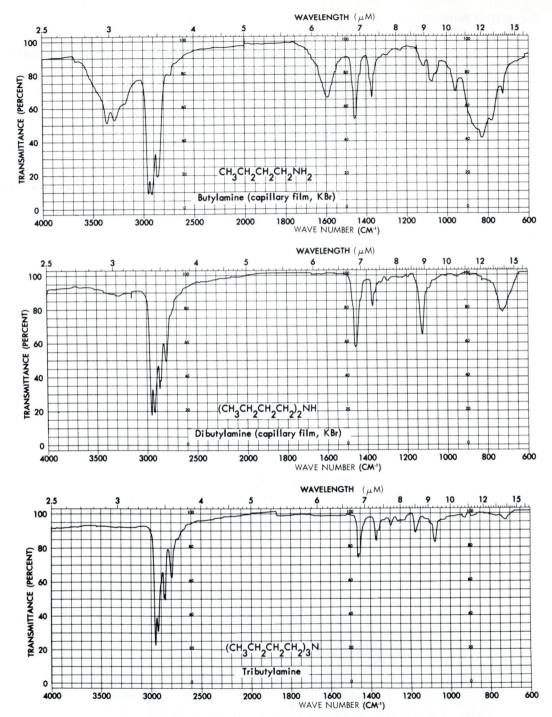

Figure 9.13 Infrared spectra of a primary, secondary, and tertiary amine. In the spectrum of the primary amine note the strong N—H stretching, bending, and deformation bands. The N—H stretching and bending bands of the secondary amines are frequently very weak and may not be observed in the usual cell thickness.

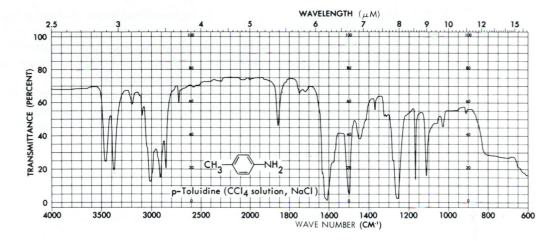

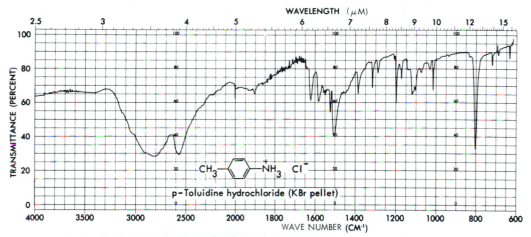

Figure 9.14 Typical infrared spectra of an aromatic amine and its hydrochloride salt. In the spectrum of *p*-toluidine note the strong 1,4-disubstitution bands in the 2000 to 1660 cm^{-1} (5 to 6 μm) region.

Tertiary amines

$$ArSO_2Cl + R_3N + NaOH \longrightarrow \text{ tertiary amine recovered}$$

PROCEDURE: HINSBERG TEST

To 0.2 g or 0.2 mL of the amine and 5 mL of 10% KOH is added 0.4 mL of benzene-sulfonyl chloride. Stopper the mixture and shake, with cooling if necessary, until the odor of the benzenesulfonyl chloride is gone. At this point, the mixture should still be strongly basic. If not, add small volumes of the base until basic.

If the mixture has formed two layers, separate the two layers and test the organic layer for insolubility in 10% HCl. Tertiary amines will be soluble. Some higher molecular weight amines form benzenesulfonamides that have low solubility in aqueous base. They may give the same solubility characteristics as the benzenesulfonamides of secondary amines. In order to differentiate between these possibilities, the separated aqueous phase is brought to pH 4. A precipitate indicates that a benzenesulfonamide of a primary amine was formed.

If the original mixture did not separate into two layers, a soluble salt of a benzenesulfonamide derived from a primary amine was formed. This can be substantiated by bringing the pH to 4, in which case the benzenesulfonamide of the primary amine will precipitate.

——— ————————————

This procedure can be scaled up and used to separate mixtures of primary, secondary, and tertiary amines. Unreacted tertiary amines can be recovered by solvent extraction or steam distillation. The sulfonamides of primary and secondary amines can be separated by taking advantage of the solubility of the primary derivative in dilute sodium hydroxide. The original primary and secondary amines can be recovered by hydrolysis of the sulfonamides with 10 parts of 25% hydrochloric acid, followed by neutralization and extraction. The sulfonamides of primary amines require 24 to 36 hours of reflux; those of secondary amines require 10 to 12 hours for hydrolysis.

For a discussion of the reaction of benzenesulfonyl chloride with tertiary amines, see C. R. Gambill, T. D. Roberts, and H. Shechter, *J. Chem. Ed.,* **49,** 287 (1972).

Nitrous Acid

CAUTION: A number of N-nitroso amines (R_2N-NO) are toxic and/or carcinogenic! Because of the high risk involved, we do not recommend treating amines or any samples of unknown structure with nitrous acid.

The following equations illustrate the reactions of nitrous acid with amines. The reactions are the basis for a classical method of distinguishing between various types of amines. The details of test procedures can be found in most organic texts and organic qualitative analysis texts.

Primary amines

$$RNH_2 \xrightarrow{\text{HONO}} R-N_2^+ \xrightarrow{-N_2} R^+ \longrightarrow \text{Carbocation products}$$

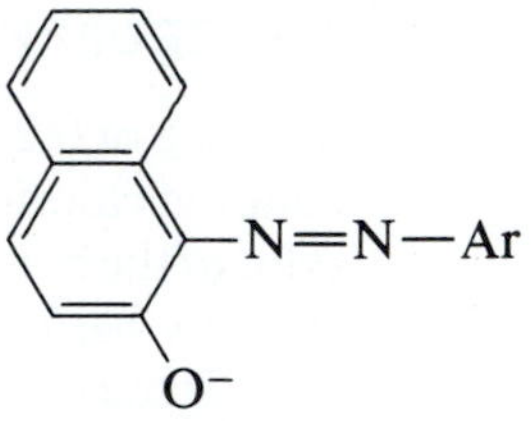

red azo dye

Secondary amines

$$R_2NH \xrightarrow{\text{HONO}} R_2N-NO$$
(yellow oils or solids)

Tertiary amines

$$R_3N \xrightarrow{\text{HONO}} R_3\overset{+}{N}-NO \xrightarrow{H_2O} R_2NNO + \text{aldehydes, ketones}$$

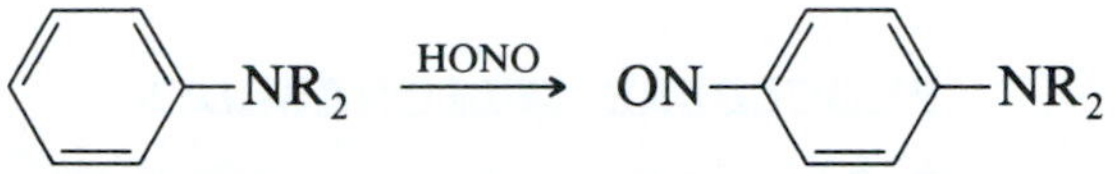

$$\text{⬡}-NR_2 \xrightarrow{\text{HONO}} ON-\text{⬡}-NR_2$$

The greatest utility is the demonstration of the presence of primary aliphatic amino group (by the loss of N_2) and primary aromatic amine group (through coupling with reactive aromatic nuclei such as β-naphthol to form azo dyes).

Characterization

Note: A number of the procedures for the preparation of derivatives of amines are quick and simple methods that can be run on a very small scale. For this reason, it is often expedient to use certain of these reactions as a combination classification-derivation procedure.

In many cases, with simple amines the combination of boiling or melting point, solubility in acid, and infrared spectrum provides sufficient information to allow the investigator to choose an appropriate derivative. In other cases, it is wise to examine selected classification tests and to obtain additional physical (e.g., pK_b) and spectral (NMR or ultraviolet) data before proceeding. See Section 6.4.1 for NH chemical shifts. In some cases the NH proton signals are broadened by quadrupole interactions with the nitrogen to the extent that the signal can barely be differentiated from the base line. In other cases, the signal is much sharper. If there is any doubt in the assignment of the peak due to protons on nitrogen, the investigator may find it helpful to exchange the nitrogen protons for deuterium (add a drop of D_2O to the NMR tube and shake) or to examine the spectrum of a salt of the amine. A simple technique for the latter is to use trifluoroacetic acid as solvent; with this solvent and tertiary amines, the $\overset{+}{>}NH$ shows up between δ 2.3 and 2.9.

In ultraviolet spectra of aromatic amines, a dramatic change is observed in going from the free base to the protonated salt.

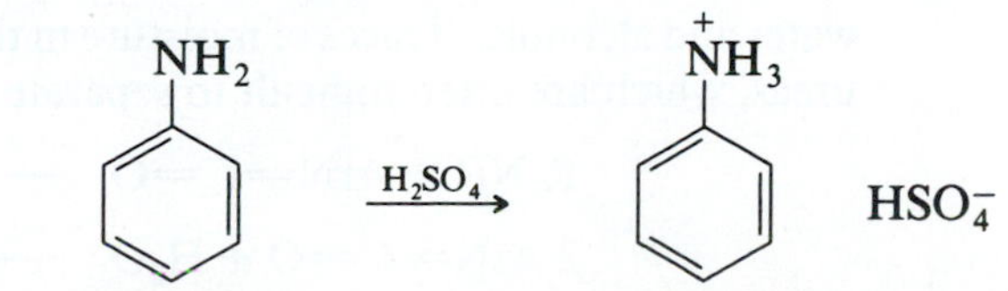

$$\underset{\lambda_{max}\ 230,\ 280\ nm}{\text{⬡}-NH_2} \xrightarrow{H_2SO_4} \underset{\lambda_{max}\ 230,\ 254\ nm}{\text{⬡}-\overset{+}{N}H_3}\ HSO_4^-$$

9.10.2   Derivatives of Primary and Secondary Amines

Amides and Related Derivatives

The most commonly employed derivatives are the *benzene-* and *p-toluenesulfonamides*. These can be made by the procedure outlined earlier for the Hinsberg test. The product should be recrystallized from 95% ethanol.

Benzyl- (α-toluene-), p-bromobenzene-, m-nitrobenzene, α-naphthalene-, and *methanesulfonamides* are also occasionally used.

PROCEDURE: SULFONAMIDES

Reflux for 5 to 10 minutes a mixture of 1 mmol of the sulfonyl chloride and 2.1 mmol of the amine in 4 mL of toluene. Allow the mixture to cool. Filter off the amine hydrochloride that precipitates. Evaporate the toluene filtrate to obtain the crude sulfonamide. Recrystallize from 95% ethanol.

Primary and secondary amines react with isothiocyanates to yield *substituted thioureas*. The most useful derivatives are obtained from phenyl- and 1-naphthyl-isothiocyanate; these reagents have the advantage of being insensitive to water and alcohols.

$$RNH_2 + ArN=C=S \longrightarrow ArNH-\overset{\overset{\textstyle S}{\|}}{C}-NHR$$

PROCEDURE: THIOUREAS

(a) Reflux a solution of 1 mmol of the isothiocyanate and 1.2 mmol of the amine in 2 mL of ethanol for 5 to 10 minutes. Add water dropwise to the hot mixture until a permanent cloudiness occurs. Cool, and scratch with a glass rod if necessary, to induce crystallization. Collect the solid. Wash with 1 to 2 mL of hexane. Recrystallize from hot ethanol.

(b) Mix equal amounts of the amine and isothiocyanate in a test tube and shake for 2 minutes. If no reaction occurs, heat the mixture gently over a small flame for 1 to 2 minutes. Cool in an ice bath, and purify by recrystallization.

Primary and secondary amines react with aryl isocyanates to produce substituted ureas. In contrast to the isothiocyanates, the isocyanates react rapidly with water and alcohols. Traces of moisture in the amines result in the formation of diaryl ureas, which are often difficult to separate from the desired product.

$$R_2NH + ArN=C=O \longrightarrow ArNHCONR_2$$

$$2\ ArN=C=O + H_2O \longrightarrow ArNHCONHAr + CO_2$$

Both aliphatic and aromatic primary and secondary amines readily react with acid anhydrides or acid chlorides to form substituted amides. The reaction of benzoyl chloride with amines under Schotten–Baumann conditions (aqueous sodium hydroxide) produces *benzamides,* which are generally excellent derivatives. The *p-nitro-* and *3,5-dinitrobenzamides* are also useful derivatives.

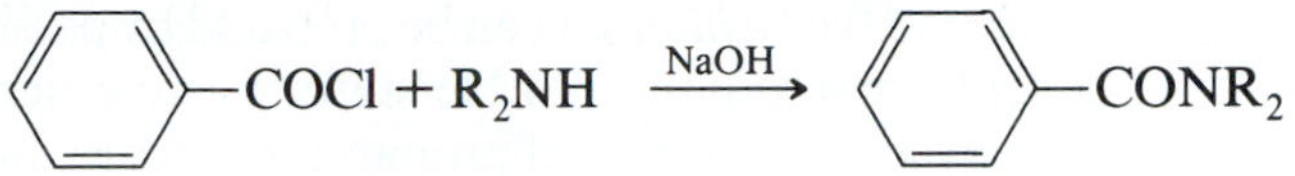

$$\text{C}_6\text{H}_5\text{—COCl} + \text{R}_2\text{NH} \xrightarrow{\text{NaOH}} \text{C}_6\text{H}_5\text{—CONR}_2$$

PROCEDURE: BENZAMIDES

C A U T I O N : Benzoyl chloride is a lacrymator and should be used and stored in a hood.

(a) About 1 mmol of the amine is suspended in 1 mL of 10% sodium hydroxide, and 0.3 to 0.4 mL of benzoyl chloride is added dropwise, with vigorous shaking and cooling. After about 5 to 10 minutes, the reaction mixture is carefully neutralized to about pH 8 (pH paper). The *N*-substituted benzamide is collected, washed with water, and recrystallized from ethanol-water.

(b) Benzoyl, *p*-nitrobenzoyl, or 3,5-dinitrobenzoyl chloride (2 to 3 mmol) is dissolved in 2 to 3 mL of toluene; 1 mmol of the amine is added, followed by 1 mL of 10% sodium hydroxide. The mixture is shaken for 10 to 15 minutes. The crude amide is obtained by evaporation of the toluene layer. Alternatively the amine and aroyl chlorides are mixed with 2 mL of pyridine, and the mixture is refluxed for 30 minutes. The reaction mixture is poured into ice water, and the derivative is collected and recrystallized from ethanol-water.

Most water-insoluble primary and secondary amines form crystalline *acetamides* upon reaction with acetic anhydride.

$$\text{RNH}_2 + (\text{CH}_3\text{CO})_2\text{O} \xrightarrow{\text{NaOH}} \text{CH}_3\text{CONHR}$$

PROCEDURE: ACETAMIDES

Dissolve about 0.2 g of water-insoluble amine in 10 mL of 5% hydrochloric acid. Add sodium hydroxide (5%) with a dropping pipette until the mixture becomes cloudy, and remove the turbidity by the addition of a few drops of 5% hydrochloric acid. Add a few chips of ice and 1 mL of acetic anhydride. Swirl the mixture and add 1 g of sodium acetate (trihydrate) dissolved in 2 mL of water. Cool in an ice bath and collect the solid; recrystallize from ethanol-water.

Salts

The amine salts of a number of proton acids are useful for purposes of identification or purification. Some salts have reproducible melting or decomposition points; others may be useful only as solid derivatives that can be carefully purified for purposes of microanalysis, spectroscopic studies, X-ray crystallography, and so on.

Hydrochlorides can be prepared by passing dry hydrogen chloride (from a tank or from a sodium chloride/sulfuric acid generator) into an ether, toluene, or isopropyl alcohol solution of primary, secondary, or tertiary amines. The hydrobromides and hydroiodides are usually more hygroscopic than the hydrochlorides.

Fluoroborates (BF_4^-) are often well-defined crystalline substances and can often be formed easily by dropwise addition of an ether solution that has been saturated with aqueous HBF_4 to an ether solution of an amine. *Picrates* are frequently found in the older literature as derivatives (salts or complexes) of amines (as well as amine oxides, aromatic hydrocarbons, and aryl ethers); picric acid and its complexes are explosive and are not recommended.

Ammonium tetraphenylborates are quite water-insoluble. The addition of an aqueous solution of sodium tetraphenylborate to a solution of an amine salt (usually the hydrochloride) causes the immediate precipitation of the amine as the tetraphenylborate salt. The method can be adapted as a means of detection of amines and amine salts, including quaternary ammonium salts. The melting points of a number of ammonium tetraphenylborates have been recorded.[18]

$$RNH_2 \xrightarrow{\text{HCl}} R\overset{+}{N}H_3\ Cl^- \xrightarrow{NaB(C_6H_5)_4} R\overset{+}{N}H_3\ B(C_6H_5)_4^-$$

$$R_4N^+X^- \xrightarrow{NaB(C_6H_5)_4} R_4N^+\ B(C_6H_5)_4^-$$

Table 9.12 lists physical properties of primary amines, secondary amines, and their derivatives.

9.10.3 Derivatives of Tertiary Amines

For solid derivatives of tertiary amines, salts are usually prepared. Quarternary ammonium salts, formed by alkylation of the tertiary amine with methyl iodide, benzyl chloride, or methyl *p*-toluenesulfonate, have been used. The most extensive data are available on the *methiodides*.

$$R_3N + CH_3I \longrightarrow R_3\overset{+}{N}CH_3\ I^-$$

PROCEDURE: METHIODIDES

C A U T I O N : Methyl iodide can cause serious toxic effects on the central nervous system. Methyl iodide should be used in an efficient hood, and care should be taken to avoid exposure of the skin to liquid methyl iodide.

---

[18] F. E. Crane, Jr., *Anal. Chem.* **28,** 1794 (1956).

TABLE 9.12 Physical Properties of Primary and Secondary Amines and Derivatives

Name of Compound	mp (°C)	bp (°C)	mp (°C) of Derivatives				
			Acetamide	Benzamide	Benzene-sulfon-amide	p-Toluene-sulfon-amide	Phenyl thiourea
Isopropylamine		33		71	26		101
t-Butylamine		46		134			120
Propylamine		49		84	36	52	63
Methylisopropylamine		50		(oil)			120
Diethylamine		56		42	42		34
Allylamine		58			39	64	98
sec-Butylamine		63		76	70	55	101
Butylamine		77					65
Diisopropylamine		86					
Pyrrolidine		89		(oil)		123	69
Pentylamine		104					
Piperidine		106	(oil)	48	94	96	101
Dipropylamine		109	(oil)		51		69
Diallylamine		112					
Ethylenediamine		116	172(di) 51(mono)	244(di) (249)	168(di)	360(di) 123(mono)	102
Hexylamine		129		40	96		77
Morpholine		130		75	118	147	136
Cyclohexylamine		134	104	149	89		148
Diisobutylamine		139	101		57		113
Heptylamine		155	86				75
Dibutylamine		159					86
Octylamine		180					
Aniline		184	114	160	112	103	54
Benzylamine		185	60	105	88	116	156
N-Methylaniline		196	102	63	79	94	87
β-Phenylethylamine		198	51	116	69		135
o-Toluidine		200	112	43(146)	124	108	136
m-Toluidine		203	65	125	95	114	
N-Ethylaniline		205	54	60	(oil)	87	89
o-Ethylaniline		216	111	147			
p-Ethylaniline		216	94	151		104	
2,4-Dimethylaniline		217	133	192	130	81	152(133)
3,5-Dimethylaniline		220	144	136			153
N-Propylaniline		222	47		54		104
o-Methoxyaniline		225	88	60(84)	89	127	136
Dibenzylamine		300		112	68	159	
N-Benzylaniline*	37	298	58	107	119	149	103
p-Toluidine	45	200	147	158	120	118	141
1-Naphthylamine*	50		159	160	167	157(147)	165
Indole	52	253	158	68	254		
4-Aminobiphenyl*	53		171	230			
Diphenylamine	53–4		101	180	124	141	152
p-Anisidine	58	240	130	157	95	114	157(171)

(table continues)

TABLE 9.12   (Continued)

Name of Compound	mp (°C)	bp (°C)	mp (°C) of Derivatives				
			Acetamide	Benzamide	Benzene-sulfon-amide	p-Toluene-sulfon-amide	Phenyl thiourea
2-Aminopyridine	60	204	71	165			
o-Nitroaniline	71		94	110(98)	104	142	
p-Chloroaniline	72		179(172)	192	122	95(119)	152
o-Phenylenediamine	102		185(di)	301(di)	185	260(di)	
2-Naphthylamine*	112		132	162	102	133	129
Benzidine*	127		317(di)	252(di)	232(di)	243(di)	
			199(mono)	205(mono)			
o-Toluidine	129		314	265			
p-Nitroaniline	147		215	199	139	191	
o-Aminophenol	174		209(201)	167	141	146	
			124(di)				
p-Aminophenol	184s†		150(di)	217			150
			168(mono)				
p-Aminobenzoic acid	188		251d	278	212		

* *These compounds have been shown to be carcinogenic!* The use and handling of these compounds have been restricted by the federal government. Extreme caution should be exercised in the handling of all suspected amines.
† Sublimes.

(a) A mixture of 0.5 g of the amine and 0.5 g of methyl iodide is warmed in a test tube over a low flame or in a water bath for a few minutes and then cooled in an ice bath. The tube is scratched with a glass rod, if necessary, to induce crystallization. Recrystallize from ethanol or ethyl acetate.

(b) A tertiary amine-methyl iodide mixture dissolved in toluene is refluxed until precipitation of the methiodide is complete.

The salts formed between tertiary amines and protonic acids form an important category of derivatives. The most extensive data available is that on the picrates that are no longer recommended due to the *explosive nature of picric acid and its complexes! Hydrochlorides* (see Section 9.10.2 on salts) are often useful.

Table 9.13 lists the physical properties of tertiary amines and their methiodides.

9.10.4   Amine Salts

Salts of amines fall into two categories — the salts of primary, secondary, and tertiary amines with proton acids, and the quaternary ammonium salts. Both types are generally water-soluble and insoluble in nonpolar solvents, such as ethers and hydrocarbons. Many are soluble in alcohols, dichloromethane, and chloroform. See Section 5.4.3 for characteristic infrared absorptions of $-\overset{+}{N}H_3$, $-\overset{+}{N}H_2$, and $-\overset{+}{N}H$.

TABLE 9.13 Physical Properties of Tertiary Amines and Derivatives

Name of Compound	bp (°C)*	Quaternary Methiodide
Trimethylamine	3	230d†
Triethylamine	89	
Pyridine	116	117
2-Methylpyridine	129	230
2,6-Dimethylpyridine	143	233
3-Methylpyridine	143	
4-Methylpyridine	143.1	
Tripropylamine	156.5	208
2,4-Dimethylpyridine	159	
N,N-Dimethylbenzylamine	181	
N,N-Dimethylaniline (mp 1.5–2.5°C)	196	228d
Tributylamine	216.5	180
	(93–5/15 mm)	
N,N-Diethylaniline	218	102
	(93–5/10 mm)	
Quinoline (mp 15.6°C)	239	133
Isoquinoline (mp 26.5°C)	243	
2-Phenylpyridine	268–9	
N,N-Dibenzylaniline (mp 72°C)		135
Tribenzylamine (mp 91°C)		184
Triphenylamine (mp 127°C)		

* bp at 760 mm Hg pressure unless otherwise noted.

† d denotes decomposition occurs at that temperature.

There are no characteristic bands for quaternary ammonium salts. Inorganic anions can be identified by the standard methods of inorganic qualitative analysis.

The *salts of proton acids* usually give slightly acidic solutions; exceptions are the salts of weak acids such as acetic, propionic, and so on. Upon treatment with dilute alkali, the amine will separate if it is insoluble or sparingly soluble in water. Even if separation does not occur, the presence of the free amine will be recognized by its characteristic odor. It is often convenient to recover the free amine from the alkaline solution by extraction with ether. In other cases, it may be more convenient to adjust the pH of the solution and make a sulfonamide, benzamide, or acetamide according to the directions given under amines.

Quarternary ammonium salts form neutral-to-basic solutions. The hydroxides and alkoxides are very strong bases. These salts can be converted to other salts that may be suitable for identification purposes by anion exchange reactions, as illustrated in the following equations.

$$R_4N^+I^- + AgBF_4 \longrightarrow R_4N^+BF_4^- + AgI$$

$$R_4N^+Cl^- + NaB(C_6H_5)_4 \longrightarrow R_4N^+B(C_6H_5)_4^- + NaCl$$

$$R_4N^+SO_4H^- \xrightarrow[\text{column}]{\text{Ion exchange}} R_4N^+OH^- \xrightarrow{\text{HBF}_4} R_4N^+BF_4^-$$

9.11 α-AMINO ACIDS

Classification

Amino acids are commonly encountered in identification work as the free amino acids, the hydrochlorides, or peptides. The free amino acids exist as the internal salts or zwitter ions, for example,

$$R-CH-COO^-$$
$$|$$
$$^+NH_3$$

They are very slightly soluble in nonpolar organic solvents, sparingly soluble in ethanol, and very soluble in water to give neutral solutions. They show increased solubility in both acidic and basic solutions, and they have melting points or decomposition points between 120 and 300°C, which depend on the rate of heating.

The principal infrared absorption bands of free amino acids are those arising from the carboxylate (asymmetric stretching near 1600 cm^{-1} and symmetric stretching near 1400 cm^{-1}) and the ammonium group ($\overset{+}{N}H_3$ stretching bands overlapping CH bands near 3000 cm^{-1}; overtones are sometimes present in the 2600 to 1900 cm^{-1} region, asymmetric bending near 1600 cm^{-1}, and symmetric bending near 1500 cm^{-1}) (see Figure 9.15). In the hydrochloride salts, the very broad —OH band of the carboxylic acid obscures the $-\overset{+}{N}H_3$ and CH stretching region. The other $-\overset{+}{N}H_3$ bands appear as above, along with carboxylic carbonyl near 1720 cm^{-1} (Figure 9.15). Because of low solubility, the free acids and salts are usually run as potassium bromide discs or as mulls. Under these conditions, the spectra of the racemic and optically active forms may show considerable differences. The solution spectra, however, are identical.

Upon warming an aqueous solution of an α-amino acid with a few drops of 0.25% *ninhydrin* (indane-1,2,3-trione hydrate), a blue-to-violet color appears. Proline and hydroxyproline give a yellow color. (Ammonium salts also give a positive test.)

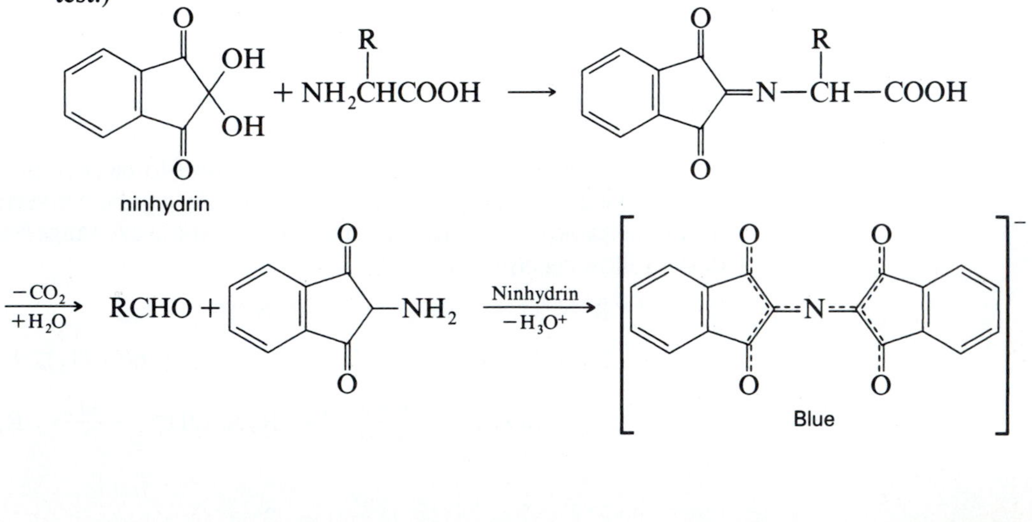

ninhydrin

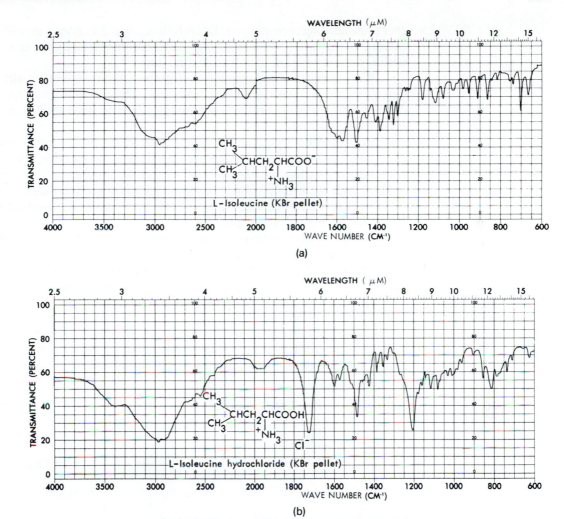

Figure 9.15 Infrared spectra of a typical amino acid and its hydrochloride salt.

Characterization

The twenty common amino acids found as constituents of proteins are optically active (except glycine) and have the L configuration. A number of less common amino acids are found elsewhere in nature, some of which have the D configuration. The specific rotations are valuable in identification.

Paper and thin-layer chromatography[19] are excellent techniques for the tentative identification of amino acids. *Handbook of Tables for Organic Compound Identification* lists the R_F values on paper using a variety of solvent systems. When-

---

[19] E. Stahl, *Thin-Layer Chromatography* (Berlin: Springer-Verlag, 1965).

ever possible, direct comparison should be made using a known, the unknown, and a mixture of the two. Two-dimensional paper chromatography has proven especially useful in the separation of amino acid mixtures.[20]

With paper or thin-layer chromatography, the spots are usually detected by developing the chromatogram with ninhydrin. The chromatogram is sprayed with a 0.25% solution of ninhydrin in acetone or alcohol, and developed in an oven at 80 to 100°C for 5 minutes.

An interesting technique for determining the absolute configuration of microgram quantities of amino acids has been developed. By spraying a paper chromatogram with a solution of D-amino acid oxidase (a commercially available enzyme isolated from pig kidneys) in a pyrophosphate buffer and incubating, only D-amino acids are destroyed;[21] the α-keto acids, thus formed, can be detected with an acidified 2,4-dinitrophenylhydrazine spray.[22]

The qualitative and quantitative composition of amino mixtures (as obtained from the hydrolysis of peptides) can be achieved by commercially available automated amino acid analyzers. These instruments employ buffer solutions to elute the amino acids from ion-exchange columns. The effluent from the column is mixed with ninhydrin, and the intensity of the blue color is measured photoelectrically and plotted as a function of time.

Crystalline derivatives of the amino acids are usually obtained by reactions at the amino group by treatment with conventional amine reagents in alkaline solution. Extensive data are available on the *N-acetyl, N-benzoyl* and *N-p-toluenesulfonyl* derivatives.

$$R-CH-COO^- + NaOH \longrightarrow R-CH-COO^-Na^+ + H_2O$$
$$\underset{NH_3^+}{\qquad} \qquad\qquad\qquad \underset{NH_2}{\qquad}$$

$$R-CH-COO^-Na^+ \xrightarrow[(2)\ HCl]{(1)\ R'COCl} R-CH-COOH$$
$$\underset{NH_2}{\qquad} \qquad\qquad\qquad \underset{NHCOR'}{\qquad}$$

PROCEDURE: *N*-ACETYL, *N*-BENZOYL, AND *N-p*-TOLUENESULFONYL DERIVATIVES

(a) Aqueous alkaline solutions of amino acids react with acetic anhydride or benzoyl chloride to yield the acetyl or benzoyl derivatives (see Sections in 9.10.2 on benzamide and acetamide procedures). In order to precipitate the derivative, the solution should be neutralized with dilute mineral acid. In the benzoyl case, any benzoic acid that precipitates can be removed from the desired derivative by washing with a little cold ether.

---

[20] R. J. Block, E. L. Durrum, and G. Zweig, *Paper Chromatography and Paper Electrophoresis,* 2nd ed. (New York: Academic Press, 1958).

[21] T. S. G. Jones, *Biochem. J.,* **42**, LIX (1948).

[22] J. L. Auclair and R. L. Patton, *Rev. Can. Biol.,* **9**, 3 (1950).

(b) The *p*-toluenesulfonyl derivatives can be conveniently made by stirring an alkaline solution of 1 equivalent of amino acid with an ether solution of a slight excess of *p*-toluenesulfonyl chloride for 3 to 4 hours. The aqueous layer is acidified to Congo red with dilute hydrochloric acid and cooled if necessary. The crude sulfonamide can be recrystallized from dilute ethanol.

The *2,4-dinitrophenyl derivatives* are yellow crystalline materials of relatively sharp melting points. Mixture of these "DNP" derivatives of amino acids can be separated by thin-layer chromatography on silica gel. This method has been used for *N*-terminal amino acid analysis in proteins and peptides; the free amino groups on the end of proteins or peptides are reacted with the DNP reagent prior to hydrolysis.[23] The DNP derivatives of simple primary and secondary amines are sometimes useful in identification work.

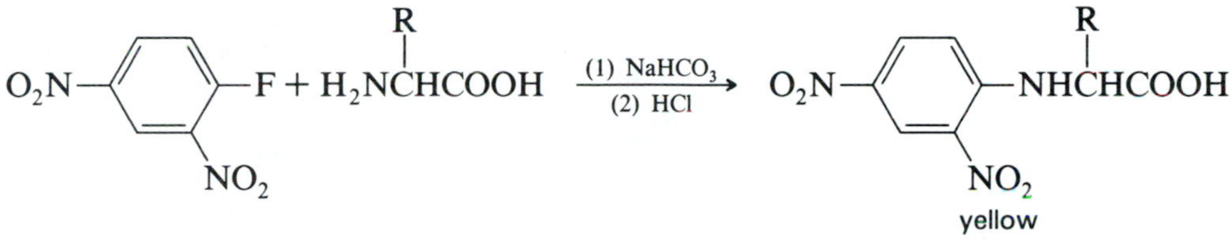

$$O_2N-\!\!\!\bigcirc\!\!\!-F + H_2N\overset{R}{\underset{|}{C}}HCOOH \xrightarrow[(2)\ HCl]{(1)\ NaHCO_3} O_2N-\!\!\!\bigcirc\!\!\!-NH\overset{R}{\underset{|}{C}}HCOOH$$

$$NO_2 \qquad\qquad\qquad\qquad\qquad\qquad\qquad NO_2$$
yellow

PROCEDURE: *N*-2,4-DINITROPHENYL DERIVATIVES

Dissolve or suspend 0.25 g of the amino acid in 5 mL of water containing 0.5 g of sodium hydrogen carbonate. Add a solution of 0.4 g of 2,4-dinitrofluorobenzene in 3 mL of ethanol. Stir the mixture for 1 hour. Add 3 mL of saturated salt solution and extract several times with small volumes of ether to remove any unchanged reagent. Pour the aqueous layer into 12 mL of cold 5% hydrochloric acid, with stirring. The mixture should be distinctly acid to Congo red. The derivative can be recrystallized from 50% ethanol. These derivatives are light-sensitive; the reaction should be run in the dark (cover flask with aluminum foil), and the products should be kept in a dark place or in dark containers. As a substitute for the 2,4-dinitrofluorobenzene, the chloro compound can be used; in this case, however, the reaction mixture should be refluxed for 4 hours.

Table 9.14 lists the physical properties of amino acids and their derivatives.

[23] F. Sanger, *Biochem. J.*, **45**, 563 (1949) and previous papers.

TABLE 9.14   Physical Properties of Amino Acids and Derivatives

Name of Compound	mp (°C)*	mp (°C) of Derivatives†			
		Acetyl	Benzoyl	p-Toluene sulfonyl	Dinitro-phenyl Derivative
L-(+)-Valine	96	156	127	147	
N-Phenylglycine	127	194	63		
L-Ornithine	140		240(mono)		
			189(di)		
o-Aminobenzoic acid	147	185	182	217	
D-(−)-Valine	157	156			
D,L-Glutamic acid	199	187.5	155	117	
β-Alanine	200		120		
D,L-Proline (monohydrate, mp 191°C)	203				
L-Arginine	207d		298(mono)		252
			235(di)		
L-Glutamic acid	213	199		131	
N-Methylglycine	213	135	104		
L-Proline	222d		156(mono)	133	137
L-Lysine	225d		150(di)		
L-Aspargine	227		189	175	
D,L-Threonine	229		145		152
L-Serine	228d				
Glycine	232	206	187.5	150	195
D,L-Arginine	238		230(di,anh)		
D,L-Serine	246d		150	213	199
D,L-Isoserine	248d		151		
D,L-Threonine	252		148		
D,L-Cystine	260d		181(di)	205	109
D,L-Aspartic acid	270d		119(hyd)		196
			177(anh)		
L-Aspartic acid	270		185	140	
L-Hydroxyproline	274		100(mono)	153	
			92(di)		
D,L-Tryptophane	275–82		176		
D,L-Methionine	281	114	145	105	117
L-Methionine	283d	99			
D,L-Phenylalanine	283d(320d)		146	165	
L-(+)-Isoleucine	283d		117	132	
L-Histidine	288(253)		230d(mono)	202–4d	
L-Tryptophane	290		183	176	175
D,L-Isoleucine	292		118	141	166
D,L-Leucine	293–5d		141		
D,L-Alanine	295	137	166	139	
D,L-Norleucine	300			124	
D,L-Valine	298d		132	110	
Creatine	303	165(di)			
Tyrosine	314–8d	148	167	188	
	(290–5d)	172(di)	212(di)	114(di)	

* d denotes decomposition.

† (anh) denotes anhydrous; (hyd) denotes a hydrate.

9.12 AMIDES AND RELATED COMPOUNDS

Classification

Amides, imides, ureas, and urethanes are nitrogen-containing compounds that exhibit a carbonyl band in the infrared and do not give a positive 2,4-dinitrophenyl-hydrazine test or show evidence for other nitrogen functions such as nitrile, amino, and nitro. Almost all are colorless crystalline solids; notable exceptions are certain *N*-alkylformamides and tetraalkylureas, which are high-boiling liquids.

Infrared spectra can be of great diagnostic value in the classification of amides (see Figure 9.16). Attention should be given to the presence of NH stretch, the exact position of the carbonyl band (amide I band), and the presence and position of the NH deformation (amide II band). Refer to Chapter 5 on infrared spectroscopy for details.

Occasionally the investigator may have difficulty in deciding from infrared analysis whether a compound in question is an amide. In such cases it may be useful to test for liberation of an amine by treatment of a small sample of the unknown with refluxing 20% sodium hydroxide or fusion with powdered sodium hydroxide. A number of color tests have been developed for various classes of amides, ureas, and so on.[24]

In the NMR, amide hydrogens exhibit broad absorptions in the region of δ 1.5 to 5.0. The spectra of tertiary amides indicate magnetic nonequivalence of the *N*-alkyl groups due to the partial double-bond character of the C—N bond.

Characterization

The only completely general method for the preparation of derivatives of amides and related compounds is *hydrolysis* and identification of the products.

$$R'CONR_2 \xrightarrow{H_2O} R'COOH + R_2NH$$
amides

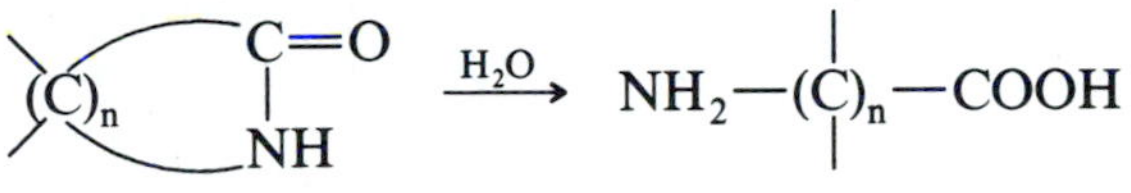

lactams

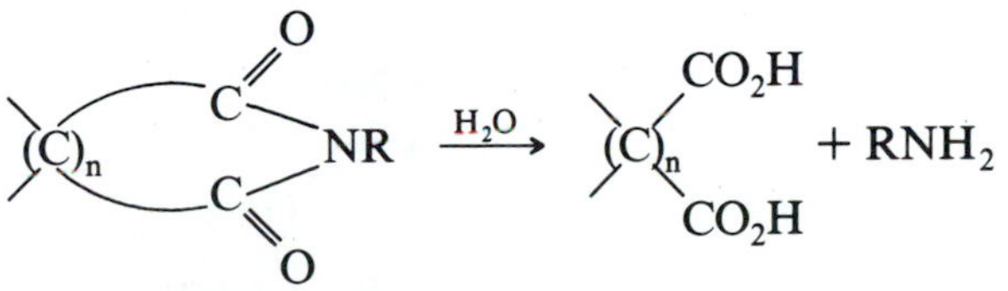

imides

---

[24] N. D. Cheronis, J. B. Entrikin, and E. M. Hodnett, *Semimicro Qualitative Organic Analysis,* 3rd ed. (New York: Interscience, 1965); F. Feigl, *Spot Tests in Organic Analysis* (Amsterdam: Elsevier, 1960).

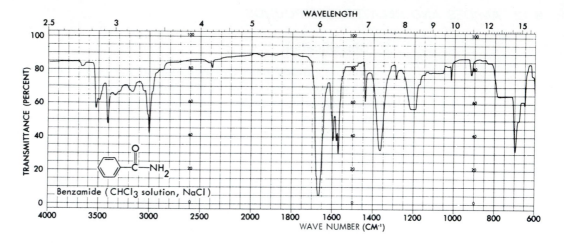

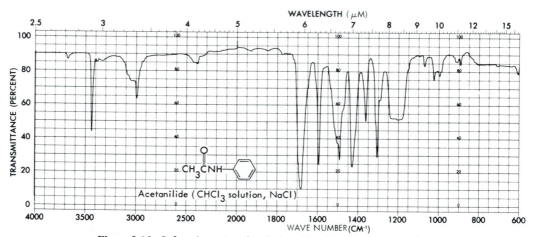

Figure 9.16 Infrared spectra of typical amides. Note the absence of the amide II band in the spectrum of the lactam.

$$\begin{array}{c} RNH \\ {\searrow} \\ C{=}O \\ {\nearrow} \\ RNH \end{array} \xrightarrow{\text{H}_2\text{O}} \ 2\,RNH_2 + CO_2$$

ureas

$$R'O\overset{\displaystyle O}{\overset{\displaystyle \|}{C}}NHR \xrightarrow{\text{H}_2\text{O}} R'OH + RNH_2 + CO_2$$

urethanes

Alkaline hydrolysis produces the free amine and the salt of the carboxylic acid; acidic hydrolysis yields the free acid and a salt of the amine. A special case is

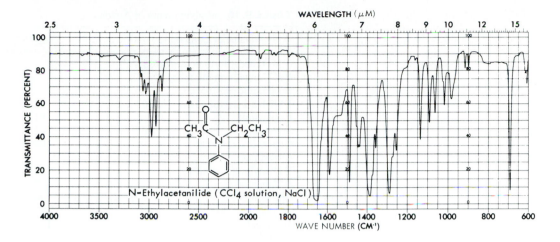

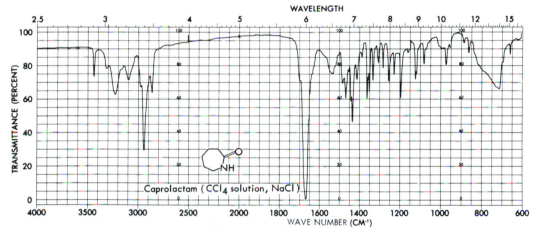

Figure 9.16 (*continued*)

presented by the lactams, which yield amino acids upon hydrolysis; these amino acids can be converted to derivatives utilizing the general principles discussed in Sec. 9.11 for the α-amino acids.

The hydrolysis can be effected by refluxing with $6N$ hydrochloric acid or 10 to 20% sodium hydroxide. The latter is usually faster. Resistant amides can be hydrolyzed by boiling with 100% phosphoric acid (prepared by dissolving 400 mg of phosphorous pentoxide in 1 g of 85% phosphoric acid), or by heating at 200°C in a 20% solution of potassium hydroxide in glycerol. Tertiary amides are readily hydrolyzed with anhydrous hydroxide/t-butoxide in ether.[25]

---

[25] P. G. Gassman, P. K. G. Hodgson, and R. J. Balchunis, *J. Am. Chem. Soc.*, **98,** 1275 (1976).

TABLE 9.15 Melting Points of Amides

Name of Compound	mp (°C)
N,N-Dimethylformamide (bp 153°C)	
N,N-Diethylformamide (bp 176°C)	
Formamide (bp 196°C)	
1-Methyl-2-pyrrolidinone (bp 202°C)	−24
Propanamide	81
Acetamide	82
4-Phenylbutanamide	84
Acrylamide	85
N-Phenylpropanamide	87
m-Toluamide	97
Heptanamide	96
Octanamide	99
Acetanilide	114
Butanamide	115
Methacrylamide	116
Phenylacetanilide	118
Adipamic Acid (monoamide)	130
Succinimide	126
Nicotinamide	128
Isobutyramide	129
Benzamide	130
o-Toluamide	143
Phthalamic Acid (monoamide)	149
Succinimide (monoamide)	157
p-Toluamide	160
Maleimide	170(190)
Benzanilide	175
Phthalamide (diamide)	220

Primary amides can be converted to acids by reaction with nitrous acid.

$$RCONH_2 + HONO \longrightarrow RCOOH + N_2 + H_2O$$

The Hofmann degradation is of occasional value in structural work. Under appropriate conditions, urethanes or symmetrical ureas can be obtained from the intermediate isocyanates.

$$RCONH_2 \xrightarrow{Br_2, NaOH} RNCO \xrightarrow{H_2O} RNH_2 + CO_2$$

Table 9.15 lists melting points of amides.

9.13 NITRILES

Classification

Aliphatic nitriles are usually liquids; the simpler aromatic nitriles are liquids or low melting solids. Both types have a characteristic odor reminiscent of cyanide. In the infrared, the nitrile absorption appears as a weak band in the 2260 to

2210 cm^{-1} region. This region somewhat overlaps that of an isocyanate, but these can be readily differentiated on the basis of their reactivity with water, alcohols, and so on, and the fact that the isocyanate absorption is much more intense than that of a nitrile. Further preliminary tests for the nitrile grouping are usually not necessary. Other texts have recommended the conversion of the nitrile to a hydroxamic acid by reaction with hydroxylamine. Hydroxamic acids can be detected by a color test with ferric chloride. Alternatively, a small amount of the supposed nitrile can be heated with potassium hydroxide in glycerol, with the evolution of ammonia being detected by odor or by the litmus or copper sulfate test (see Section 9.10 on amines). Both of these methods, however, depend on the prior establishment of the absence of other functional groups such as esters and amides.

Characterization

Nitriles can be *hydrolyzed to acids* under acidic or alkaline conditions. If the resulting acid is solid, it can be employed as such a derivative; if it is a liquid or water-soluble, it is usually converted directly to a solid derivative.

$$RCN + 2\,H_2O \longrightarrow RCOOH + NH_3$$

PROCEDURE: HYDROLYSIS TO ACIDS

(a) *Acid hydrolysis.* In a small flask equipped with a reflux condenser, place 4 mL of 85% phosphoric acid, 1 mL of 75% sulfuric acid, and 0.2 to 0.5 g of nitrile. Add a boiling chip and gently reflux the mixture for 1 hour. Cool the mixture and pour onto a small amount of crushed ice. If the acid is a solid, it may precipitate at this point. It should be collected on a filter. If amide impurities are present, the precipitate can be taken up in base, filtered, and reacidified to recover the acid. If the acid does not precipitate, the reaction mixture should be extracted several times with ether; the ether is evaporated to recover solid acid. If the acid is a liquid, it can be converted to conventional acid derivatives such as the *p*-toluidide, *p*-bromobenzyl ester, or *S*-benzylthiuronium salt.

(b) *Alkaline hydrolysis.* In a small flask equipped with a reflux condenser, place 2 g of potassium hydroxide, 4 g of glycerol or ethylene glycol, and 250 to 500 mg of nitrile. Reflux for 1 hour. Dilute with 5 mL of water, cool, and add several milliliters of ether. Shake and allow the layers to separate. Decant the ether layer and discard. Acidify the aqueous solution by slow addition of 6N hydrochloric or sulfuric acid. Extract the acid with several portions of ether, and handle the extract as indicated above under acid hydrolysis.

For those nitriles that yield solid, water-insoluble amides, partial *hydrolysis to amides* is often a very satisfactory method of derivatization.

$$RCN \xrightarrow[\text{H}_2\text{O}]{\text{H}_2\text{SO}_4} RCONH_2$$

PROCEDURE: HYDROLYSIS TO AMIDES

A solution of the nitrile (0.1 g) in 1 mL of concentrated sulfuric acid is warmed on the steam bath for several minutes. The mixture is cooled and poured into cold water. The precipitated crude amide is collected and suspended in a small volume of bicarbonate solution to remove any acid formed. The insoluble amide is recollected and recrystallized (aqueous ethanol). The amide can be further converted to the acid by action of nitrous acid.

Nitriles, especially aromatic nitriles, can be smoothly converted to amides by *alkaline hydrogen peroxide.*

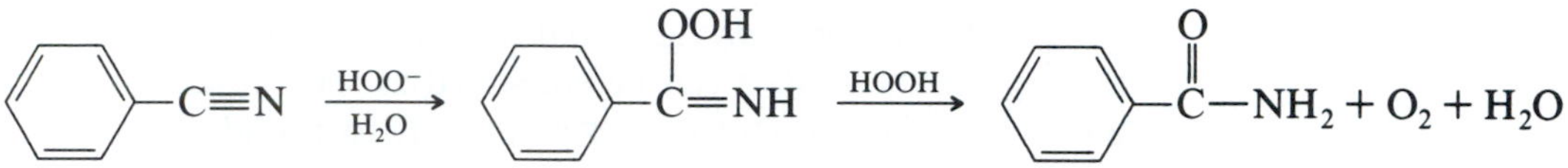

**PROCEDURE: CONVERSION TO AMIDES
BY ALKALINE HYDROGEN PEROXIDE**

In a small test tube, place 0.1 g of nitrile, 1 mL of ethanol, and 1 mL of $1N$ sodium hydroxide. To this mixture add dropwise (cooling if the reaction foams too vigorously) 1 mL of 12% hydrogen peroxide. After addition is complete, maintain the solution at 50 to 60°C in a water bath for an additional 30 minutes to 1 hour. Dilute the reaction mixture with cold water, and collect the solid amide. Recrystallize from aqueous ethanol.

Table 9.16 lists physical properties of nitriles.

9.14 NITRO COMPOUNDS

Classification

Most nitroalkanes are liquids that are colorless when pure, but they develop a yellow tinge on storage. The liquid nitro compounds have characteristic odors; they are insoluble in water and have densities greater than unity. Many of the aromatic compounds are crystalline solids. The aromatic nitro compounds are generally yellow, the color increasing markedly on polynitration.

When the nitro group is the primary functionality in the molecule, its presence is readily revealed on casual inspection of the infrared spectrum; strong bands appear

TABLE 9.16 Physical Properties of Nitriles

Name of Compound	bp (°C)	mp (°C)
Acrylonitrile	78	
Acetonitrile	81.6	
Propanonitrile	97.3	
2-Methylpropanonitrile	104	
Butanonitrile	117.4	
trans-2-Butenonitrile	119	
3-Methylbutanonitrile	130	
Pentanonitrile	141	
Hexanonitrile	165	
Benzonitrile	190.2	
o-Toluonitrile	205	
Ethylcyanoacetate	207	
m-Toluonitrile	212	
p-Toluonitrile	217	27
Phenylacetonitrile	234	
Cinnamonitrile	256	20
Hexanedinitrile	295	21
Malononitrile	219	31
1-Naphthonitrile		32
o-Chlorobenzonitrile	232	43
Succinonitrile		57
Cyanoacetic acid		66
p-Chlorobenzonitrile		96
o-Nitrobenzonitrile		110
p-Nitrobenzonitrile		147

near 1560 and 1350 cm^{-1} (see Figure 9.17). More frequently, the group appears as a substituent in other classes of compounds, and its presence is revealed by the physical constants of the compound and its derivation and/or by more careful inspection of the infrared spectrum. Like halogen substituents, when the nitro group is present along with a more reactive functionality, for example, carbonyl, hydroxyl, and so on, derivatization is made at the more reactive site.

The high electronegativity of the nitro group results in extensive deshielding of adjacent aliphatic hydrogens in NMR spectra; nitromethane appears at δ 5.72, RCH_2NO_2 at δ 5.6, and R_2CHNO_2 at δ 5.3. The characteristic chemical shift, along with coupling patterns, allows distinction to be made between primary and secondary aliphatic nitroalkanes.

Characterization

Many of the simple nitroalkanes can be identified by reference to physical and spectral properties. Relatively few vinyl nitro compounds are known; nitroethylene and other lower nitroalkenes readily polymerize.

Both aromatic and aliphatic nitro compounds can be converted to *primary amines by reduction with tin and hydrochloric acid*. The reduction can also be

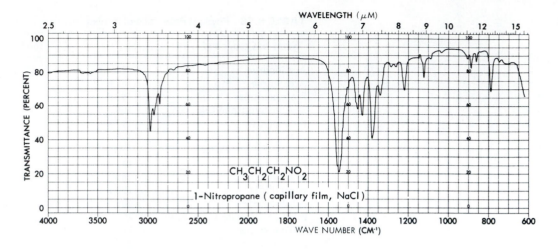

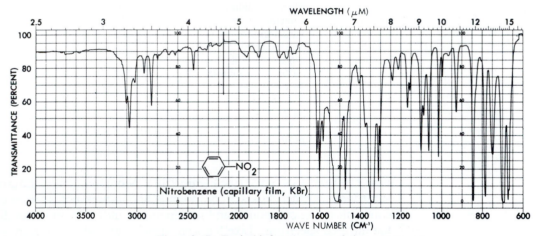

Figure 9.17 Typical infrared spectra of nitro compounds.

effected by catalytic hydrogenation. The amine thus obtained is identified by standard procedures. This provides the only general method for the conversion of nitroalkanes to solid derivatives.

$$RNO_2 \xrightarrow{\text{Sn/HCl}} RNH_2$$

PROCEDURE: REDUCTION WITH TIN AND HYDROCHLORIC ACID

In a 50-mL round-bottomed flask fitted with a reflux condenser, place 1 g of the nitro (nitroso, azo, azoxy, or hydrazo) compound, and 2 to 3 g of granulated tin. If the compound is highly insoluble in water, add 5 mL of ethanol. Add 20 mL of 10%

hydrochloric acid in small portions, shaking after each addition. After the addition is complete, warm the mixture on the steam bath for 10 minutes. Decant the hot solution into 10 mL of water and add sufficient 40% sodium hydroxide to dissolve the tin hydroxide. Extract the mixture several times with ether. Dry the ether over magnesium sulfate; remove the ether and identify the amine by conversion to one or more crystalline derivatives.

For highly volatile amines, the procedure must be modified. As the solution is made alkaline, the amine can be distilled into dilute hydrochloric acid.

Primary and secondary nitroalkanes can be *oxidized to the corresponding carbonyl compounds* by potassium permanganate. Magnesium sulfate solution is added to act as a buffering agent.[26]

$$R-CH_2NO_2 + KMnO_4 \xrightarrow[\text{MgSO}_4]{} RCHO$$

Aromatic mononitro compounds can be characterized by conversion to *di-* or *trinitro derivatives* (follow procedures for aromatic hydrocarbons, Sec. 9.1.4). Many polynitro aromatics form stable *charge-transfer complexes* with naphthalene and other electron-rich aromatic compounds.[27]

Nitro derivatives of alkylbenzenes can be oxidized to the corresponding nitrobenzoic acids with dichromate (preferred) or permanganate (follow procedures for alkyl benzenes, see Section 9.1.4). Bromination of the ring often affords a suitable derivative.

Table 9.17 lists the physical properties of nitro compounds.

9.15 SULFUR COMPOUNDS

A comprehensive discussion of the methods of detection and identification of the many types of organic sulfur compounds is far beyond the scope of this text. In general, organic sulfur chemistry follows along the lines of organic oxygen chemistry; many of the same general types of functional groups occur, but the variety is greatly increased by the variable oxidation states of sulfur, as indicated by the following examples.

Oxygen series

| ROH | Alcohol | ROR | Ether | ROOR | Peroxide |

---

[26] H. Shechter and F. T. Williams, Jr., *J. Org. Chem.*, **27**, 3699 (1962).
[27] O. C. Dermer and R. B. Smith, *J. Am. Chem. Soc.*, **61**, 748 (1939).

TABLE 9.17   Physical Properties of Nitro Compounds

Name of Compound	bp (°C)	mp (°C)
Nitromethane	101	
Nitroethane	114	
2-Nitropropane	120.3	
1-Nitropropane	131.6	
2-Nitrobutane	140	
1-Nitrobutane	153	
2-Nitropentane	154	
1-Nitropentane	173	
Nitrobenzene	210	
o-Nitrotoluene	221.7	
o-Nitroanisole	265	10
m-Nitrobenzyl alcohol		27
o-Chloronitrobenzene		32
2,4-Dichloronitrobenzene	258	33
2-Nitrophenol		37
4-Chloro-2-nitrotoluene	240	38
p-Nitrotoluene	238	52
p-Nitroanisole		52.5
1-Nitronaphthalene*		57
2,4,6-Trinitroanisole		68
2,4-Dinitrotoluene		70
p-Nitrobenzyl chloride		72
2,4-Dinitrobromobenzene		72
o-Nitrobenzyl alcohol		74
2-Nitronaphthalene*		78
2,4,6-Trinitrotoluene		80.1
2,4,6-Trinitrochlorobenzene		83
p-Chloronitrobenzene		84
m-Dinitrobenzene		90
2,4-Dinitroanisole		95
4-Nitrobiphenyl*		114

Compound and/or reduction product ($-NH_2$) is carcinogenic.

Sulfur series

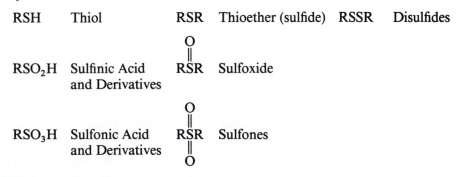

TABLE 9.18 Physical Properties of Sulfides, Sulfones, Sulfoxides, and Thiols

Sulfide	bp (°C)	Sulfoxide (mp, °C)	Sulfone (mp, °C)
Dimethyl	38	18.4 (bp 189°C)	109 (bp 238°C)
Ethyl methyl	67		36
Thiophene	84		
Diethyl	91		71
Methyl propyl	95.5		
Ethyl propyl	118.5		
Tetrahydrothiophene	119		27 (bp 285°C)
Diisopropyl	119		36
Dipropyl	142.8		30
Dibutyl	189	34 (bp 250°C)	45 (bp 287–295°C)
Methyl phenyl	188		
Ethyl phenyl	204		42
Diphenyl	296	70.4	128
	mp (°C)		
Benzyl phenyl	41	123	146
Dibenzyl	50	134.8	151.7
Di-*p*-tolyl	57	93	158

		mp (°C) of Derivatives	
Thiol	bp (°C)	2,4-Dinitrophenyl thioether	2,4-Dinitrophenyl sulfone
Methanethiol	6	128	189.5
Ethanethiol	36	115	160
2-Propanethiol	56	94	140.5
2-Methyl-1-propanethiol	88	76	105.5
2-Propen-1-thiol	90	72	105
Butanethiol	97	66	92
Pentanethiol	126	80	83
Hexanethiol	151	74	97
Cyclohexanethiol	159	148	172
Benzenethiol	169	121	161
Phenylmethanethiol	194	130	182.5
1-Phenylethanethiol	199	89	133.4

In this chapter, only the most commonly occurring organic sulfur functions will be discussed. The most comprehensive listing of the properties of organic bivalent sulfur compounds is found in the series by Reid.[28]

Table 9.18 lists the physical properties of thiols, sulfoxides and sulfones.

---

[28] E. E. Reid, *Organic Chemistry of Bivalent Sulfur,* Vols. I–VI (New York: Chemical, 1958–1966).

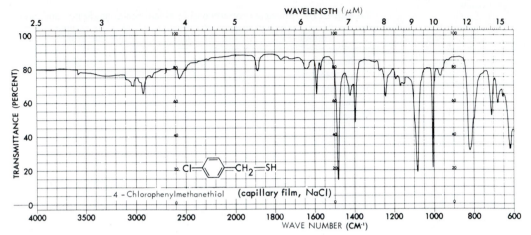

Figure 9.18 Typical infrared spectrum of a thiol. Note the very weak S—H band at 2560 cm^{-1}.

9.15.1   Thiols (Mercaptans)

The classification of a compound as a thiol seldom awaits the interpretation of the infrared spectrum. For the uninitiated, most thiols have an objectionable odor reminiscent of hydrogen sulfide. Thiols are generally insoluble in water, but they react with aqueous alkali to form soluble salts.

$$RSH + OH^- \longrightarrow RS^- + H_2O$$

In the infrared, the S—H band appears in the 2600 to 2550 cm^{-1} region (see Figure 9.18). This is a relatively weak band and in some cases may be overlooked, especially if the spectrum is run as a dilute solution. This may result in misassignment of the compound as a sulfide. Sulfides and thiols can be differentiated on the basis of solubility of the latter in dilute alkali and by the *lead mercaptide test;* the former is insoluble in base and does not precipitate in the lead mercaptide test.

$$2 RHS + Pb^{++} + 2 H_2O \longrightarrow Pb(SR)_2 \downarrow + 2 H_3O^+$$

PROCEDURE: LEAD MERCAPTIDE TEST

Add two drops of the thiol to a saturated solution of lead acetate in ethanol; yellow lead mercaptide precipitates.

———————————————

Perhaps the most useful derivatives of the alkyl and aryl thiols are the *2,4-dinitrophenyl sulfides* and the corresponding sulfones.

$$RSH \xrightarrow{\text{NaOH}} RSNa$$

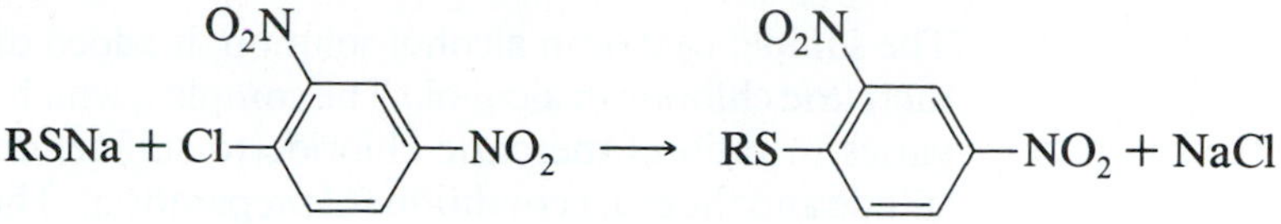

RSNa + Cl—[2,4-dinitrophenyl]—NO$_2$ ⟶ RS—[2,4-dinitrophenyl]—NO$_2$ + NaCl

PROCEDURE: 2,4-DINITROPHENYL SULFIDES

In a test tube dissolve 250 mg of the thiol in 5 mL of methanol. Add 1 mL of 10% sodium hydroxide solution. Add this solution to 3 mL of methanol containing 0.5 g of 2,4-dinitrochlorobenzene in a small flask arranged for reflux. Reflux the mixture for 10 minutes, cool, and recrystallize the product from methanol or ethanol.

Occasionally it may be expedient to identify thiols by conversion to simple sulfides or sulfones. The methyl and benzyl sulfides, produced by alkylation of the sodium thiolates with methyl iodide or benzyl bromide in alcohol, are the most useful. These sulfides, as well as the 2,4-dinitrophenyl sulfides, can be oxidized to sulfones by procedures outlined under sulfides in the next section. The 2,4-dinitrophenyl sulfones are particularly valuable because they exhibit a wide range of melting points.

9.15.2 Sulfides and Disulfides

Infrared spectroscopy alone usually cannot provide sufficient information to adequately classify a compound as belonging to the sulfide or disulfide class. Both these classes of compounds exhibit a carbon-sulfur stretching frequency in the 800 to 600 cm^{-1} region, which is of little utility because it is easily confused with other bands in that region and, furthermore, is frequently obscured by the solvent. The sulfur-sulfur stretching frequency is a weak band in the 550 to 450 cm^{-1} region, and it is thus inaccessible on most instruments.

Aliphatic and most aromatic sulfides and disulfides have unpleasant odors somewhat resembling that of hydrogen sulfide. Compounds with such odors, which are not mercaptans or sulfoxides (as determined by the absence of characteristic infrared bands), should be strongly suspect as sulfides or disulfides. Most aliphatic sulfides and disulfides are liquids. The aromatic sulfides are liquids or low melting solids; aromatic disulfides are low melting solids. A number of polysulfides of general formula R—S$_n$—R are also known.

Most sulfides give *crystalline complexes with mercuric chloride.*

$$R_2S + nHgCl_2 \longrightarrow R_2S(HgCl_2)_n \xrightarrow{\text{NaOH}} R_2S$$

PROCEDURE: MERCURIC CHLORIDE COMPLEXES

The sulfide, neat or in alcohol solution, is added dropwise to a saturated solution of mercuric chloride in alcohol. The complex, which usually precipitates immediately, varies in ratio of mercuric chloride to sulfide, depending on the structure of the sulfide and the exact conditions of preparation. The sulfides can be regenerated from the complex by aqueous sodium cyanide or dilute hydroxide. The procedure can be used as a test for sulfides or as a method for obtaining a solid derivative.

The most useful solid derivatives of the sulfides are the *sulfones* produced by oxidation with permanganate or peroxides.

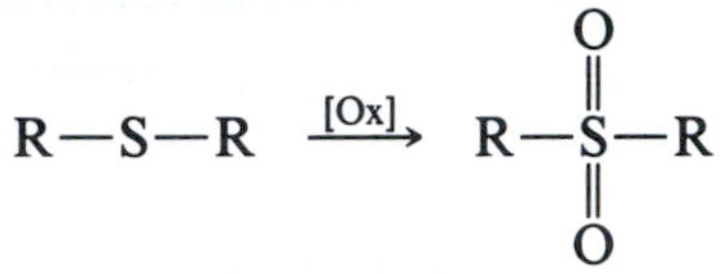

$$R-S-R \xrightarrow{[Ox]} R-\overset{\overset{\displaystyle O}{\|}}{\underset{\underset{\displaystyle O}{\|}}{S}}-R$$

PROCEDURE: OXIDATION TO SULFONES

(a) Dissolve the sulfide in a small volume of warm glacial acetic acid. Shake and add 3% potassium permanganate solution in 2- or 3-mL portions as fast as the color disappears. Continue the addition until the color persists after shaking for several minutes. Decolorize the solution by addition of sodium hydrogen sulfite. Add 2 to 3 volumes of crushed ice. Filter off the sulfone and recrystallize from ethanol.

(b) Dissolve 1 mmol of the sulfide in 3 mL of toluene and 3 mL of methanol. Add 0.5 to 1 mmol of vanadium pentoxide and heat to 50°C with stirring. Slowly add a slight excess of *t*-butyl hydroperoxide until the excess can be detected by acidified starch iodide paper. Add solid sodium hydrogen sulfite in small portions, and stir for 5 to 10 minutes. Filter and concentrate to a heavy oil. Add a small amount of water, extract with methylene chloride or other suitable solvent, and dry over magnesium sulfate.

(c) Dissolve the sulfide in glacial acetic acid. Add excess hydrogen peroxide and allow to stand overnight. Warm the solution to 50 to 60°C for 5 to 10 minutes. Pour into ice water and collect the crystalline sulfone.

N-p-Toluenesulfonyl sulfilimines are readily prepared by shaking a suspension of the sulfide in an aqueous solution of Chloramine-*T*.

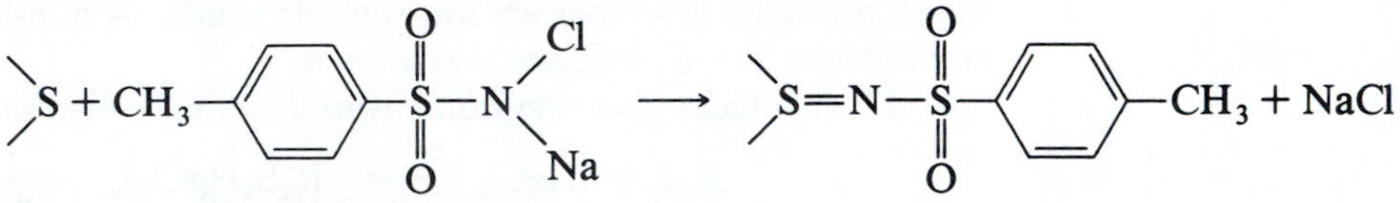

Chloramine-*T*

Disulfides can be reduced to thiols by the action of zinc and dilute acid, or they can be oxidized to sulfonic acids by potassium permanganate.

$$RSSR \xrightarrow[H^+]{Zn} 2\ RSH$$

$$RSSR \xrightarrow{KMnO_4} 2\ RSO_3H$$

9.15.3 Sulfoxides

Sulfoxides are usually low melting, highly hygroscopic solids with a noticeable odor similar to that of sulfides — which, in fact, is often due to trace amounts of contaminating sulfide. The sulfoxide grouping is pyramidal and configurationally stable; hence sulfoxides can exist as enantiomeric and/or diastereomeric pairs. Sulfoxides are weak bases; isolable salts are formed with some strong acids, including nitric acid.

C A U T I O N : Sulfoxides react explosively on contact with perchloric acid.

The sulfoxide grouping is readily noted by the characteristic strong infrared band at 1050 to 1000 cm^{-1} (see Figure 9.19). This band is quite sensitive to hydrogen bonding; the addition of methanol to the solution causes a significant shift to lower frequency. Likewise, a shift to lower frequency is observed in going from carbon tetrachloride to chloroform.

In the NMR, hydrogens alpha to sulfoxide appear in the δ 2.0 to 2.5 region. Methylene protons adjacent to sulfoxide often appear as an *AB* system.

Sulfoxides can be reduced to sulfides by reduction with sodium borohydride[29] or aqueous sodium hydrogen sulfite.[30] The yields are high and the sulfides are obtained in essentially pure form.

9.15.4 Sulfones

Sulfones are colorless, odorless, solids; however, a number of the dialkyl sulfones have low melting points.

When the sulfone is the primary functionality in the molecule, the two strong sulfone bands at 1325 and 1175 cm^{-1} are the most prominent in the infrared spectrum (Figure 9.19).

Sulfones are relatively inert substances; the group is not readily subject to chemical alteration. Some sulfones can be reduced to sulfide by lithium aluminum hydride. The sulfone group markedly increases the acidity of α-hydrogens. Sulfone anions, readily formed by action of strong bases (e.g., butyllithium), undergo the expected reactions of halogenation, alkylation, and acylation.

Infrared, nuclear magnetic resonance and mass spectral data (the SO_2 grouping

---

[29] D. W. Chasar, *J. Org. Chem.*, **36**, 613 (1971).

[30] C. R. Johnson, C. C. Bacon, and J. J. Rigau, *J. Org. Chem.*, **37**, 919 (1972).

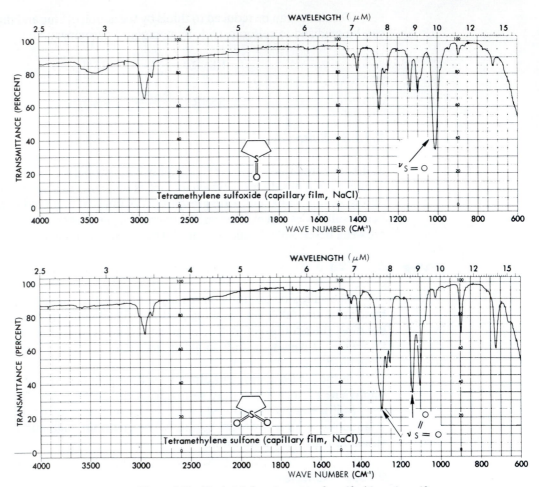

Figure 9.19 Typical infrared spectra of a sulfoxide and a sulfone.

is transparent down to 185 nm in the ultraviolet), along with physical constants, constitute the best method of identification.

Aromatic sulfones yield phenols when fused with solid potassium hydroxide.

$$ArSO_2R \xrightarrow[\text{(2) } H_2O^+]{\text{(1) KOH, }\Delta} ArOH$$

9.15.5 Sulfonic Acids; RSO₃H

Sulfonic acids are strong acids, comparable to sulfuric acid. The free acids and their alkali metal salts are soluble in water and insoluble in nonpolar organic solvents (ether, etc.).

The infrared spectrum of sulfonic acids exhibits three SO bands—1250 to 1160 cm^{-1}, 1080 to 1000 cm^{-1}, and 700 to 610 cm^{-1}. The first of these is the most intense.

The free acids are very hygroscopic and do not give sharp melting points. The acids are listed in *Handbook of Tables for Organic Compound Identification* in order of increasing melting point of the sulfonamide.

The preparation of *S-benzylthiuronium sulfonates* from free sulfonic acids and their salts is highly recommended because of the ease with which they are made and because of the extensive data available.

$$RSO_3Na + C_6H_5CH_2SC(NH_2)_2^+Cl^- \longrightarrow C_6H_5CH_2SC(NH_2)_2^+ RSO_3^- + NaCl$$

PROCEDURE: *S*-BENZYLTHIURONIUM SULFONATES

To a solution of 0.5 g of the salt in 5 mL of water, add several drops of 0.1 N hydrochloric acid. Add a cold solution of 1 g of *S*-benzylthiuronium chloride in 5 mL of water. Cool in an ice bath and collect and recrystallize the derivative from aqueous ethanol.

If the free acid is available, dissolve 0.5 g in 5 mL of water and neutralize to phenolphthalein endpoint with sodium hydroxide. Then add two to three drops of 0.1 N hydrochloric acid and proceed as before.

Probably the most commonly employed method for the derivation of sulfonic acids and salts is the *conversion to sulfonamides* via the sulfonyl chlorides.

$$RSO_3Na + PCl_5 \longrightarrow RSO_2Cl + POCl_3 + NaCl$$

$$RSO_2Cl + 2 NH_3 \longrightarrow RSO_2NH_2 + NH_4Cl$$

PROCEDURE: SULFONAMIDES

Place 1.0 g of the free acid or anhydrous salt and 2.5 g of phosphorous pentachloride in a small, dry flask equipped with a reflux condenser. Heat in an oil bath at 150°C for 30 minutes. Cool and add 20 mL of toluene; then warm and stir to effect complete extraction. (If desired, the sulfonyl chloride can be isolated by removal of the solvent. Recrystallize from petroleum ether.) Add the toluene solution slowly, with stirring, to 10 mL of concentrated ammonia solution. If the sulfonamide precipitates, isolate it by filtration. Otherwise it can be isolated by evaporation of the organic layer. Recrystallization can be accomplished from water or ethanol.

Substituted sulfonamides can be obtained by treatment of the foregoing sulfonyl chloride with amines in the usual manner.

9.15.6 Sulfonyl Chlorides; RSO₂Cl

The simple aliphatic and a number of the aromatic sulfonyl chlorides are liquids. High molecular weight aliphatic and substituted aromatic sulfonyl chlorides are solids. They are insoluble in water and soluble in most organic solvents. Their vapors have a characteristic irritating odor and are lacrymatory. In the infrared, strong bands are observed at 1360 and 1170 cm⁻¹.

Sulfonyl chlorides are easily converted to amides or substituted amides by reaction with concentrated ammonia solution or with amines in the presence of aqueous alkali (see Section 9.10.2 on amines).

9.15.7 Sulfonamides; RSO₂NH₂

Most sulfonamides are colorless crystalline solids of high melting point that are insoluble in water. The primary and secondary sulfonamides are soluble in aqueous alkali (Hinsberg test). In the infrared, sulfonamides exhibit the appropriate N—H stretching and deformation bands, as well as asymmetric (1370 to 1330 cm⁻¹) and symmetric (1180 to 1160 cm⁻¹) stretching of the S=O.

The only general method for characterization of sulfonamides is *hydrolysis to the sulfonic acid and amine.* The hydrolysis can be effected in a number of ways. Refluxing with 25% hydrochloric acid requires 12 to 36 hours. Eighty percent sulfuric acid or a mixture of 85% phosphoric and 80% sulfuric can also be used. The latter method is particularly efficient.

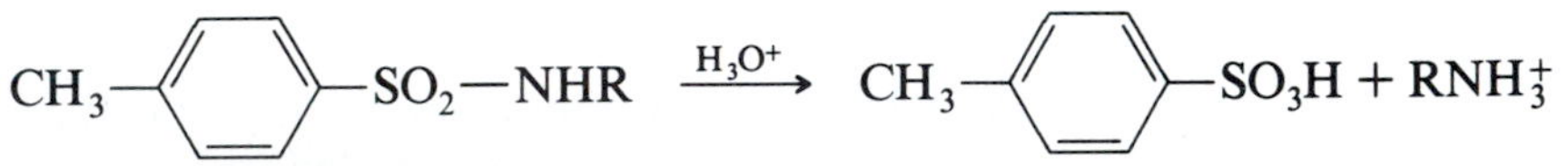

$$CH_3-\!\!\left\langle\bigcirc\right\rangle\!\!-SO_2-NHR \xrightarrow{\;H_3O^+\;} CH_3-\!\!\left\langle\bigcirc\right\rangle\!\!-SO_3H + RNH_3^+$$

PROCEDURE: HYDROLYSIS OF SULFONAMIDES

In a test tube mix 1 mL of 80% sulfuric acid with 1 mL of 85% phosphoric acid. Add 0.5 g of the sulfonamide and heat the mixture to 160°C for 10 minutes or until the sulfonamide has dissolved. A dark, viscous mixture will result. Cool and add 6 mL of water. Continue cooling and, while stirring, render the solution alkaline with 25% sodium hydroxide solution. Separate the amine by extraction or distillation. Prepare an appropriate derivative of the amine (benzoyl) and of the sodium sulfonate (*S*-benzylthiuronium salt).

10

Laboratory Experiments

10.1 INTRODUCTION

Experience is the best teacher. This chapter presents a number of practical experiments in organic chemistry with emphasis on the "hands-on" approach. The techniques described in the earlier chapters are elaborated in a laboratory environment. Early experiments focus on gaining experience with laboratory equipment and technique. Later experiments combine several techniques and provide experience with functional group modification, functional group interconversions, and bond-forming and bond-breaking reactions. A number of concepts of organic chemistry are tested experimentally. Mechanistic and synthetic studies are developed from the elementary levels to those sophisticated enough for a modern research laboratory. For full experience and practicality, the scale of the laboratory experiments varies from microscale to the miniscale most commonly used in research laboratories. Before beginning any of the following experiments, you should review the safety procedures described in Section 1.2 and the procedures for handling and disposal of chemicals in Section 1.3.

10.2 MELTING POINTS

The term *melting point* has been defined and discussed in Section 3.2. It should be reemphasized that in a laboratory situation, a *melting range* is observed. The melting range of a crystalline substance is an easily determined characteristic that provides an

TABLE 10.1 Melting Point Standards

Compound	mp (°C)
p-Nitrotoluene	51–52
Benzil	93–94
Urea	133–135
Adipic acid	152–153
p-Anisic acid	182–184

TABLE 10.2 Alternate Melting Point Standards

Compound	mp (°C)
Vanillin	81–83
Acetanilide	114–116
Acetophenetidin	134–136
Sulfanilamide	164.5–166.5
Sulfapyridine	190–193
Caffeine	235–237.5

indication of the identity and/or purity of the substance. The types of apparatus and procedure for the determination of melting "points" have also been described in Section 3.2. However, any data, including melting points, obtained by any type of instrumentation must be calibrated. Calibration of your thermometers is especially important since it will be used to determine melting points and boiling points of your starting materials, reagents, products, and unknowns. Your thermometer(s) should be calibrated at several temperatures in the range that you will normally be taking measurements. To allow this calibration and to help you perfect your melting point technique, you must have pure compounds available as standards. Table 10.1 provides a list of readily available compounds that adequately cover the temperature range normally encountered in the laboratory. Alternatively, Table 10.2 provides a list of standards available from the Arthur H. Thomas Company, Philadelphia, Penn., which are also provided with a Thomas–Hoover melting point apparatus (see Figure 3.3). The Aldrich Chemical Company also provides 26 melting point standards that cover the 50° to 300°C range.

EXPERIMENT 1

Thermometer Calibration

Check your thermometer(s) and make certain that the mercury column is not separated. Place a small amount of each standard in separate melting point capillary tubes or between glass plates (for a Fisher–Johns apparatus) as described in Section 3.2. Using the available apparatus (Figures 3.2 to 3.5), carefully determine the melting range of each standard. In most types of apparatus you may insert two or three capillary tubes at once. Remember that the rate of heating should not exceed 1 to 2°C per minute while traversing the melting point range. If any melting points differ by more than 1 or 2°C from the expected value, repeat those determinations carefully to be certain that your technique is correct. If your thermometer appears to be reproducibly very far off, it should be checked by someone else. If the problem is confirmed, a new thermometer should be obtained. When you are confident of your observed values, plot the high point of the observed melting point ranges (X-axis) versus the high point of the melting ranges (y-axis) of the standards provided in Tables 10.1 and 10.2 or the literature. *Use this calibration curve to correct all future melting points or boiling points that you may obtain.* Obviously the calibration must be done carefully for each thermometer since it will affect all your subsequent temperature determinations.

EXPERIMENT 2

Melting Point of an Unknown

You will be given a small amount of a pure, unknown solid. Record the sample number or code name. Determine the melting range of the solid very carefully. Correct the values you obtain using your calibration curve. Record the values in your notebook. If you are given enough sample, you may save a considerable amount of time by first taking the melting point rapidly to get an approximate indication of the melting point and then repeating the determination at the proper heating rate (1 to 2°C per minute) over the melting range. Compare your determined and recorded values to those on the list of possible unknowns[1] provided by your instructor. Decide which of the possible listed substances your unknown might be. Since the list may contain several substances with similar melting points, obtain a small sample of each compound with a reported melting point near that which you observed for your unknown (being careful not to contaminate the bottles of comparison compounds). Separately mix each sample with a nearly equal amount of your unknown. Mix the samples very well by grinding them together with a spatula or pestle. Measure and record the melting ranges of the mixtures you have made. Compounds which have

Compound	Approximate mp (°C)
Naphthalene	80
Phenylacetic acid	80
Vanillin	80
Glutaric acid	100
Oxalic acid dihydrate	100
o-Toluic acid	100
Acetanilide	113
Fluorene	113
p-Nitrophenol	113
o-Aminophenol	172
Hydroquinone	172
d-Tartaric acid	172

[1] Instructor's note: The list above provides suitable unknowns with approximate melting points.

been mixed with your unknown that are different from the unknown will depress the melting point and often expand the melting range. However, a mixture of your unknown compound with a compound with which it is identical will result in no melting point depression or melting range expansion. Record the results in your notebook and indicate that compound which you believe is the same as your unknown.

10.3 MICRO BOILING POINTS

As indicated in Section 3.4, ultramicro boiling points can conveniently be determined on 3 to 4 μL of liquid in standard glass capillary tubes. This process is routinely more effective than larger-scale or even microscale boiling point determinations. The process should first be practiced using several of the (pure) solvents or other liquids available in the laboratory. Ethyl acetate (bp 77°C), toluene (bp 111°C), and related organic liquids work well. Water (bp 100°C) has a large surface tension and is difficult to insert in the glass capillary.

E X P E R I M E N T 3

Microscale Boiling Point Determination

Obtain a small sample of an unknown liquid in a vial from your instructor. Record the sample number or code name. Do not open the vial until you are ready to begin the experiment. Because of their volatility, liquid samples should only be handled in a hood or ventilated area. *(CAUTION: Do not attempt to identify your unknown by its odor. Even small amounts of organic vapors may be harmful if inhaled.)* Open the sample vial and withdraw several microliters with a *clean and dry* syringe. (Air or gas bubbles may form in the syringe when the plunger is withdrawn. These can be eliminated by rapidly pushing the plunger in and out several times while the tip of the syringe needle is immersed in the liquid sample.) Insert the syringe needle as far into a melting point capillary as possible and gently depress the plunger. If the liquid does not flow to the bottom of the capillary, the capillary should be placed in a small centrifuge tube and briefly centrifuged. A fine capillary is prepared by heating a piece of 1- to 3-mm soft glass tubing with a small burner and, immediately after removing it from the flame, drawing it out to a diameter small enough to fit into the melting point tube containing the liquid sample (see Section 3.4). A small length (5 mm) of the drawn-out capillary is cut and sealed at one end by touching it to the flame of the small burner. (Preferably the inner capillary should be prepared in advance or in an area remote from that where flammable organic liquids will be handled.) Place the small capillary *open end first* into the larger one containing the sample. If the inner capillary does not fall to the bottom, it may be centrifuged carefully.

Insert the assembled system into your melting point apparatus. Raise the temperature rapidly to 15 to 20°C below the anticipated boiling point of your sample. Then adjust the rate of the temperature increase to about 2°C per minute until a fine stream of bubbles is emitted from the inner capillary. Adjust the heat control until the temperature begins to decline. Take the boiling point as the temperature where the last escaping bubble collapses. This is the point where the vapor pressure of the substance equals the atmospheric pressure (boiling point definition). If necessary, repeat the process on the same sample several times to confirm the results. Record the data in your notebook. Compare your boiling point with those of several possibilities provided by your instructor and predict the identity of your unknown.

10.4 RECRYSTALLIZATION

Recrystallization (see Section 2.2) is one of the oldest, most widely used, and convenient means of purifying a solid. In the following experiments, you will recrystallize known compounds by standard procedures (see Section 2.2) to learn the techniques and then apply what you have learned to the purification of an unknown.

EXPERIMENT 4

Recrystallization of Acetanilide from Water

acetanilide (*N*-phenylacetamide)

Weigh a 1- to 5-g sample of impure[2] acetanilide and place it in a 50- to 250-mL Erlenmeyer flask, with the size of the flask chosen to be directly proportionate to the amount of the crude sample. Heat water to boiling (use a burner or hot plate, since a steam bath will not boil water) and add a minimum amount of the hot water to dissolve the acetanilide and soluble impurities while continuously heating the mixture at the boiling point. Start with about 20 to 100 mL of water, depending on the initial amount of crude acetanilide used. Continue to add hot water until no more material appears to dissolve. Do not continue to add solvent in attempts to dissolve the insoluble contaminants. Note that the acetanilide will melt in the hot solution,

---

[2] Instructor's note: Impure acetanilide may be prepared by thoroughly mixing acetanilide with various amounts of ordinary table sugar (a soluble impurity) and insoluble impurities such as sand and/or sawdust.

forming an oily layer, and enough boiling water should be added to dissolve the oil. Using a short stem or stemless funnel, gravity filter (Figure 2.1) the hot solution into an appropriately sized beaker. Cool the clear filtrate by allowing it to stand in a water bath. After about 30 minutes collect the crystals by suction filtration (Figure 2.3). Wash the crystals with a few mL of *cold* water and partially dry them by pressing them on the Büchner funnel with a spatula or small beaker. Allow the crystals to dry on a watch glass or filter paper until the next laboratory period.

Transfer the total filtrate from the recrystallization to an Erlenmeyer flask and concentrate the solution to about one-fourth of the original volume by boiling it vigorously *(CAUTION: Use a boiling chip)*. Cool the concentrate to obtain a second crop of crystals. If the concentrate is noticeably colored, the crystals obtained from it will also be colored unless the solution is first clarified. For this purpose, remove the solution from the heat source. Allow it to cool slightly below the boiling point and add a "pinch" (a small amount on the end of a spatula) of decolorizing charcoal. Be certain not to add the charcoal while the solution is boiling since the solution would then rapidly boil over the top of the flask. Reheat the solution to boiling and gravity filter it while it is hot. Cool the clear filtrate. Collect the crystals and dry them as before.

Separately weigh the two samples of dry crystals. Determine the melting points (corrected ~ 114 to 116°C) of your purified samples and calculate the mass recovery. If the melting point of the second crop of crystals is identical to that of the first crop, the samples may be combined. If the melting point of the second crop is lower than that of the first, the sample may be recrystallized again from hot water. Compare the melting points to that determined for pure acetanilide in Experiment 1. Calculate the mass recovery of pure acetanilide from the starting impure mixture. Record all the data in your notebook. Store the purified acetanilide in a tared (preweighed) and labeled vial for inspection by your instructor.

EXPERIMENT 5

Recrystallization of Benzoin from Organic Solvents

benzoin
(2-hydroxy-1,2-diphenylethanone)

Selecting a solvent. Many single or mixed organic solvents may be used for recrystallizing organic solids. The choice of the solvent system depends on the relative solubility of the solid in the solvent at high and low temperature. Thus, when an ample supply of material is available, the solvent choice is made by performing several small-scale trial recrystallizations. Place a small amount of crude benzoin (about the size of a pea) and add a few drops of 95% ethanol (just enough to cover the crystals). Very little of the material should dissolve. Heat the mixture to boiling using a steam bath or preheated sand bath. Do not distill the solvent away from your sample, but keep it boiling gently and watch carefully to see if all the benzoin dissolves. Impurities such as sawdust, sand, or broken glass obviously will never dissolve, so try to distinguish these impurities visually from the benzoin itself. If undissolved benzoin remains in the boiling mixture after 30 to 60 seconds, add more 95% ethanol dropwise while keeping the mixture boiling. When all the benzoin has dissolved, stop adding solvent. Only if necessary, gravity filter the hot solution to remove any insoluble impurities. Cool the mixture first with a water bath and then with an ice bath. If necessary, scratch the inside of the cooled tube to induce crystallization. Note whether most, little, or none of the starting benzoin has crystallized. Estimate the volume ratio of ethanol used relative to the amount of starting benzoin (1:1, 2:1, 10:1?).

Try this same test tube experiment using pentanes (Skellysolve F, low-boiling petroleum ether) as the solvent. (*CAUTION: These solvents are highly flammable.*) How does the use of pentanes compare to the use of 95% ethanol as a solvent for the recrystallization of benzoin?

Preparative recrystallization. Weigh out about 1 to 5 g of crude benzoin (exact weight? melting point range?) and place it in a 25- to 125-mL Erlenmeyer flask (with the size chosen proportionate to the amount of benzoin) along with a boiling chip or a wooden applicator stick (which will serve the same purpose and is easier to remove later). Add enough preheated ethanol to cover the crystals (~ 4 to 20 mL) and heat the mixture to boiling on a steam bath. While keeping the mixture boiling gently, slowly add small portions of 95% ethanol *just* until all the benzoin, but not the insoluble impurities, dissolves. If *all* the material dissolves and the solution is colorless, skip the steps described in the next paragraph.

If the solution is colored, note the volume and add about 50 to 100% more ethanol. Cool the solution to just below the boiling point and add a small amount ($\sim \frac{1}{2}$ of a small spatula full) of activated (decolorizing) charcoal to adsorb the colored impurities. Boil the suspension briefly. Gravity filter (Figure 2.1) the hot solution into a beaker using a short stem or stemless funnel. [If the recrystallation was started with a small amount of benzoin (~ 1 g) and a correspondingly small amount of solvent, a filter pipette (Figure 2.1c) may be more appropriate.] During the filtration, some of the hot ethanol may evaporate from the funnel and crystals may begin to form in the funnel itself. This can be minimized by placing a watch glass over the funnel and heating the receiving beaker on the steam bath during the filtration. Once all the solution has been filtered, add a small amount of hot ethanol to the original flask to rinse it and pour this rinse through the funnel. Concentrate the resulting filtrate by boiling off some of the solvent under the hood until only the original minimal volume of hot solvent required for dissolving all the benzoin remains.

Cover the hot beaker or flask containing the dissolved benzoin with a watch glass and allow the solution to cool slowly to room temperature. If crystals do not begin forming within a few minutes, scratch the flask or seed the solution with a tiny crystal of benzoin. If the product oils out, add slightly more ethanol and reheat the mixture to dissolve the oil. Again, cool and scratch or seed the solution to induce crystallization. Once crystals begin to form and/or the solution cools to room temperature, chill it further in an ice bath for 5 to 10 minutes. Suction filter (Figure 2.3) and wash the crystals with two small (1- to 5-mL portions) of cold ethanol. Draw air through the crystals for several minutes, then spread them out on a piece of dry filter paper to air dry until near the end of the laboratory period or until the next period. If the crystals are stored, place them in a loosely capped vial or cover the filter paper with an inverted beaker. Determine the weight and melting range (~ 134 to $136°C$) of the dry crystals. Calculate the mass recovery and record all your results in your notebook.

EXPERIMENT 6

Microscale Recrystallization of Benzoin

Using a Craig tube (Figure 2.4) and sand bath as the heating source, carefully recrystallize 50 mg (0.05 g) of benzoin. Note that you may need to filter twice through a heated filter pipette (Figure 2.1c) if you need to use decolorizing carbon. Isolate your product by centrifugation as described in Section 2.2.1 and in Figure 2.4. Allow the product to dry and record the yield and melting range. Compare this data with that obtained from the large-scale recrystallization.

EXPERIMENT 7

Recrystallization of Naphthalene
from a Two-Solvent System

Test tube experiments. Place a small amount of crude naphthalene (about the volume of a pea) in a test tube and add just enough acetone dropwise to dissolve the solid at room temperature. Heat the solution to boiling with a steam bath or preheated sand bath. Slowly add hot water dropwise until the solution turns cloudy. To the turbid boiling solution add just enough acetone (one or two drops) to clarify the solution. Cool the solution first with a water bath and then with an ice bath. Scratch the inside of the tube if necessary to induce crystallization. If the solution oils out, reheat it until it clears or boils, and once boiling add just enough acetone to effect clearing. Cool the solution again to induce crystallization. Estimate the approximate ratio of acetone to water used.

Preparative recrystallization.   Dissolve about 1 to 5 g (record the exact weight and melting range) of crude naphthalene in the *minimum* amount (5 to 30 mL, depending on the amount of naphthalene used) of warm acetone (steam bath or sand bath, *use no flames since acetone is very flammable*).   Heat the solution to boiling and slowly add hot water portionwise with swirling until a slight cloudiness persists.   Add acetone dropwise to clarify the solution.   Remove the solution from the heat source and allow it to air cool slowly with scratching or seeding if required to initiate crystallization.   After the solution has cooled to room temperature, cool it in an ice bath.   Suction filter the crystals and wash them with a few mL of a cold mixture of acetone and water (~ 1 : 1).   Dry the crystals.   Determine the weight of the crystals and their melting point.   Store the pure product in a labeled and tared vial.   Record all the data in your notebook.

EXPERIMENT 8

Microscale Recrystallization of Naphthalene

Using a Craig tube (Figure 2.4), carefully recrystallize 50 mg of naphthalene using the acetone-water solvent system.   Isolate the product by centrifugation.   Dry the product and record the yield and melting point.   Compare the data with that obtained from the large-scale recrystallization.

EXPERIMENT 9

Recrystallization of an Unknown

Obtain an impure sample of a solid unknown compound from your instructor. Record the sample number or code name.   Determine the weight of the sample and its melting point.   Determine a suitable solvent or solvent mixture for recrystallization of your unknown.   Common and useful solvents include 95% ethanol, methanol, ethyl acetate, acetone, pentanes, hexanes, toluene, and water.   Note that recrystallizations from pure water require heating with a flame to achieve boiling.

 C A U T I O N :   Be extremely careful that no flammable solvents are nearby before using a flame, and *never* heat a flammable solvent in a test tube, beaker, or other open flask with a flame.

Recrystallize your entire sample (you might want to do this in two portions) using an appropriately sized apparatus and technique.   Note that not all unknowns will require the use of decolorizing carbon or hot filtration.   After you have obtained crystals, dry and weigh them.   Calculate and record the mass recovery.   Determine the melting point of the pure material and compare it with the melting point of the starting material. Record all the data in your notebook. Store the pure product in a labeled and tared vial.

Questions for Experiments 4 to 9

1. What properties are necessary and desirable for a single solvent in order that it be well suited for recrystallizing a particular organic compound?

2. Which of the following mixtures could not be used for two-solvent systems and why?
   (a) acetone-water   **(b)** dichloromethane-water   **(c)** ethanol-water
   (d) ether-water   **(e)** hexane-water   **(f)** toluene-water

3. Suppose that the naphthalene in Experiment 7 or 8 contained colored impurities. How would you modify the recrystallization procedure to provide easy and efficient decolorizing of the sample?

4. Why is suction filtration preferable to ordinary gravity filtration for separating purified crystals from the mother liquor? Why should the vacuum be released at the apparatus before turning off the aspirator?

10.5 DISTILLATION AND GAS CHROMATOGRAPHIC ANALYSIS

In this section you will perform simple, fractional, and microscale distillations. You will also monitor the progress and efficiency of distillations by gas chromatography (gc). The material in Section 2.3 should be reviewed before beginning any of the experiments.

EXPERIMENT 10

Simple Distillation

Place 25 mL of methanol and 25 mL of water[3] in an appropriately sized round-bottomed flask and add a boiling chip. Attach a distillation head and a condenser. Place a thermometer in the distillation head so that the mercury bulb is slightly below the joint leading to the condenser. Heat the liquid to the boiling point with an appropriate heat source (see Section 1.7.1 and avoid the use of flames, since methanol is flammable). Collect the distillate in a graduated cylinder. Separately collect single drops of the distillate periodically for later analysis by gc (see Experiment 15). The first drop and then single drops at about every 10 mL of distillate will be sufficient. Record the temperature as a function of the volume of distillate collected at 2-mL

---

[3] Alternatively, a 50% mole mixture of acetone and 1,1,1-trichloroethane (~ 30 mL of acetone and 40 mL of 1,1,1-trichloroethane) works well. However, since many halogenated materials are toxic, this distillation should only be done in a well-ventilated hood.

intervals. A good rate of distillation will result in the collection of one drop of distillate every 2 to 5 seconds. Construct a graph of temperature versus volume for the simple distillation.

EXPERIMENT 11

Fractional Distillation

Place 25 mL of methanol and 25 mL of water in a round-bottomed flask and add a boiling chip. Place a fractionating column (Figure 2.8) between the flask and the distillation head. Again record the boiling point as a function of volume of distillate collected at 2-mL intervals and periodically save drops of distillate for gc analysis (see Experiment 15). Construct a graph of temperature versus volume and compare it to that obtained from the simple distillation.

EXPERIMENT 12

Distillation of an Unknown

Obtain a 30- to 50-mL sample of an unknown binary mixture from your instructor. Record the unknown number or code name. Save a few mL of the unknown for use later in a semimicro fractional distillation (see Experiment 14). Fractionally distill the rest of the unknown. Record the boiling point as a function of the volume collected. Save early and late samples for gc analyses.

EXPERIMENT 13

Microscale Distillation

Practice a microscale distillation with a Hickman still (Figure 2.9) using a common organic solvent. Methanol or ethyl acetate works well. Then obtain a single component unknown (100 to 500 μL or more depending on the size of the Hickman still head) from your instructor. Note that the unknown may be contaminated with a number of colored, relatively nonvolatile impurities. Record the number or code name of the unknown. Carefully distill it from a small round-bottomed flask, or preferably from a conically shaped reaction vial fitted with a small stir vane and the Hickman still. A sand bath is a convenient heat source. A thermometer may be suspended in the Hickman still near the point of vapor condensation to obtain an approximate boiling point of the liquid.

EXPERIMENT 14

Fractional Semimicroscale Distillation

Place 2 to 3 mL of your original unknown from Experiment 12 into a 5-mL conical vial fitted with a stir vane and distill approximately one-half of the solution using a Hickman still head. Collect the distillate with a syringe. Save a small sample (appropriately labeled) for later gc analysis. Transfer the remaining distillate to a new 2- to 3-mL vial and distill it again using a Hickman still head. Save the appropriately labeled second distillate for gc analysis.

EXPERIMENT 15

Gas Chromatographic Analysis of Distillation Fractions

Practice your gc (see Section 2.6.7) and syringe techniques with a known liquid. Routine laboratory solvents, including methanol or ethyl acetate, work well for this practice. Recall that air or gas bubbles trapped in a syringe can often be eliminated by repetitive "pumping" of the syringe barrel while the tip of the needle is immersed in the liquid. Care should be taken not to bend the syringe needle or barrel during use. The glass barrel of the syringe should not be allowed to touch the hot injection port of the gc while an injection is being made through the septum, since the glass may fracture.

Analyze a sample of your original unknown (Experiment 12) by gc, using an appropriate column and conditions (see Section 2.6.7). Note how easily this technique can separate the components of the unknown. Analyze the fractions previously set aside from the simple and fractional distillations of your unknown. Use the results to compare the efficiency of simple and fractional distillation. Similarly analyze the first and second distillates from the fractional semimicro distillation. Compare all your results.

Questions for Experiments 10 to 15

1. Define the term *boiling point.* How does a reduction of the external pressure influence the boiling point?
2. What is an azeotrope?
3. What is the advantage of a fractional distillation over a simple distillation?
4. Why is gas chromatography so efficient relative to the distillations described?
5. Describe the components of a gas chromatograph and indicate the purpose of each.

10.6 *EXTRACTION*

Separation of organic products from inorganic impurities and other components after an organic reaction can quite often be done by extraction (see Section 2.5). Extractions can be performed on any scale, micro to industrial, depending on the need and the availability of appropriate immiscible solvents. This section will describe experiments that employ common extraction techniques used in teaching and research laboratories.

 C A U T I O N : Many extractions require the use of strong acid and base solutions. Avoid contact with skin and clothing. Wear rubber gloves and frequently vent the separatory funnel as described in the following sections.

EXPERIMENT 16

Preparation of a Standard Acid Solution

Make a stock acid solution by dissolving 5.0 mL of glacial acetic acid in 100 mL of distilled water. (Other acids, such as propionic acid, can also be used, for a comparison of the distribution changes based on structure.) Using phenolphthalein solution (two drops) as an indicator, titrate a 5-mL aliquot with standard sodium hydroxide solution (0.10 to 0.15N solutions work best) to determine the exact acetic acid concentration.

EXPERIMENT 17

Simple Extraction: Determination of Distribution Coefficients

Place 15 mL of the acetic acid solution and 15 mL of diethyl ether in a small separatory funnel. Hold the separatory funnel carefully with one hand on the stopcock and the other on the stopper area. Invert the funnel carefully and swirl it. Tip the stopcock end up and vent the funnel by opening the stopcock. Close the stopcock and shake the funnel more vigorously. Vent again. Mount the separatory funnel on a ring stand and open the top. Separate the layers by draining the aqueous layer out the bottom stopcock. Measure the new volume of the aqueous phase. Determine the new concentration of the acetic acid in the aqueous phase by titrating a 10-mL aliquot with standard sodium hydroxide solution. Calculate the concentration of acetic acid in the aqueous phase, the amount extracted into the ether phase, and use the results to determine the distribution coefficient as described in Eq. (2.4) (see Section 2.5).

EXPERIMENT 18

Multiple Extractions

Place another 15 mL of the acetic acid stock solution and 15 mL of diethyl ether into a separatory funnel and shake with venting. Separate the layers and discard the ether layer. Extract the aqueous phase with a second 15-mL portion of ether and again discard the ether. Do a third extraction with 15 mL of ether. Withdraw the aqueous layer and titrate the entire solution with standard sodium hydroxide solution. Calculate the amount of acetic acid remaining in the aqueous phase and the amount extracted into the ether phase. Express the results in terms of percentages and compare the simple extraction with the multiple extraction.

Repeat the multiple extractions with dichloromethane (density = 1.33 g/mL) and with hexanes (density < 1). Compare the distribution results for the different solvents.

EXPERIMENT 19

Microscale Extraction and Determination of Distribution Coefficient

Carefully weigh 100 to 500 mg of benzoic acid into a tared vial. Record the exact weight. With a pipette, add 2 mL of dichloromethane and 2 mL of water. Swirl or shake the vial until all the benzoic acid has dissolved. Carefully separate the two layers by removing the top aqueous layer with a disposable pipette. Evaporate the remaining dichloromethane layer directly in the vial using a rotary evaporator if the vial is fitted with a ground-glass joint, or by warming in a water bath and blowing a gentle stream of nitrogen into the vial. If neither a rotary evaporator nor a nitrogen source is available, place the tip of a disposable pipette just above the surface of the solution and attach it with tubing through a trap previously attached to an aspirator. Turn on the aspirator to aspirate air past the solution to aid evaporation. (Evaporation with a stream of air from compressed air in most laboratories should be avoided since many pipes in air lines contain traces of oil that will contaminate the extraction sample unless the air is appropriately filtered.) *(CAUTION: This evaporation should be done in a fume hood.)* After the solvent has been completely removed, weigh the vial containing the residual benzoic acid. Determine the partition coefficient of benzoic acid between water and dichloromethane.

Repeat the reaction, but replace the water with two 2-mL portions of 10% KOH solution. In this case, the benzoic acid will react with the KOH to form potassium benzoate that distributes exclusively into the aqueous phase. Little or no benzoic acid should be recovered in the organic phase.

EXPERIMENT 20

Separation of Unknown Acids and Bases by Extraction

Obtain a sample of an unknown from your instructor. The unknown consists of a binary mixture of approximately equal weights of crude acetanilide (a neutral component) and *either* m-nitroaniline (a basic component) or benzoic acid. Weigh the sample accurately before you begin. Determine whether the sample contains the acid or basic component by thoroughly mixing a small amount ($\sim$ 50 to 200 mg, or the amount that will fit on the end of a spatula) with approximately the same amount of sodium hydrogen carbonate (sodium bicarbonate) in a watch glass or small test tube. Add a few drops of water, stir, and note if there is any effervescence to indicate the reaction of benzoic acid with the bicarbonate to form carbon dioxide, a gas, and sodium benzoate. A positive test can be confirmed by acidifying the test mixture with a few drops of 1 M HCl to precipitate benzoic acid, which is only slightly soluble in cold water. If benzoic acid is present, proceed with Part A. If benzoic acid is apparently not present, go on to Part B.

NHCOCH$_3$ CO$_2$H   NH$_2$

O$_2$N

acetanilide   benzoic acid   m-nitroaniline

Part A. Calculate *two* times the theoretical volume of 1.5N sodium hydroxide to convert the benzoic acid (half the weight of the sample) to sodium benzoate. Dissolve the sample in 60 mL or less or dichloromethane (dichloromethane, density > 1). If insoluble impurities are present, filter the solution. Transfer the solution to a 125-mL separatory funnel. Extract the solution twice, each time using half the volume of the 1.5N NaOH calculated above. *Remember to hold the separatory funnel with both hands and to vent it frequently with the lower end pointed upward and away from other persons.* Separate and save the combined aqueous layers and the organic layer.

Extract the aqueous portion again with two 15-mL portions of dichloromethane to ensure that any neutral compound in the water layer will be completely removed (this is referred to as a back extraction). Combine *all* the dichloromethane layers and predry the solution by extracting it once with 10 to 15 mL of saturated aqueous sodium chloride solution. Dry the organic layer further over anhydrous sodium sulfate, using just enough so that upon swirling some of the drying agent is easily suspended in the solution (somewhat like snowflakes) and does not remain clumped in the solution. After standing for about 1 minute, the dichloromethane solution should be clear. If not, add more anhydrous sodium sulfate and swirl again. Gravity filter the dry dichloromethane solution containing the acetanilide into a tared, round-bottomed flask and evaporate the solvent on a rotary evaporator. Obtain the weight and melting point of the crude product. If necessary, recrystallize the acetanilide as described in the section on recrystallization.

Place the aqueous solution containing the sodium benzoate in a beaker or an Erlenmeyer flask and cool it in an ice bath. Acidify the sodium benzoate solution carefully by adding a slight excess of concentrated ($\sim 12N$) hydrochloric acid. Check the resulting suspension with pH paper to make certain that it is acidic. Cool the suspension for a few more minutes and then collect the benzoic acid by suction filtration on a Büchner funnel. Wash the product with a few mL of *cold* water. Air dry the benzoic acid and determine its weight and melting point. If necessary, the benzoic acid may be crystallized from water using about 10 to 15 mL of boiling water per gram of sample.

Part B. Calculate 20 times the theoretical volume of $6N$ hydrochloric acid to convert the *m*-nitroaniline (half the weight of the original sample) to its hydrochloride salt. (The excess acid ensures complete protonation of the amine and thereby avoids any equilibrium problems.) Dissolve the sample in 60 mL of dichloromethane. Gravity filter the solution if insoluble impurities are present. Transfer the solution to a separatory funnel. Extract the solution *three* times, each time using one-third of the volume of the $6N$ HCl calculated above. (*Note:* Wear rubber gloves.) The hydrochloride salt of the *m*-nitroaniline may precipitate from the aqueous layer. If so, add small amounts of water portionwise to redissolve the precipitate. Remember that the dichloromethane will be the lower layer and that you should vent the funnel frequently. Separate and save the combined aqueous layers *and* the organic layer. Back extract the combined aqueous layers to ensure that any neutral compound in the water layer will be completely removed. Combine *all* dichloromethane layers. Dry the dichloromethane layer first by extracting it with 10 to 15 mL of saturated aqueous sodium chloride solution and then by the addition of anhydrous sodium sulfate. Swirl the resulting suspension and check for the "snowstorm effect." Allow the suspension to settle for 1 minute and check if the dichloromethane solution is clear. If it is not clear, add more drying agent portionwise as necessary. Gravity filter the dry suspension into a tared, round-bottomed flask and evaporate the solvent on a rotary evaporator. Obtain the weight and melting point of the crude product. If necessary, recrystallize the acetanilide as described earlier.

Place the aqueous solution containing the *m*-nitroaniline hydrochloride in a beaker or an Erlenmeyer flask and cool the container in an ice bath. Basify the cold *m*-nitroaniline hydrochloride solution by slowly and carefully adding an excess of a concentrated aqueous solution of sodium hydroxide. (Prepare the sodium hydroxide solution by dissolving a 10 to 20% excess, relative to that needed to neutralize the acid of solid NaOH in 15 to 20 mL of water and cool the solution in an ice bath.) Check the solution with pH paper to be certain that it is basic. Cool the resulting suspension and collect the precipitated *m*-nitroaniline by suction filtration on a Büchner funnel. Wash the solid on the funnel with a few mL of *cold* water. Air dry the sample and determine its weight and melting point. If necessary, recrystallize the sample using about 40 to 60 mL of hot water. Determine and record the weight, yield, and melting point of the purified *m*-nitroaniline.

Extraction Flow Sheet

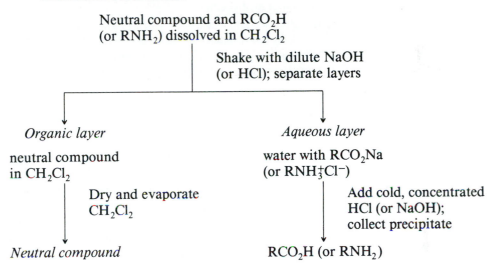

EXPERIMENT 21

Extraction of a Three-Component Mixture Containing a Neutral, Acidic, and Basic Component

Combination of the procedures described above can also be used to isolate each of the components of a ternary mixture containing neutral, acidic, and basic compounds. A particularly nice example[4] employs 1 g each of colored compounds: azobenzene (neutral, orange), benzoic acid (acid, white), and *m*-nitroaniline (base, yellow). Describe what you would do and see during the separation of this mixture.

Questions for Experiments 16 to 21

1. Does acetic acid or propionic acid partition more effectively into ether when extracted from water? Rationalize your answer.

---

[4] G. D. Hobbs and J. D. Woodyard, *J. Chem. Ed.,* **59,** 386 (1982).

2. The solubility of caffeine in water is 2.5 g/100 mL at room temperature and 45 g/100 mL at 60°C. The solubility in chloroform is 20 g/100 mL at room temperature and 90 g/100 mL at 60°C. At what temperature will the larger fraction of caffeine be extracted into the chloroform phase? Show using calculations.

3. Can ethanol be used to extract acetic acid from water? Why or why not?

4. Suppose an aqueous reaction product mixture containing 30 g of malononitrile, $CH_2(CN)_2$, was diluted to 300 mL with water. If the solubility of malononitrile at room temperature is 100 mg/mL in water and 200 mg/mL in diethyl ether, what weight of malononitrile would be recovered (i.e., removed from the aqueous solution) by extraction at room temperature with (a) three 100-mL portions of diethyl ether and (b) one 300-mL portion of diethyl ether?

5. If one is uncertain which of the two layers in a separatory funnel is the aqueous layer, what is an easy test to determine which is which without looking up the density of the solvent?

6. Why should you remove the top stopper from the separatory funnel before opening the bottom stopcock to drain out the bottom layer?

10.7   CHROMATOGRAPHY

Paper chromatography is a type of partition chromatography that is especially adapted to the separation of ultra-small quantities of materials. The cellulose of filter paper contains absorbed water, which is the stationary phase. The sample is applied as a spot to the filter paper, and the developing solvent is allowed to pass through the filter paper by capillary action. As it does, the components of the sample move at different rates through the paper. The different rates are proportional to the partition coefficients of the components between the absorbed water and the solvent used for elution.

Paper chromatography has been a very valuable tool in biochemistry, especially for separating mixtures of amino acids derived from hydrolysis of proteins. In the present experiment, paper chromatography will be used to illustrate the separation of some synthetic food dyes and to determine whether a particular dye is homogeneous.

EXPERIMENT 22

Paper Chromatography of a Food Dye

Fold an 18.5-cm sheet of Whatman No. 1 filter paper approximately 2 to 3 cm from the bottom and cut the lower section to give a flat bottom. Repeat on the top end of the paper. Now cut up from the ends of the flat bottom to give a rectangular sheet for your chromatogram. During the handling of the paper, try to avoid getting finger-prints on it. Draw a horizontal thin pencil line 1 to 2 cm from the bottom of the paper and place "x"s about 1.5 to 2 cm apart along the line. Label them Y, B, G, R, and M (representing the four food dyes — yellow, blue, green, red — and a mixture).

Using a separate drawn-out melting point capillary for each sample, place a small spot (<0.1 cm) of yellow food dye (diluted ~1 : 10 with water) at position Y; and at B, G, R, and M, place similar spots of diluted blue, green, red, and the mixture of food dyes. Allow the spots to dry thoroughly or they will appear as smears at the end of the experiment.

In a jar or 500-mL Erlenmeyer flask, place enough solvent (n-propyl alcohol-water, 2 : 1 volume/volume) to just cover the bottom. Position the filter paper in the flask so that the paper stands vertically with the spots towards the bottom. If the paper does not stand, it can be hung from the top of the chamber with a paperclip and wire. Cover the flask and allow the solvent to rise up the paper. This may take 2 to 3 hours, so the experiment should be run to overlap with another. When the solvent has risen almost to the top of the paper, remove the paper and mark the upper limit of the solvent front. Allow the paper to dry and calculate the R_f values (see Section 2.6.4) of the spots.

Questions for Experiment 22

1. What factors determine the rate at which the components of the samples move up the paper? In other words, why is the R_f of the yellow dye different from the R_f of the blue dye?

2. Why are the spots generally larger on the finished chromatogram than at the origin?

3. Which dyes are homogeneous?

4. More than one compound can have a given R_f in a particular chromatography system. Can you really be 100% certain, then, that any of the dyes are homogeneous? If two samples on the same paper chromatogram give different R_f values, can they be composed of the same compound?

Thin-layer chromatography is also a very useful technique for affecting the separation of ultra-small quantities of materials. The following experiment will demonstrate the use of this technique, along with the use of micro column chromatographic techniques.

EXPERIMENT 23

Separation of the Components of Black Ink

Separation of the components of black inks by TLC. Obtain silica gel thin-layer chromatography plates from your instructor and *lightly* draw a pencil line about ¼ in. above the bottom so as not to disturb the adsorbent. This line will correspond to the "origin." (Pencil must be used since the dyes in an ink pen will be soluble in solvents to be used to develop the plates.) The plates should be a minimum of 2 cm by 7 cm. Spot the plates with a variety of ink samples according to the following directions.

Several different inks can be spotted on a given plate if the spots are light and placed at least $\frac{1}{2}$ cm apart. If a felt-tipped pen is available, lightly touch the tip to the line previously drawn on the silica gel plate. If more than one variety of felt-tipped pen is available, try spotting each side by side (again, at least $\frac{1}{2}$ cm apart). Ink from ballpoint pens or other sources can also be used.

Extract some of the ink from the pen cartridge with a syringe and place about 20 to 100 μL or a small drop in a small test tube or vial containing 1 mL of 95% ethanol. Alternatively, try to dissolve some ink in ethanol by immersing the tip of the pen in about 1 mL of ethanol in a small test tube or vial. Draw out a fine capillary from a melting point tube as described earlier (see Section 2.6.4). Insert the tip of the capillary in the ethanolic solution of the ink sample and then *lightly* touch the capillary to the premade line on the silica gel plate. Prepare the developing chamber (a jar with a cap will do) by cutting a piece of filter paper to line about one-half to two-thirds of the inside of the jar. Place about $\frac{1}{8}$ in. of developing solvent in the chamber. Cap the chamber and shake it to saturate the filter paper and the atmosphere in the chamber with the developing solvent.

The best solvent for separating the individual dye components of the inks used can be determined by trial and error. Most effective solvents will be quite polar and consist of mixtures of ethyl acetate, ethanol, or other alcohols and water in various ratios. Extensive studies have been made to separate the dyes of some common inks. A $1:1:1:1:1$ (volume/volume) mixture of ethyl acetate, sec-butyl alcohol, n-propyl alcohol, absolute ethyl alcohol, and water or a $60:10:20:0.5$ (volume/volume) mixture of n-butyl alcohol, absolute ethyl alcohol, water, and acetic acid works well for most inks.[5]

Carefully place the TLC plate in the developing chamber so that you can see the surface of the plate through the side of the chamber and replace the cap. Allow the solvent to move up the plate by capillary action. Note the process of separation of the colored dyes and remove the plate from the chamber just before the solvent front reaches the top of the plate. Allow the plate to dry in the hood and measure the R_f values of the component dyes. Which inks do you think have one or more common dyes? Record your results in your notebook.

In most cases, compounds that are separated by thin-layer or paper chromatography are not visible, but must be visualized after the separation has been completed. Section 2.6.4 includes appropriate procedures. Some of these visualizations will be used in subsequent experiments.

Separation of black inks by micro column chromatography. Column chromatography can also be practiced with the ink samples. Carefully score a disposable pipette with a file about 1 to 2 in. below the constriction of the pipette and carefully break off the long bottom portion of the pipette. Place a small amount of glass wool or cotton in the pipette from the top so that it is gently wedged in the constricted portion. Pour enough silica gel into the pipette to give a column about 2 in. high. Be careful not to

---

[5] B. Olesen and D. Hopson, *J. Chem. Ed.,* **60,** 232 (1983).

breathe the silica "dust." Based on the TLC experiment in the previous section, layer the same, or a somewhat less polar solvent, on top of the silica gel column and either let it percolate through the silica or gently pressurize the top of the column with a pipette bulb ("micro flash chromatography"). Once the solvent has saturated the entire column, allow the solvent to drain through the column until the solvent level nearly reaches the top of the silica gel. At this point, your column should look like a "micro" version of that shown in Figure 2.15.

As in normal column chromatography (see Section 2.6.2) never let the solvent level go below the top of the adsorbent in the column. If a pipette bulb was used to push the solvent through the silica gel, gently remove it by pushing it sideways. If it is simply pulled from the top of the pipette, the column bed will be disturbed and it will be ineffective for the subsequent separation. Apply a few μL (no more, as the column can be easily overloaded) of the ethanolic solution of the black ink prepared previously to the top of the column. Allow the ink solution to drain into the silica and rinse it in with a few more μL of ethanol or the eluting solvent. Repeat the rinse procedure until all the ink is adsorbed onto the top of the column. Apply the eluting solvent to the top of the column and allow the elution to begin by gravity flow, or under the *gentle* pressure of a pipette bulb applied to the top of the column. Ideally the components of the ink should begin to separate into distinct bands as shown in Figure 2.16. However, depending on the brand of ink and eluent, the separations are usually not so distinct. Instead, a "rainbow" pattern of colors is often observed in the column. Add more solvent to the top of the column as necessary. Elute the various dyes and collect the dye fractions in separate test tubes. To further confirm that the fractions are homogeneous (contain only one dye), analyze by TLC aliquots of the individual fractions.

Several other interesting TLC experiments have been reported.[6,7]

10.8 CHEMISTRY OF ALKANES AND ALKENES

Alkanes are aliphatic hydrocarbons that only contain single σ bonds. Alkenes, or olefins, are hydrocarbons that also contain double π bonds. The greater chemical reactivity of the double bonds in alkenes can be used to distinguish alkenes from alkanes (review Section 9.12). Even when both classes of compounds can react with the same chemical reagents, the processes (or mechanisms) and products are often different.

---

[6] For an interesting study on the separation of a series of phenols see M. J. Kurth, *J. Chem. Ed.,* **63,** 360 (1986).

[7] For a description of TLC separation and identification of the analgesics aspirin, phenacetin, acetaminophen, and the stimulant caffeine, see V. T. Lieu, *J. Chem. Ed.,* **48,** 479 (1971).

Alkanes and alkenes react with halogens such as bromine by the processes shown below. In the presence of a free-radical initiator such as light or peroxides, alkanes react by a radical chain process to give alkyl halides. The net result is a *substitution* of a halide for a hydrogen on the starting alkane. In contrast, alkenes undergo a net electrophilic *addition* of halogens to the double bond by way of a polar mechanism without the need for an initiator. In either case, the characteristic red/brown color of bromine dissipates during the reactions. This color change serves as the visual test for the reactions. Other functional groups may interfere with the visual tests. For example, carbonyl groups (ketones, aldehydes, and esters) with α-protons can enolize and react with bromine to give the α-bromo carbonyl compound. Since this is also a net substitution reaction, HBr is liberated and can be detected. In contrast to the reactions of alkanes with Br_2, which also liberate HBr, no initiator is required for the reaction of α-hydrogen carbonyl compounds.

Free-radical alkane bromination

$$Br_2 + Initiator \longrightarrow 2Br\cdot$$

$$Br\cdot + R_3C{-}H \longrightarrow R_3C\cdot + HBr$$

$$R_3C\cdot + Br_2 \longrightarrow R_3C{-}Br + Br\cdot$$

Bromination of alkenes

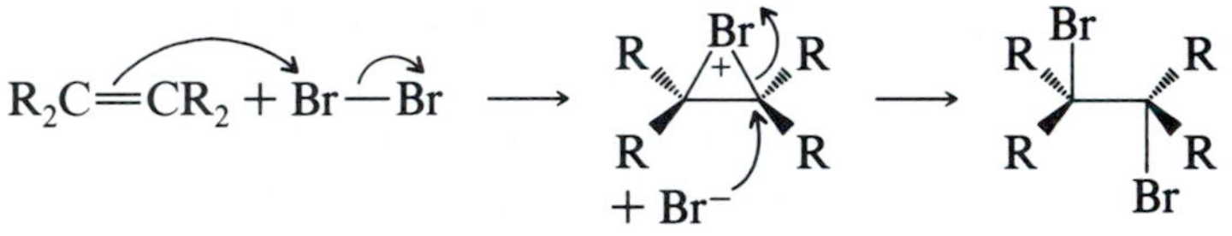

Bromination of ketones

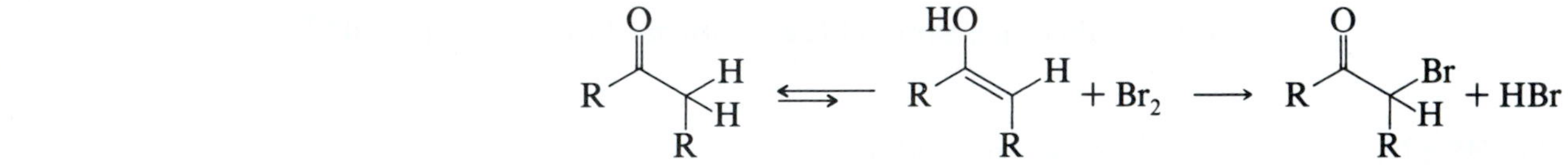

EXPERIMENT 24

Reaction of Alkanes and Alkenes with Bromine Water

Add ~ 1.5 mL of a 3% aqueous bromine solution (*CAUTION: Bromine reacts with skin*) to each of three test tubes. Add 0.5 mL of hexanes (petroleum ether or Skelly-solve) to two of the tubes and add 0.5 mL of cyclohexene to the third. Stopper the tubes lightly. Shake the tubes and record your initial observations (any color changes in any of the tubes). Place one of the tubes containing the hexanes in a dark place or wrap it with aluminum foil so it will not be exposed to any light. Meanwhile, expose the other tube containing hexanes to sunlight, a sun lamp, or a light bulb. After an hour or so compare the two tubes containing the hexanes. Remove the stoppers from all the tubes and separately hold a piece of moist pH paper over the top of the tubes. Record all of your observations in your notebook. Interpret your observations mechanistically based on the equations shown above.

EXPERIMENT 25

Reaction of Alkanes and Alkenes with Bromine in Nonaqueous Solutions

In the hood place 0.5 mL of hexanes or pentanes and 0.5 mL of cyclohexene in separate test tubes. Add 2 to 3 drops of a 3% solution of bromine in carbon tetrachloride to each tube *(CAUTION: Both bromine and carbon tetrachloride are toxic)*. Note the persistence or disappearance of the bromine color immediately after the additions. Hold a moist piece of pH paper over the tops of the tubes to check for the evolution of HBr. Record all your observations and interpret the results mechanistically.

EXPERIMENT 26

Reaction of Alkenes with Pyridinium Hydrobromide Perbromide

pyridinium bromide perbromide

Pyridinium hydrobromide perbromide is a commercially available, solid, odorless complex that serves as a source of molecular bromine (Br_2) in the presence of an alkene or other electrophilic substrate even though it is relatively insoluble in most organic solvents. The complex is reasonably stable upon storage and is much more easily handled than bromine itself.

Suspend 320 mg (1 mmol) of pyridinium bromide perbromide in 2 mL of acetic acid *(CAUTION: The reagent causes severe burns to the skin)* in a 10-mL Erlenmeyer flask or a 5-mL vial. Swirl or magnetically stir the mixture and add cyclohexene dropwise with a calibrated syringe or pipette until the reagent appears to be consumed. Note the exact amount of cyclohexene needed and calculate the number of mg of cyclohexene (molecular weight 82) used. Dilute the solution with water. Can you distinguish the product from the starting material?

Many other electrophilic reagents, including hypohalous acids (HOX, where X = Cl or Br) and sulfuric acid, can also add to alkenes. Concentrated sulfuric acid is strong enough to protonate alkene double bonds to give alkyl hydrogen sulfates, but cannot react with an alkane. Alkenes can also be oxidized to diols with permanganate and other oxidizing agents. Some of these reactions will be demonstrated in the following experiments.

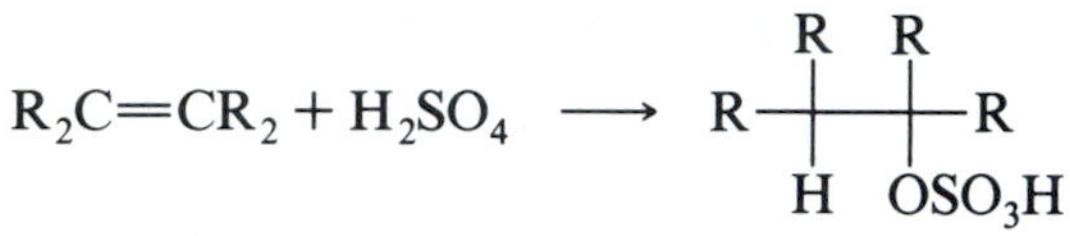

$$R_2C{=}CR_2 + H_2SO_4 \longrightarrow R\underset{H}{\overset{R}{-}}{\underset{OSO_3H}{\overset{R}{-}}}R$$

EXPERIMENT 27

Reaction of Alkanes and Alkenes with Sulfuric Acid

In separate test tubes or vials, add 0.5 to 1 mL of hexanes and cyclohexene. Cool the samples in an ice bath and add 1.5 to 3 mL of concentrated sulfuric acid *(CAUTION: Sulfuric acid causes severe burns to the skin)*. Stir or shake the samples. Note which sample undergoes an exothermic reaction by carefully touching the side of the containers. One of the mixtures should become homogeneous, while the other should continue to exist as two layers. Explain the results.

N-Bromosuccinimide (NBS) reacts with water to provide a convenient source of hypobromous acid (HOBr). The other product, succinimide, is water soluble. Subsequent reaction of the solution containing HOBr with alkenes produces a bromohydrin. Two experiments will illustrate the sequence. The first is the reaction with cyclohexene and the second with a solid, less-volatile alkene, 3-sulfolene (mp 65 to 66°C).

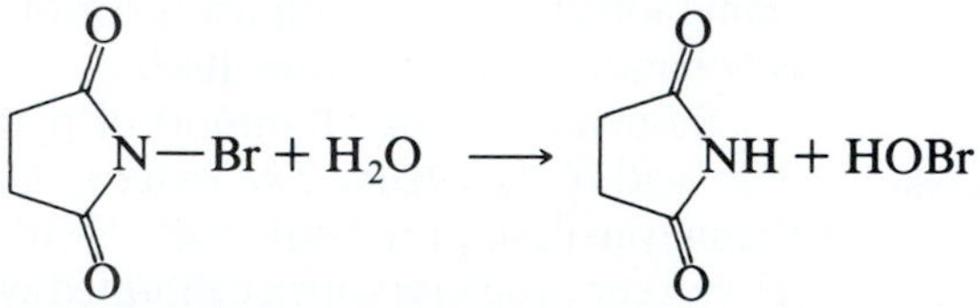

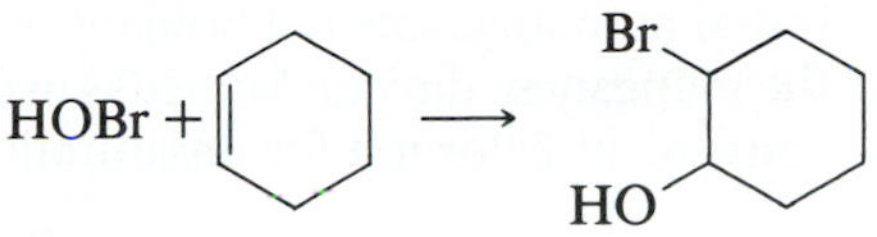

$$HOBr + \text{[cyclohexene]} \longrightarrow \text{[2-bromocyclohexanol]}$$

EXPERIMENT 28

Reaction of Alkenes
with Hypobromous Acid

Dissolve 178 mg (1 mmol) of N-bromosuccinimide in 0.5 mL of *p*-dioxane *(CAU-TION: See Table 1.1)* or tetrahydrofuran (THF) in a test tube or vial. Add a solution of 1 mmol of cyclohexene (determined from the experiment with pyridinium bromide perbromide, or by first checking the density of cyclohexene and determining the volume corresponding to 1 mmol in 0.5 mL of dioxane or THF. In a separate tube or vial, add ~0.05 mL (50 μL, ~1 drop) of concentrated sulfuric acid to 0.2 mL of cold water. Transfer the acidic aqueous solution to the first solution and observe the result. Dilute the resulting solution with water until the product separates.

Reaction with 3-sulfolene. Place about 1 g (record exact weight) of 3-sulfolene (2,5-dihydrothiophene-1,1-dioxide) in a 125-mL Erlenmeyer flask. Add 100 mol % (an equivalent number of moles) of solid N-bromosuccinimide (NBS) and 15 mL of water. Heat the mixture on a steam bath or a hot plate for 30 minutes while maintaining the temperature at 85 to 95°C. Swirl the flask occasionally. (While the reaction is proceeding, verify the presence of a carbon-carbon double bond in 3-sulfolene by the procedure described in Experiment 25.) Cool the flask in an ice bath and collect the precipitated crude product. The succinimide is soluble even in cold water. Air dry the product and determine its melting range. Pure product has a melting range of 190 to 191°C. If necessary, recrystallize the product from a minimum amount of water. Record the yield and melting point. What product would have been obtained if an alcohol (methanol or ethanol) had been used instead of water as the solvent?

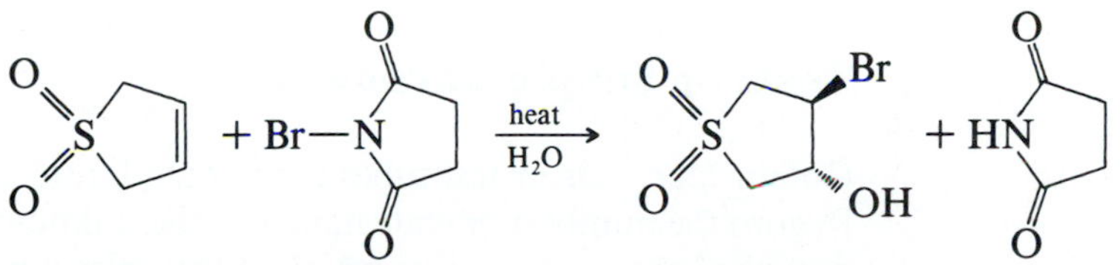

3-sulfolene   (NBS)   (plus mirror image)

Permanganate reacts with alkenes to form intermediate manganate esters that hydrolyze to the diols in water. During this process, the manganese is reduced. The initial permanganate is a brilliant purple, whereas after the reaction with the alkene the manganese dioxide formed is usually a brown precipitate. The overall result is a convenient color test for unsaturation (alkenes in this case).

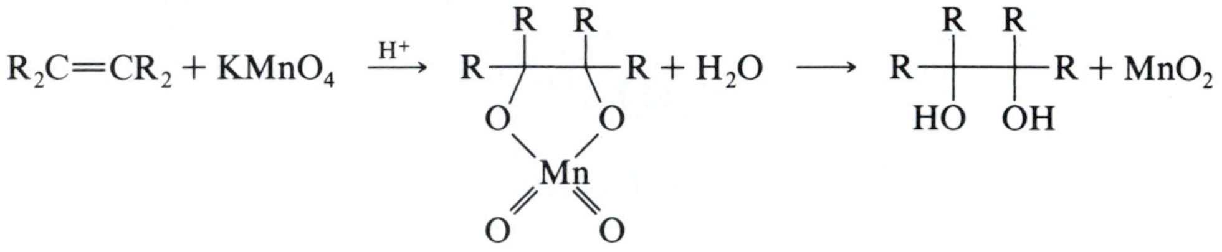

$$R_2C{=}CR_2 + KMnO_4 \xrightarrow{H^+} R\!-\!\underset{\substack{O \\ \diagdown \\ }}{\overset{R}{\underset{}{|}}}\!-\!\underset{\substack{O \\ }}{\overset{R}{\underset{}{|}}}\!-\!R + H_2O \longrightarrow R\!-\!\underset{HO}{\overset{R}{\underset{}{|}}}\!-\!\underset{OH}{\overset{R}{\underset{}{|}}}\!-\!R + MnO_2$$

EXPERIMENT 29

Reaction of an Alkene with Permanganate (Baeyer Test for Unsaturation)

A. Reaction of permanganate with cyclohexene.   Place 0.5 mL of hexanes and cyclohexene in separate test tubes or vials. Add a drop of a 1 to 2% aqueous potassium permanganate solution to each sample (see also Section 9.1.2). Shake or stir the reactions. If the solutions decolorize, add additional drops of the permanganate solution. Note the results and interpret them based on the generalized chemical reaction shown above.

B. Reaction of permanganate with 3-sulfolene.   Perform the permanganate test with sulfolene by dissolving 100 to 500 mg of 3-sulfolene in a minimum amount of hot water in a test tube or reaction vial. Add a drop of a 2% aqueous potassium permanganate solution to the sample. Shake or stir the reactions and note the results. Draw the structure of the organic product. How many possible stereo-isomers could be obtained?

EXPERIMENT 30

Tests on Simple Unknowns

Obtain four vials or test tubes containing liquid unknowns from your instructor.[8] Record the numbers or code names of the unknowns. Using the tests and reactions described above, determine which of the unknowns are alkanes, alkenes, or contain

---

[8] Instructor's note: Liquid samples of 0.1 to 1.0 mL are sufficient. Small amounts of the samples can be conveniently transferred to conical vials or small test tubes with a syringe for the tests. All syringes used must be clean and dry to avoid interfering results. Cyclohexene is a convenient alkene. Hexanes or cyclohexane serve well as alkanes and acetone can be used as a ketone to illustrate a bromine reaction without an initiator that produces HBr.

other functional groups. Clean, dry syringes work well for transferring small samples to be tested.

Cyclohexene and the other alkenes used in the previous experiments were all symmetrical. Thus, only one addition product could be obtained upon reaction with a given electrophilic reagent. Most alkenes are not symmetrical. Often they may have different substituents at the ends of the double bond. The hydration of 1-hexene is an example of a reaction with an unsymmetrical alkene. Isolation, distillation, and determination of the boiling point of the product allow assignment of the structure of the product and, thus, indicate the direction of addition of sulfuric acid across the unsymmetrical double bond to confirm Markovnikov's rule. Alternatively, the faces of the double bond may be sterically differentiated. This latter situation can be demonstrated by electrophilic addition reactions of bicyclic alkenes such as bicyclo[2.2.1]heptene (norbornene). In this experiment, norbornene will be treated with concentrated sulfuric acid to form a sulfate. This initial reaction can conceivably occur at either of the two faces (exo or endo) of the norbornene. Base hydrolysis converts the sulfate to the corresponding alcohol. The bicyclic alcohol can be isolated and purified by sublimation. The identity of the product alcohol can be determined by comparing its melting point with that of the known two isomeric norborneols.

EXPERIMENT 31

Hydration of 1-Hexene[9]

$$CH_3(CH_2)_3CH{=}CH_2 \xrightarrow{H_2SO_4} \left\{ \begin{array}{c} CH_3(CH_2)_3HC{-}CH_3 \\ | \\ OSO_3H \\ \text{or} \\ CH_3(CH_2)_3H_2C{-}CH_2 \\ | \\ OSO_3H \end{array} \right\} \xrightarrow[100°C]{H_2O} \begin{array}{c} CH_3(CH_2)_3HC{-}CH_3 \\ | \\ OH \\ \text{bp } 136°C \\ \text{or} \\ CH_3(CH_2)_3H_2C{-}CH_2 \\ | \\ OH \\ \text{bp } 156.5°C \end{array}$$

Carefully add 15.2 mL of concentrated (98%) sulfuric acid to an ice-cooled flask containing 5 mL of water (*CAUTION: Handle carefully and always add acid to water, not water to acid*). Transfer the resulting 85% sulfuric acid solution to a small separatory funnel and add 10 mL of 1-hexene. Shake the mixture with frequent venting of the separatory funnel until the mixture is homogeneous (~ 5 minutes). If

---

[9] J. R. McKee and J. M. Kauffman, *J. Chem. Ed.*, **59**, 695 (1982).

the separatory funnel becomes uncomfortably hot, remove the stopper and set the funnel aside for a minute or so and then continue to shake it. Add another 10 mL of 1-hexene (the total amount is then about 0.16 mol) and shake again until homogeneous. Allow the reaction to continue for another 5 minutes without shaking. Drain the mixture into a 250-mL round-bottomed flask containing about 50 mL of water. Rinse the separatory funnel with another 20 mL of water and add the rinse to the previous mixture. The resulting diluted mixture should form two layers. Attach a reflux condenser (Figure 1.8), add a boiling chip, and heat the mixture to boiling using an oil bath. Continue to boil the solution for 5 minutes to complete the hydrolysis of the sulfate to the alcohol. Cool the solution in an ice bath and transfer the solution to the separatory funnel. Separate the layers and save the upper organic layer (15 to 20mL). Wash the organic layer with 10.0 mL of 5% sodium hydroxide. Separate the layers and dry the organic layer over ~ 1 g of anhydrous sodium sulfate. Carefully decant the liquid into a 25-mL round-bottomed flask. Distill the liquid using a simple distillation setup. The early fractions (bp ~60 to 136°C) may be discarded. The product (~ 10 mL) should then distill. Determine the boiling point during the distillation or determine the ultramicroboiling point on a fraction obtained midway through the distillation.

Based on the boiling point of the product, determine the structure of the product and comment on the applicability of Markovnikov's rule. The structure of the product can also be verified by an iodoform test (see Section 9.5.4). This chemical test works well on secondary methyl alcohols, but not on primary alcohols (except ethanol). Perform the iodoform test on the hydration product and interpret the results.

EXPERIMENT 32

Microscale Hydration of 1-Hexene

Perform the hydration reaction of 1-hexene by reducing all the proportions by a factor of 20. Prepare the 85% sulfuric acid solution in a 5-mL conical reaction vial that has a ground-glass joint or other connection for eventual attachment to a microscale condenser, and add 0.5 mL of 1-hexene. Magnetically stir or shake the mixture vigorously until it becomes homogeneous. Add another 0.5 mL of 1-hexene and stir again until homogeneous. Allow the reaction to continue for another 5 minutes with slow or no stirring. Carefully add 2 to 2.5 mL of water. Very gently stir the nearly full vial. Attach a microscale condenser and while magnetically stirring, heat the mixture to boiling for about 5 minutes with a preheated sand bath. Cool the solution in an ice bath and carefully remove the top organic layer with a disposable pipette. Dry the product by passing the liquid through another pipette containing a 0.5- to 1-cm layer of anhydrous sodium sulfate supported by a plug of glass wool or cotton. The liquid should be drained directly into a tared (preweighed) 2-mL conical reaction vial. Use a pipette bulb to force the liquid that may be adsorbed on the drying agent through the column into the vial. Check to be sure that the liquid in the vial is clear and homogeneous. If it is turbid, add a small amount of sodium sulfate to the vial and stir the mixture for a few minutes. Then transfer the liquid to another clean dry reaction vial with the aid of a disposable pipette containing a small amount of glass wool or cotton in the tip (to filter any drying agent). Add a stir vane and attach a Hickman still to the vial and, while magnetically stirring, distill the product using a sand bath preheated to about 10 to 20°C higher temperature than the expected boiling point. Perform an ultramicro boiling point determination on the distillate. Perform an iodoform test on a portion of the distillate. Record your results and predict the structure of the hydration product of 1-hexene.

EXPERIMENT 33

Hydration of Norbornene

Norbornene is volatile (bp 96°C) and has a very unpleasant odor, so it should be handled in a hood. Obtain a solution of approximately 2.5 g of norbornene in about 7.5 mL of hexanes from your instructor. Under the hood, transfer the solution to a 50-mL Erlenmeyer flask. Immediately stopper the flask and cool it in an ice bath. Separately prepare a mixture of 4 mL of concentrated sulfuric acid and 2 mL of water *(CAUTION: Handle carefully and always add acid to water, not water to acid).* Cool the acid solution in the ice bath. Remove the stopper from the flask with the organic solution and add the acid mixture to it. Loosely restopper the flask and swirl it with brief ice cooling to control the initial exothermic reaction. Remove the flask from the ice bath and allow it to react at room temperature for 10 minutes with occasional swirling. Meanwhile, prepare a solution of 3 g of KOH in 15 mL of water. Cool both solutions in an ice bath and then add the base solution slowly to the reaction mixture. Swirl it for a few minutes to complete the hydrolysis of the sulfate.

Transfer the reaction mixture to a separatory funnel and extract it with two portions of ether (30 to 40 mL, then 10 mL). If three phases form during the attempted extraction, add a few drops of water. Combine the two ether layers and sequentially extract them with 5 mL of water and 10 mL of saturated sodium carbonate solution *(be careful of a pressure build-up from the evolution of carbon dioxide).* Pour the ether solution out the top of the separatory funnel into an Erlenmeyer flask containing a small amount of anhydrous magnesium sulfate. Filter the solution into a tared 125-mL suction filtration flask. This flask will also serve as a sublimation chamber (see Figure 10.1). (If desired, the sublimation technique can be practiced using camphor as described below.)

Carefully evaporate the ether under reduced pressure (aspirator vacuum, using a water trap!) while keeping the flask near room temperature with the aid of a water bath. Do not heat the flask much above room temperature or some product will be lost. (Note that using a water-aspirator vacuum at full water flow can effectively lower the boiling point of solvents by about 100°C.) Weigh the crude material by weighing the flask and subtracting its tared weight.

EXPERIMENT 34

Sublimation of Norbornanol

Insert a test tube through a rubber filter adapter or through a stopper bored with a hole through which the test tube will fit snugly. Insert the modified test tube into the suction filtration flask. Cool the center of the test tube (condenser) with a gentle stream of air through a disposable pipette or by placing crushed ice in the test tube. Heat the suction filtration flask containing the crude product gently with a steam bath or hot plate while applying a vacuum with an aspirator. If most of the product tends to sublime to the sides of the flask instead of the test tube, warm the sides gently. If a steam bath is used as the heat source, the flask can be partially lowered into the bath. If a hot plate is used, a warm-air hood can be created by loosely wrapping the flask with aluminum foil, keeping it $\sim\frac{1}{2}$ in. away from the flask at the bottom. After most of the product has sublimed onto the center tube, remove the

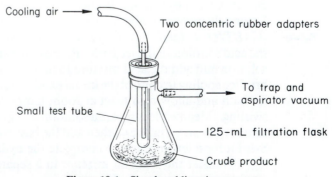

Figure 10.1 Simple sublimation apparatus.

flask from the heat source and gently release the vacuum (break the vacuum before turning off the aspirator to avoid water back-up into the trap and the sublimation flask). Remove the test tube from the flask and scrape the sublimed material into a flask or onto a piece of weighing paper. Determine the melting point on a small portion and store the remaining product in a well-sealed vial. Compare the melting range to the melting points listed above near the structures and determine the structure of the product. Account for the facial selectivity of the reaction (molecular models will be very helpful).

_____

Acid-catalyzed dehydration of alcohols is a common method for the preparation of alkenes. The dehydration process has several limitations. The reaction is an equilibrium, so the desired alkene can be rehydrated under the same conditions unless the product can be removed from the reaction mixture. Performing the reaction on an unsymmetrical alcohol results in the production of a mixture of alkenes in which the major product is usually the thermodynamically most stable alkene. In many cases rearrangements of the intermediate carbocations also occur. The two experiments that follow provide examples of alkene synthesis from a symmetrical alcohol (cyclohexanol) and a unsymmetrical alcohol (3,3-dimethyl-2-butanol).

Treatment of cyclohexanol with a strong acid establishes an equilibrium with the protonated alcohol. Loss of water produces the secondary carbocation intermediate that can either lose a proton to give the alkene, react with the conjugate base of the acid used, or with the starting alcohol to produce an ether. Removal of one of the components of this equilibrium mixture will shift the equilibrium. Cyclohexene is the lowest boiling component of the equilibrium mixture and can be distilled from the reaction as it is formed. Thus, cyclohexene can be formed in good yield.

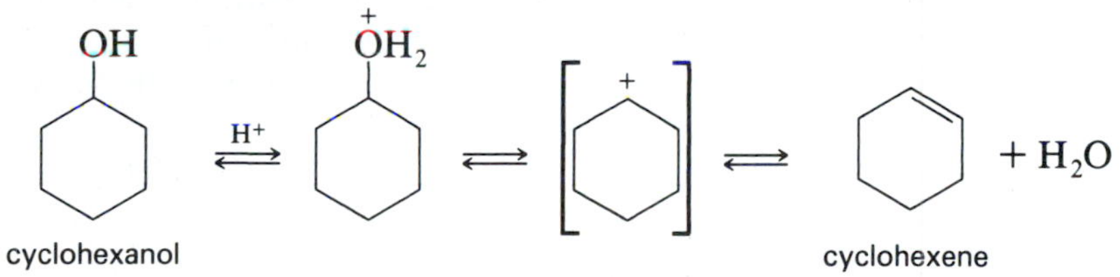

cyclohexanol   cyclohexene

EXPERIMENT 35

Preparation of Cyclohexene from Cyclohexanol

Place 10 to 20 mL of cyclohexanol (bp 161°C, mol. wt. 100, $d = 0.962$) and 2.5 to 5 mL of 85% phosphoric acid (H_3PO_4) and boiling chips in a 25- to 50-mL round-bottomed flask. Use this flask as the distillation flask and, in the hood, set up the rest of the apparatus needed for a simple distillation (see Section 2.3), including a 25- to 50-mL receiving flask. Immerse the receiving flask up to its neck in an ice bath to avoid loss of any of the product (bp 83°C). Heat the distillation flask carefully so that a steady distillation rate is obtained. Do not allow the temperature of the distilling vapors to exceed 103°C. Stop heating the reaction mixture when only a few mL of liquid remain in the distilling flask or when the vapors exceed 103°C.

Saturate the distillate by adding solid sodium chloride portionwise with swirling. Try to keep the excess sodium chloride to a minimum. Now, neutralize any acid in the receiving flask by adding just enough 10% aqueous sodium carbonate solution to the mixture with swirling to make the aqueous layer basic (use pH paper). Transfer the mixture to a separatory funnel and drain off the aqueous layer. Pour the organic layer into a small Erlenmeyer flask and dry it with just enough anhydrous magnesium sulfate to create a fine suspension when the mixture is swirled. Gravity filter the product into a clean, dry, round-bottomed flask and redistill the dry cyclohexene into an ice-cooled tared receiver using an anhydrous calcium chloride drying tube attached to the outlet adapter to avoid condensation of moisture in the distillate. Record the boiling point and weight of the product.

If time allows, perform some of the tests for alkenes described earlier on your product. Describe how you would use chemical tests to distinguish the cyclohexene product from the starting cyclohexanol.

———————————————

3,3-Dimethyl-2-butanol is an unsymmetrical alcohol. Acid-catalyzed dehydration initially produces a secondary carbocation that can either lose a proton to form a terminal alkene or undergo rearrangement to a more stable, tertiary carbocation and then lose either of two different protons to produce a different terminal alkene or a tetrasubstituted alkene (tetramethylethylene).

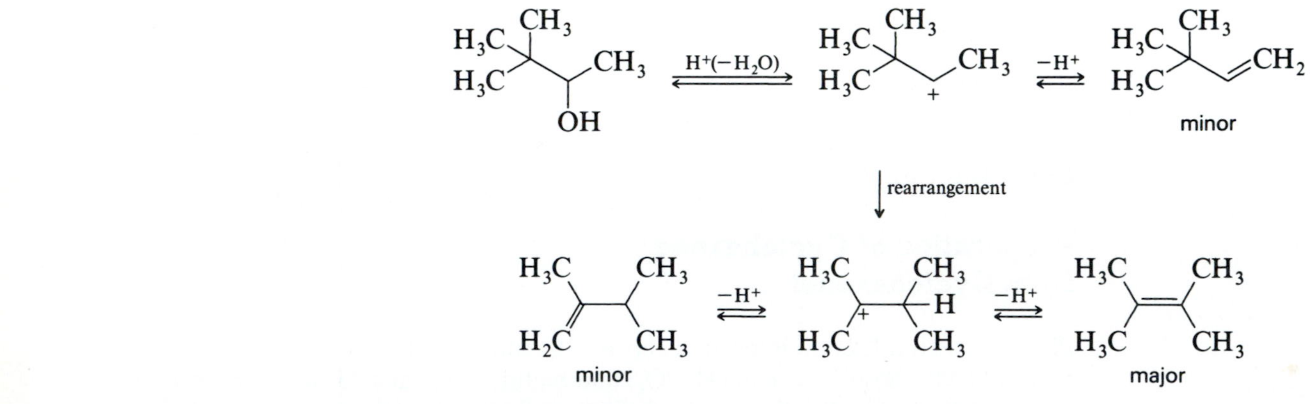

EXPERIMENT 36

Dehydration of 3,3-Dimethyl-2-butanol

To approximately 20 g of crushed ice in a 100-mL beaker, slowly add 15 mL of concentrated sulfuric acid with stirring. Transfer the cold solution to a 100-mL round-bottomed flask and cautiously add 12.3 g (~15 mL) of 3,3-dimethyl-2-butanol. (This alcohol is commercially available, but because of its rather high cost, it can be prepared from the corresponding ketone as discussed in Experiment 46.) Add a boiling chip or stir bar and set up a fractional distillation in the hood, using the initial 100-mL round-bottomed flask as the distilling flask, with an empty fractionating column. Cool the receiving flask in an ice bath. Heat the distillation flask so that the distillation rate is about one drop every 5 seconds.

The major portion of the alkene product mixture should distill at 65 to 75°C. Stop the distillation when it becomes difficult to maintain a constant temperature at the *top* of the fractionating column below 80°C. Transfer the cold distillate to a small separatory funnel and gently shake it with 10 mL of 5% sodium hydroxide solution. Drain off the aqueous layer and pour the organic layer into a small Erlenmeyer flask. Dry the liquid over a small amount of anhydrous magnesium sulfate or sodium sulfate. Decant the volatile liquid into a dry distillation flask and redistill it into a tared round-bottomed flask using a simple distillation setup. Record the volume and mass recovered, the yield, and the boiling range of the isomeric alkenes. Analyze the isomeric mixture by gas chromatography (see Section 2.6.7 and Experiment 15). Record the nuclear magnetic resonance (NMR) spectrum of the isomeric mixture. From the gc and NMR data determine the relative amount of the major alkene product.

Rotation about the carbon-carbon double bond of alkenes is not possible without temporarily converting the double bond to a single bond. This can be accomplished by several methods, including acid catalysis, or in some cases by photolysis. In general, acyclic *trans* alkenes are more stable than *cis* alkenes, and under equilibrating conditions the *trans* isomer will be preferentially formed. This experiment will illustrate the acid-catalyzed conversion of maleic acid, a *cis* alkene, to fumaric acid, the isomeric *trans* alkene. Maleic acid is readily available, but can also be easily generated by the hydrolysis of maleic anhydride.

maleic anhydride → maleic acid → fumaric acid

EXPERIMENT 37

Isomerization of Maleic to Fumaric Acid

Heat 10 mL of water to boiling in a small Erlenmeyer flask, or a 25-mL round-bottomed flask, and then add 8 g of maleic anhydride. The low-melting anhydride will first melt then dissolve in the water as it hydrolyses to maleic acid. Cool the solution to allow some of the maleic acid to crystallize out of solution. When crystals no longer form, collect the maleic acid on a Büchner funnel using aspirator suction. Do not wash the material because it is very soluble in water. Allow the material to dry and determine its melting range ($\sim$ 136 to 139°C).

Meanwhile, place the filtrate, which still contains a considerable amount of maleic acid, in a 25-mL round-bottomed flask. Add 8 mL of concentrated HCl and fit the flask with a reflux condenser (Figure 1.8). Using a Bunsen burner or steam bath, heat the solution at reflux for about 10 minutes. Cool the solution and collect the crystals of fumaric acid that form. If necessary, recrystallize a small portion of the fumaric acid using a Craig tube and then determine its melting range ($\sim$ 298 to 300°C; do not use an oil bath). As further evidence that the two products, maleic acid and fumaric acid, are indeed different, perform a mixed melting point.

Free-radical catalyzed isomerization. An alternative to the acid-catalyzed isomerization is a free-radical process. Dissolve 3 g of maleic acid in 20 mL of water in a 50-mL Erlenmeyer flask. With caution, add one drop of bromine and place the flask under a sun lamp in the hood for about 15 to 20 minutes. Suction filter the solid fumaric acid that precipitates. Dry the product. Determine the melting range and compare it to that obtained for the previous products.

The following experiments demonstrate some of the possible stereochemical consequences of electrophilic additions to alkenes.[10] Acenaphthylene is a planar molecule in which the geometry of the alkene portion is determined by fusion to a rigid aromatic (naphthalene) system. Addition of electrophilic reagents to the alkene gives relatively rigid products with restricted rotation about the resulting carbon-carbon single bond. Thus, the net mode of the addition (*syn* or *anti*) can be directly determined by analysis of the products. A convenient analysis uses the difference in the coupling constant of the vicinal hydrogens in the NMR spectrum of the product. The Karplus equation (see Section 6.5 and Figure 6.4) predicts that a relatively low coupling constant $J = 0$ to 1 Hz) will be observed from the vicinal coupling constant of the nonaromatic protons of the *anti* addition product since the dihedral angle of these vicinal protons is about 100°C. Alternatively, the coupling constant for the

---

[10] E. F. Silversmith, *J. Chem. Ed.,* **59,** 346 (1982).

nonaromatic protons of the *syn* addition product should be about 8 Hz, corresponding to a dihedral angle of near 0°C.

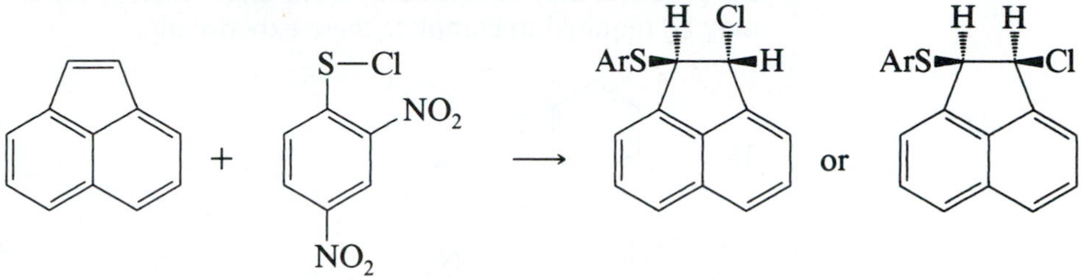

| acenaphthylene | 2,4-dinitrobenzene-sulfenyl chloride | *anti*-addition product | *syn*-addition product |

EXPERIMENT 38

Addition of 2,4-Dinitrobenzenesulfenyl Chloride to Acenaphthalene

Place 0.5 g of 2,4-dinitrobenzenesulfenyl chloride and 0.35 g of acenaphthylene (recrystallized from 95% ethanol, mp 86 to 89°C) in a 10-mL Erlenmeyer flask or a 10-mL round-bottomed flask and, with caution, add 2 to 5 mL of glacial acetic acid. Heat the mixture on a hot plate, sand bath, or an oil bath at 75 to 85°C in the hood for about 20 minutes or until testing of a small drop with moist starch iodide paper is negative. Allow the solution to cool to room temperature and collect the product by suction filtration. Wash the product (~0.5 g) on the suction filtration funnel with a few mL of ether and air dry it. Record the melting range of the yellow-brown solid. If it is not sharp and near 186 to 187°C, the compound may be recrystallized from 1-butanol using a Craig tube (see Figure 2.4). Record the yield, melting range, and the NMR spectrum of the product. [Deuterated DMSO (dimethylsulfoxide) is a convenient solvent. Other solvents may be used that do not have signals in the δ 4.5 to 6.0 region to be observed.] Determine coupling constants for the nonaromatic protons from the NMR spectrum and comment on the mode (*syn* or *anti*) of the electrophilic addition.

Write a mechanism that accounts for the results.

The reaction of *meso*-stilbene dibromide with silver acetate in acetic acid and with silver acetate in aqueous acetic acid provides a striking example of how neighboring groups and solvent may influence the stereochemical outcome of a reaction.

The first part of this experiment involves the preparation of *meso*-stilbene dibromide from *trans*-stilbene and pyridinium bromide perbromide (Experiment

26). The next steps involve the conversion of the product, *meso*-stilbene dibromide, into both *meso*- and *dl*-1,2-diphenylethane-1,2-diol. The stereochemical purity of the products may be estimated from their NMR spectra. Three laboratory periods may be required to complete these experiments.

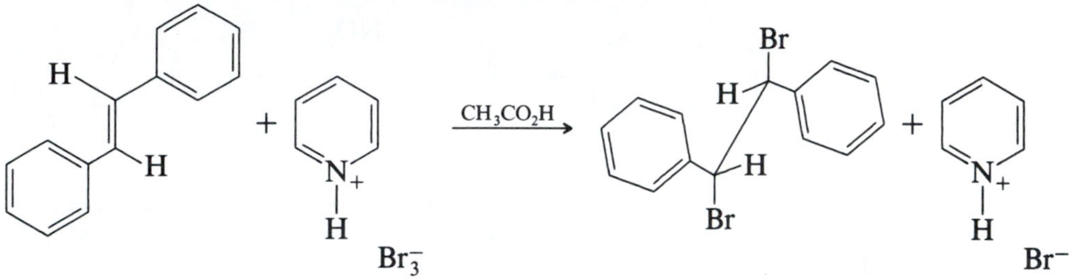

trans-Stilbene   *meso*-Stilbene dibromide

EXPERIMENT 39

Preparation of *Meso*-Stilbene Dibromide

Place 4 g of *trans*-stilbene and, with caution, 80 mL of glacial acetic acid in a 250-mL Erlenmeyer flask. Dissolve the solid by heating the mixture on a steam bath or hot plate in the hood. When the solution is homogeneous, add 8 g of pyridinium bromide perbromide (the reagent may be freshly recrystallized from acetic acid before use). Stir, or swirl, the contents of the flask while heating for 1 to 2 minutes. If some crystals cling to the sides of the flask, rinse them down with a small amount of acetic acid. Cool the mixture in a cold-water bath or in a stream of water from a cold-water tap. (Note that acetic acid freezes at ~ 17°C.) Collect the product by suction filtration and wash it on the filter funnel with a small amount of methanol. The yield of colorless crystals with a melting point of about 237°C should be about 6.4 g.

EXPERIMENT 40

Preparation of *dl*-1,2-Diphenylethane-1,2-diol

Place 2 g of *meso*-stilbene dibromide, 2 g of silver acetate, 25 mL of glacial acetic acid, and 1 mL of water in a 50-mL Erlenmeyer flask. Heat the uniform white suspension for 10 minutes on a steam bath or hot plate with frequent swirling. During the heating process, the suspension should change to a curdy dense precipitate of silver bromide. A pink color may develop, and then the solution should become clear and colorless. Cool the mixture to room temperature and suction filter it to remove the silver bromide. Wash the silver bromide with a small amount of ether. Extract a suspension prepared from the combined filtrate and 60 mL of water with 40 mL of ether. Wash the ether extract twice with 25-mL portions of water and once with 25 mL of 10% aqueous sodium hydroxide. Dry the solution over potassium carbonate, gravity filter the solution, and evaporate the ether under reduced pressure on a rotary evaporator, if one is available, or by the use of a steam bath in a hood.

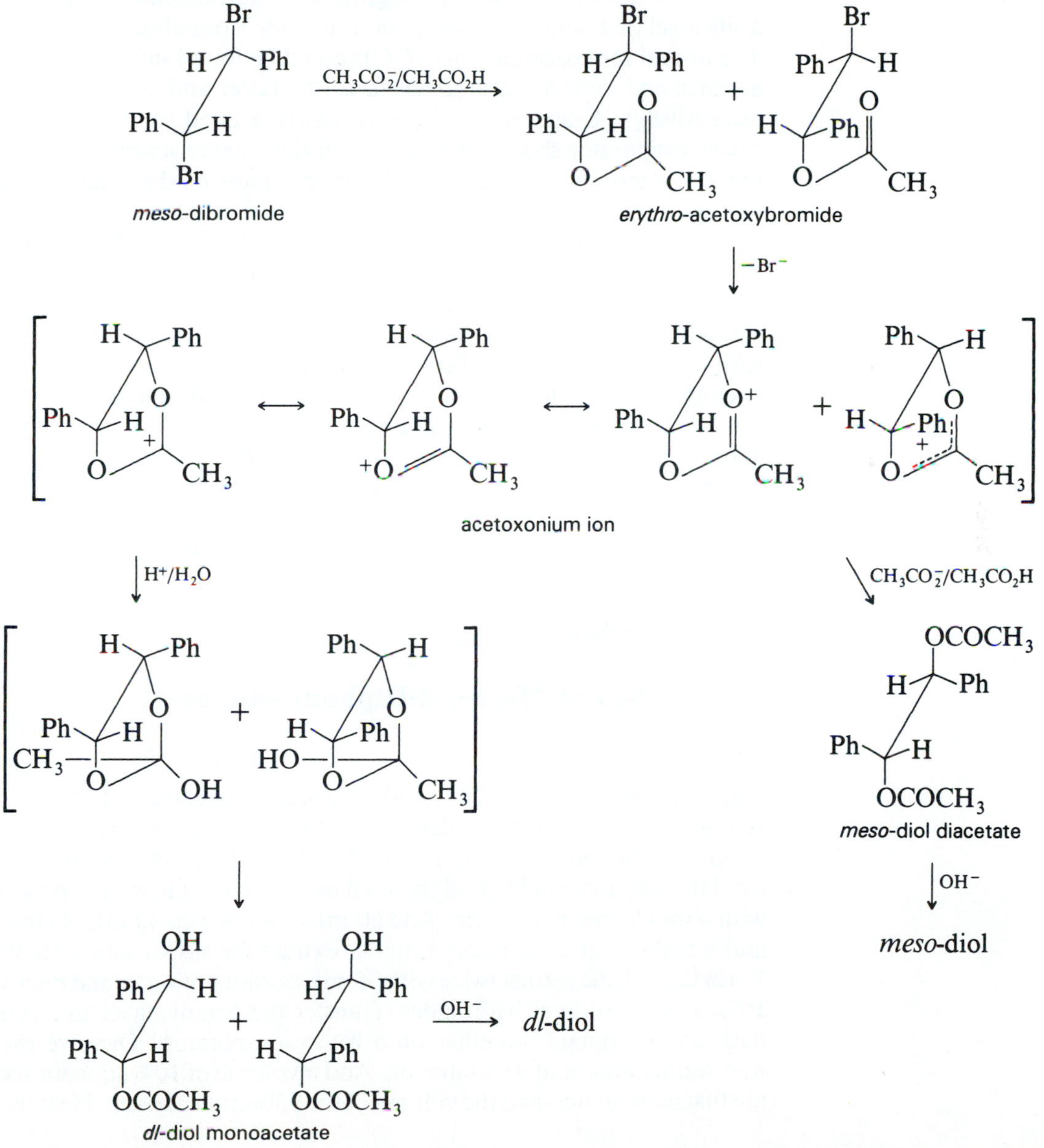

$CH_3CO_2^-/CH_3CO_2H$
meso-dibromide
erythro-acetoxybromide
$-Br^-$
acetoxonium ion
H^+/H_2O
$CH_3CO_2^-/CH_3CO_2H$
meso-diol diacetate
OH^-
meso-diol
dl-diol monoacetate
OH^-
dl-diol

Dissolve the oil obtained in a minimum amount of ethyl acetate or dichloromethane. Using a drawn-out capillary, remove a small sample for analysis by thin-layer chromatography (see TLC, Section 2.6.4). Spot the solution of the product on a silica gel or alumina TLC plate and develop it with mixtures of ethyl acetate and hexanes. Start with 5% ethyl acetate in hexanes and if necessary increase the polarity (amount of ethyl acetate) to optimize the separation of the products. Using the optimum solvent system as indicated by the degree of separation of the components of the mixture, or preferably one slightly less polar, chromatograph the mixture using a silica gel or a neutral alumina column made from about 25 to 50 g of adsorbent. The *dl*-diol monoacetate (mp 87°C) should be found in intermediate fractions (diacetates and diols are also obtained as the faster and slower eluting components, respectively). Evaporation of the solvent in a tared round-bottomed flask on the rotary evaporator should produce an oil that slowly solidifies upon standing overnight. Determine the weight and melting point of the product. Verify that the sample is the *dl*-diol monoacetate by NMR spectroscopy.

Dissolve all the *dl*-diol monoacetate in 10 mL of 95% ethanol. Add 5 mL of 10% aqueous sodium hydroxide and heat the solution on a steam bath or hot plate (90 to 100°C internal temperature) for 10 minutes to hydrolyze the acetate. Cool the solution to room temperature and dilute it with 30 mL of water. Extract the solution with two 25-mL portions of ether. Combine the ether extracts and dry the solution over anhydrous potassium carbonate. Filter the solution and concentrate the solution on the rotary evaporator to a volume of about 10 mL. Add 15 mL of hexanes and concentrate the solution on the rotary evaporator, without heating, until crystals just begin to separate. Cool the solution in an ice bath and collect the crystals by suction filtration. The yield is usually low (~0.35 g). Record the melting point (~120°C). Obtain and interpret the NMR spectrum of the diol.

EXPERIMENT 41

Preparation of *Meso*-1,2-Diphenylethane-1,2-diol

Place 2 g of *meso*-stilbene dibromide, 2 g of silver acetate, and 25 mL of glacial acetic acid in a 50-mL Erlenmeyer flask. Heat the suspension on a steam bath or hot plate (internal temperature ~85 to 100°C) for about 10 minutes or until it becomes clear. Cool the solution and filter off the solid using suction filtration. Wash the silver salts with a small amount of ether. Add 60 mL of water and 40 mL of ether to the filtrate and transfer it to a separatory funnel. Extract the suspension with 40 mL of ether. Wash the organic extract twice with 25-mL portions of water and once with 25 mL of 10% aqueous sodium hydroxide. Transfer the organic layer to a round-bottomed flask and evaporate the ether on a rotary evaporator. Dissolve the residue in a minimum amount of 95% ethanol. Add a volume of 10% aqueous sodium hydroxide that is about one-half the volume of the 95% ethanol used. Heat the solution on a steam bath or hot plate for about 10 minutes. Cool the mixture and dilute it with two equivalent volumes of water. Scratch the sides of the flask to induce crystallization. Cool the suspension in an ice bath and collect the product by suction filtration. The product should have a melting point near 137°C. Record the melting point and yield. Obtain and interpret the NMR spectrum. Compare the spectrum of the product with that obtained in the previous part of this experiment.

Many carbenes are electrophilic and are capable of adding to alkenes. Carbenes (R_2C:) are conveniently prepared by an α-elimination process in which both of the electrophilic and nucleophilic groups depart from a common carbon atom.

$$R_2C\begin{smallmatrix} \nearrow E \\ \searrow N \end{smallmatrix} \longrightarrow R_2C\text{:} + E^+ + N^-$$

$E^+ =$ electrophile
$N^- =$ nucleophile

Most carbene-forming reactions involve a base-catalyzed elimination. In fact, one of the simplest methods for the formation of dichlorocarbene is the reaction of aqueous or ethanolic base with chloroform. However, the use of hydroxide and hydroxylic solvent leads to further detrimental reactions of the carbene and decreases the efficiency of the process. Groups other than a proton can be used to promote the α-elimination and eventual formation of a carbene. For example, trichloroacetic acid can be converted to sodium trichloroacetate that upon heating undergoes decarboxylation followed by loss of chloride ion to produce dichlorocarbene. Under these relatively nonbasic conditions the side reactions with the dichlorocarbene are minimized.

Once generated, carbenes can undergo a variety of reactions including (1) insertion into C—H bonds, (2) addition (insertion) into multiple bonds, (3) rearrangement reactions to produce alkenes, and (4) dimerization to form alkenes. Dichlorocarbene is a relatively unreactive carbene and usually only undergoes addition reactions with alkenes to form 1,1-dichlorocyclopropanes. The parent carbene, methylene (CH_2:), and other alkylcarbenes are extremely reactive and undergo several of the other reactions listed above.

$$Cl_3CH + OH^- \xrightarrow{-H_2O} Cl_3C^- \xrightarrow{-Cl^-} Cl_2C\text{:}$$

Since carbenes can be considered both electrophiles and nucleophiles, they are usually generated in inert solvents. However, the following experiment demonstrates that carbenes can be formed in the presence of water, if a phase-transfer catalyst is used. In this case, benzyltriethylammonium chloride is the catalyst, although many others work well.

benzyltriethylammonium chloride

EXPERIMENT 42

Addition of Dichlorocarbene to Cyclohexene

Add 40 to 50 mg of benzyltriethylammonium chloride to a preformed mixture of 2 mL (1.64 g, 20 mmole) of cyclohexene, 1.7 mL (2.5 g, 21 mmole) of chloroform *(CAUTION: Cancer suspect agent)* and 4 mL of 50% aqueous sodium hydroxide in a 25- to 50-mL round-bottomed flask. Swirl or vigorously stir the mixture with a magnetic stirrer to produce a sudsy emulsion. Suspend a thermometer in the flask and monitor the temperature which should first gradually, and then rapidly rise. Maintain the temperature between 50 and 60°C with an ice bath, if necessary and continue to swirl, or stir, the thick paste (magnetic stirring may no longer be practical). After about 10 minutes, upon completion of the exothermic reaction, allow the mixture to cool slowly and then dilute the mixture with 10 mL of water. Transfer the biphasic solution to a small separatory funnel. Rinse the reaction flask with 5 mL of methylene chloride and add the rinse to the funnel. Recall that the organic layer should now be on the bottom. Separate the layers. Extract the aqueous layer once with 5 mL of methylene chloride. Combine this extract with the original organic layer. Extract the combined organic portions with 5 mL of water. Pour the organic solution into a small Erlenmeyer flask and dry it with anhydrous sodium sulfate. Gravity filter the solution into a 25-mL round-bottomed flask and carefully remove the methylene chloride and excess chloroform and cyclohexene on a rotary evaporator. Distill the product (bp ~ 190 to 200° C) in portions using a Hickman still, heating with a sand bath to obtain enough product for analysis. The NMR spectra of the starting cyclohexene and the *1,1*-dichlorocyclopropane product can be conveniently recorded in deuterochloroform ($CDCl_3$).

Questions for Experiment 42

1. Dichlorocarbene can also be generated in the presence of tetrachloroethylene used as one of the solvents, yet not react with this alkene. Why?

2. Draw the two possible structures of the product from the reaction of the dichlorocarbene with *cis,cis*-1,5-cyclooctadiene.

3. Actually, only one isomer of the two drawn in question 2 is formed. Which one is it? Rationalize your answer.

10.9 CHEMISTRY OF ALKYNES

Like alkenes, alkynes (acetylenes) can undergo addition reactions. With alkynes, however, the product after the addition still contains a double bond. The fate of that double bond depends on its reactivity toward the initial addition reagent or its tendency toward other reactions. Mercuric ion-catalyzed hydration of an alkyne proceeds by the addition of water to the triple bond of an alkyne. In this case, the initial product is an unsaturated alcohol (enol) that tautomerizes to produce a carbonyl compound. With disubstituted alkynes, the product of hydration is always a ketone. With a terminal alkyne, addition could conceivably produce either a ketone or an aldehyde, depending on whether addition of the hydroxyl group occurred at the internal or terminal carbon of the alkyne. The direction of the addition is determined by the stability of the intermediates. Reaction of the mercuric ion with the alkyne proceeds by preferential addition to give products apparently arising from the secondary vinyl cation, rather than the less stable primary vinyl cation, as expected for a Markovnikov addition. Actually, an intermediate cyclic mercuronium species is probably involved, but the observed regioselectivity is consistent with the cationic intermediates. Subsequent addition of water produces the enol corresponding to the ketone and not the aldehyde. This mode of addition will be confirmed in Experiment 43.

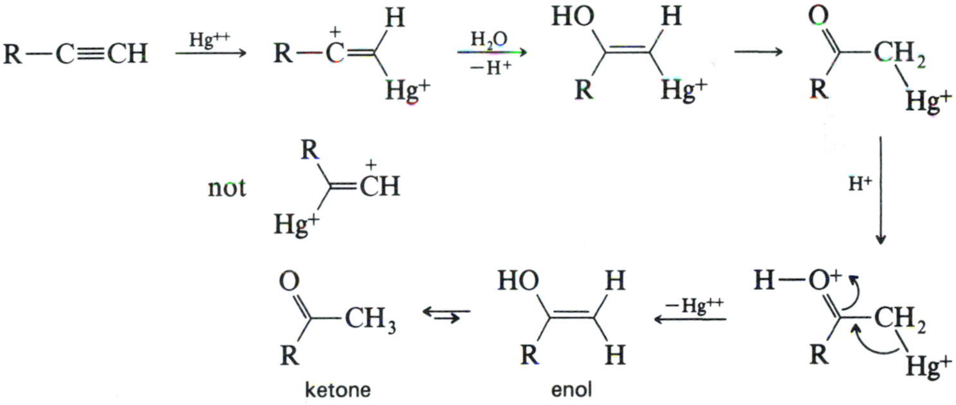

In this experiment you will determine if addition of a hydroxyl substituent on the carbon adjacent to an alkyne will inhibit or change the mode of addition of water to the alkyne under mercuric ion-catalyzed conditions. You will test for the formation of a carbonyl group (aldehyde or ketone) by attempting to form a semicarbazone derivative of the carbonyl compound. An iodoform test on the purified hydration product will also allow you to distinguish between the potential aldehyde and the methyl ketone products. An additional experiment, the periodic acid test, will confirm or deny the possibility that the product is an α-hydroxy ketone. Before determining the structure of the product, you will steam distill (see Section 2.3.4) the crude product from the solution containing the mercuric ion catalyst. The steam will be

generated internally by adding water to the reaction flask before heating to distill the
product.

(mp 30° c)   ketone   aldehyde

EXPERIMENT 43

Hydration of 1-Ethynyl-1-cyclohexanol

Obtain about 12 g of the 3-hydroxy-1-alkyne (1-ethynyl-1-cyclohexanol) from your
instructor. Add 9 mL of concentrated sulfuric acid *(CAUTION: Concentrated sulfuric acid is a strong acid)* to 60 mL of water in a 250-mL round-bottomed flask.
Dissolve 0.5 g of mercuric oxide *(toxic)* in the acid solution. The temperature of the
initial acid solution will be elevated from mixing the acid with water. Cool the
mixture to 50°C with an ice bath and attach a reflux condenser. Carefully add about
6 g (about half) of the alkyne through the top of the condenser with a disposable
pipette. Mix vigorously by swirling (or with a magnetic stirrer, if possible). The
initially formed precipitate (a mercury-alkyne complex) should begin to dissolve.
After 2 minutes, heat the mixture to reflux with a burner. As soon as the mixture
begins to reflux, stop heating and cool the flask back to 50°C. Add the remainder of
the alkyne, mix, and heat through a simple distillation apparatus (see Section 2.3),
until almost all the organic material has been distilled (between 35 to 75 mL total
volume). Transfer the solution to a separatory funnel. Add potassium carbonate to
the distillate until saturated to salt out the organic product and separate the layers.
Remove the organic layer and dry it with about 4 g of anhydrous potassium carbonate. Filter the solution through a small cotton plug and redistill the product from a
100-mL flask (to avoid foaming). Record the yield and boiling point of the product.

Determination of the structure of the hydration product. Prepare the semicarbazone derivative as described by the general procedure in Section 9.5.5. A positive
result with this test will confirm that you have either a ketone or an aldehyde, but it
will not differentiate between them unless you can compare the melting point of the
product with that of the semicarbazone derivatives of the authentic aldehyde and
ketone.

    The iodoform test (see Section 9.5.4) can be used to determine if your product is
a methyl ketone and thereby distinguish between the possible aldehyde and ketone
product. Formation of a yellow precipitate of iodoform within 15 minutes is a
positive test. It is good practice to isolate some of the product, wash it with water, and
ascertain whether it melts near 119 to 120°C, which is the correct melting point for
authentic iodoform.

Questions for Experiment 43

1. What is the structure of the hydration product that you obtained in this experiment?
2. Write the mechanism for the formation of the semicarbazone from your product.
3. The iodoform test also gives a positive result with certain alcohols. Why would traces of your starting alcohol not interfere with the iodoform test?
4. Why should you not clean your test tube with acetone or ethanol before trying the iodoform test?

10.10 CHEMISTRY OF ALCOHOLS

Alcohols can be prepared by a variety of reactions from a number of other functional groups. Preparation of an alcohol by hydration of alkenes has already been described (Experiments 31 and 33). Hydroboration of alkenes followed by an oxidative work-up is a very efficient method for the preparation of alcohols by a net anti-Markovnikov addition of water to the alkene. A complimentary reaction, oxymercuration followed by reductive removal of the mercuri group, provides an efficient method for the preparation of alcohols by a net Markovnikov addition of water to alkenes. Solvolysis of alkyl halides also produces alcohols. Reduction of carbonyl compounds is one of the most common methods of preparing alcohols. *Reduction of aldehydes, esters, or carboxylic acids produces primary alcohols, whereas reduction of ketones produces the corresponding secondary alcohol.* Secondary alcohols can also be prepared by the reaction of aldehydes with Grignard reagents, and tertiary alcohols can be prepared by the reaction of ketones or esters with Grignard reagents.

Alcohols can also serve as precursors to a number of other functional groups. Dehydration of alcohols to alkenes has already been described (Experiments 35 and 36). They can also be converted to sulfonate esters, halides, esters, acetals, and ketals. Primary alcohols can also be oxidized to aldehydes or acids. Secondary alcohols can be oxidized to ketones.

EXPERIMENT 44

Reduction of Cyclohexanone to Cyclohexanol with Sodium Borohydride[11]

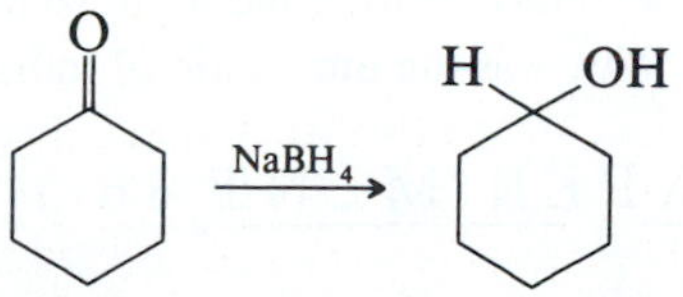

This reaction evolves hydrogen gas and should be carried out in a hood. Place a magnetic stir bar in a 125-mL Erlenmeyer flask containing 30 mL of water and 0.25 mL (about five drops) of 10% aqueous sodium hydroxide. Cool the solution in

---

[11] N. M. Zaczek, *J. Chem. Ed.,* **63,** 909 (1986).

an ice-water bath and while stirring add 1.5 g (0.04 mol) of sodium borohydride. Continue stirring and add 9.8 g (0.1 mol, 10.3 mL) of cyclohexanone dropwise while maintaining the temperature under 40°C. After the addition is complete (5 to 10 minutes), remove the ice bath and stir the solution for another 10 minutes. Continue stirring and slowly add 10 mL of 6 M HCl. Hydrogen gas evolution may be vigorous, but can be controlled somewhat by the rate of addition of the acid.

After all the acid has been added, check the solution with pH paper to confirm that it is acidic. Saturate the solution by adding about 10 g of solid sodium chloride with stirring. Transfer the solution to a separatory funnel. Rinse the reaction flask with 15 mL of ether and pour the rinse into the separatory funnel. Stopper the separatory funnel, invert it, and swirl gently with frequent venting. Separate the layers and extract the aqueous layer with two more 15-mL portions of ether. Combine the ether extracts in an Erlenmeyer flask. Dry the ether solution over anhydrous magnesium sulfate, and gravity filter the solution into a round-bottomed flask. Carefully distill the mixture of ether and product, cyclohexanol, and collect the fraction boiling between 155 to 165°C. Alternatively, carefully evaporate the ether and then distill the residue.

Record the infrared and NMR spectra of the starting cyclohexanone and cyclohexanol. Comment on the differences in the spectra of starting materials and products. Submit the product to your instructor. The product can be used for the dehydration experiment described in Experiment 35.

EXPERIMENT 45

Microscale Reduction of Cyclohexanone

This reaction can also be performed on one-tenth the scale in a reaction vial. The extraction should then be done by transferring the layers with a disposable pipette and the final product should be distilled with a Hickman distillation head.

Questions for Experiments 44 and 45

1. Sodium hydride is also a source of hydride (H$^-$). Why is it not an effective reducing agent for the conversion of cyclohexanone to cyclohexanol? What other reaction might occur upon reaction of sodium hydride with cyclohexanone?

2. Lithium aluminum hydride is a very effective reagent for the reduction of ketones to alcohols. Why could it not be used under the experimental conditions described above?

3. Why was the mole ratio of sodium borohydride and cyclohexanone not unity?

EXPERIMENT 46

Reduction of 3,3-Dimethyl-2-butanone with Sodium Borohydride

Combine 2 g (2.5 mL, 20 mmol) of 3,3-dimethyl-2-butanone (pinacolone, t-butyl methyl ketone) and 6 mL of reagent grade methanol in a 25-mL single-necked, round-bottomed flask. Cool the mixture to slightly below 20°C with the aid of an ice bath and add 0.4 g of sodium borohydride in portions with swirling or magnetic stirring over a 10-minute period. During the addition, cool the solution as required to maintain the temperature of the reaction mixture below 20°C and minimize hydrogen gas evolution. After the addition is complete, attach a reflux condenser and slowly heat the mixture to 40 to 50°C for 10 minutes with a steam bath, warm-water bath, or preheated sand bath.

Cool the reaction mixture to about 20°C and carefully transfer it to a separatory funnel containing about 25 mL of diethyl ether. Saturate the aqueous layer by adding solid sodium chloride with gentle swirling. Separate the aqueous layer and wash the ether layer with two 10-mL portions of water followed by three 5-mL portions of saturated sodium chloride solution. (Some salt may separate during each washing.) Dry the ether layer over anhydrous magnesium sulfate and gravity filter the solution into a round-bottomed flask. Carefully remove the ether using a rotary evaporator. Cooling the receiving flask on the rotary evaporator will reduce exposure to ether vapors. Distill the residue and collect only the material boiling *above* 110°C. Alternatively, the ether solution of the crude product can be distilled directly with collection of only the higher boiling (>110°C) fractions. 3,3-Dimethyl-2-butanol has a boiling point of 120°C. Record the boiling range, mass yield, and percent of the theoretical yield of your product. Record and compare the NMR and IR spectra of the starting ketone and product alcohol. Submit your product to your instructor.

EXPERIMENT 47

Reduction of 2-Butanone to 2-Butanol
(A Volatile Alcohol)

Start with 1.4 g (1.74 mL, 20 mmol) of 2-butanone and use the procedure described in Experiment 46 to reduce 2-butanone to 2-butanol. Be especially careful during the final evaporation of the diethyl ether since the product, 2-butanol, boils at 99 to 100°C.

Questions for Experiments 46 and 47

1. What is the purpose of the five washes with saturated sodium chloride during the work-up of the sodium borohydride reduction?

2. Identify the organic product obtained in each of the following reactions after the usual work-up.
   (a) CH_2=CH—CH_2—$COCH_3$ + $NaBH_4$
   (b) $HCOCH_2CH_2CH_3$ + $NaBH_4$
   (c) $ClCOCH_2$—CH=CH—CH_2—CO_2CH_3 + $NaBH_4$

3. Outline another synthesis of 3,3-dimethyl-2-butanol.

The reaction of an alcohol with a hydrohalic acid is a standard method for the preparation of alkyl halides. It is used here to prepare several pentyl halides. Before coming to the laboratory, prepare a table of properties of reagents, reactants, products, and byproducts. Include data on molecular weights, densities of liquids, boiling or melting points of organic compounds, and solubilities in water.

EXPERIMENT 48

Competitive Conversion of 1-Pentanol to 1-Bromo and 1-Chloropentane

Place 30 g of crushed ice in a 250-mL round-bottomed flask, and carefully add 25 mL of concentrated sulfuric acid. Mix the solution by swirling. Add 0.12 mol of ammonium chloride, 0.12 mol of ammonium bromide, and a boiling chip. Attach a reflux condenser, and provide a trap for acidic gases (see Section 1.12). Warm until the solids are dissolved (swirling the contents of the flask may help). *If the solids do not dissolve within 10 to 15 minutes, continue on.* Cool the flask slightly, and pour 0.20 mol of 1-pentanol through the top of the condenser. Heat the reaction mixture to maintain gentle reflux for 1 to 1 1/2 hours (see Experiment 49). Do not overheat.

After refluxing, cool the reaction mixture and transfer it to a separatory funnel in the hood. Separate the layers, and wash the organic layer twice with 10 mL portions of concentrated sulfuric acid (*CAUTION: Wear rubber gloves during the of extraction procedure*), once with 50 mL of water, and once with about 50 mL of 5% aqueous sodium bicarbonate (*CAUTION: Extensive gas evolution will occur*). Transfer the organic material to a 50-mL Erlenmeyer flask containing 2 to 3 g of anhydrous calcium chloride. Cork the flask, and swirl for a few minutes. Gravity filter the product, collecting it in a sample bottle. Estimate the ratio of 1-chloropentane to 1-bromopentane in the product by gas chromatography. This gc analysis should be performed on the same day as the chemical reaction. If this is not possible, the reaction product should be stored in a tightly stoppered container in a refrigerator. (All refrigerators used for storage of organic compounds should be explosion-proof.)

EXPERIMENT 49

Competitive Reaction of 1-Pentanol and 2-Methyl-2-butanol with Hydrochloric Acid

While refluxing the reaction mixture in Experiment 48, place 10 mL of 2-methyl-2-butanol and 10 ml of 1-pentanol in a separatory funnel. Add 100 mL of concentrated hydrochloric acid, and shake vigorously for 10 minutes. During the shaking period, invert the funnel several times and release the pressure developed by carefully opening the stopcock (in the hood!). Separate the layers, and wash the organic layer several times with an equal volume of cold water. Dry the organic layer over about 1 g of anhydrous magnesium sulfate, and filter into a 100-mL round-bottomed flask. In the hood, distill the product, collecting the product that boils from 83 to 88°C. The product is quite volatile, and care should be exercised during the experiment to keep evaporation loss to a minimum. Analyze the product by GC, comparing the chromatogram of the product with that of an authentic mixture of 1-chloropentane and 2-chloro-2-methylbutane.

Questions for Experiments 48 and 49

1. Which is the better nucleophile, chloride ion or bromide ion? Explain on the basis of your experimental observations in Experiment 48.
2. Is the product in Experiment 49 1-chloropentane or 1-chloro-1,1-dimethylpropane? Explain why only one product was formed.

The rate of hydrolysis of organic halides to alcohols depends significantly on the structure of the halide. This multistep experiment will demonstrate the hydrolysis of two different halides.[12] The kinetics of the hydrolyses will be determined that will allow conclusions to be made about the mechanism for the hydrolysis reactions. The substrates for the kinetic experiments will be 2-chloro-2-methylpropane (*t*-butyl chloride), which undergoes a first-order solvolysis, and 3-chloro-3-methyl-1-butyne, which undergoes solvolysis in a second-order process in basic solution. The latter substrate may first be prepared for a large group from the corresponding alcohol in a single large-scale reaction. The entire sequence, including the preparation of 3-chloro-3-methyl-1-butyne and the kinetics studies, requires about four (3- to 4-hour) laboratory periods. However, less total laboratory time can be used if the students work in pairs and each pair performs only some of the kinetic experiments. The data from all the experiments should be analyzed by each of the students.

EXPERIMENT 50

Preparation of 3-Chloro-3-methyl-1-butyne

In a 500-mL Erlenmeyer flask in the hood, cool approximately 165 to 170 mL (~2 mol) of concentrated hydrochloric acid (12N) to 7 to 10°C with the aid of an ice bath. Add a trace of copper-bronze powder (~20 mg) and then in one portion add 42 g (0.5 mol; 49 mL) of 2-methyl-3-butyn-2-ol. Cool the solution to 15°C with stirring and immediately add 35 to 38 g of calcium chloride. While stirring, allow the temperature to rise to 20 to 25°C but be careful not to allow it to rise above 25°C. Maintain the reaction temperature between 15 and 25°C by cooling with stirring as required for 40 to 50 minutes or until all the calcium chloride is dissolved and the upper oily product layer appears fully developed.

---

[12] J. A. Duncan and D. J. Pasto, *J. Chem. Ed.,* **52**, 666 (1975).

At this time, transfer the reaction mixture to a separatory funnel and remove the aqueous layer. (You may wish to retain this layer since if the layers were separated prematurely any additional product may be recovered later.) Wash the product layer with two 25-mL portions of cold water and once with 25 mL of cold 5% sodium carbonate solution. Dry the crude product[13] (~ 47 to 53 g of straw-colored liquid) over 2 to 4 g of anhydrous potassium carbonate for a minimum of 24 hours.

Rapidly distill the crude product from a clean flask containing 0.5 g of fresh anhydrous potassium carbonate through an empty fractionating column and collect that portion of the distillate boiling between about 60°C and 92°C.[14] (This distillation may be accompanied by a slight evolution of hydrogen chloride. Should hydrogen chloride evolution become copious near the end of this distillation, terminate the distillation process immediately and proceed with the redistillation.)

In the same fashion redistill the fraction obtained above, again from 0.5 g of anhydrous potassium carbonate, and collect and retain the fraction with boiling point 72 to 79°C (mostly 75 to 77°C). Determine percent yield.

Questions for Experiment 50

1. Show how you would prepare 3-methyl-4-hexyn-3-ol from readily available reagents.

2. Account mechanistically for the observation that the reaction of 2-methyl-3-butyn-2-ol with HCl affords reasonable quantities of 3-methyl-3-buten-1-yne and 1-chloro-3-methyl-1,2-butadiene in addition to 3-chloro-3-methyl-1-butyne.

3. Account for the rather unique reactivity of 3-chloro-3-methyl-1-butyne in which it affords both ethers with alcohols and alkenylidene cyclopropanes with alkenes.

EXPERIMENT 51

Solvolysis of 2-Chloro-2-methylpropane

Note: For Experiment 51, *work in pairs.* Each pair (*xx*) will collaborate with one other pair (*yy*). Pair *xx* from each group of four will perform experimental variations 1 to 3, while pair *yy* performs variations 4 to 6. The data obtained should be exchanged within each group of four. Each pair should perform Experiment 52.

Prepare a constant temperature bath by allowing a 1000-mL beaker of water to equilibrate at the desired temperature. The bath should be brought to and maintained at this temperature by addition of small amounts of ice or hot water as

---

[13] bp 75.5° at 745 torr; n_D^{25} 1.4145; 0.90. For information concerning the chemistry of this halide consult *J. Amer. Chem. Soc.,* **72**, 3542 (1950); **73**, 4735 (1951); **75**, 1653 (1953); **79**, 2142 (1957); **82**, 4908 (1960); *J. Org. Chem.,* **26**, 725 (1961); **26**, 2677 (1961); **34**, 1319 (1969).

[14] In addition to 3-chloro-3-methyl-1-butyne, this crude product may contain varying amounts of the following substances, which can successfully be removed by careful distillation:

$$HC\equiv C-\underset{\underset{CH_3}{|}}{C}-CH=CH_2 \qquad ClCH=C=\underset{\underset{CH_3}{|}}{C}-CH_3 \qquad CH_2=\underset{\underset{Cl}{|}}{C}-\underset{\underset{CH_3}{|}}{C}=CH_2 \qquad ClCH=CH-\underset{\underset{CH_3}{|}}{C}=CH_2$$

and several dichlorides with molecular formula $C_5H_8Cl_2$.

required. Pipette 50.0 mL of the ethanol-water solvent system (see variations 1 to 6 and use only the prepared solutions) into a 125-mL Erlenmeyer flask, stopper with a cork, and allow this mixture to equilibrate in the bath at the specified temperature.

While waiting for the bath to equilibrate, fill a 50-mL burette with $0.02N$, standardized sodium hydroxide solution, noting the exact concentration. Using a graduated cylinder, measure 20 mL of 95% ethanol into a 125-mL Erlenmeyer flask, add six drops (from a disposable pipette) of methyl red indicator solution (0.02 g of solid indicator in 100 mL of 60% ethanol-water), and cool the solution in an ice-water bath.

When all is in readiness, pipette exactly 1.0 mL of 2-chloro-2-methyl propane (using a rubber pipette bulb; the halide should be at 20 to 25° for this measurement) into the 50-mL reaction-solvent system, swirl the flask, and *very quickly* remove a 5.0-mL sample aliquot with a pipette and quench it by adding it to the cold ethanol solution. Note the *exact time* that this operation was performed. Titrate the quenched solution *in an ice bath* with the standard sodium hydroxide solution to the methyl red endpoint (red/yellow). Record the burette reading to ascertain the "blank" volume that should be used to correct each subsequent titration volume. (*Note:* Keep the reaction flask well stoppered with the cork when not withdrawing samples.)

In the same manner, titrate five or six additional 5.0-mL aliquots over a $2\frac{1}{2}$- to 3-hour period at gradually lengthening intervals (the reaction clearly will slow down as the reactants are used up), removing the first sample after 5 to 10 minutes. The time should be noted at the point when the 5-mL pipette is approximately half empty. Each solution must be titrated in an ice-water bath immediately to the methyl red endpoint. A clean flask containing 20 mL of fresh, cold 95% ethanol, to which has been added six drops of indicator, must be used to titrate each sample aliquot.

Experimental variations (solvent systems and temperatures)

1. 60% ethanol-water (by volume) at 15°C
2. 60% ethanol-water at 25°C
3. 60% ethanol-water at 35°C
4. 70% ethanol-water at 25°C
5. $0.184M$ sodium chloride in 60% ethanol-water at 25°C
6. $0.184M$ lithium perchlorate in 60% ethanol-water at 25°C

EXPERIMENT 52

Solvolysis of 3-Chloro-3-methyl-1-butyne

Prepare a constant-temperature water bath in a 1000-mL beaker as described in Experiment 50 and equilibrate at 25°C. Pipette exactly 50.0 mL of the $0.175N$ sodium hydroxide in 60% ethanol-water solution into a 125-mL Erlenmeyer flask, stopper, and equilibrate at 25°C in the water bath.

Charge a burette with standardized 0.02N hydrochloric acid solution. Place 15 mL of 95% ethanol in a 125-mL Erlenmeyer flask. Add four drops of the phenolphthalein indicator and cool the solution in an ice bath.

When all is in readiness, pipette exactly 1.0 mL of 3-chloro-3-methyl-1-butyne into the basic reaction solvent system and swirl the flask (time = 0 is taken as the time when the pipette is half empty). Quickly remove a 5.0-mL sample aliquot and quench it by adding it to the cold ethanol solution (time = t is taken as the time when the pipette is half empty). Titrate the quenched solution in an ice bath with the standard 0.02N hydrochloric acid solution to the phenolphthalein endpoint.

In exactly the same manner, titrate five additional 5-mL aliquots over a $1\frac{1}{2}$- to $2\frac{1}{2}$-hour period at gradually lengthening intervals. Once again the time should be noted at the point when the 5-mL pipette is half empty and each solution must be titrated in an ice bath to the indicator endpoint in a clean 125-mL Erlenmeyer flask containing 15 mL of 95% ethanol and four drops of phenolthalein solution.

Notes—Experiment 51

1. The density of 2-chloro-2-methylpropane used is approximately 0.85 g/mL at 20 to 25°, and when 50.0 mL of 60% ethanol-water and 1.0 mL of 2-chloro-2-methylpropane are mixed a total volume of 50.8 mL is obtained.

2. The product composition for the solvolysis of 2-chloro-2-methylpropane in various solvent systems is given in Table 10.3.

TABLE 10.3 Solvolysis Products of 2-Chloro-2- methylpropane in Several Solvent Systems

Solvent Composition			Product Composition		
%C_2H_5OH-H_2O (by volume)	Mole %				
	H_2O	C_2H_5OH	Alkene	Ether	Alcohol
60	68.3	31.7	13	15.5	71.5
70	58.2	41.8	15	21	64
60 (0.8M NaOH added)			28	6	66

Note—Experiment 52

The density of 3-chloro-3-methyl-1-butyne is approximately 0.90 g/mL at 20 to 25°C, and when 50.0 mL of 0.175N sodium hydroxide in 60% ethanol-water and 1.0 mL of 3-chloro-3-methyl-1-butyne are mixed a total volume of 50.7 mL is obtained.

Calculations

1. For experimental variations 1 to 6 of Experiment 51, determine in each case the first-order rate constant for the solvolysis of 2-chloro-2-methylpropane using the integrated rate expression for first-order reactions (see item 3 below).

2. Calculate $\Delta H^{\neq}$ and $\Delta S^{\neq}$ for the solvolysis of 2-chloro-2-methyl-propane in 60% ethanol-water.

3. In Experiment 51, determine the kinetic order of the reaction of 3-chloro-3-methyl-1-butyne in 60% ethanol-water 0.175N in sodium hydroxide. Integrated rate expressions for first- and second-order reactions follow:

$$\text{First Order:} \qquad k_1 = \frac{2.303}{t} \log \frac{a}{a-x}$$

$$\text{Second Order } (a \neq b): \qquad k_2 = \frac{2.303}{t(a-b)} \log \frac{b(a-x)}{a(b-x)}$$

$$\text{Second Order } (a = b): \qquad k_2 = \frac{1}{t} \cdot \frac{x}{a(a-x)}$$

In each equation, k is the rate constant, x is the concentration of substrate reacted at time t, and a and b are the initial ($t = 0$) concentrations of reactants.

4. Determine the value of the appropriate rate constant for the reaction in Experiment 52.

5. *Results and Discussion:* Provide important data results and discuss them fully. Include a discussion of the following:
   (a) The implications the $\Delta H^{\neq}$ and $\Delta S^{\neq}$ values have with respect to the reaction mechanism.
   (b) The effect of changing the reaction solvent.
   (c) The effect of adding sodium chloride.
   (d) The effect of adding lithium perchlorate.
   (e) The difference in reactivity of the two substrates (2-chloro-2-methylpropane and 3-chloro-3-methy-1-butyne) employed.

A number of useful, selective reagents have been developed for the oxidation of hydroxyl groups to carbonyl groups. Under appropriate conditions, primary alcohols can be oxidized to aldehydes or to the corresponding carboxylic acids. Secondary alcohols can be oxidized to ketones. Some the of most common and selective oxidants are derived from chromium (VI) compounds. However, significant effort has been made to avoid their use because of the toxicity of chromium salts. The following experiment will use sodium hypochlorite (NaOCl) in acetic acid to oxidize a secondary alcohol to a ketone. Secondary alcohols react with the hypochlorous acid (HOCl) generated to form an alkyl hypochlorite ($R_2CH-O-Cl$). Subsequent E2 elimination of HCl provides the ketone.

$$R_2CHOH + HOCl \longrightarrow H_2O + R_2CH-O-Cl \longrightarrow R_2C=O + HCl$$

Primary alcohols oxidize to aldehydes, initially, but upon reaction with water they form the hemiacetal, which are further oxidized to carboxylic acids. Tertiary alcohols do not oxidize easily, since they do not have the essential hydrogen on the hydroxyl-bearing carbon.

$$RCH_2OH + HOCl \longrightarrow H_2O + RCH_2OCl \longrightarrow RCHO$$

$$\Big\downarrow H_2O$$

After the oxidation of an alcohol with hypochlorite is completed, excess hypochlorite can be detected by a starch iodide test for the oxidant. Excess hypochlorite can be destroyed by reduction with sodium bisulfite.

$$NaOCl + 2\,NaI + H_2O \longrightarrow NaCl + 2\,NaOH + I_2$$

$$NaOCl + NaHSO_3 \longrightarrow NaCl + NaHSO_4$$

In Experiment 53, you will be given an unknown secondary alcohol to be oxidized to the corresponding ketone with hypochlorite. Typical unknowns include cyclohexanol, 4-methylcyclohexanol, cyclopentanol, 2-pentanol, and 3-pentanol. The products will be removed from the reaction mixture by a steam distillation in which the steam will be generated internally. Subsequent salting out of the product followed by extraction with ether, drying, and distillation should provide pure ketone product. Characterization of the product by its boiling point and the melting point of its 2,4-dinitrophenylhydrazone (Section 9.5.1) will allow unambiguous determination of the structure of the ketone product and the corresponding starting alcohol.

EXPERIMENT 53

Oxidation of an Unknown Alcohol

Obtain 10 g of an unknown alcohol from your instructor (see list above) and place it along with 25 mL of acetic acid in a 250-mL Erlenmeyer flask. Cool the flask in an ice-salt bath (do not freeze the acetic acid). Place 75 mL of aqueous sodium hypochlorite solution (minimum 11% by weight) in a separatory funnel and suspend the funnel above the reaction flask. Add the hypochlorite solution dropwise to the flask at a rate to maintain the internal temperature of the reaction mixture between 30 to 35°C. The addition will require 30 to 45 minutes. When the addition is complete, test the resulting greenish-yellow solution for excess oxidant with the starch-iodide test (Section 1.15.8). If the test is negative, add more sodium hypochlorite (1 to 3 mL) to the mixture to give a positive test and impart the greenish-yellow color to the reaction mixture. Finally add 5 more mL to be sure that an excess of hypochlorite is present. Swirl or magnetically stir the solution for 15 minutes at room temperature. Add saturated sodium bisulfite solution (1 to 5 mL) until the mixture turns colorless and gives a negative potassium iodide-starch test.

Transfer the mixture to a 250-mL distilling flask, add 60 mL of water, and distill the aqueous solution (internally generated steam distillation). Collect the first 45 mL of distillate in a graduated cylinder. This distillate consists of the product ketone, water, and acetic acid. Transfer the distillate to a 125-mL flask. Rinse the graduated cylinder with 5 mL of water and add the rinse to the distillate. Neutralize the acetic acid by carefully adding solid anhydrous sodium carbonate (6.5 to 7.0 g *total*) in portions until the evolution of carbon dioxide ceases. Then saturate the aqueous solution by adding 10 g of solid sodium chloride and stir for 15 minutes. Decant the mixture into a separatory funnel and separate the two layers. Transfer the crude ketone (top layer) to a 50-mL Erlenmeyer flask. Pour the aqueous layer back into the separatory funnel and extract it with 25 mL of ether. Separate the two layers and add the ether extract to the ketone. Dry the solution over anhydrous magnesium sulfate. Decant or gravity filter the solution into a dry 100-mL round-bottomed flask. Distill the dry ketone containing ether solution directly, saving only the higher boiling fractions. From your distillation determine and record the boiling point of the ketone. Determine and record the weight of the product.

Prepare a 2,4-dinitrophenylhydrazone derivative (Section 9.5.1) of the ketone product by dissolving 1 mL (about 20 drops) of the product in 1 mL of ethanol and adding this to 5 mL of the 2,4-dinitrophenylhydrazine solution. Recrystallize your derivative (try 95% ethanol first) and determine the melting range of the pure dry material. Compare the boiling point of your ketone and the melting point of the 2,4-DNP derivative with those listed below or those provided by your instructor and decide which alcohol you started with.

Alcohol	Ketone	Ketone bp (°C)	2,4-DNP mp (°C)
Cyclohexanol	Cyclohexanone	~151	~156
4-Methylcyclohexanol	4-Me-cyclohexanone	~166	~137
Cyclopentanol	Cyclopentanone	~130	~140
2-Pentanol	2-Pentanone	~100	~140
3-Pentanol	3-Pentanone	~100	~156

Hypochlorite has been used to oxidize many other alcohols to ketones.[15,16] An interesting oxidation of a series of secondary alcohols with chromium (VI) has been reported that includes a titrimetric method for the determination of the pseudo first-order rate of the oxidations. The results provide an estimation of the relative steric strain within a series of cyclic or acyclic alcohols.[17]

---

[15] J. W. Hill, J. A. Jenson, C. F. Henke, J. G. Yaritz, and R. L. Pedersen, *J. Chem. Ed.,* **61,** 1118 (1984).

[16] R. A. Perkins and F. Chau, *J. Chem. Ed.,* **59,** 981 (1982).

[17] T. J. Mason, *J. Chem. Ed.,* **59,** 980 (1982).

TABLE 10.4 Approximate Boiling Point and Fragrance of Some Esters

Ester	Approximate bp (°C)	Fragrance
Methyl butyrate	102	Apple
n-Propyl acetate	105	Pear
Ethyl butyrate	121	Pineapple
Isobutyl propionate	137	Rum
Isoamyl acetate	147	Banana
Isopentenyl acetate	153	Juicy fruit
Octyl acetate	204	Orange
Benzyl acetate	218	Peach
Ethyl phenylacetate	227	Honey
Methyl anthranilate	135 (15mm)	Grape

Esters are an important class of organic molecules that are formally derived from alcohols and carboxylic acids. Esters occur widely in nature and are often an important component of many compounds used in the flavor and fragrance industries. As illustrated in Table 10.4, simple structural variation of the alcohol or carbonyl portions of many esters markedly affects the fragrance of the ester.

Formation of an ester from an alcohol and a carboxylic acid (i.e., acetic acid) can be catalyzed by the addition of strong acids such as sulfuric or anhydrous hydrogen chloride. As is evident from the reaction sequence illustrated below, esterification is a reversible process. Esters can be hydrolyzed with water generated in the reaction. The equilibrium can be shifted to favor the ester by using excess of either the carboxylic acid or the alcohol, but purification of the product is made more difficult.

$$R\overset{\displaystyle O}{\underset{\displaystyle }{\|}}\!\!-\!OH + R'OH \underset{}{\overset{H^+}{\rightleftharpoons}} R\overset{\displaystyle O}{\underset{\displaystyle }{\|}}\!\!-\!OR' + H_2O$$

$$K_{eq} = \frac{[RCOOR']\,[H_2O]}{[RCOOH]\,[R'OH]}$$

An alternative solution is not to use the acid itself, but some more reactive derivative of the acid that has an equilibrium in its reaction with alcohols shifted far towards ester formation. Both acid chlorides and acid anhydrides are suitable reagents for inexpensive acids. The reaction of an alcohol with an acid anhydride (unlike the reaction of the free acid itself) is a highly exothermic process, strongly favoring the ester product. The reaction of alcohols with anhydrides can also be catalyzed by strong acids.

$$ROH + (CH_3CO)_2O \longrightarrow ROCOCH_3 + HO_2CCH_3$$

In this experiment you will obtain an unknown alcohol from your instructor and convert it into one of the acetate-ester natural products shown in the list by reaction with acetic anhydride. Sulfuric acid will be used as a catalyst. A slight excess of the anhydride will be used to ensure complete reaction of the starting alcohol. Separation of the acetic acid byproduct will be accomplished by extraction. The final ester will be purified by distillation. The identity of the ester, and consequently the starting alcohol, can be determined from its fragrance.

EXPERIMENT 54

Esterification of an Unknown Alcohol

Place 5 mL[18] of your unknown alcohol, along with a boiling chip, in a single-necked round-bottomed flask. Attach a Claisen adapter to the flask. Place a thermometer adapter with a thermometer in the main arm of the Claisen adapter so that the thermometer bulb is immersed in the liquid. Place a condenser in the side arm of the Claisen adapter. Add 6 mL of acetic anhydride *(CAUTION: Acetic anhydride is a severe skin irritant)* through the condenser and swirl or magnetically stir the resulting solution. Monitor the temperature for the first 5 minutes. Then remove the thermometer and adapter and add three drops of concentrated sulfuric acid. Replace the thermometer adapter and thermometer and again swirl or magnetically stir the solution while monitoring the temperature of the reaction mixture. When the internal temperature has returned to room temperature or after 20 minutes, transfer the mixture to a separatory funnel using a 25-mL ether rinse to aid the transfer. Extract the mixture with two or three 10-mL portions of *ice cold* 5% NaOH solution. Check the last extract with pH paper to be certain that it is basic. Wash the organic portion with 15 mL of a saturated sodium chloride solution. Pour the organic layer into an Erlenmeyer flask and dry it over anhydrous magnesium sulfate. Gravity filter the organic solution into a clean, dry, round-bottomed flask. Carefully evaporate the ether on the rotary evaporator or, preferably, distill the ether and product *(CAUTION: Make sure that there are no open flames near you)*. Collect the appropriate fraction ($> 100°C$) and record its boiling point and weight. Carefully wave your hand over the mouth of the distillate receiving flask and note the fragrance. Predict the structure of your ester and then deduce the structure of the starting alcohol. Once you have predicted the structure, calculate the yield of the reaction.

The structure of the ester may be verified by recording its NMR spectrum in $CDCl_3$. Alternatively, a gas chromatographic analysis of your ester and alcohol can be performed followed by a coinjection with authentic materials obtained from your instructor. In the gc experiment, first inject the ester and alcohol separately and determine their retention times. Next inject the authentic samples and note whether

---

[18] This reaction can also be performed conveniently on one-tenth the scale using microscale glassware and using a disposable pipette for the extraction transfers.

the retention times are the same or considerably different. If the results are inconclusive, mix approximately equal volumes of the unknown ester or alcohol with the corresponding authentic compounds. Inject the samples and note whether one or two peaks are found for the alcohol combination and the ester combination. If single peaks are observed in each case, it is reasonably safe to conclude that you have an alcohol and ester identical to the authentic samples.

Questions for Experiment 54

1. Draw structures of all 10 esters listed in the table for this experiment.
2. If you were esterifying an expensive alcohol with an expensive carboxylic acid, would it be wiser to use the acid chloride or acid anhydride reagent? Why?
3. Show the intermediates in the acid-catalyzed reaction of an alcohol with an acid anhydride, indicating the involvment of the strong acid catalyst in the reaction.
4. Why is so much NaOH used in the extraction since only three drops of sulfuric acid were used as a catalyst?
5. Why would the gc coninjection experiment with your product and an authentic sample not be totally conclusive evidence that you had correctly determined the structure of the product ester?

As indicated above, esters can be hydrolyzed to the corresponding acids and alcohols by treatment with aqueous acids. Alternatively, treatment of esters with strong base (saponification) is also used to generate the constituent alcohol and salt of the acid. This is commonly done to facilitate the structure determination of complex ester-containing molecules. In this experiment an ester (a triglyceride) will be isolated from nutmeg. From its melting point and that of the carboxylic acid that is obtained from the ester after alkaline hydrolysis, the ester can be identified.

$$
\begin{array}{ccc}
\mathrm{H_2C-O_2CR} & & \mathrm{H_2C-OH} \\
| & \xrightarrow[\text{2. neutralize}]{\text{1. KOH}} & | \\
\mathrm{HC-O_2CR} & & \mathrm{HC-OH} + 3\ \mathrm{RCO_2H} \\
| & & | \\
\mathrm{H_2C-O_2CR} & & \mathrm{H_2C-OH} \\
\text{triglyceride} & & \text{glycerol} \quad\quad \text{carboxylic acid}
\end{array}
$$

EXPERIMENT 55

Hydrolysis of Nutmeg

Place about 5 g of ground nutmeg in a 100-mL round-bottomed flask in the hood. Add 50 mL of ether, attach a reflux condenser, and heat the mixture (no flames!) under reflux for 15 minutes to extract the organic soluble triglyceride. Cool the flask to room temperature and gravity filter the ether solution directly into a round-bottomed flask. Evaporate the ether on a rotary evaporator to obtain a yellow oil or an off-white solid residue.

Dissolve the residue by adding 25 mL of warm acetone to the flask. Gravity filter the mixture into a small Erlenmeyer flask or round-bottomed flask. Allow the flask to stand for 15 minutes. Then cool the flask in an ice bath for 15 minutes. If crystals do not form, scratch the inside of the flask. Collect the white crystalline product by suction filtration. Wash the product with 1 mL of *cold* acetone. Air dry the product by drawing air through it on the filter funnel. Save enough of the product to determine its melting point. This product is not responsible for the flavor or odor of nutmeg, but is part of the fat content of the nut. If you want to find the odor chemical, evaporate the filtrate from your recrystallization.

Place 0.5 g of the recrystallized product (a triglyceride) from the extraction in a 100-mL round-bottomed flask and add 20 mL of a 2% ethanolic potassium hydroxide solution. Attach a water-cooled condenser and heat the mixture to reflux with a steam bath or oil bath for 45 to 60 minutes. Cool the mixture, dilute it with 20 mL of *distilled water,* and acidify it by adding about 20 mL of 10% hydrochloric acid containing an equal volume of ice. A white solid should separate during the acidification. Continue to add the hydrochloric acid solution until the precipitation is apparently complete. Dilute the mixture with another 50 mL of distilled water and suction filter the product. Wash the solid residue with two 5-mL portions of distilled water and air dry it on the suction filter. Recrystallize a portion of the solid from methanol-water using a Craig tube.

Record the melting points of both the triglyceride and the acid obtained from the saponification reaction.

Questions for Experiment 55

1. What are the structures of the triglyceride and the acid obtained in these experiments? (Look up "nutmeg" in the Merck Index or elsewhere for further clues.)

2. If the hydrolysis were carried out with standardized base on a known weight of the triglyceride, the saponification equivalent of the fat could be obtained. What is the saponification equivalent?

3. The triglyceride and the acid have similar melting points. What evidence can be cited to demonstrate that the acid was actually obtained rather than simply recovering the triglyceride after the saponification?

4. What chemical species is present in solution after treatment of the triglyceride with 2% ethanolic potassium hydroxide? Why is it soluble in water? Why is water added before acidification of this solution?

10.11 CHEMISTRY OF CARBONYL COMPOUNDS

Carbonyl groups (aldehydes, ketones, acids, esters, amides, and related derivatives) are among the most versatile functional groups in organic chemistry. Several of these groups can be prepared by oxidation of the corresponding alcohols or activated hydrocarbons (benzylic or allylic) by reactions with inorganic and organic oxidants. Permanganate can effectively oxidize side chains of alkyl benzenes to carboxylic acids. In Experiment 56 either *o*- or *p*-chlorotoluene are oxidized to the corresponding benzoic acid.

| *o*-chlorotoluene | *o*-chlorobenzoic acid | *p*-chlorotoluene | *p*-chlorobenzoic acid |

EXPERIMENT 56

Oxidation of *o*- or *p*-Chlorotoluene with Permanganate[19,20]

Mix 25 mL of water and 6.5 mL of 6N potassium hydroxide in a 100-mL round-bottomed flask and add 1.3 g of potassium permanganate. Stir or swirl the flask for 1 minute and add 0.5 g of either *o*- or *p*-chlorotoluene. Attach a reflux condenser to the flask and heat the mixture at reflux temperature for 2 hours. Destroy any excess permanganate by *cautiously* adding ethanol ($\sim$ 5 to 10 drops) through the top of the condenser *after removing the heat source.* Cool the mixture to room temperature or below with an ice bath. Carefully add 10% sulfuric acid until the solution is acidic as determined with pH paper. Gently boil the acidified mixture with stirring and add successively very small portions of solid sodium hydrogen sulfite to destroy the manganese dioxide. If necessary, add small amounts of water portionwise to make the mixture homogeneous. If the mixture is colored, add a small amount of decolorizing carbon. (Cool the solution to below the boiling point first to avoid boiling over when the charcoal is added.) Gravity filter the hot solution and cool the filtrate. Collect the crystals of the crude product by suction filtration and recrystallize the product from a suitable solvent or solvent pair. Suitable solvents for recrystallizing aryl carboxylic acids include water, aqueous ethanol, toluene, and ethyl acetate–hexanes. At 25°C, the solubility of *o*-chlorobenzoic acid is 210 mg/100 mL in water and *p*-chlorobenzoic acid is 8 mg/100 mL in water. Determine and record the mass yield, mole percent yield, and melting point of the purified product. The melting point of *o*-chlorobenzoic acid is 139 to 140°C and that of *p*-chlorobenzoic acid is 238

---

[19] H. T. Clarke and E. R. Taylor, *Organic Syntheses,* Coll. Vol. II, A. H. Blatt, ed. (New York: John Wiley, 1943), p. 135.

[20] D. J. Pasto and C. R. Johnson, *Organic Structure Determination* (Englewood Cliffs, N.J.: Prentice Hall, 1965), p. 343.

to 239°C. Based on your melting point data, which isomeric chlorotoluene did you start with? The corresponding methyl ester can also be prepared by the procedure described in Experiment 61.

Questions for Experiment 56

1. Write a balanced chemical equation for the oxidation of *p*-chloroethylbenzene with potassium permanganate.
2. Write a balanced equation for the reaction with ethanol that results in the destruction of the excess potassium permanganate in this experiment.
3. What would be the product of the reaction of 2,4-dimethylchlorobenzene with excess potassium permanganate?

Several permanganate oxidations are often facilitated by the presence of phase-transfer catalyst (PTC) reagents. For example, tetraalkylammonium chlorides have been used to assist the oxidation of terminal aliphatic alkenes to the corresponding carboxylic acids with one less carbon.[21] A nice sequence of reactions that also relies on a phase-transfer catalyzed permanganate oxidation is the preparation of vanillin from eugenol, a readily available natural product.[22] In this series, eugenol is first isomerized to isoeugenol with base and the phenolic hydroxyl group is protected as an acetate before the phase-transfer catalyzed permanganate reaction is used to cleave the alkene to an aldehyde oxidatively. Acid hydrolysis of the acetate produces vanillin in modest yield.

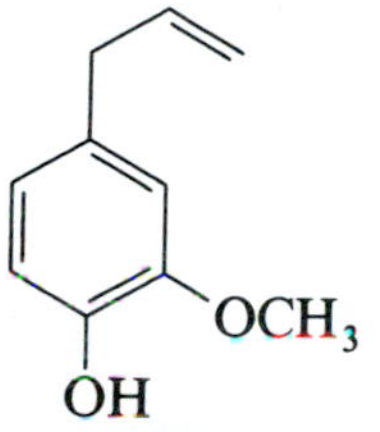

Phase-transfer catalyzed oxidation of aliphatic alkenes to carboxylic acids

$$CH_3(CH_2)_nCH{=}CH_2 + KMnO_4 + PTC \longrightarrow CH_3(CH_2)_nCO_2H$$

Phase-transfer catalyzed oxidation as the key step in the synthesis of vanillin from eugenol

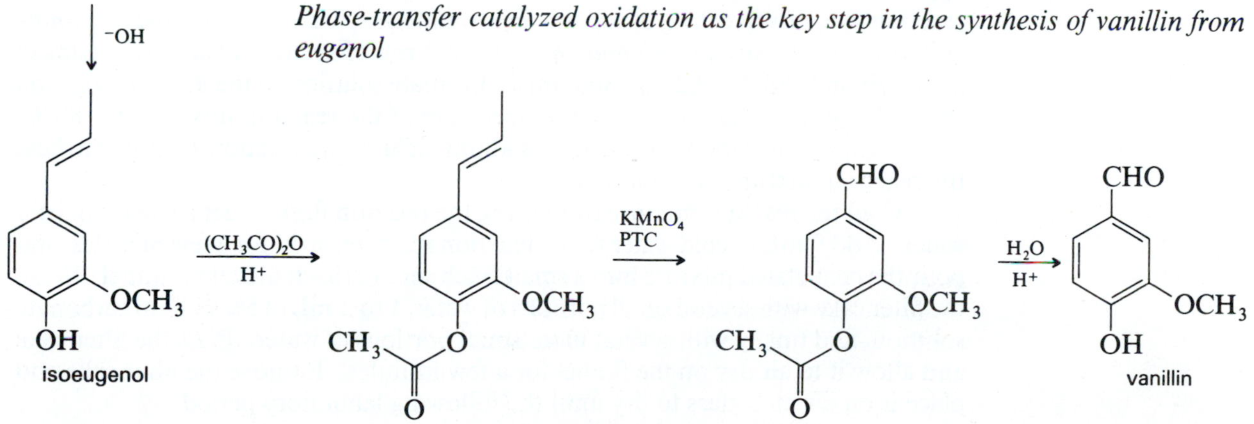

[21] K. C. Brown, V. S. Chang, F. H. Dar, S. E. Lamb, and D. G. Lee, *J. Chem. Ed.*, **59**, 696 (1982).
[22] G. M. Lampman and S. D. Sharpe, *J. Chem. Ed.*, **60**, 503 (1983).

The oxidation of fluorene to fluorenone can be accomplished by a number of oxidants, including oxygen in the air. The reaction represents another situation in which an activated hydrocarbon can be selectively oxidized. In this experiment, fluorene will be oxidized to the ketone, fluorenone, with sodium dichromate in acetic acid. The oxidant and conditions are chosen intentionally to result in an incomplete reaction. Separation of the starting material and product will then be accomplished by column chromatography. During and after the oxidation process, while the product mixture is drying, various types of chromatography that will be employed should be practiced. Review Sections 2.6 and 10.7 before beginning these experiments.

fluorene, mp 114°C $\xrightarrow[\text{CH}_3\text{CO}_2\text{H}]{\text{Na}_2\text{Cr}_2\text{O}_7}$ fluorenone, mp 82°C

EXPERIMENT 57

Oxidation of Fluorene to Fluorenone and Chromatographic Separation of the Fluorene and Fluorenone

Dissolve 0.3 g of fluorene in 1.5 mL of glacial acetic acid in a small flask by heating at approximately 80°C on a steam bath or sand bath in the hood. In another flask or reaction vial, dissolve 0.9 g of sodium dichromate dihydrate (*CAUTION: Chromate salts are very strong oxidants and are toxic*) in 3 mL of acetic acid also by heating at approximately 80°C. Add the sodium dichromate solution to the fluorene solution *in small portions,* maintaining the temperature of the reaction mixture at ~80°C. After completion of the addition of the sodium dichromate solution, continue to heat the reaction mixture for 30 minutes.

Cool the reaction mixture by swirling the reaction flask under a stream of cold water. Add 9 mL of cold water to the reaction mixture, swirl for a few minutes, and pour the coagulated mixture into a small Büchner or Hirsch filtration funnel. Wash the filter cake with several small portions of water, 1 to 2 mL of 5% sodium carbonate solution, and finally with several more small portions of water. Press the filter cake and allow it to air dry on the funnel for a few minutes. Remove the filter cake and place it on a watch glass to dry until the following laboratory period.

Analysis and separation of the fluorene-fluorenone mixture will be accomplished by thin-layer chromatography (TLC) and column chromatography. The analysis and separation are more effectively carried out when students work in pairs.

Using at least two different solvent systems (first try $9:1$ hexanes-ethyl acetate, $8:2$ hexanes-ether, or $8:2$ hexanes-dichloromethane), qualitatively analyze the crude reaction mixture, comparing it with authentic samples of fluorene and fluorenone on silica gel TLC plates. Comparisons should always be made on the same TLC plate. Try to find a solvent system that separates the two components with a difference in the R_f of at least 0.2 with the faster-moving component having an R_f value of about 0.3 to 0.5, but no greater. Chromatograms can be visualized with short wavelength ultraviolet light (~ 254 nm) (*CAUTION: Avoid looking directly at the ultraviolet light*) or iodine vapor. The TLC system may be worked out before the reaction product has thoroughly dried.

Using the best solvent system found for the TLC analyses (or one slightly less polar),[23] prepare a column (Section 2.6) using approximately 10 to 20 g of alumina or silica gel. Place a solution of approximately 50 to 200 mg of the fluorene-fluorenone mixture (determine the exact weight) in the minimum amount of the same solvent or a slightly more-polar solvent (dichloromethane or toluene work well) and apply it to the column. Elute the column first with a slightly less-polar solvent than used for the TLC analysis. Monitor the progress of the column chromatography by TLC analysis. Once the less-polar band (fluorene) has eluted, increase the polarity to elute the fluorenone. Evaporate the fractions containing the two compounds separately in tared round-bottomed flasks on the rotary evaporator. Determine the melting points of the products. Calculate the material balance. If possible, record the IR (KBr pellet) and NMR ($CDCl_3$) spectra of the two compounds and point out the characteristic differences.

Questions for Experiment 57

1. When an excess of both sodium dichromate and acetic acid is used, the oxidation of 2-acetylfluorene affords fluorenone-2-carboxylic acid and carbon dioxide. The dichromate is reduced to Cr^{+3}. Write a balanced equation for this transformation using H^+ to represent acetic acid.

2. What effect does using a more-polar solvent have on the R_f value in normal silica gel chromatography?

10.12 GRIGNARD REACTIONS

Grignard reactions are very useful for the formation of carbon-carbon bonds. The versatility of Grignard reactions was recognized early by the award of the Nobel Prize in chemistry to Victor Grignard in 1912. A few of the many possible applications of Grignard reactions are summarized below. Alkyl iodides react to form Grignard reagents more easily than do bromides or especially chlorides, which are formed with

---

[23] For a detailed description of the chromatographic separation of a fluorene-fluorenone mixture, see G. T. Miller and L. Milakofsky, *J. Chem. Ed.,* **59,** 1072 (1982).

difficulty. Bromides are most commonly used because they are readily available and more stable than the corresponding iodides. Experimental procedures for several representative examples of Grignard reactions follow.

$$R\text{—}Br + Mg \longrightarrow R\text{—}MgBr$$

$$\xrightarrow{\quad H_2O \quad} R\text{—}H, \text{ a hydrocarbon}$$

$$\xrightarrow{\quad D_2O \quad} R\text{—}D, \text{ a deuterated hydrocarbon}$$

$$\xrightarrow{\quad 1.\ R'\text{—}Br \quad} R\text{—}R', \text{ coupled product}$$

$$\xrightarrow[\text{2. H}^+/\text{H}_2\text{O}]{\quad 1.\ CO_2 \quad} R\text{—}CO_2H, \text{ a carboxylic acid}$$

$$\xrightarrow[\text{2. H}^+/\text{H}_2\text{O}]{\quad 1.\ R'CHO \quad} R\underset{\underset{H}{|}}{\overset{\overset{OH}{|}}{-}}R'$$

$$\xrightarrow[\text{2. H}^+/\text{H}_2\text{O}]{\quad 1.\ R'_2CO \quad} R\underset{\underset{R'}{|}}{\overset{\overset{OH}{|}}{-}}R'$$

$$\xrightarrow[\text{2. H}^+/\text{H}_2\text{O}]{\quad 1.\ R'CO_2R'' \quad} R\underset{\underset{R}{|}}{\overset{\overset{OH}{|}}{-}}R' + R''OH$$

EXPERIMENT 58

Preparation of a Grignard Reagent

Obtain an unknown bromide from your instructor. The unknown will be one of the following: bromobenzene, 2-bromotoluene, 3-bromotoluene, 4-bromotoluene, or 4-bromoanisole. In a hood, charge a clean, dry, 250-mL round-bottomed flask with 2.4 g (100 mmol) of magnesium turnings. Insert a Claisen adapter and attach both a dropping or addition funnel (see Figure 1.2) and a reflux condenser to the Claisen adapter. Attach a calcium chloride drying tube to the top of the condenser. All the joints should be *lightly* greased. *(CAUTION: Check to be certain that no flammable materials are in the area.)* Remove the condenser and heat the remaining apparatus with a gentle burner flame until all the water has been removed. The heating should be started at the bottom of the apparatus and slowly moved to the top. Be careful not to heat extensively near joints or clamps. If the condenser is left in the apparatus, moisture may collect inside of it and later drip into the reaction mixture to quench the reaction. Allow the apparatus to cool slightly, but while still warm attach the condenser with the calcium chloride tube. If the condenser is inserted into a joint while it is still hot, the joint will later contract slightly and make it impossible to remove the condenser. When the assembled apparatus is cool to the touch, add approximately 15 mL of fresh *anhydrous* ether and all 10 mL of your unknown into the dropping funnel. Stopper the dropping funnel.

Have an additional 15 mL of anhydrous ether ready in a dry stoppered flask. Start the water running through the condenser, and then add about 3 to 4 mL of the halide solution to the magnesium turnings. It may be necessary to temporarily remove the stopper from the addition funnel during the addition. If so, replace the stopper as soon as possible. Watch for the evolution of heat as evidenced by the boiling of the ether or just a slightly warming of the flask. Also watch for development of cloudiness (turbidity) in the reaction mixture. These are signs that the Grignard reaction has been initiated. If nothing happens after several minutes, try (1) adding a drop of methyl iodide (*CAUTION: Methyl iodide is a powerful alkylating agent*) to the flask, (2) adding a small crystal of iodine, (3) crushing a piece of Mg carefully against the side of the flask with a glass stir rod (being careful not to keep the reaction exposed to the atmosphere too long), or (4) immersing the flask in a sonicating bath and sonicating the solution for a few minutes. The use of a sonicating bath is preferred, if one is available. After the reaction has been initiated and appears to be well under way, add the second 15-mL portion of dry ether through the condenser (*not* through the drying tube) to moderate the reaction. If the reaction is too slow, gently warm it with a warm-water bath. Then add the remainder of the halide solution dropwise from the funnel over about a 15-minute period so as to maintain a steady reflux. When the addition is complete, wait for the reflux to slow (15 to 30 minutes) and then warm the flask with a hot-water bath to continue the reflux for another 10 minutes. Cool the reaction mixture to room temperature.

EXPERIMENT 59

Reaction of a Grignard Reagent with Carbon Dioxide

While the Grignard reaction is cooling, weigh out approximately 25 to 35 g of solid carbon dioxide (Dry Ice) in a small plastic bag (to minimize water condensation on the surface of the Dry Ice). Wrap the bag with a towel and crush the Dry Ice by hammering with a mallet, hammer, or other device. (Be careful not to break the plastic bag and expose the Dry Ice to the atmosphere, thereby allowing moisture to condense on the carbon dioxide.)

Quickly add the crushed Dry Ice to a dry 500-mL Erlenmeyer flask and add the cooled solution of the Grignard reagent slowly while gently stirring the resulting mixture. Slowly add 75 mL of technical-grade ether and cover the flask with two or three sheets of filter paper and an inverted small beaker. Allow the excess carbon dioxide to escape in a hood while covered. Experiments 58 and 59 must be completed to this point in one laboratory period.

Once carbon dioxide no longer appears to be bubbling out of the reaction mixture, carefully add 10 to 15 g of ice followed by 15 to 20 mL of water. Follow this with a mixture of 15 to 16 mL of concentrated sulfuric acid and about 35 g of ice. Cool the reaction mixture and dilute the ether layer to at least 100 mL with more technical-grade ether until all the solid present, if any, dissolves. Mix the solution thoroughly by swirling while keeping it cold. Pour the biphasic solution into a separatory funnel and separate the layers. Extract the aqueous layer with two 20-mL portions of ether. Combine the ether extracts.

Extract the acid product from the ether solution with three 25-mL portions of 5% aqueous sodium hydroxide. If necessary, add a small amount of decolorizing charcoal to the combined aqueous layers, stir, gravity filter, and cool. While stirring, precipitate the free carboxylic acid by dropwise addition of cold concentrated hydrochloric acid, until the milky suspension is strongly acidic to pH paper. Suction filter the mixture and wash the product with a small amount of *cold* water. Recrystallize the product from water, ethanol, methanol, or mixtures of these solvents as determined by small test tube trials. Dry the sample. Determine the weight, percent yield, and melting point. Compare your melting point with those of the benzoic acids listed below and identify your acid and starting aromatic bromide. Record and compare the IR spectra of your starting bromide (neat film) and acid product (KBr). Note the characteristic peaks corresponding to the carboxyl group (Sections 5.2.2 and 5.2.7).

Compound	Approximate mp (°C)
o-Toluic acid	100
m-Toluic acid	110
Benzoic acid	119
p-Toluic acid	178
p-Methoxybenzoic acid	183

EXPERIMENT 60

Reaction of a Grignard Reagent with Deuterium Oxide (D$_2$O)

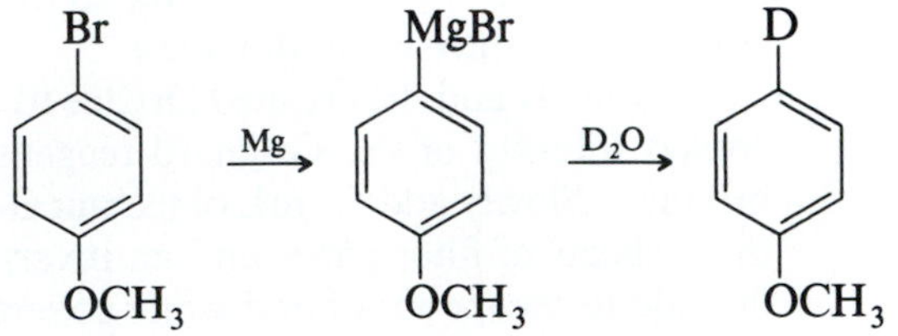

This experiment can be performed as an option. It will demonstrate the ease of quenching Grignard reactions with water and also provide an example of preparing a deuterium-labeled product. To facilitate isolation, Grignard reactions should be done starting with arylbromides that will eventually give a relatively high boiling product. Thus, p-methoxybromobenzene is a good choice of starting material.

Prepare p-methoxyphenylmagnesium bromide as indicated in Experiment 58. Instead of adding Dry Ice to the mixture, quench the Grignard by slowly adding 2 to 3 mL of D_2O down the top of the condenser to the ice-cooled Grignard reagent. A vigorous reaction may occur. Stir or swirl the reaction mixture for 10 to 15 minutes and then add 5 g of ammonium chloride in 15 to 20 mL of ice-cold ordinary water followed by an additional 15 to 20 mL of water. Stir or swirl the mixture to dissolve the solids.

Add about 50 mL of technical-grade ether to the mixture and transfer it to a separatory funnel. Separate the aqueous layer and extract the ether layer with another 10 to 15 mL of water. Wash the ether layer with 10 to 15 mL of saturated aqueous sodium chloride. Pour the ether layer into an Erlenmeyer flask containing a small amount of anhydrous sodium sulfate or magnesium sulfate. Gravity filter the solution into a round-bottomed flask and carefully evaporate the ether on a rotary evaporator or fractionally distill off the ether. Distill the product (or a portion of it with a Hickman still) and determine its boiling point. Record the NMR spectrum of the deuterated product and compare the spectrum to that of authentic methoxybenzene (anisole) and to that of the starting p-methoxybromobenzene.

Because it is difficult to exclude water completely while attempting to quench the Grignard with D_2O, deuterium incorporation in the product is seldom complete. However, as indicated in the following figures, the amount of deuterium incorporation can be readily determined by NMR spectroscopy. Figure 10.2a shows the H^1 NMR spectrum of methoxybenzene (anisole). Figure 10.2b is the same spectrum expanded in the aromatic region. Figures 10.2c and 10.2d show the corresponding normal and expanded proton decoupled ^{13}C NMR spectra. Figure 10.2e shows the ^1H NMR spectrum of the product after formation of the Grignard from p-methoxybromobenzene and quenching with D_2O. Expansion of the spectrum and integration in the aromatic region (see Figure 10.2f) allows determination of the ratio of deuterated to nondeuterated product (anisole, methoxybenzene), which is 78 : 22 in this case. The ^{13}C NMR (see Figure 10.2g) of the product mixture also reveals it to be a mixture. Expansion of a portion of the spectrum (see Figure 10.2h) further reveals the mixture. It is also interesting to note that the C_4 carbon attached to the deuterium is coupled to the deuterium and appears as a triplet with its chemical shift slightly offset from that of the undeuterated material because of the presence of the deuterium (isotope-induced shift).

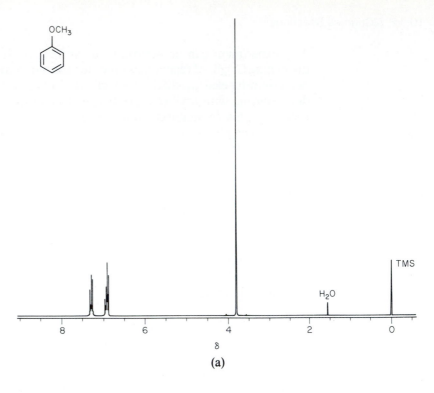

(a)

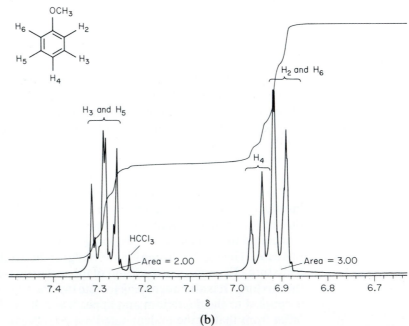

(b)

Figure 10.2 (a) ^{1}H NMR spectrum of methoxybenzene (anisole). (b) Expansion of the aromatic region of (a).

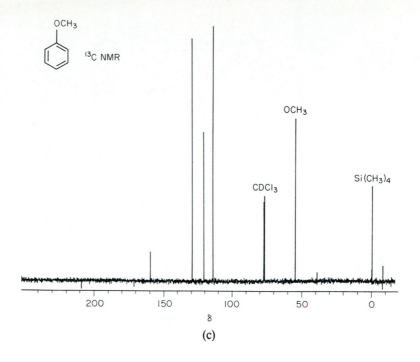

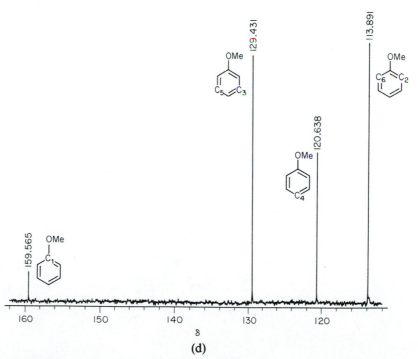

Figure 10.2 (c) Proton decoupled ^{13}C NMR spectrum of methoxybenzene. (d) Expanded region of (c).

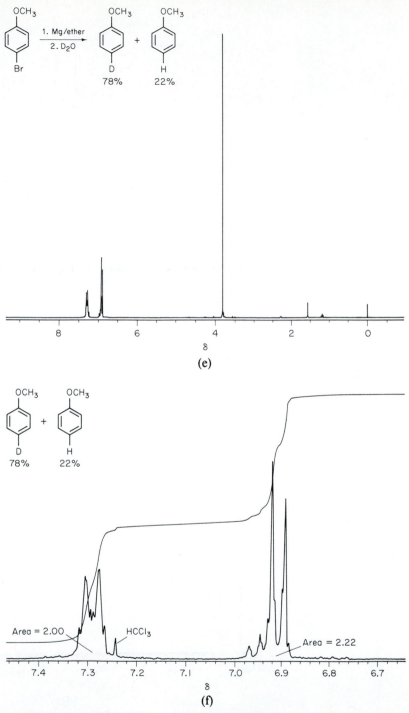

Figure 10.2 (e) ^{1}H NMR spectrum of the product derived from D$_2$O quench of the Grignard reagent. (f) Expansion and integration of the aromatic region of (e).

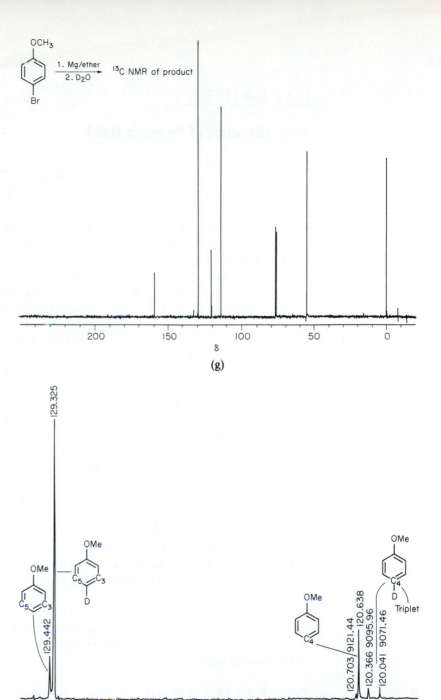

Figure 10.2 (g) Proton decoupled ^{13}C NMR spectrum of product from the Grignard reagent. (h) Expansion of (g).

EXPERIMENT 61

Esterification of Benzoic Acid

The following procedure describes an esterification on a small (1 g) scale. The process can be easily scaled up if a subsequent Grignard reaction is planned using the ester as a substrate (Experiment 62).

Place 1 g of benzoic acid or the substituted benzoic acid prepared in Experiment 59 and 5 mL of methanol in a round-bottomed flask and add 0.6 mL of concentrated sulfuric acid *slowly* to the mixture. Swirl or magnetically stir to mix the solution thoroughly. Attach a reflux condenser and reflux the mixture for 30 minutes using a preheated ($\sim$ 70 to 80°C) sand bath. Cool the solution and transfer it to a separatory funnel with the aid of 10 to 15 mL of ether. Add 5 mL of water and shake the separatory funnel. Discard the aqueous layer and wash the organic phase with a second 5-mL portion of water and then with three 5-mL portions of 5% sodium carbonate solution or until carbon dioxide evolution ceases. Wash with three 5-mL portions of saturated sodium chloride solution and dry the organic layer over anhydrous sodium sulfate in a small flask. Gravity filter the organic layer into a round-bottomed flask (rinse the salts with additional ether, filter, and add the rinses to the original organic layer). Evaporate the ether using a rotary evaporator. Dry the residual oil by adding $\sim$0.1 g of anhydrous sodium sulfate and swirl the mixture to remove any droplets of water. Transfer the crude methyl ester to a small conical vial by filtration through a cottom plug in a disposable pipet and distill it using a Hickman still head topped with a condenser.

Since the boiling points of the aromatic methyl esters are relatively high (methyl benzoate, bp = 198 to 200°C), preheat the sand bath, using a *minimum* amount of sand, to about 200°C, being careful not to touch the hot bath or hot plate. Alternatively, the methyl ester can be distilled at a lower temperature using vacuum distillation with a short-path distillation head (Figure 2.9a), a bulb-to-bulb distillation apparatus or Hickman still. Record the yield and boiling point of the product (if necessary, obtain the boiling point after the distillation by the ultramicro boiling point process described in Section 10.3). Obtain the IR spectrum of the ester and compare it to the IR spectrum of the starting acid.

EXPERIMENT 62

Preparation of an Amide from a Carboxylic Acid

Prepare the anilide from your benzoic acid by first making the acid chloride and reacting it with aniline as described in Section 9.7.1. Determine the melting point of the amide product. Record the IR spectrum of the amide and compare it to the spectra obtained for the acid and ester.

10.13 PREPARATION AND REACTIONS OF TRIPHENYLCARBINOL

Triphenylcarbinol is chemically a very interesting compound. It is easily prepared by the reaction of two equivalents of phenylmagnesium bromide with methyl benzoate.

$$2 \text{ PhMgBr} + \text{PhCO}_2\text{CH}_3 \longrightarrow (\text{Ph})_3\text{COH}$$

The following experimental description is based on starting with 4 g of methyl benzoate. If necessary, adjust the scale to the amount of methyl benzoate or other aryl ester available (from Experiment 61 or from your instructor) and prepare the appropriate amount of phenylmagnesium bromide by the procedure described in Experiment 58. Note that 2 mole of the phenylmagnesium bromide will be required for each mole of ester used.

EXPERIMENT 63

Preparation of Triphenylcarbinol

Place 4 g of methyl benzoate and 20 mL of anhydrous ether into the addition funnel of the previously assembled apparatus used for the preparation of the phenylmagnesium bromide. Cool the flask containing the phenylmagnesium bromide solution briefly in an ice bath. Open the stopcock of the addition funnel and add the methyl benzoate solution slowly with cooling only as necessary to control the mildly exothermic reaction. An intermediate addition compound may separate as a white solid. Swirl or vigorously stir magnetically the mixture until it cools to room temperature and the reaction has obviously subsided. Complete the reaction by either refluxing the mixture for 30 minutes or storing the mixture until the next laboratory period. *The experimental sequence should not be stopped before this point.*

Pour the reaction mixture into a 250-mL Erlenmeyer flask containing 50 mL of 10% sulfuric acid and about 25 g of ice and use a few mL of both technical-grade ether and 10% sulfuric acid to rinse the flask. Swirl or stir the mixture well to promote hydrolysis of the addition compound. During this process, basic magnesium salts are converted into water-soluble neutral salts and triphenylcarbinol is distributed into the ether layer. If a suspension results, add a small additional amount of ether. Pour the mixture into a separatory funnel. Rinse the flask with a few more mL of ether and add the rinse to the separatory funnel. Carefully shake the funnel (with proper venting) and allow the layers to separate. Remove the aqueous layer and extract the ether layer in the separatory funnel with about 5 mL of 10% sulfuric acid. Remove the acid layer and wash the ether layer with 10 mL of an aqueous saturated sodium chloride solution. Separate the layers and pour the ether layer into an Erlenmeyer flask containing a small amount of anhydrous sodium sulfate as a drying agent. Swirl the flask for a minute and then gravity filter the solution into a tared round-bottomed flask. Evaporate the ether on the rotary evaporator. The residual solid or thick oil is a mixture of triphenylcarbinol and biphenyl, a side product resulting from coupling of the Grignard reagent with some of the starting bromobenzene or magnesium-induced coupling of two bromobenzene molecules. Recrystallize the crude triphenylcarbinol from hexanes, ether-hexanes, or toluene. Collect the product by suction filtration and determine its weight and melting point. Triphenylcarbinol melts at 164 to 165°C. Save the product for later studies of the chemistry of triphenylcarbinol.

EXPERIMENT 64

Alternate Preparation
of Triphenylcarbinol from Benzophenone

benzophenone   phenylmagnesium   triphenylcarbinol
                    bromide

Starting with 7.9 g (0.05 mol) of bromobenzene and 1.4 g (0.55 mol) of magnesium turnings, prepare phenylmagnesium bromide as described in Experiment 58. Place 9.1 g (0.05 mol) of benzophenone in 50 mL of dry ether into the addition funnel of the Grignard apparatus. Cool the flask containing the Grignard solution with an ice-water bath to about 10 to 15°C and add the benzophenone solution slowly (dropwise) over about 10 to 15 minutes. Remove the cooling bath and heat the resulting mixture to reflux for 20 minutes. Cool the solution to room temperature and remove the condenser, funnel, and Claisen adapter. Pour the solution into a 500-mL round-bottomed flask that contains about 50 mL of crushed ice. Stir the mixture and carefully add 6 to 8 mL of concentrated hydrochloric acid to dissolve the basic magnesium salts. Separate the ether layer, dry it over anhydrous magnesium sulfate, and evaporate the ether on a rotary evaporator. Cool the flask and collect the crude, solid triphenylcarbinol by suction filtration. Air dry the solid and then recrystallize it from hexanes, ether-hexanes, or toluene. Cool the recrystallization solution and collect the product by suction filtration. Air dry the product and record its weight and melting point. Authentic triphenylcarbinol melts at 164 to 165°C. Save some of the product for later studies of the chemistry of triphenylcarbinol.

Questions for Experiments 58 to 64

1. If methyl iodide was used to help start the Grignard reaction, what unwanted acid byproduct would be formed upon later reaction with carbon dioxide? At what point of the purification would most of this acid be separated from the desired product?

2. Explain with the aid of balanced equations how the addition of iodine and 1,2-dibromo-ethane can aid in the initiation of a Grignard reaction.

3. How does sonication aid in the initiation of a Grignard reaction?

4. Provide a balanced chemical equation for the predominant side reaction (formation of biphenyl) that occurs during the preparation of phenylmagnesium bromide from bromo-benzene and magnesium in the absence of any oxygen and water.

5. What would be the product of the reaction of phenylmagnesium bromide with acetone? With ethyl acetate?

The following experiments with the triphenylcarbinol obtained from the Grignard experiments carried out earlier or from your instructor will illustrate the reactivity of triphenylcarbinol. As illustrated by these experiments, the chemistry of triphenylcarbinol is dominated by the stability of the triphenylmethyl carbocation.

EXPERIMENT 65

Conversion of Triphenylcarbinol to Triphenylmethane

Dissolve 0.5 g of triphenylcarbinol in 10 mL of warm glacial acetic acid. Add 2 mL of 57% HI. Heat the mixture on a steam bath or with a preheated sand bath for 1 hour in the hood. Note the formation of the iodine color indicating an oxidation of iodide to iodine and a reduction to give triphenylmethane. Cool the reaction and add 1 g of sodium bisulfite and 20 mL of water. Collect the precipitate by suction filtration. Wash the product on the filter funnel with water and recrystallize it from about 15 mL of methanol. Record the melting point of the product (93 to 94°C).

EXPERIMENT 66

Conversion of Triphenylcarbinol to Triphenylmethyl Acetate

Dissolve 0.5 g of triphenylcarbinol in 10 mL of warm acetic acid. Add 2 mL of an acetic acid solution containing 5% of chloroacetic acid and 1% of sulfuric acid. Heat the mixture for 5 minutes and then add 5 to 6 mL of water and let the mixture stand until the product crystallizes out. Collect the product by suction filter and wash the crystals with a 1 : 1 solution of methanol and water. Dry the crystals and record the melting point of the product (95 to 97°C). Perform a mixed melting point with triphenylmethane to confirm the difference between the two compounds.

EXPERIMENT 67

Conversion of Triphenylcarbinol to Triphenylmethyl Chloride

Dissolve 0.5 g of triphenylcarbinol in 10 mL of warm acetic acid. Add a hot solution of 2 g of stannous chloride in 5 mL of concentrated HCl. Heat the mixture for 1 hour on a steam bath or with a sand bath at 90 to 100°C. Cool the reaction mixture and collect the crystalline product by suction filtration. Wash the crystals with methanol and air dry them. Record the melting point of the product (110 to 112° C).

EXPERIMENT 68

Conversion of Triphenylcarbinol to Triphenylmethyl Bromide

Dissolve 1 g of triphenylcarbinol in 25 mL of methanol and add 1 mL of 48% hydrobromic acid. Heat the mixture gently for 15 minutes on a steam bath. Cool the solution and collect the product by suction filtration. Wash the product on the suction filter with methanol and dry it. Record the melting point of the product (152 to 154°C).

Question for Experiments 65 to 68

1. Why is triphenylcarbinol so susceptible to reaction with acids?

10.14 AROMATICITY AND THE CHEMISTRY OF AROMATIC COMPOUNDS

The properties and chemistry of aromatic compounds are different from those of other unsaturated systems. Very early on, many of these compounds were characterized by their odor. Later it was recognized that the same types of compounds are unusually stable, typically undergoing substitution reactions rather than addition reactions characteristic of alkenes. We now know that the structures of aromatic compounds all have similar properties. The compounds are cyclic, planar, and have $4n + 2\,\pi$ electrons. Another indication of aromaticity is the ability to sustain a ring current that significantly affects the chemical shifts of aromatic protons in the nuclear magnetic resonance spectrum. Several experiments described in this section will illustrate the tendency of aromatic compounds to undergo both electrophilic and nucleophilic substitutions. These fundamental reactions are extremely important.

Alkylation of aromatic compounds is usually accomplished by the reaction of alkyl halides with Lewis acids to generate carbocation intermediates that are the actual alkylating agents. The process involves a typical electrophilic aromatic substitution. The main problem with the reactions are rearrangements of the carbocation intermediates. This type of Friedel–Crafts reaction is often demonstrated by the alkylation of benzene. The cumulative toxicity of benzene has discouraged its use. The same type of chemistry can be demonstrated nicely by the double alkylation of less-volatile biphenyl with the t-butyl cation generated from t-butyl chloride.[24]

$$(CH_3)C_3-Cl + FeCl_3 \longrightarrow (CH_3)_3C^+$$

$$2(CH_3)_3C^+ \; + \; \text{biphenyl} \longrightarrow (CH_3)_3C-\!\!\text{—}\!\!-C(CH_3)_3$$

biphenyl   4,4′-di-t-butylbiphenyl

EXPERIMENT 69

Friedel–Crafts Alkylation: Preparation of 4,4′-Di-t-butylbiphenyl

Commercially available t-butyl chloride can be used or it can be generated from t-butyl alcohol just before use by the following process. Cool a 25- to 50-mL round-bottomed flask or Erlenmeyer flask containing 12 mL of concentrated hydrochloric acid in an ice bath in a hood. Add 4 mL of t-butyl alcohol and magnetically stir or swirl the mixture for 20 minutes. Carefully, withdraw the product t-butyl chloride layer with a disposable pipette or transfer the mixture to a small separatory funnel and separate the layers. Dry the halide over $CaCl_2$ in a vial or round-bottomed flask. Distill the product *in the hood* with a short path still, or with a Hickman still, collecting several fractions boiling between 49 to 50°C. *(**CAUTION:** Avoid contact or inhalation of this alkyl halide!)*

Place 1 g (6 mmol) of biphenyl in a 25-mL round-bottomed flask in the hood. As with many aromatic compounds, biphenyl should be handled with extreme care. Although it is not volatile like benzene, contact with the skin should be avoided. *Wear rubber gloves.* Add 5 mL of dichloromethane and a small stir bar and magnetically stir the mixture until all the biphenyl has dissolved. Add 2 mL (18 mmol) of t-butyl chloride with a syringe or graduated pipette. Continue to stir the solution and add 40 mg of anhydrous ferric chloride. To the top of the flask, immediately attach a T-tube connected to a gas trap containing 10% aqueous sodium hydroxide (Figure 1.14). Stir the solution for 20 to 30 minutes at room temperature or slightly higher (with the aid of a warm-water bath). Transfer the reaction solution into another small round-bottomed flask or a small separatory funnel containing 4 to 5 mL of 10%

[24] D. A. Horn, *J. Chem. Ed.*, **60**, 246 (1983).

HCl. Stir the mixture and separate the layers. If the reaction mixture had been transferred to a flask rather than a separatory funnel, the bottom organic layer can be removed by carefully inserting a disposable pipette through the top aqueous layer and withdrawing the bottom organic layer. Extract the organic layer with several more portions of 10% HCl. Dry the organic layer over anhydrous calcium chloride or sodium sulfate. Gravity filter the solution, or transfer the solution to another tared round-bottomed flask using a disposable pipette containing a cotton plug to remove the drying agent (Figure 2.1c). Remove the solvent with a rotary evaporator and determine the crude yield. Recrystallize the residual 4,4'-di-*t*-butylbiphenyl from a minimum amount of 95% ethanol. Determine the yield and melting point. The authentic product has a melting point of 127 to 128°C. Record the IR spectra (KBr pellet) of the starting material and product. What characteristic bands, typical of a monosubstituted benzene ring, allow you to distinguish the starting material from the product, a *p*-disubstituted aromatic system? Record the NMR spectra of both the starting material and product in $CDCl_3$ and describe how they can be used to distinguish the two.

Electrophilic aromatic substitution reactions on benzene or its derivatives often produce mixtures of isomeric products. Mesitylene is a symmetrical trisubstituted aromatic system. Thus, substitution at any of the remaining three positions provides the same product. Subsequent dinitration is unlikely because the introduction of the first nitro group deactivates the ring toward further electrophilic substitution.

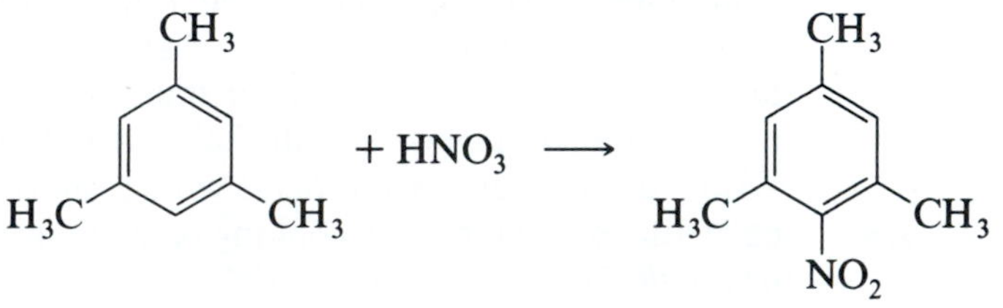

EXPERIMENT 70

Nitration of 1,3,5-Trimethylbenzene (Mesitylene)

Being very careful to avoid contact with your skin *(wear gloves!)* or clothing, place 1 mL of concentrated nitric acid in a 5-mL conical reaction vial containing a stir wedge, or in a small (10- to 25-mL) round-bottomed flask with a stir bar, set up in a hood. Cool the flask in an ice bath and while magnetically stirring, *carefully* add

1 mL of concentrated sulfuric acid with a pipette. (*CAUTION: This mixture will cause a severe reaction with skin and clothing; be especially careful.*) Suspend a thermometer in the solution so the bulb is immersed as far as possible without interfering with the action of the stir bar. While stirring, add 0.6 g of mesitylene dropwise with a disposable pipette or syringe while maintaining the temperature at or near 35°C with the ice bath. When all the mesitylene has been added, continue to stir the mixture for 30 minutes and then *carefully* transfer it with a disposable pipette to another flask containing about 10 g of crushed ice. When all the ice has melted, collect the product by suction filtration on a small Büchner funnel. Determine the melting point of the crude product. Recrystallize the product from 95% ethanol using a Craig tube. Determine the melting point of the pure product and compare it with that of the crude product. Record the IR spectra of the starting mesitylene (as a neat film between salt plates) and the nitrated product (as a KBr pellet). Note the difference the nitro group makes. Record the NMR spectra of the starting material and product in $CDCl_3$. The NMR spectra should reflect the symmetry of the starting material and product, yet allow you to distinguish the two by a predictable chemical shift difference in the aromatic region, and the presence of the methyl peaks at higher field.

Questions for Experiments 69 and 70

1. Another method of generating the *t*-butyl cation for Friedel–Crafts alkylation is to treat isobutylene (2-methylpropene) with a strong acid such as sulfuric acid in the presence of an aromatic compound. Show how this reaction works.

2. Realizing that carbocations are susceptible to rearrangement, what would be the major product from the Friedel–Crafts alkylation of benzene with 1-chloro-2,2-dimethylpropane?

3. Write the mechanism for the conversion of *t*-butyl alcohol to *t*-butyl chloride with HCl.

4. Show the products which would be obtained from the mononitration of toluene.

The utility of Friedel–Crafts alkylations and acylations as well as aromatic nitration reactions is well established. A nice combination of these two processes has been reported to facilitate the synthesis of 2-acetyl-1,4-naphthoquinone from 1-naphthol.[25] The reactions are shown below. Friedel–Crafts acylation of 1-naphthol with acetic anhydride and zinc chloride produces 2-acetyl-1-naphthol in good yield. Nitration with fuming nitric acid produces 2-acetyl-4-nitro-1-naphthol. Subsequent reduction of the nitro group gives the corresponding amine that readily oxidizes to 2-acetyl-1,4-naphthoquinone. With careful use of the corrosive reagents, this series of reactions can be easily carried out in the undergraduate laboratory.

---

[25] I. R. Green, *J. Chem. Ed.,* **59**, 698 (1982).

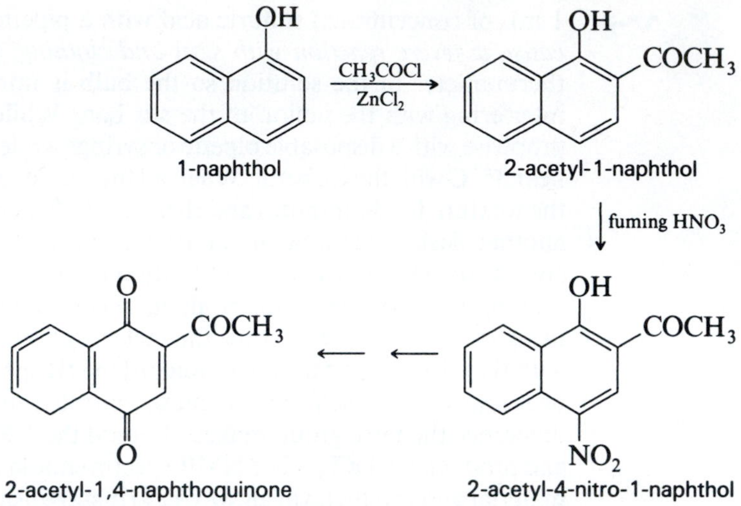

1-naphthol → 2-acetyl-1-naphthol (via CH_3COCl, $ZnCl_2$)

fuming HNO_3

2-acetyl-1,4-naphthoquinone ← ← 2-acetyl-4-nitro-1-naphthol

A single electrophilic substitution on toluene can produce three products, the *ortho-*, *meta-*, or *para*-toluenesulfonic acid. In practice, *para*-toluenesulfonic acid is the main product because of the steric hindrance at the *ortho* positions and deactivation towards *meta* substitution. (*CAUTION: Strong acid is also used in this experiment, so use extreme care and work in your hood.*)

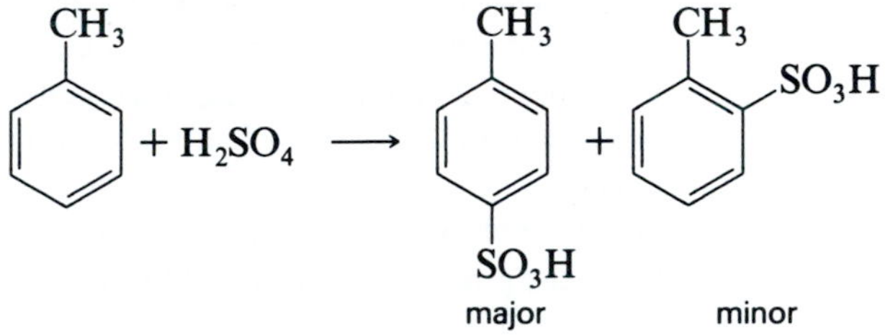

toluene + H_2SO_4 → (para, major) + (ortho, minor)

EXPERIMENT 71

Sulfonation of Toluene

Carefully place 8 mL of toluene and 5 mL of concentrated sulfuric acid in a 25-mL round-bottomed flask containing a $\frac{1}{2}$-in. magnetic stir bar. Attach a reflux condenser and stir the two-phase mixture vigorously while gently heating the flask with a preheated sand bath until the toluene begins to boil. Reflux the solution for about 20 minutes or until just a thin or no layer of toluene remains. Remove the flask from the heat source and while still warm pour the mixture into 25 mL of distilled water in a 100-mL beaker. (*CAUTION: Be careful, wear your gloves!*) Partially neutralize the strongly acidic solution by adding 4 g of sodium bicarbonate portionwise. Addition of excess bicarbonate will be detrimental. The solution may foam during this process and should be stirred gently during and after the addition of each portion of the bicarbonate. Add 10 g of solid sodium chloride to the solution to "salt out" the product. During the addition, the p-toluenesulfonic acid should begin to separate as white plates. Stir or swirl (magnetic stirring of the suspension may not be effective) the mixture until all the granular salt has dissolved and a maximum amount of the sulfonic acid appears to have separated. Cool the flask in an ice bath and collect the product by suction filtration on a small Büchner funnel. Rinse the flask with a few mL of saturated sodium chloride and use the rinse to wash the crystals of product. *(CAUTION: Do not wash the product with water since it will dissolve!)*

Dissolve the crude product in a small amount (15 to 25 mL) of boiling water. Add 4 to 5 g of solid sodium chloride portionwise and stir or swirl until a homogeneous solution is obtained. Cool the solution thoroughly in an ice bath. If necessary, scratch the inside of the flask to induce crystallization. Collect the product by suction filtration on a small Büchner funnel. Wash the crystals with a few mL of cold methanol. The product is reasonably hydroscopic so extended air drying may be detrimental. Alternatively, a portion of the product can be placed in a round-bottomed flask on a rotary evaporator and evacuated with gentle warming. Record the melting point of the product. Comparison of the infrared spectra of the starting material (neat film between salt plates) and product (KBr pellet) will illustrate the diagnostic sulfonic acid bands.

Questions for Experiment 71

1. Why would addition of excess bicarbonate be detrimental during the described partial neutralization?
2. Write the mechanism for the sulfonation.
3. Sulfonation of aromatic compounds is often reversible. Write the mechanism for the reverse reaction.

Electrophilic aromatic substitution reactions have often been used as the key step in the synthesis of important therapeutic agents. In this experiment you will synthesize sulfanilamide **9**, a compound commonly used for the treatment of bacterial infections in humans before the discovery of more-effective antibiotics. The synthesis is multistep and will not only introduce new reactions but also protecting group strategy that is often required in organic synthesis. With efficient work, the entire process can be completed in a single 3-hour laboratory period.

The first step involves the synthesis of trifluoroacetanilide **5** followed by its reaction with an electron-deficient species **3** to effect the electrophilic aromatic substitution. This reaction is facilitated by the lone pair of electrons of the nitrogen of **5** that, through resonance, helps stabilize the intermediate **6**. Reaction of the sulfonyl chloride **7** with ammonia gives the sulfonamide **9** by reaction at both the sulfonyl chloride and the trifluoroacetyl groups. The overall reaction is usually depicted as proceeding according to the scheme. The electrophile **2** is generated as shown in Equation (10.1).

$$H^+ + ClSO_3H \longrightarrow ClSO_2OH_2^+ \longrightarrow ClSO_2^+ + H_2O \qquad (10.1)$$

$$\underset{\mathbf{1}}{} \qquad\qquad \underset{\mathbf{2}}{} \qquad\qquad \underset{\mathbf{3}}{}$$

Scheme

$$\mathbf{10} + \mathbf{10} \longrightarrow NH_2\text{—}\!\!\bigcirc\!\!\text{—}SO_2NH\text{—}\!\!\bigcirc\!\!\text{—}SO_2Cl$$

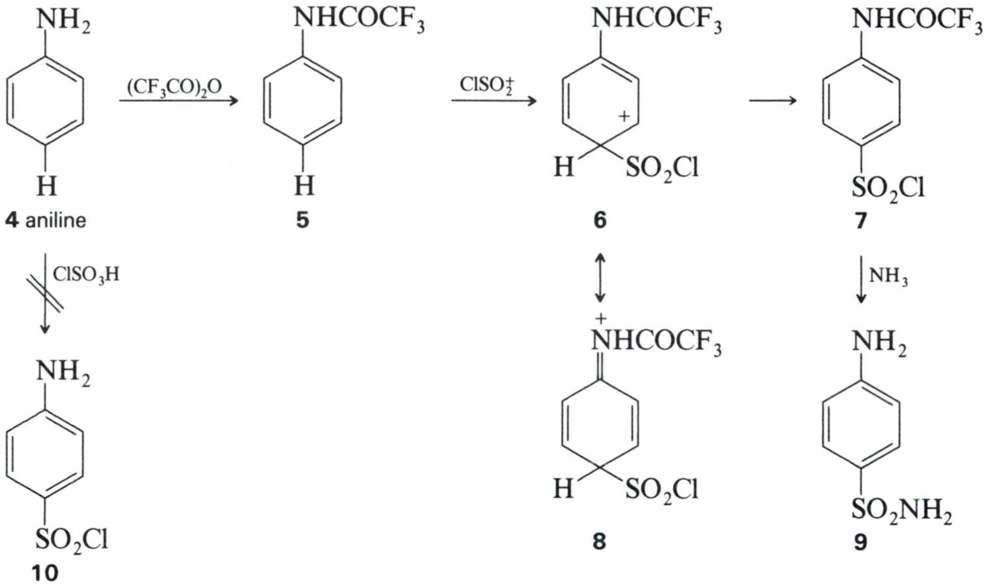

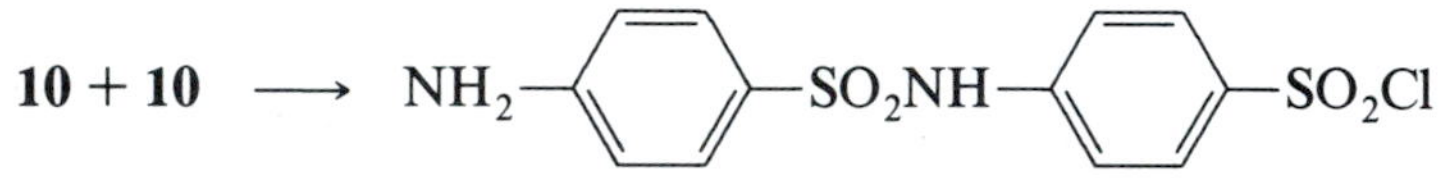

11

You should note that the complete synthetic scheme includes aniline **4** as the precursor to your key reactant, trifluorocetanilide. Aniline also contains a nitrogen with a lone pair of electrons attached to the ring and, therefore, should also facilitate the electrophilic aromatic substitution. Why, then, prepare trifluoroacetanilide since the trifluoroacetyl group is removed later?

The reaction of aniline with chlorosulfonic acid could proceed directly. Conceptually, simple reaction of **10** with ammonia should produce sulfanilamide **9**. However, **10** also contains a free amino group (such as ammonia) that can react either with chlorosulfonic acid or the sulfonyl chloride **10** produced to form **11**. Further reaction would result in a polymeric mess. Also, the amino group of aniline would be protonated, decreasing the reactivity of the ring and directing substitution to the *meta* position. Therefore the amino group of aniline must be *protected* by first converting it to an amide (trifluoroacetanilide **5**) that is not nucleophilic enough to react with the sulfonyl chloride. This concept of using a protecting group for one functional group while reacting another functional group is commonly used in organic synthesis.

C A U T I O N : *Chlorosulfonic acid is extremely corrosive.* The chlorosulfonic acid will be dispensed from a special apparatus (a commercial repipetter works well). Wear your gloves at all times. Do not breathe any of the fumes that might be generated from the reagent. If the reagent comes in contact with your skin or clothing wash thoroughly and immediately with cold water. Should chlorosulfonic acid come in contact with your gloves, carefully, but rapidly, remove the glove and wash your hand. Do not use disposable gloves, since if you try to rinse the acid from these gloves while wearing them, the resulting exothermic reaction [Equation (10.2)] will melt the glove and allow the acid to burn your skin.

$$H_2O + ClSO_3H \longrightarrow HCl + H_2SO_4 + heat \tag{10.2}$$

Be especially careful. Any glassware containing chlorosulfonic acid should be inverted over an ice bath and the acid allowed to drop into the ice. Do not add water to any container having even traces of the acid. An explosive reaction may result between the chlorosulfonic acid and water.

EXPERIMENT 72

Preparation of Sulfanilamide[26]

*(**CAUTION:** This experiment must be carried out in a hood.)* To a flame-dried 5-mL flask connected with a T-tube to a fume trap (see Figure 10.3) add aniline (0.23 mL) and dichloromethane (0.25 mL) using a syringe. Cool this solution in an ice bath and rapidly add trifluoroacetic anhydride (0.50 mL) diluted in dichloromethane (0.25 mL) with a pipette. After standing for 1 minute, replace the ice bath with a hot-water bath and continue to warm the flask and evaporate the dichloromethane and trifluoroacetic acid through the fume trap until a solid residue **(5)** remains.

Cool the flask containing the crude solid **(5)** in the ice bath and then remove the T-tube and slowly add chlorosulfonic acid (0.9 mL) with a syringe. Immediately

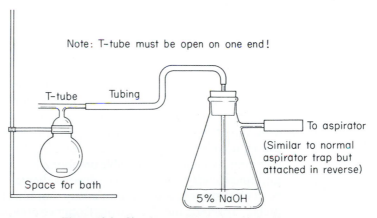

Note: T-tube must be open on one end!

T-tube   Tubing

To aspirator

(Similar to normal aspirator trap but attached in reverse)

Space for bath

5% NaOH

Figure 10.3   Simple apparatus to trap evolved gases.

[26] E. C. Hurdis and J. W. Yang, *J. Chem. Ed.,* **46,** 697 (1969).

replace the T-tube connection to the fume trap and allow the mixture to stand in the ice bath for 10 minutes. Replace the ice bath with a preheated sand bath (100°C) or a steam bath and heat the mixture for 10 minutes. Cool the mixture in an ice bath and carefully pour it into 5 mL of crushed ice in a small beaker. Filter off the resulting solid (7) and wash it with 2 to 3 mL of ice water. Air dry the solid for 10 minutes. A typical yield is about 0.45 g (62%) of a pink solid (mp 142 to 145°C).

Place the solid (7) in a 5-mL flask and add 1.25 mL of concentrated ammonium hydroxide and 1.25 mL of water. Attach a condenser and reconnect a fume trap. Heat the mixture with a preheated sand bath (100°C) or a steam bath until the solid dissolves. Heat an additional 1 minute, then cool the solution in an ice bath for about 15 minutes. The resulting crystals (9) should be filtered and air dried for 10 minutes. A typical crude yield is 0.18 g (42%); mp 154.5 to 157°C. Recrystallization from 1 to 2 mL of water using a Craig tube should provide about 0.12 g (28%) of the pure product (9) with mp 159 to 163°C (lit. mp 163 to 164°). Record the melting point and yield. Record the IR spectrum (KBr) and note the position of the sulfonamide bands relative to those of the sulfonic acid prepared in Experiment 71.

Questions for Experiment 72

1. Draw three other resonance structures of 6.
2. Draw four resonance structures for the initial intermediate derived from *ortho* attack of $ClSO_2^+$ upon trifluoroacetanilide. Why does *para*- and not *ortho*-substituted product form in this experiment?
3. Draw the major resonance structures for the initial intermediate derived from *meta* attack of $ClSO_2^+$ upon trifluoroacetanilide. Why does *para*- and not *meta*-substituted product form in this experiment?
4. The sulfonyl chloride 7 can hydrolyze by reaction with water to a sulfonic acid $ArSO_3H$. Since aqueous ammonia was used in the reaction of 7, why does the amide 9 and not the sulfonic acid form?

There are many other types of organic compounds that possess chemical properties very similar to benzene, naphthalene, and so forth and their derivatives. They are generally referred to as nonbenzenoid aromatic compounds. The following experiments will illustrate how aromatic properties of nonbenzenoid compounds can be used in organic and organometallic chemistry. Dicyclopentadienyl iron, or ferrocene, is a remarkably stable organometallic compound that consists of a ferrous ion sandwiched between two cyclopentadienide anions. The bonding involves the six π electrons of each anion in a way that binds every carbon atom equally to the metal. Together, the 12 electrons from two cyclopentadienide anions and 6 electrons from the ferrous cation result in the stable 18-electron configuration of the inert gas krypton. Ferrocene is stable to 450°C and is soluble in organic solvents.

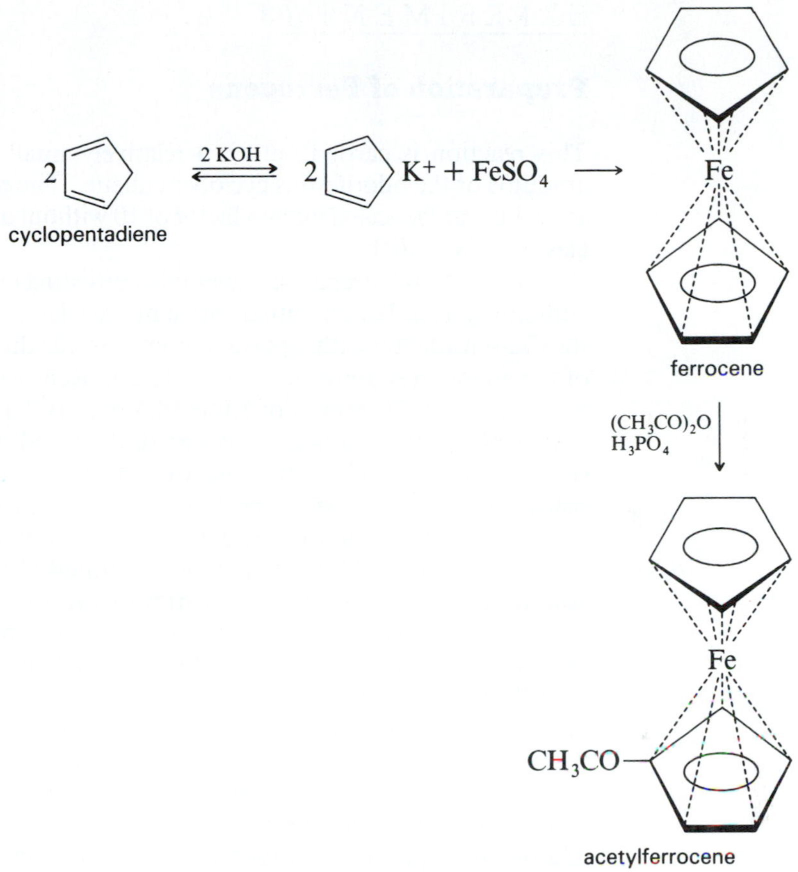

cyclopentadiene

ferrocene

acetylferrocene

The chemistry of ferrocene suggests that it has aromatic properties. It undergoes electrophilic substitution reactions with ease. However, it is susceptible to air oxidation to the blue ferricinium cation. Thus, the preparation and reactions of ferrocene should be carried out with minimal exposure to air.

The first step in the procedure will be to prepare the anion of cyclopentadiene. While most hydrocarbons and alkenes are not very acidic, deprotonation of cyclopentadiene leads to the formation of a planar cyclic molecule with 6 π electrons. Thus, the acidity of cyclopentadiene ($pK_a = 15.5$), due to the formation of the relatively stable cyclopendienyl anion, emphasizes again the importance of aromaticity. Under purely aqueous conditions, reaction of KOH would produce only a small amount of the cyclopentadienyl anion in equilibrium with the cyclopentadiene itself. Other polar, highly solvating solvents such as 1,2-dimethoxyethane (glyme) and dimethyl sulfoxide (DMSO) also work well.

EXPERIMENT 73

Preparation of Ferrocene

This reaction is carried out on a relatively small scale to avoid the use of large amounts of the odoriferous cyclopentadiene. The reaction works quite well on this scale, but can be scaled up by a factor of 10 without a problem if only large laboratory glassware is available.

In the hood, prepare an assembly consisting of a 25-mL round-bottomed flask containing a stir bar and fitted with a micro Claisen adapter. Close the top joints of the Claisen adapter with septa and insert a needle through the septum on the side arm of the adapter (as a pressure-release tube). Remove the Claisen adapter and in the flask place 4 g (70 mmol) of KOH (flakes or pellets), followed by 0.8 g (12 mmol, 1 mL) of freshly distilled cyclopentadiene[27] and 10 mL of 1,2-dimethoxyethane (glyme). Immediately reattach the Claisen adapter and stir or swirl the mixture as the black solution of potassium cyclopentadienide is formed.

Meanwhile, in a small Erlenmeyer flask prepare a solution of 1.2 g (7.4 mmol) of ferrous chloride dihydrate in 6 mL of dimethyl sulfoxide (*CAUTION: DMSO is readily absorbed by the skin),* by stirring or swirling the flask over a gentle steam bath for a few minutes (not more than 5 minutes; also be certain that no moisture from the steam bath is carried over into the very hygroscopic DMSO solution). Cool the solution to 15 to 20°C in an ice bath and transfer it portionwise by syringe to the previously assembled flask containing the potassium cyclopentadienide. (Inject the syringe through the septum on the main arm of the Claisen adapter and add the ferrous chloride solution dropwise.) The entire ferrous chloride solution should be added over a 10-minute period while stirring or swirling vigorously. Stir or swirl an additional 5 minutes and then pour the reaction mixture into a 100-mL beaker containing 20 g of ice and 17.5 mL of 6N HCl (concentrated HCl is 12N). Rinse the reaction flask with about 2 mL of water and add the rinse to the beaker and stir with a glass rod until the ice has melted. The experiment may be interrupted at this point, but *not earlier.*

Suction filter the reaction mixture using coarse filter paper in a Hirsch funnel, and maintain liquid in the funnel until all the mixture has been introduced. Wash the crude ferrocene on the filter with about 2 mL of water to remove the blue color (from oxidation of ferrocene to the blue cation $[Fe(C_5H_5)_2]^+$).

Dry the product by drawing air through the filter. The product is readily purified by sublimation (Section 2.4), especially if a small sublimation apparatus is constructed from a test tube and a 125-mL suction flask (see also Experiment 33 and Figure 10.1). Careful recrystallization from hexanes or methanol also effects purification. Small samples (< 50 mg) can also be purified by chromatography on silica gel

---

[27] Fresh cyclopentadiene is prepared by carefully fractionally distilling ("cracking") dicyclopentadiene (see Experiment 85). This can be done by one person for a whole class if the cyclopentadiene is stored cold and used the same day.

in a small column prepared from a disposable pipette (see also micro flash chromatography in Experiment 23). Larger samples of 0.5 to 1.0 g can be purified by modified flash chromatography with a 100-mm column of silica gel in a 300- $\times$ 19-mm column using dichloromethane as the eluent and balloons as the pressure source.[28] This method is especially effective for separating ferrocene from the acetyl ferrocene after the acetylation reaction described next. Record the melting point and NMR spectrum (in $CDCl_3$) of the purified ferrocene. The melting point should be taken in a sealed capillary melting point tube to avoid oxidative decomposition and sublimation during the determination.

EXPERIMENT 74

Acetylation of Ferrocene

Fit a 10-mL flask with a magnetic stir bar and a drying tube filled with anhydrous calcium chloride. In the flask place 0.1 g (0.54 mmol) of purified ferrocene and add exactly 1 mL (10.5 mmol) of acetic anhydride. To the reaction mixture, *carefully* add 0.2 mL of 85% phosphoric acid with stirring. With the drying tube attached, heat the flask in a preheated sand bath at 100°C for about 10 minutes. Allow the reaction mixture to cool to room temperature and pour it into about 4 to 5 g of ice in a beaker. Gradually and with thorough stirring, add small quantities of crushed solid NaOH until the solution is neutral to pH paper. When the mixture has again cooled to 20°C, collect the product by suction filtration on a small Büchner or Hirsch funnel. Draw air through the product for about 15 minutes to dry it. Purify the crude acetylferrocene by small-scale column chromatography with dichloromethane as the eluent on silica gel as described in Experiment 23. Alternatively, the product can be chromatographed on a neutral alumina column using about 2.5 g of adsorbent. In the latter case, elute with a mixture of 25% ether and 75% hexanes until the red-orange band of acetylferrocene has passed through the column. Recrystallize the chromatographed acetylferrocene from hexanes and record the yield and melting point. Record the NMR spectrum of acetylferrocene in $CDCl_3$ and compare it to that of ferrocene. Account for the differences in the spectra.

Question for Experiment 74

1. Explain why only one peak at δ 4.1 is observed in the NMR spectrum of ferrocene.

Cyclopentadiene used in the synthesis of ferrocene (preceding section) is unusually acidic because the conjugate base is aromatic. Indene ($pK_a = 20$) and fluor-

---

[28] W. L. Bell and R. D. Edmondson, *J. Chem. Ed.*, **63,** 361 (1986).

ene ($pK_a = 23$) are also relatively acidic cyclic hydrocarbons that yield aromatic anions by deprotonation. Like cyclopentadiene, the anions can be induced to undergo a number of useful reactions. While the cyclopentadienyl anion reacts readily with the carbonyl groups of aldehydes and ketones, the corresponding reactions with aromatic anions from indene and fluorene often require vigorous conditions or phase-transfer catalysis (PTC). The condensations involve nucleophilic attack of the carbanion at a carbonyl carbon followed by dehydration to give a carbon-carbon double bond. Thus, the following experiment illustrates carbanion formation, aromatization and the ($4n + 2$) π-rule, an aldol-like condensation, dehydration, and alkene synthesis.[29]

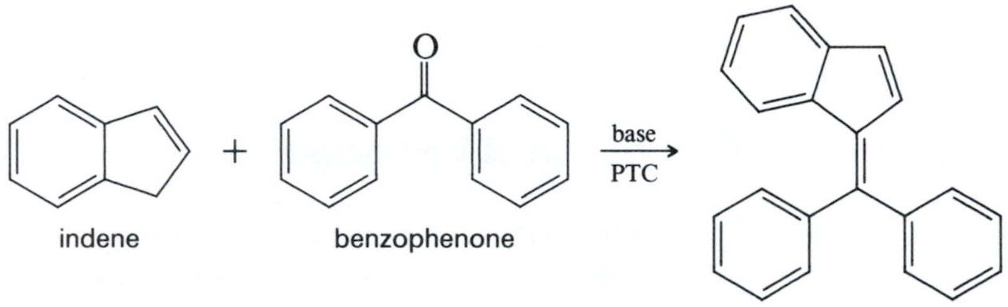

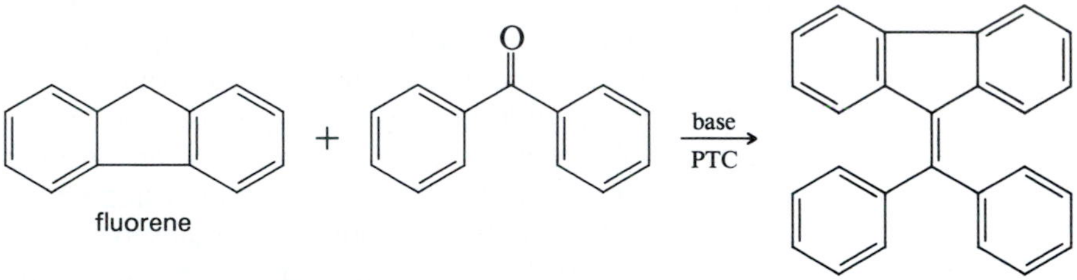

EXPERIMENT 75

Preparation of a Fulvene

In the hood, dissolve 1.16 g (10 mmol) of indene in 5 mL of xylene and add the solution to 1.2 g (30 mmol) of sodium hydroxide pellets in a 25-mL round-bottomed flask containing a $\frac{1}{2}$-in. magnetic stir bar. Carefully add 0.34 g (1 mmol) of tetrabutylammonium hydrogen sulfate (the phase-transfer catalyst) followed by a small drop of water and 1.82 g (10 mmol) of benzophenone. Attach a reflux condenser and heat the mixture with a preheated sand bath while stirring vigorously. The initial temperature of the sand bath should be 100 to 150°C. However, as soon as the reaction appears to start it becomes exothermic and the flask must be removed from the sand bath. This can be conveniently done if the sand bath and magnetic stirrer are

[29] Adapted from J. W. Hill, J. A. Jensen and J. G. Yaritz, *J. Chem. Ed.,* **63,** 916 (1986).

mounted on a laboratory jack that can be lowered once the reaction starts. If necessary, moderate the exothermic reaction with a cool-water bath or ice bath. The solution should be dark green from the indenyl anion. Allow the reaction to proceed until the temperature decreases by itself. Cool the flask and carefully add 6N HCl until the aqueous layer is acidic to pH paper. Separate the aqueous layer by extending a disposable pipette through the top organic layer and carefully withdrawing the acid solution. Evaporate the solvent on the rotary evaporator and recrystallize the crude fulvene derivative (benzhydrylideneindene, mp 113 to 114°C) from 95% ethanol or 2-propanol. Record the melting point and the NMR and IR spectra.

Substitution of fluorene for indene produces benzhydrylidenefluorene (mp 222 to 224°C). Reaction of fluorene with benzaldehyde under the same conditions produces benzylidene fluorene (mp 73 to 75°C).

Pyrrole is isoelectronic with the cyclopentadienyl anion (Experiment 73) and is aromatic. However, since it is a neutral compound, rather than an anion, it is not as basic. In fact it reacts with electrophiles, even a proton, by typical ring substitution processes rather than reaction at the nitrogen. As can be shown by molecular orbital calculations and demonstrated chemically, the α-carbons of pyrrole are the most electron-rich. Consequently, electrophiles tend to react with pyrrole regioselectively at the α-positions. Since there are two α-positions on pyrrole itself and the molecule is symmetrical, reaction can occur on either side or both sides of the nitrogen. An interesting example is the reaction of pyrrole with benzaldehyde. The multiple condensation of pyrrole with benzaldehyde followed by air oxidation produces *meso*-tetraphenylporphyrin remarkably efficiently. Porphyrins are extremely im-

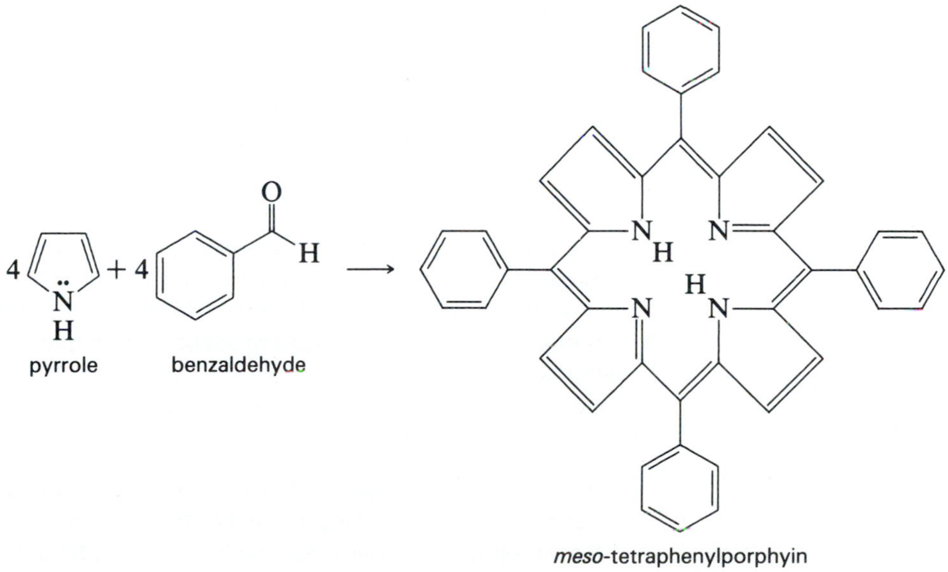

meso-tetraphenylporphyin

portant biological pigments. A similar porphyrin, protoporphyrin IX, binds iron in the center to form heme, the prosthetic group of hemoglobin. Other porphyrin derivatives and their metal complexes play a variety of important roles in physiological processes. Some of the simpler synthetic derivatives, such as tetraphenylporphyrin,[30,31] and their metal complexes have been used as models to help elucidate the chemistry and biology of this vital group of compounds.

EXPERIMENT 76

Preparation of *Meso*-Tetraphenylporphyrin

In the hood, place 25 to 30 mL of propionic acid *(CAUTION: Propionic acid is a strong organic acid and causes burns)* in a 50- to 100-mL round-bottomed flask containing a magnetic stir bar or boiling chips. Attach a reflux condenser and heat the acid to boiling. Cool the solution to a few degrees below the boiling point. Carefully add 0.5 mL (7.14 mmol) of freshly distilled pyrrole and 0.72 mL (7.2 mmol) of reagent-grade benzaldehyde either directly through the top of the condenser with a long pipette or by temporarily removing the condenser. Reheat the flask to boiling and reflux the dark, essentially black, mixture for 30 minutes. Cool the mixture to room temperature or below with an ice bath and collect the beautiful purple crystals of *meso*-tetraphenylporphyrin by suction filtration on a small Büchner or Hirsch funnel. Wash the crystals thoroughly with methanol. Dissolve a few of the tiny crystals in a drop of chloroform, or other solvent, and using a drawn-out capillary spot some of the solution on a piece of filter paper. Allow the solvent to evaporate in the hood and then shine a *long wavelength* ultraviolet light on the filter paper. Observe the intense fluorescence. The fluorescence is quenched when porphyrins bind metal ions. Several metal complexes can be made by literature procedures.[32] The copper complex is the easiest to prepare.

Questions for Experiment 76

1. Write a mechanism for the initial condensation of a single pyrrole molecule with one molecule of benzaldehyde. Continue to write a mechanism for the condensation of another pyrrole molecule with the product from the first condensation.
2. Trace out as many aromatic rings as possible in the porphyrin structure (ignore the phenyl substituents).

---

[30] A. D. Adler, F. R. Longo, and J. D. Finarelli, *J. Org. Chem.,* **32,** 476 (1967).
[31] R. E. Bozak and C. L. Hill, *J. Chem. Ed.,* **59,** 36 (1982).
[32] A. D. Adler, F. R. Longo, F. Kampas, and J. Kim, *J. Inorg. Nucl. Chem.,* **32,** 244 (1970).

The syntheses of many heterocyclic aromatic compounds are based on simple reactions such as condensation of carbonyl groups with amino and activated methylene groups in bifunctional molecules. The next experiment demonstrates how a number of the fundamental heterocyclic rings can be built from a single starting material, acetylacetone. The 1,3-diketone system in acetylacetone is largely enolized, and the compound and its reactions should be represented with this tautomeric form. The enolic form contains two electrophilic centers and a highly reactive nucleophilic central carbon. Condensations can occur between any two of these functional groups and two complementary groups in another molecule or molecules, as indicated schematically in **12** and **13**.

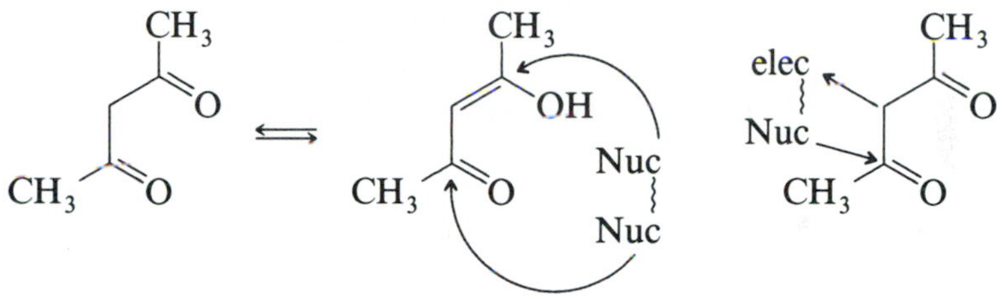

The syntheses to be carried out are condensations of type **12**, in which two nucleophilic centers in the second component condense with the dicarbonyl system, eliminating 2 mol of water. Typical examples are the formation of the pyrazole **14** and the pyrimidine **15** illustrated from dibenzoylmethane.

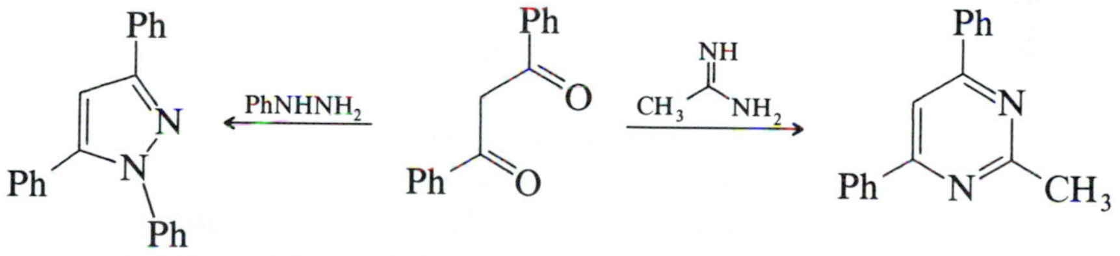

Condensations of this type are very simple to carry out, and with the very reactive acetylacetone, these preparations involve little more than mixing the reagents in an appropriate medium. When the solubilities of starting materials and products permit, water is a very effective solvent. The rate of these carbonyl condensations in aqueous solutions is pH dependent, a slightly basic solution being optimum for the reactions studied here.

Procedures are indicated in the following paragraphs for reactions of acetylacetone with the two reagents listed alphabetically in Table 10.5. The names of the products are given, also in alphabetical order, in Table 10.5. Before proceeding with the experiments, determine the product that will be obtained from each reagent. Only minimum directions are given; use your own judgment and experience in isolating the products. Determine the melting point and identity of the products.

TABLE 10.5 Reactions of Acetylacetone with Cyanoacetamide and Guanidine

Starting Materials	Formula	Products
Cyanoacetamide	$NCCH_2CONH_2$	3-Cyano-4,6-dimethyl-2-pyridone, mp 290 to 291°C (from H_2O)
Guanidine	$(NH_2)_2C\!\!=\!\!NH$	2-Amino-4,6-dimethylpyrimidine, mp 156 to 157°C (from MeOH)

EXPERIMENT 77

Preparation of a 2-Pyridone

Dissolve 0.84 g of cyanoacetamide and 0.5 g of sodium carbonate in 15 mL of water and add 1 mL of acetylacetone. Collect the solid product by suction filtration. Air dry the product and determine its melting point.

EXPERIMENT 78

Preparation of a Pyrimidine

Dissolve 1.0 g of guanidine carbonate and 1 g of sodium acetate in 5 mL of water and add 2 mL of acetylacetone. Collect the product by suction filtration. After completely air drying the product, purify it using vacuum sublimation. Use an aspirator for vacuum and a hot-water bath for heat. If droplets of water condense on the condenser when you begin to heat the sample, stop and dry off the condenser, then start again. Repeat if needed. Do *not* heat the solid to the melting point; if it starts to turn to a liquid, remove it from the heat and proceed at a lower temperature.

In addition to electrophilic substitution reactions, certain aromatic compounds also undergo nucleophilic substitution reactions. The ability of substituted benzene derivatives to undergo nucleophilic substitutions is determined by the type, number, and position of the electron-withdrawing substituents on the ring. The greater the number of electron-withdrawing groups, the faster the nucleophilic reaction, especially when the attack occurs *ortho* or *para* to the electron-withdrawing groups. The mechanism for a typical nucleophilic aromatic substitution is shown for the reaction of a 2,4-dinitrophenyl halide (X = halide). In this case, the electron-withdrawing nitro groups stabilize the carbanion intermediate.

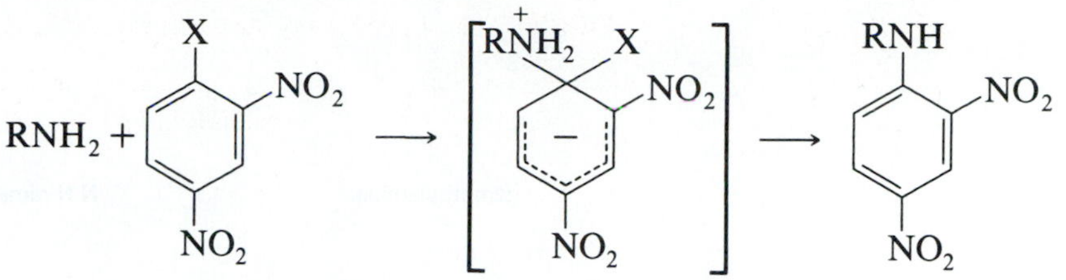

One important application of this reaction was developed by Sanger (see page 367), who reacted 2,4-dinitrofluorobenzene (DNFB) with the N-terminal amino acid of various proteins (essentially amino acid polymers). Normally the bond between the carbonyl carbon and the amine nitrogen (the peptide bond) in a protein is susceptible to acid hydrolysis by $6N$ HCl. However, the DNFB-labeled amino acid is stable. Thus, by reacting a small amount of protein with DNFB and hydrolyzing the product with $6N$ HCl it was possible to determine the identity of the amino acid at the N-terminus of the protein. Repeating this procedure with various overlapping segments of fragments of the protein allows for the determination of the complete amino acid sequence of the protein. One of the most important medical accomplishments in recent times was the determination, by this technique, of the complete amino acid sequence of insulin. Insulin is now prepared commercially by recombinant DNA technology for the treatment of diabetes.

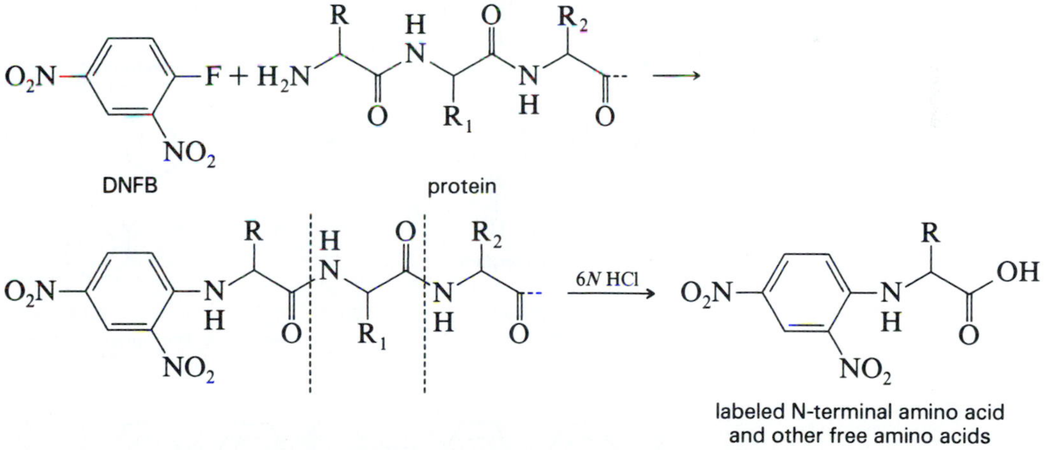

labeled N-terminal amino acid
and other free amino acids

In this experiment, an unknown compound[33] (see Table 10.6) will be reacted with 2,4-dinitrofluorobenzene to obtain the 2,4-dinitrobenzene derivative.

---

[33] Instructor's note: Morpholine and pyrrolidine work especially well in this experiment. If the experiment is done without incorporating the unknown aspect, use of either of these amines will be effective.

TABLE 10.6 DNFB Derivatives of Some Common Amines

Unknowns:

1. $CH_3-NH-CH_3 \longrightarrow$

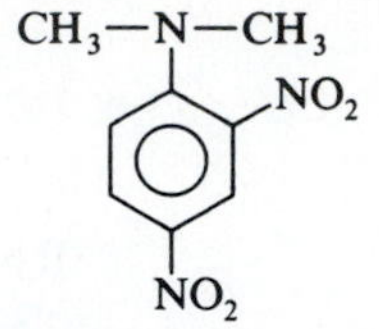

dimethylamine

N,N-dimethyl-2,4-dinitroaniline
−75°C

2. $\longrightarrow$

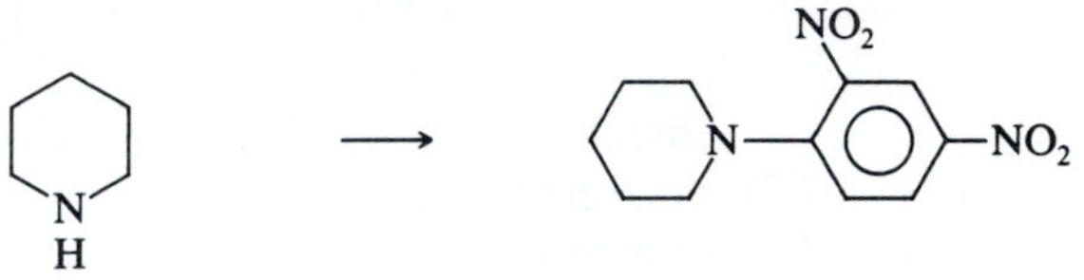

piperidine

N-(2,4-dinitrophenyl)piperidine
−90°C

3. $\longrightarrow$

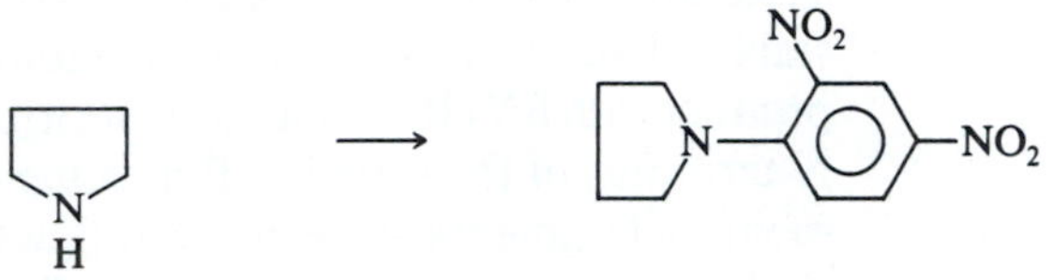

pyrrolidine

N-(2,4-dinitrophenyl)pyrrolidine
−110°C

4. $\longrightarrow$

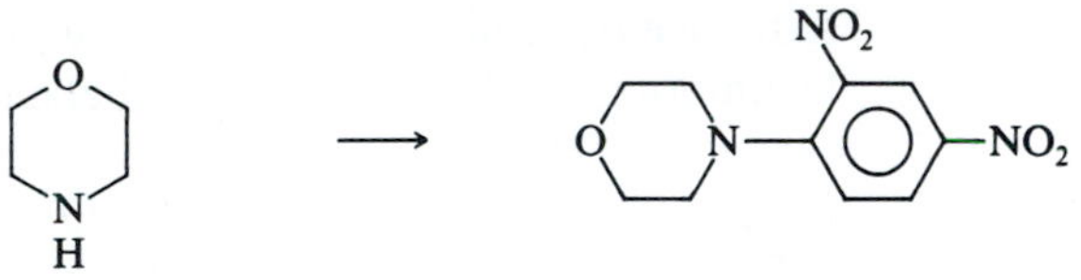

morpholine

N-(2,4-dinitrophenyl)morpholine
−120°C

5. $\longrightarrow$

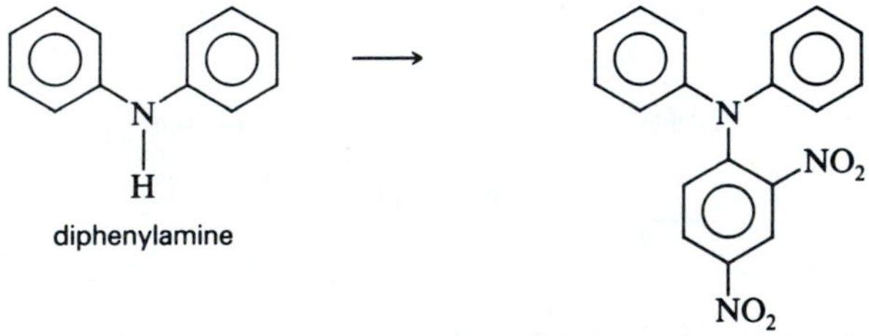

diphenylamine

2,4-dinitrophenyldiphenylamine
−130°C

6. $-NH_2 \longrightarrow$

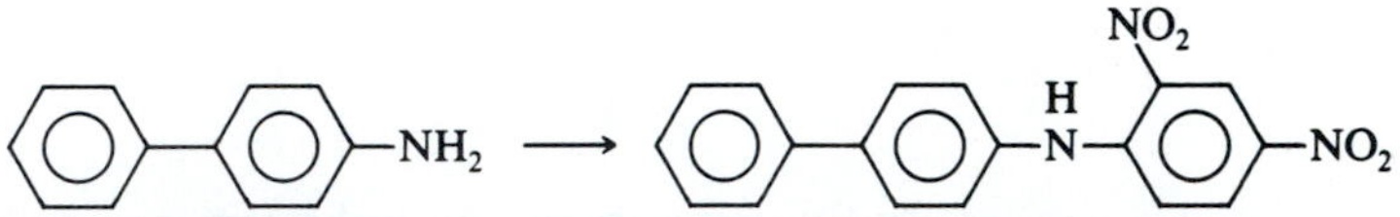

biphenylamine
(p-aminobiphenyl)

2,4-dinitrophenylbiphenylamine
−145°C

7. $\longrightarrow$

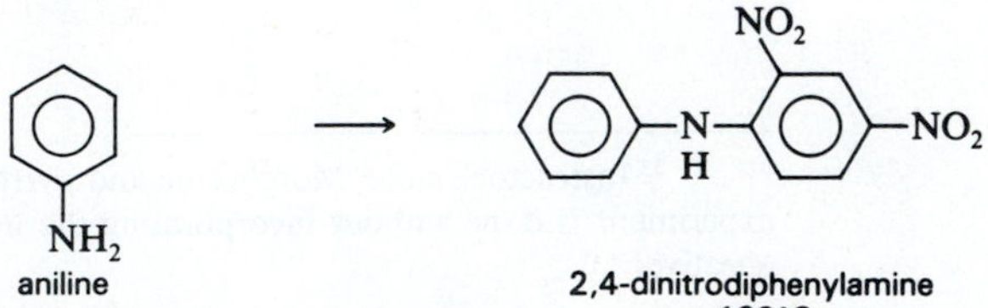

aniline

2,4-dinitrodiphenylamine
−160°C

EXPERIMENT 79

Nucleophilic Aromatic Substitution

Measure out 2.5 mL of a solution of 2,4-dinitrofluorobenzene (DNFB) dissolved in absolute ethanol (0.08 g/mL) and pour it into a 10-mL round-bottomed flask containing a stir bar. *(CAUTION: DNFB reacts readily with proteins, so avoid contact with the skin. Also, 2,4-dinitrophenol, which would be obtained from competitive hydrolysis of the reagent, is toxic.)*

Set up a reflux condenser (water running) on a ring stand. To the top of the reflux condenser attach a $CaCl_2$ drying tube containing *fresh* $CaCl_2$. Obtain an unknown (0.2 g of material dissolved in ethanol) and add it to the 10-mL round-bottomed flask and attach the flask to the reflux condenser. Note the color change, if any. At this point a few crystals may appear. Reflux the mixture gently using a preheated sand bath ($\sim 100°C$) on the stirrer hot plate, noting if more crystals begin to appear in the flask. After refluxing, allow the flask to *air* cool (with occasional swirling) for about 10 minutes. If no crystals have appeared, try adding ice-cold water dropwise until crystallization begins (a few drops should be sufficient). Once crystallization has begun, cool the flask in an ice bath for 10 to 15 minutes (again, with occasional swirling), or until a large crop of crystals is evident. Collect the product by suction filtration on a small Büchner or Hirsch funnel. Note that it may be necessary to use a double layer of filter paper to increase retention of the very fine crystals. Also note the shape of the crystals.

If a gummy residue remains in the flask, dissolve it in a few mL of absolute ethanol with heat from a steam bath or the sand bath. Air cool the solution with occasional swirling and collect the crystals as described above. Recrystallize the product using a Craig tube. Suggested solvents include ethanol, ethyl acetate, and methanol. Note that a relatively large amount of solvent may be required for the recrystallization. If so, try doing the recrystallization using < 0.5 g of product at a time. Avoid ice cooling during recrystallization, if possible, in order to minimize impurities in the final product. Allow the crystals to dry completely. Record the melting point of the product and determine the identity of the unknown amine from the melting point data given in Table 10.6.

Calculate a yield based on the known starting amount of 2,4-dinitrofluorobenzene and assuming 0.2 g of your unknown.

As an alternative, more biologically relevant experiments form the 2,4-dinitrobenzene derivatives of amino acids as described on page 367.

Questions for Experiment 79

1. Why was absolute ethanol used in this reaction?
2. Which reactant was used in excess and why?
3. Are the products colored? If so, why, since the starting materials were practically colorless?

Especially reactive toward nucleophilic substitution are the aryl diazonium cations. Primary aromatic amines react with nitrous acid to give diazonium salts that are usually stable in aqueous solution at temperatures below 5°C. Thus, when a cold aqueous solution of aniline hydrogen sulfate is treated with sodium nitrite, it gives phenyldiazonium hydrogen sulfate [Equation (10.3)]. Aromatic diazonium salts are soluble in water and acetone but not in hydrocarbons and ethers. Since they are potentially explosive, diazonium salts are almost never isolated but they are usually converted into the desired derivative while still in solution.

$$\underset{}{\text{C}_6\text{H}_5\text{NH}_2} \; + \; \text{NaNO}_2 + 2\,\text{H}_2\text{SO}_4 \; \longrightarrow \; \underset{\text{HSO}_4^-}{\text{C}_6\text{H}_5\overset{+}{\text{N}}\!\equiv\!\text{N}} \; + \text{NaHSO}_4 + 2\,\text{H}_2\text{O} \qquad (10.3)$$

$$\underset{\text{HSO}_4^-}{\text{C}_6\text{H}_5\overset{+}{\text{N}}\!\equiv\!\text{N}} \; + \; \text{KI} \; \longrightarrow \; \underset{\text{HSO}_4^-}{\text{C}_6\text{H}_5\text{I}} \; + \text{N}_2 + \text{KHSO}_4 \qquad (10.4)$$

The experimental procedure for the preparation of aromatic diazonium compounds is quite straightforward. First the amine is dissolved in a suitable volume of dilute mineral acid (with warming if necessary) and the resulting solution of amine salt is cooled to 0 to 5°C with the aid of an ice-salt bath or by the addition of a few lumps of ice. The amine salt usually crystallizes from the solution at this point and an aqueous solution of sodium nitrite is added dropwise and with stirring until a slight but persistent excess of nitrous acid is present (starch iodide test). During the addition of the nitrite solution, the solid amine salt gradually dissolves in the reaction mixture, which progressively takes on a dark yellow, orange, or brown color. At the end of the diazotization, excess nitrous acid is decomposed by the addition of urea.

Aromatic diazonium compounds can be converted into a wide variety of compounds, for example, halides, nitriles, and nitro compounds, merely by treating their aqueous solutions with the appropriate salts. Iodides are the easiest to prepare; it is only necessary to add a solution of potassium iodide to the diazonium solution to effect conversion [Equation (10.4)]. Bromides and chlorides can only be prepared by treating diazonium salts with cuprous salts (cuprous ion catalyzes the reaction) and by excluding from the reaction mixture all anions that could compete with the bromide or chloride. The difference in reactivity between iodine and chloride is so great that it is possible to carry out the conversion of diazonium salts into iodides in dilute hydrochloric acid.

Diazonium salts couple readily with aromatic amines in two ways, such as with the formation of an N—N bond [Equation (10.5)] or a C—N bond [Equation (10.6)]. Since amine salts tend to dissociate slightly into their components [Equation (10.7)], it is necessary to have substantial excess of acid present in the diazotization reaction mixture in order to suppress dissociation and prevent coupling.

The reaction mixture must be kept cold (0 to 5°C) during all phases of the work, otherwise some of the diazonium salt will be converted into the corresponding phenol [Equation (10.8)]. Nitrous acid has oxidizing properties; consequently, an excess is to be avoided, especially in diazotizing highly reactive compounds (e.g., *o*- and *p*-aminophenol). Furthermore, in the preparation of iodobenzene, an excess of nitrous acid will destroy some of the potassium iodide [Equation (10.9)].

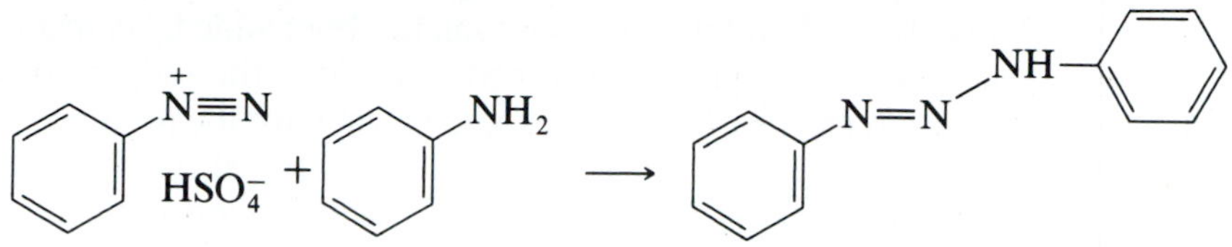

$$(10.5)$$

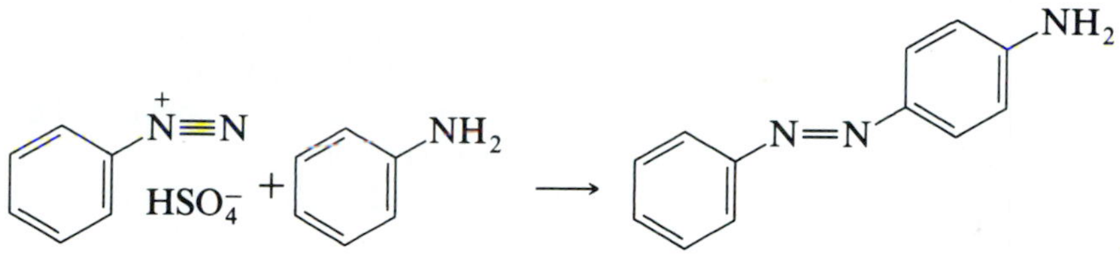

$$(10.6)$$

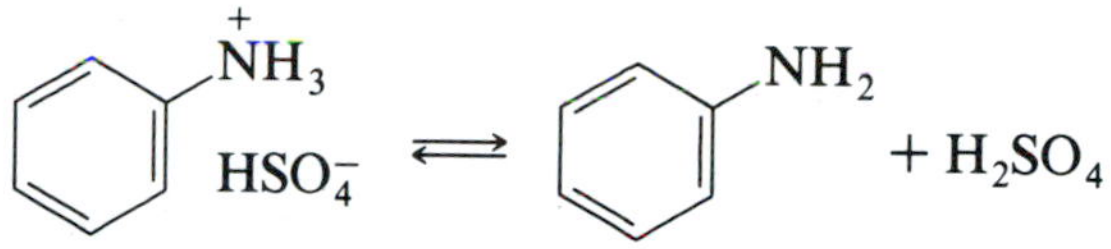

$$(10.7)$$

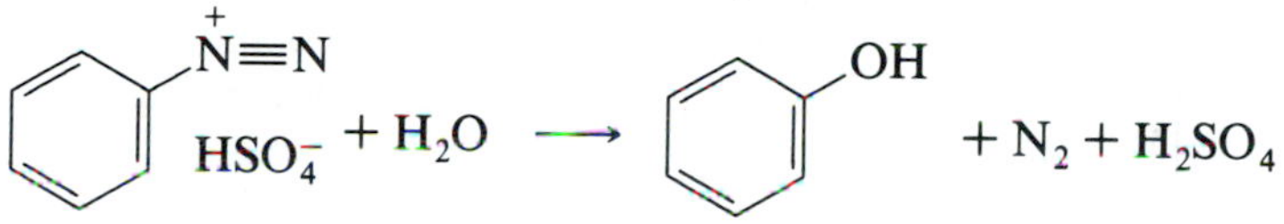

$$(10.8)$$

$$2\,NO_2^- + 2\,I^- + 4\,H^+ \longrightarrow 2\,NO + I_2 + 2\,H_2O \qquad (10.9)$$

EXPERIMENT 80

Preparation of Iodobenzene

Add 7.5 mL of concentrated sulfuric acid to 40 mL of water in a tared 250-mL beaker in the hood. While the acid solution is still hot, add 7.4 mL of aniline and stir until the aniline has dissolved. Finally, add 50 g of ice, immerse the reaction mixture in a salt-ice bath, and stir it vigorously with a glass stirring rod until the temperature drops to 0 to 5°C and recrystallization of anilinium hydrogen sulfate is complete.

Calculate the weight of sodium nitrite required to react with the amount of aniline and then calculate the weight of a 10% excess. Dissolve the required amount (110%) of sodium nitrite in 25 mL of water. Prepare a salt-ice bath by filling the bottom part of a steam bath (or other suitable container) half full with chipped ice and sprinkling technical-grade sodium chloride (i.e., rock salt) on the ice while stirring it with a glass rod. Dissolve 15 g of potassium iodide in 40 mL of water in a 200-mL round-bottomed flask; chill the solution in an ice bath.

Support a 60-mL separatory funnel on a ring stand above the anilinium hydrogen sulfate solution and place the sodium nitrite solution in the funnel. Add the nitrite solution to the anilinium hydrogen sulfate slurry dropwise with rapid stirring at such a rate that the temperature does not exceed 5°C. The temperature of the diazonium solution must be kept below 5°C at all times. When all but approximately 5 mL of the nitrite solution has been added, begin adding the nitrite solution in increments of five to six drops each. After the addition of each increment, stir the reaction mixture for 2 minutes, then test it for the presence of excess nitrous acid by touching the tip of the stirring rod to a piece of starch-iodide test paper. Addition should be stopped when an excess is reached. Destroy excess nitrous acid by adding a small amount of urea to the reaction mixture. Only an instantaneous (about 1 second) blue coloration on the starch-iodide test paper is indicative of excess nitrous acid. The diazonium salt solution itself will produce a gradual blue coloration (about 3 seconds) even when nitrous acid is absent. Towards the end of the reaction, the uptake of nitrous acid becomes very slow.

Add the phenyldiazonium hydrogen sulfate solution prepared above in increments of 5 to 10 mL to a cold solution of 15 g of potassium iodide (an excess) in 40 mL of water in a 500-mL round-bottomed flask. Return the flask to the ice-salt bath after each addition and swirl the reaction mixture occasionally. *(CAUTION: Frothing may occur from the nitrogen that is evolved.)* The preparation of iodobenzene may be interrupted at this point. Stopper the flask containing the iodobenzene reaction mixture with a wad of cotton and support it upright in a beaker for storage until the next laboratory period. Allowing the phenyldiazonium salt to decompose slowly in this way results in a slightly higher yield of iodobenzene.

C A U T I O N : Attempts to heat the reaction mixture at this point may cause an explosion!

To continue, make the reaction mixture strongly basic with 20% sodium hydroxide solution. Attach a distilling head and condenser to the flask. Add several boiling chips to the mixture and heat it to boiling using a sand bath or water bath warmed by a hot plate. During this steam distillation, collect the distillate in a flask. Continue heating and collecting until the distillate is no longer milky. Extract the distillate with 30 mL of diethyl ether. Dry the ether extract over a small amount of anhydrous calcium chloride. Remove the calcium chloride by gravity filtration into a tared round-bottomed flask. Evaporate the ether on the rotary evaporator. Since iodobenzene boils at 80°C at 15 mm, be careful not to heat the water bath on the rotary evaporator. Record the IR (neat film) spectrum of the iodobenzene.

10.15 CONCERTED REACTIONS

Concerted reactions are those that produce products from starting material without proceeding through intermediates. Many cycloaddition reactions and rearrangements are concerted. Perhaps the most common and useful concerted reaction is the

Diels–Alder cycloaddition reaction. The Diels–Alder reaction is the 1,4-addition of a conjugated diene to a molecule containing an active ethylenic or acetylenic bond (the dieneophile) to form an adduct that is a six-membered ring containing a double bond.

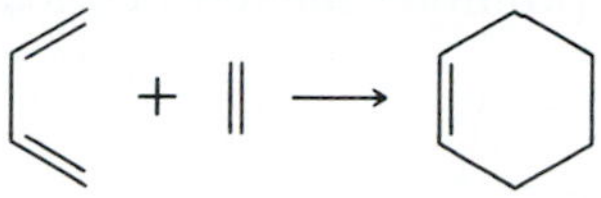

In the next experiment, 4-cyclohexene-1,2-dicarboxylic acid anhydride will be derived from the Diels–Alder reaction of maleic anhydride and butadiene. Since butadiene is a gas at room temperature, it will be generated in the reaction mixture by the thermal decomposition (chelatropic elimination) of its sulfur dioxide adduct, 3-sulfolene, or butadiene sulfone. Since sulfur dioxide is a noxious and toxic gas (but a common pollutant), it must be trapped and neutralized as it is evolved from the reaction apparatus.

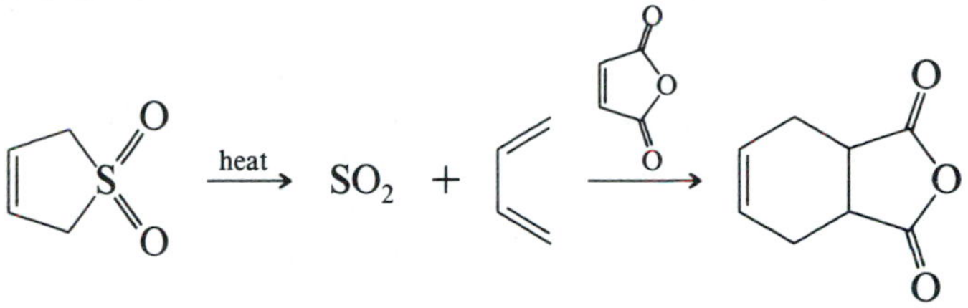

EXPERIMENT 81

Diels–Alder Reaction of Maleic Anhydride with 1,3-Butadiene

To a 100-mL round-bottomed flask in the hood, attach a reflux condenser fitted with a T-tube with tubing leading from one end to a trap containing 5% NaOH (Figure 1.14). Temporarily remove the condenser and add 10 g of butadiene sulfone and 5 g of powdered maleic anhydride to the flask. Add about 10 mL of xylene and replace the condenser. Heat the reaction mixture carefully to reflux watching for evolution of gas (SO$_2$) before boiling starts. Continue to reflux the solution for 45 minutes. Cool the mixture to room temperature and add 50 mL of toluene. Add a small amount of decolorizing charcoal, warm the mixture, and gravity filter the solution into a 125-mL Erlenmeyer flask. Then add 25 to 30 mL of hexanes and heat the solution to about 80°C. Cool the solution slowly to room temperature and then cool it further in an ice bath. Collect the resulting crystals by suction filtration. Determine the weight, percent yield, and melting point of the product.

The Diels–Alder reaction is often performed with cyclopentadiene as the diene and maleic anhydride as the dienophile. In this case, cyclopentadiene must be generated shortly before use by the thermal retro Diels–Alder reaction of its dimer cyclopentadiene. This can be done easily by fractionally distilling dicyclopentadiene that promotes the retro cycloaddition (see Experiment 85 for details).

dicyclopentadiene   cyclopentadiene

Questions for Experiment 81

1. Only dienes that can adopt an eclipsed *(cisoid)* conformation react in Diels–Alder reactions. Why?

2. Only one *(endo)* of two possible products *(endo* and *exo)* from the reaction of cyclopentadiene with maleic anhydride is shown above. Draw the other product and explain why its formation is not favored.

3. Draw the major product that you would expect to form on heating maleic anhydride with cyclohexadiene; with 2-methyl-1,3-butadiene.

4. Why was xylene rather than benzene used as the solvent in the Diels–Alder reaction?

Benzyne is extremely reactive in cycloaddition reactions, reacting with dienes to produce benzene-containing cycloadducts. Benzyne is a very reactive species and must be used as it is generated. In this experiment, benzyne will be generated *in situ* by diazotization of anthranilic acid with isoamyl nitrite that initially produces benzenediazonium-2-carboxylate *(CAUTION: This salt should never be isolated as it is explosive).* Loss of nitrogen and carbon dioxide generates benzyne. In the presence of anthracene, benzyne undergoes a Diels–Alder reaction to produce tryptycene. First, however, the anthranilic acid will be prepared by saponification (base hydrolysis) of phthalimide followed by a Hofmann degradation. The Hofmann degradation involves the reaction of an amide with bromine and base to produce an N-bromoamide initially. Ionization of the N-bromoamide initiates a rearrangement to an isocyanate. The isocyanate reacts with water to form a carbamic acid. The carbamic acid spontaneously decarboxylates to give anthranilic acid. The entire Hofmann rearrangement occurs easily in a one-flask reaction.

Alkyl nitrites are potent heart stimulants, so caution must be used to avoid exposure to the volatile isoamyl nitrite. To ensure that no excess nitrite remains, excess anthranilic acid and anthracene will be used. Any excess anthracene not used in the reaction will be consumed by the addition of maleic anhydride that will undergo a Diels–Alder reaction with the anthracene. Subsequent treatment with base will hydrolyze the anhydride to the diacid that can be removed in aqueous base.

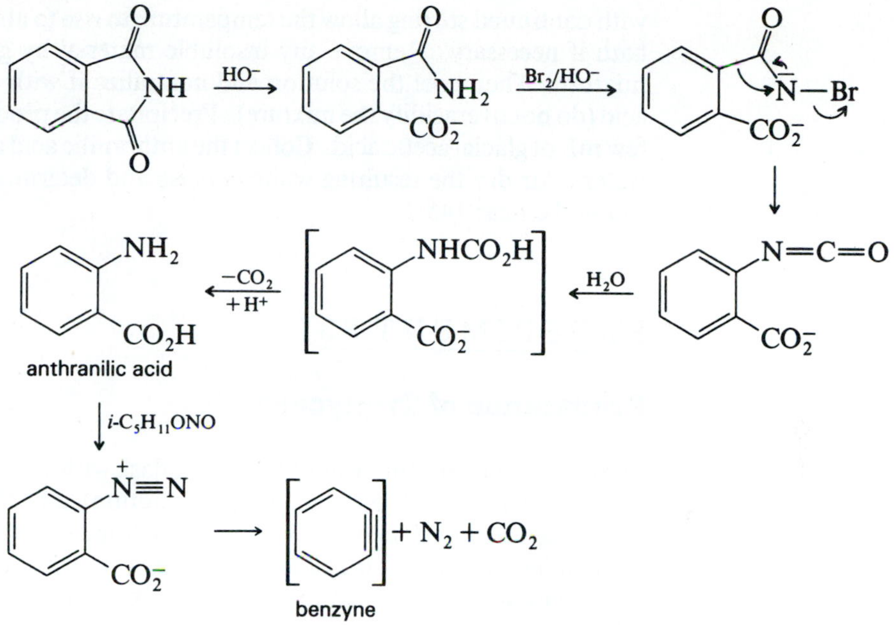

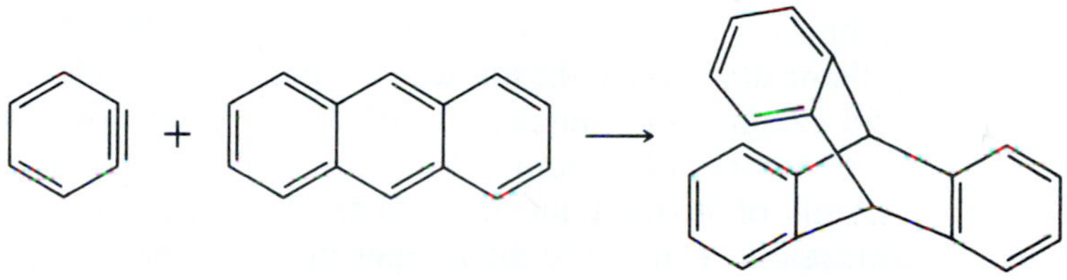

benzyne + anthracene → triptycene

EXPERIMENT 82

Preparation of Anthranilic Acid

Note: This part can be deleted if commercially available anthranilic acid is available.

C A U T I O N : This sequence of experiments must be carried out in a hood!

In a 250-mL Erlenmeyer flask dissolve 24 g (0.6 mol) of sodium hydroxide in 100 mL of water and cool the solution to 0°C in an ice bath. Transfer half of this solution to another 250-mL Erlenmeyer flask and add 16 g (0.1 mol) of bromine *(CAUTION: Avoid contact with the skin)* with rapid and continuous stirring while cooling in an ice bath. Add 14.7 g (0.1 mol) of phthalimide and the remaining portion of aqueous sodium hydroxide with stirring. Remove the cooling bath and with continued stirring allow the temperature to rise to about 80°C. Heat on a steam bath if necessary. Remove any insoluble material by gravity filtration of the hot mixture. Then cool the solution and neutralize it with concentrated hydrochloric acid (do not overacidify the mixture). Precipitate the product amino acid by adding a few mL of glacial acetic acid. Collect the anthranilic acid and recrystallize it from hot water. Air dry the resulting white crystals and determine their melting point that should be near 145°C.

EXPERIMENT 83

Preparation of Triptycene

In the hood, fit a 100-mL round-bottomed flask with a reflux condenser and add 1.5 g of anthracene, 1.5 mL of isoamyl nitrite (caution), and 15 mL of 1,2-dimethoxyethane. Insert a short stem funnel containing a fluted piece of filter paper into the top of the condenser. Place 2.0 g of anthranilic acid in the filter paper and gently pack it down with a spatula. Heat the reaction mixture in the flask until it just begins to reflux. Some of the anthracene will not immediately dissolve. Slowly dissolve the anthranilic acid in the filter paper into the reaction flask by adding 1,2-dimethoxyethane dropwise to the funnel. Attempt to completely transfer the solid anthranilic acid over a 15-minute period with about 7.5 mL of solvent.

Temporarily remove the funnel from the condenser and add an additional 1.5 mL of isoamyl nitrite through the condenser. Add an additional 1.9 g of anthranilic acid to the filter paper in the funnel and percolate it into the reaction mixture as before over another 15 minutes.

After the addition of all the anthranilic acid is complete, reflux the mixture for another 10 minutes. Then add 10 mL of 95% ethanol and a solution of 2.5 g of sodium hydroxide in 30 mL of water to produce a brownish suspension. Cool the mixture and suction filter the product in the hood. Wash the product on the filter funnel with cold 80% methanol/20% water. Transfer the moist triptycene-anthracene mixture to a tared 100-mL round-bottomed flask and dry it by evaporation on a rotary evaporator. To consume the remaining anthracene, add 0.8 g of maleic anhydride and 15 mL of triethylene glycol dimethyl ether (diglyme, bp 222°C) and heat the mixture at reflux for 5 minutes. Cool the mixture to 100°C. Add 8 mL of 95% ethanol and a solution of 2.5 g of sodium hydroxide in 30 mL of water. Cool the mixture, collect the product, and wash it with cold 80% methanol-water.

The crude product may contain some insoluble material. Dissolve the product in excess dichloromethane (~8 mL) and gravity filter it to remove the insoluble material. Add two volumes of methanol and concentrate the solution to about one-third of its total volume. If crystals begin to separate during this concentration, heat the mixture to boiling and then allow it to cool slowly to recrystallize the product. Record the melting point, which should be near 255°C. Record and interpret the NMR spectrum of the product.

Questions for Experiments 82 and 83

1. Explain why anthracene reacts with benzyne at its 9,10-position and not others. (*Hint:* Consider resonance energy of reactants and products.)
2. Write a balanced chemical equation for the diazotization of anthranilic acid with isoamyl nitrite and propose a mechanism for the process.
3. Draw the structure of the product of anthracene with maleic anhydride before it is hydrolyzed and after it is hydrolyzed with hydroxide.

Cycloaddition reactions are stereospecific: *trans*-disubstituted dienophiles producing only *trans*-disubstituted cycloadducts, and *cis*-disubstituted dienophiles producing only *cis*-disubstituted cycloadducts. In the next experiment this stereospecificity will be demonstrated in the cycloaddition reactions of *cis*- and *trans*-dibenzoylethylene. In addition, kinetic and thermodynamic control in product formation will be demonstrated.

Trans-dibenzoylethylene (1,4-diphenyl-2-butene-1,4-dione) can be photoisomerized to the *cis*-isomer. Both the *trans*- and *cis*-isomers can serve as dienophiles in Diels–Alder reactions. In this experiment they will be separately treated with cyclopentadiene (generated from its dimer) to give the cycloaddition products. Each of the cycloadducts will be treated with sodium ethoxide in equilibration experiments to determine which isomer is the more thermodynamically stable product. Acid-catalyzed isomerization of the *cis*-dibenzoylethylene will also be demonstrated.

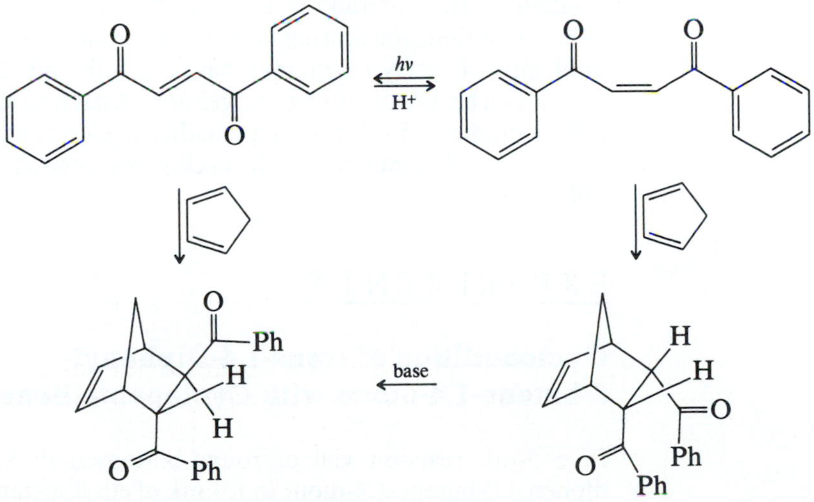

EXPERIMENT 84

Photochemical Isomerization
of *trans*-1,4-Diphenyl-2-butene-1,4-dione

Dissolve 0.3 g of *trans*-1,4-diphenyl-2-butene-1,4-dione in 12 mL of 95% ethanol in a large test tube (previously washed with dilute ammonia in ethanol) by heating on a steam bath. While still hot, stopper the tube loosely with a cork and irradiate the contents with an unfiltered 275-watt sun lamp at a distance of from 4 to 6 in. for about 6 to 8 hours. (Several tubes can be successfully irradiated simultaneously with one lamp if the lamp and tubes are surrounded with sheets of aluminum foil to reflect the light to the tubes.) Cool the tube in an ice bath and induce crystallization by scratching with a stirring rod if necessary. Collect the white solid and recrystallize it from 95% ethanol if necessary. The product should have a melting point of about 136°C.

EXPERIMENT 85

Cyclopentadiene by Thermal
Cycloreversion of Dicyclopentadiene

This experiment should be carried out in a hood in groups of five or more persons. For each cracking procedure, charge a 50-mL distillation flask with 10 to 20 mL of dicyclopentadiene and attach it to a fractional distillation column packed with stainless steel wool. Immerse the receiving flask (25 mL) in an ice bath and heat the distillation flask carefully with a sand bath at ~180°C until the contents begin to reflux. Continue the heating such that the cyclopentadiene distills over at 40 to 42°C until approximately 1.5 mL of it has been collected. Stopper the receiving flask and store it in a beaker of ice or a refrigerator. Attach another 25-mL receiving flask and collect another 1.5 mL of cyclopentadiene. Continue until each person has received 1.5 mL of cyclopentadiene. The cyclopentadiene should be used immediately or it will dimerize.

EXPERIMENT 86

Cycloaddition of *trans*-1,4-Diphenyl-
2-butene-1,4-dione with Cyclopentadiene

In a 5-mL reaction vial or round-bottomed flask, dissolve 0.2 g of *trans*-1,4-diphenyl-2-butene-1,4-dione in 1.0 mL of ethyl acetate by heating on a steam bath or sand bath. Add 0.25 mL (~0.2 g) of cyclopentadiene and after the initial exothermic reaction has subsided, heat the vial on a steam bath or with a sand bath for 2 hours. Stopper the vial and store it until the next laboratory period.

Remove most of the ethyl acetate on the rotary evaporator or, if a vial was used, by heating the vial on a steam bath in the hood. Also use a T-tube and a water aspirator if necessary to assist the solvent removal. Add 0.5 mL of 95% ethanol and cool the flask or vial and its contents until crystals form (scratch the inside of the flask if necessary). Collect the crystals by suction filtration and recrystallize from methanol using a Craig tube. The product should have a melting point near 78 to 79°C.

EXPERIMENT 87

Cycloaddition of *cis*-1,4-Diphenyl-2-butene-1,4-dione with Cyclopentadiene

Dissolve 0.1 g of *cis*-1,4-diphenyl-2-butene-1,4-dione in 3 mL of absolute ethanol in a 5-mL reaction vial or a small round-bottomed flask. Add 0.25 mL (~0.2 g) of cyclopentadiene. Attach a condenser and heat the mixture at reflux for $3\frac{1}{2}$ hours or until the end of the laboratory period. Cool the vial. Stopper it and store it until the next laboratory period.

If crystals have formed on standing, cool the flask momentarily in an ice bath. Then collect the crystals and recrystallize them from ethanol. In the event that no crystals have formed, evaporate approximately half of the solvent on a steam bath or with the rotary evaporator and proceed as above. The product should have a melting point of about 160 to 161°C.

EXPERIMENT 88

Acid-Catalyzed Isomerization of *cis*-1,4-Diphenyl-2-butene-1,4-dione

Dissolve 0.02 to 0.1 g of *cis*-1,4-diphenyl-2-butene-1,4-dione in 1 to 5 mL of 95% ethanol in a small flask or reaction vial by heating on a steam bath or in a sand bath. Add one drop of concentrated hydrochloric acid and boil the mixture for 5 minutes. Add 1 to 5 mL of water to the hot reaction mixture and cool the mixture. Collect and air dry the product. Determine its melting point and identify the product.

EXPERIMENT 89

Base-Catalyzed Isomerization of the Cycloaddition Products

Prepare dilute solutions of the two cycloadducts from Experiments 86 and 87 in ethyl acetate ($\sim$ 5 mg of adduct/0.1 to 0.2 mL of ethyl acetate) and a 1 : 1 mixture of the two adducts in ethyl acetate. Place very small drops of each of the three solutions near the bottom of a single silica gel TLC plate using drawn-out capillaries. Develop the TLC plate with ethyl acetate-hexanes (try 1 : 1 first). Air dry the plate and visualize the spots by placing the plate into a screw-capped bottle containing several crystals of iodine. Note and record the R_f values of the spots.

In each of two reaction vials containing 3 mL of absolute ethanol, cautiously add two small pieces of sodium metal (about one-tenth the size of a pea), one at a time. When the sodium has dissolved, add 50 mg of the cycloadduct prepared in Experiment 86 to one flask and 50 mg of the cycloadduct obtained in Experiment 87 to the other. Stopper each flask loosely with a cork and heat both on a steam bath. With emphasis on noting the formation of a different isomer in each case, analyze each reaction by TLC, using the same developing system and visualization method found effective above. Analyze at 10- to 15-minute intervals for at least 30 minutes or until equilibration is apparently established (no further change, if any, detected). Critically judge the outcome of the equilibrations by making sufficient comparisons of the equilibrated solutions with the solutions of the cycloadducts previously prepared, together on the same TLC plate.

To each reaction mixture, add 3 mL of water and heat on a steam bath (to effect homogeneity it may be necessary to add a small amount of ethanol). Slowly cool the clear solutions to room temperature and then further in an ice bath. Collect the crystalline products and recrystallize from methanol. Record their melting points and determine mixture melting points of each product with the initial cycloadducts. Record and interpret the NMR spectra ($CDCl_3$) of the products.

Questions for Experiments 81 to 89

1. Explain why *cis*-dibenzoylethylene is colorless and the *trans*-isomer is light yellow. Use this answer to help with the next question.
2. Why is the thermodynamically more stable *trans*-1,2-dibenzoylethylene converted to the *cis*-isomer during the photolysis?
3. Write a mechanism for the photochemical transformation.
4. Which is the thermodynamically more stable of the two isomeric cycloadducts? How did the experiment show this?
5. Write a mechanism for the isomerization of the cycloadducts with base.
6. The cycloaddition of cyclopentadiene with the *trans*-alkene occurred under milder conditions than did that with the *cis*-alkene. Why?
7. Why was a large excess of cyclopentadiene used in the cycloaddition reactions?

10.16   REARRANGEMENTS

Rearrangements can occur under a variety of conditions. Some rearrangements can be thermally induced concerted processes. Others may require prior modifications of substrates by introduction of good leaving groups or temporary modifications by catalytic processes. This section includes several examples of rearrangements that are commonly used in organic synthesis. Mechanistic and theoretical studies related to these rearrangements have also provided significant insight into many organic reactions.

Substituted allyl phenyl ethers undergo thermally induced concerted Claisen rearrangements to 6-allyl-2,4-cyclohexadienones that rearomatize to *o*-allyl phenols. The process is classified as a [3.3]-sigmatropic rearrangement.

The requisite allyl phenyl ethers are readily prepared by alkylation of phenoxide with an allyl halide. The phenoxide anion is generated by reaction of phenol (pK$_a$ ~ 9) *(CAUTION: Concentrated or neat phenol can cause severe skin burns; wear gloves)* with potassium carbonate in acetone solution. Other bases and solvents could be used, but allyl bromide is exceptionally reactive toward nucleophilic displacement by most other bases (e.g., sodium ethoxide in ethanol, *t*-butoxide in *t*-butyl alcohol, or amine bases). Potassium carbonate is an insoluble base and the carbonate anion is a poor nucleophile. However, carbonate is a strong enough base to deprotonate phenol in nonaqueous solvents.

During the available laboratory time, neither the formation nor the rearrangement of allyl phenyl ether may proceed to completion.[34] Separation of the starting phenol, allyl phenyl ether, and the product *o*-allyl phenol can be accomplished by simple extraction techniques. The *o*-allyl phenol will also be derivatized for characterization by conversion to the corresponding 2-allylphenoxyacetic acid.

---

[34] Instructor's note: Because of the short setup times, but long reaction times, required for this experiment, it can often be performed simultaneously with another experiment.

EXPERIMENT 90

Preparation of Allyl Phenyl Ether[35]

In a 25-mL round-bottomed flask in the hood, place 0.94 g (10 mmol) of phenol, 1.21 g (0.86 mL, 10 mmol) of allyl bromide *(CAUTION: Allyl bromide is a very toxic, powerful alkylating agent),* 1.4 g of finely ground potassium carbonate, and 2 to 5 mL of acetone. Use a mortar and pestle to pulverize the carbonate. Attach a reflux condenser and reflux the solution for up to 3 hours using a preheated sand bath (~ 80 to 100°C). Allow the mixture to cool and then pour it into 10 mL of water in a small separatory funnel. Extract the mixture with two 10-mL portions of ether. Combine the ether extracts and extract twice with 5-mL portions of 10% sodium hydroxide to remove the unreacted phenol. Pour the ether layer into a small Erlenmeyer flask containing a layer of solid potassium carbonate (~ 1 to 2 g) to dry it. If the ether solution is colored, add a small amount of decolorizing carbon and stir or swirl the suspension for a few minutes. Gravity filter the organic layer into small round-bottomed flask and evaporate the ether and acetone on a rotary evaporator.[36]

EXPERIMENT 91

Rearrangement of Allyl Phenyl Ether

Save a small drop of the crude allyl phenyl ether for later qualitative test comparisons and spectroscopic analysis. Place the remaining crude allyl phenyl ether in a small round-bottomed flask or a reaction vial in the hood. Attach a reflux condenser and gently reflux the solution for three hours using a preheated sand bath with a bath temperature > 200°C. Allow the reaction mixture to cool and dissolve the residue in 15 to 20 mL of ether. Transfer the ether to a small separatory funnel and extract it twice with 5-mL portions of 10% sodium hydroxide. *This aqueous base contains the salt of the rearrangement product. Save it!* Dry the ether layer over magnesium sulfate and gravity filter it into a small, tared, round-bottomed flask. Evaporate the ether on the rotary evaporator and record the weight of recovered material.

Carefully adjust the pH of the combined aqueous sodium hydroxide extracts to pH 4 to 6 with dropwise addition of 30% sulfuric acid (prepared by slowly pouring 10 mL concentrated sulfuric acid over ~ 25 mL of ice). If necessary, cool the resulting solution by adding ice chips and extract it with two 10- to 15-mL portions of

---

[35] D. S. Tarbell, *Organic Reactions,* Vol. II, R. Adams, ed. (New York: John Wiley, 1944), p. 26.

[36] Note that failure to remove all the acetone will make the subsequent rearrangement impossible since the reflux temperature may only be that of boiling acetone.

ether. Combine these two ether extracts, dry the resulting solution over magnesium sulfate, gravity filter the solution into a tared round-bottomed flask, and evaporate the solvent. Record the weight of the crude product. Also record and interpret the NMR and IR spectra of this final product and its precursor. Compare the spectra and show how they distinguish between starting material and the final product. Note the characteristic differences in the aromatic substitution pattern in the NMR spectra.

Perform and interpret the ferric chloride test (see Section 9.3.2) on phenol, allyl phenyl ether, and the rearrangement product. Prepare the solid aryloxyacetic acid derivative of 2-allylphenol (see Section 9.3.2) and determine its melting point (~150°C).

Questions for Experiments 90 and 91

1. What diagnostic differences did you notice between the IR spectra of the starting allyl phenyl ether and the rearranged product?

2. Is there a single conformation for the transition state, or is there more than a single conformation for the transition state? Draw the conformations you might consider.

3. It is conceivable that the Claisen rearrangement can occur by a two-step dissociation-recombination mechanism (by a radical process). Suggest an experiment designed to differentiate between a concerted and a two-step mechanism.

4. What would be the structures of the corresponding products derived from 2-methylphenol and 3-bromo-1-butene?

Another class of rearrangement reactions is known as electron-deficient rearrangements. These rearrangements are induced by the formation of a cationic center, or partially formed cationic center, followed by migration of a group from an adjacent atom to form a more stable cation. The Wagner–Meerwein rearrangement of carbocations is an example of such a rearrangement. Rearrangements involving cationic centers on other than carbon atoms are also known, an example being the Beckmann rearrangement of oximes.

Oximes can be prepared readily from aldehydes and ketones by reaction of carbonyl compounds with hydroxylamine (see Section 9.5.5). Conversion of the hydroxyl group of aldoximes to a good leaving group can result in an elimination to produce the corresponding nitrile. Conversion of the hydroxyl group of ketooximes to a good leaving group results in a Beckmann rearrangement to the corresponding amide in aqueous solution. For symmetrical ketones and the corresponding oximes, migration from either side of the oxime produces the same product. In unsymmetrical ketones, two oximes (*E*- or *Z*-stereochemistry) can form. Rearrangement of the unsymmetrical oximes can be stereospecific. The overall process is a remarkable and efficient conversion of a ketone to an amide.

In the following experiment, the oxime of a symmetrical ketone, fluorenone (Experiment 57), will be prepared and subjected to the Beckmann rearrangement.

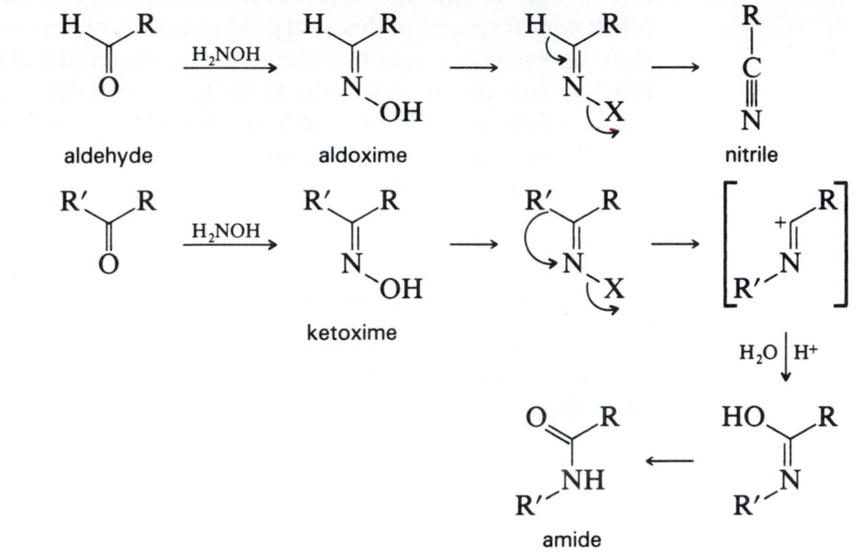

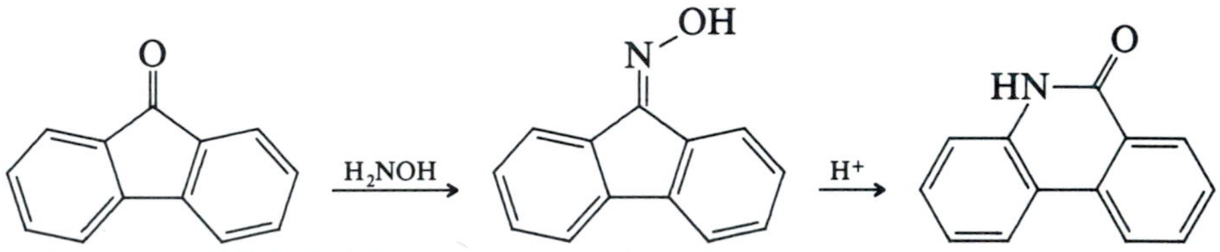

EXPERIMENT 92

Beckmann Rearrangement of Fluorenone Oxime

The oxime of fluorene may be prepared by the general procedure described for oximes (Section 9.5.5, mp ~196°C). Place 390 mg (2 mmol) of fluorenone oxime in a 25-mL round-bottomed flask fitted with a reflux condenser and a stir bar. Add 10 to 12 mL of polyphosphoric acid and heat the mixture in a sand bath at 175 to 190°C for about 10 minutes. Cool the solution to room temperature and add it to about 50 mL of water. Collect the precipitated product by suction filtration, wash it with cold water, and air dry the product. If possible determine the mp (~286 to 289°C). If necessary, recrystallize the product from ethyl acetate-hexanes. Record the IR spectra of the product and fluorenone. Compare the carbonyl stretching frequencies.

Questions for Experiment 92

1. The rate of the reaction of hydroxylamine with acetone is very pH dependent. The reaction is slow at low pH and high pH, but very fast near pH 4 to 6 (near the pK_a of hydroxylamine). Rationalize this pH-rate dependence.

2. Explain the difference in the carbonyl stretching frequency of fluorenone and the amide produced by Beckmann rearrangement of the fluorenone oxime.

As indicated above, unsymmetrical ketooximes can rearrange stereospecifically during a Beckmann process. An interesting experiment that clearly demonstrates this is the Beckmann rearrangement of the *(E)-* and *(Z)-*oximes of phenylacetone. The amide products can be readily distinguished by their NMR spectra and by independent syntheses of the amides. The details of this very instructive experiment have been published.[37]

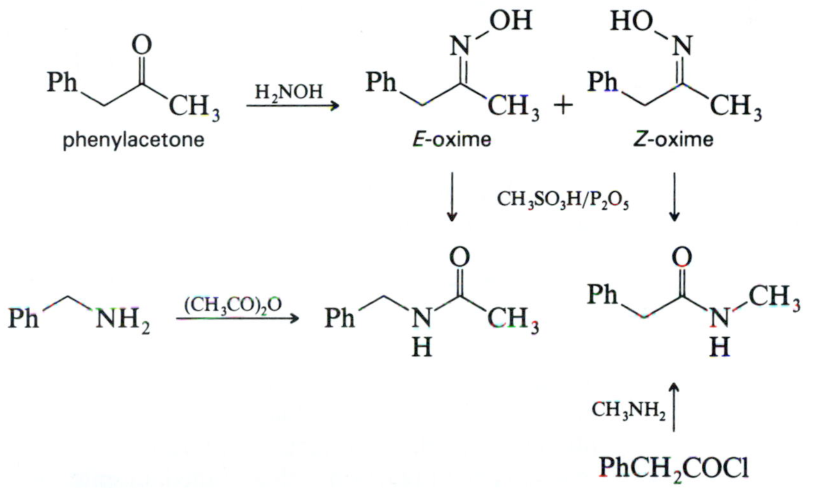

Another very common electron-deficient rearrangement that is useful for the synthesis of aldehydes and ketones is the pinacol rearrangement. The pinacol-pinacolone rearrangement is shown as the basic example. In this case, all the potential migrating groups are identical.

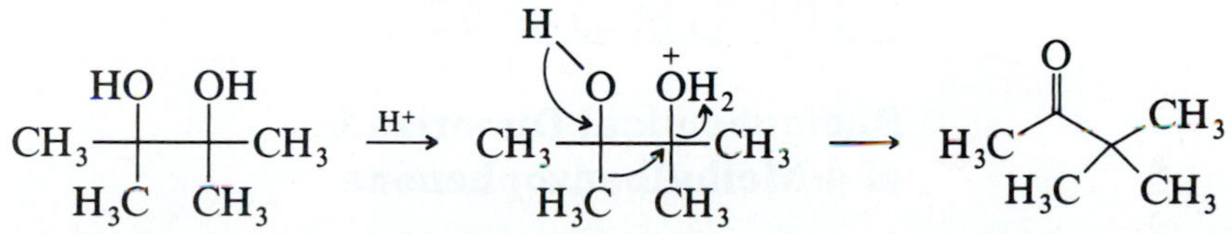

In many cases the potential migrating groups are different and the relative ability to migrate is variable. In this experiment, 4,4'-dimethylbenzopinacol **(16)**, prepared by photochemical dimerization of 4-methylbenzophenone, is subjected to

---

[37] S. S. Stradling, J. L. Hornick, and J. Riley, *J. Chem. Ed.,* **60,** 502 (1983).

dehydration conditions that result in its rearrangement to the benzopinacolones **17** and **18**, in unequal amounts.

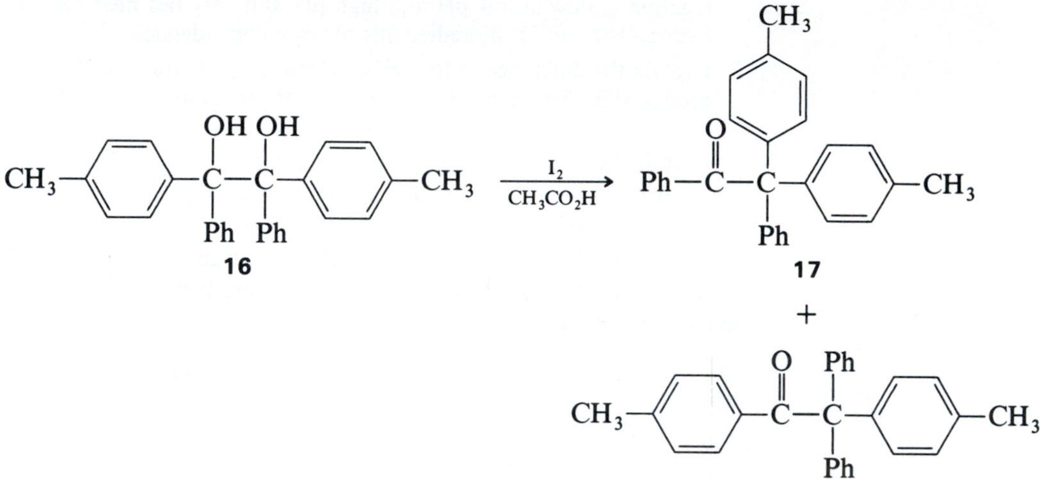

These products are not isolated; instead the product mixture is subjected to degradation using potassium *t*-butoxide in dimethyl sulfoxide (DMSO):

$$\textbf{17} + \textbf{18} \xrightarrow[\text{2. H}^+]{\text{1. K}^+ {}^-\text{OC(CH}_3)_3/\text{DMSO}} PhCO_2H + p\text{-CH}_3C_6H_4CO_2H$$

$$+ \ p\text{-CH}_3C_6H_4CHPh_2 + (p\text{-CH}_3C_6H_4)_2CHPh$$

The acid products are then separated from the hydrocarbons and converted to methyl benzoate and methyl *p*-toluate. The ester mixture is analyzed by gas chromatography and the relative migratory aptitude of the phenyl and tolyl groups during the rearrangement reaction is determined, despite the fact that the overall yield for the reaction sequence is often very low.

EXPERIMENT 93

Photochemical Dimerization of 4-Methylbenzophenone

Place in a clean 24- × 150-mm test tube 4.0 g of 4-methylbenzophenone, 25 mL of isopropyl alcohol, and two drops of glacial acetic acid and warm the mixture on a steam bath until solution is attained. Cover the test tube with aluminum foil and irradiate the solution with a 275-watt sun lamp, at an approximate distance of 15 cm, for three days. Several tubes can be simultaneously irradiated with a single lamp, especially if the group of tubes is surrounded with a reflecting sheet of aluminum foil.[38]

Cool the reaction mixture in an ice-water bath and collect the yellow to white solid obtained. Recrystallize the solid from 1-propanol-water using approximately 50 mL of the alcohol. The photodimer has a melting point of 162 to 164°C.

EXPERIMENT 94

Rearrangement of 4,4'-Dimethylbenzopinacol

To a 100-mL round-bottomed flask add 1.60 g of 4,4'-dimethylbenzopinacol, 15 mL of glacial acetic acid, and sufficient 2% iodine in acetic acid to impart to the solution a pale orange color. Heat the solution at reflux for approximately 10 minutes; add additional iodine-acetic acid solution if the orange-yellow color fades. Evaporate the solution to a viscous oil under reduced pressure on a rotary evaporator or by fitting the flask with a thermometer adapter and glass tubing and connecting the glass tube to a vacuum trap assembly that is connected to a water aspirator with thick-walled tubing. Heat the flask on a steam bath (it may prove more efficient to submerge the flask and cover it with a towel) and cool the trap in an ice-water bath. To remove all the iodine completely, treat the residue with three consecutive 5-mL portions of glacial acetic acid and each time evaporate all volatile material under reduced pressure.

EXPERIMENT 95

Degradation of the Mixture of Pinacolones

To the residue in Experiment 94, add 10 mL of dimethyl sulfoxide and warm on a steam bath to effect solution. Also place in a small Erlenmeyer flask 10 mL of dimethyl sulfoxide, 0.38 mL of water (from a small graduated pipette or syringe), and 4.0 g of potassium *t*-butoxide.

 C A U T I O N : Potassium *t*-butoxide is highly corrosive and should not be allowed to come in contact with the skin; flush skin with plenty of water should accidental contact be made. Since this reagent is also highly hydroscopic it should be weighed and dissolved as rapidly as possible, and the bottle should be recapped immediately following each use.

---

[38] If this photochemical reaction is not practical, commercially available 4,4'-dimethylbenzopinacol can be used for the next step.

Add the warm solution of the pinacolones to the slurry in the Erlenmeyer flask, swirl the resulting highly colored mixture, and cool to room temperature by briefly placing it in an ice-water bath, and then extracting it with two 20-mL portions of dichloromethane. Discard these extracts.

Acidify the aqueous layer in the separatory funnel with concentrated hydrochloric acid to approximately pH 1 (test with pH paper), and extract it with three 15-mL portions of dichloromethane. Wash the combined organic extracts with two 10-mL portions of water and one 10-mL portion of saturated aqueous sodium chloride. Dry the dichloromethane solution over magnesium sulfate, gravity filter into an appropriately sized round-bottomed flask, and evaporate to dryness using the method outlined in Experiment 94.

Transfer the residue to a 50-mL round-bottomed flask with two 10-mL portions of approximately 3% methanolic hydrogen bromide, and heat the solution at reflux for $1\frac{1}{2}$ hours. Proceed to the next section.

Analysis of products by GLC.   Evaporate the methanol solution from the last section to a volume of approximately 0.1 to 0.25 mL using the previously prescribed method. Analyze this solution of methyl benzoate and methyl p-toluate by gas chromatography as described by the instructor. Assume that the gc detector responds equally to the same mole amounts of each ester product. Determine which of the peaks corresponds to methyl benzoate by injecting an authentic sample of methyl benzoate (made up by dissolving one drop of methyl benzoate in 10 to 20 drops of methanol). If necessary, coinject a few microliters each of the reaction product mixture with the authentic methyl benzoate solution to confirm which peak corresponds to methyl benzoate. Determine the relative migratory aptitude of the phenyl group versus the tolyl group by integrating the areas for the methyl benzoate and methyl p-toluate. Provide a detailed explanation for the experimental outcome in your report.

Questions for Experiments 93 to 95

1. Suggest a synthesis of 4-methylbenzophenone employing, in addition to benzene, only other readily available organic compounds containing one carbon atom per molecule and any necessary inorganic reagents.
2. Predict the predominant product formed from dehydration-rearrangement of each of the following glycols:
   (a) 1,2-propanediol
   (b) 2-methyl-1,2-propanediol
   (c) 1-phenyl-1,2-ethanediol
   (d) 1,1-diphenyl-1,2-ethanediol
   (e) 1-phenyl-1,2-propanediol
   (f) 2,2-dimethyl-1,1-diphenyl-1,2-ethanediol
   (g) 2-methyl-1,1,2-triphenyl-1,2-ethanediol
3. How many stereoisomers of 4,4'-dimethylbenzopinacol are formed in the dimerization of 4-methylbenzophenone? Draw stereoformulas for each.

4. Provide mechanisms for the rearrangement of 4,4′-dimethylbenzopinacol and for the cleavage of the ketone products with base.

Other Rearrangements

Many other rearrangements are frequently observed in organic chemistry. Some further examples are provided in other sections of this chapter.

The Hofmann rearrangement, which is representative of a number of amide rearrangements to isocyanates (shown below), was demonstrated by the conversion of phthalimide to anthranilic acid (Experiment 82).

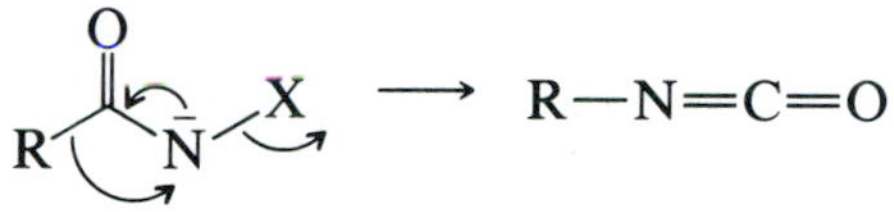

Hofmann rearrangement, X = Br
Curtius rearrangement, X = N₂
Lossen rearrangement, X = OR′
Wazonek rearrangement, X = N⁺R₃

The benzil to benzilic acid rearrangement is described in Experiment 102.

10.17 CONDENSATIONS AND RELATED REACTIONS

One of the most important processes in organic chemistry is the formation of carbon-carbon bonds. In most cases such bond formation requires the formation of a carbon nucleophile (usually a carbanion) and its reaction with an electron-deficient carbon (an electrophile). This section will provide several experiments that allow for the formation of carbon-carbon bonds by such condensation processes. Elaboration of some of the condensation products will also be demonstrated.

The benzoin condensation involves the reaction of two nonenolizable aldehydes in such a manner that the final product results from direct attachment of the original two aldehyde carbonyl carbons. This net result initially appears to violate the commonly accepted mode of formation of carbon-carbon bonds in which one component has negative character and the other has positive character, since both aldehyde carbonyl carbons bear a partial positive charge. In fact, the condensation requires a temporary reversal of polarity at one of the carbonyl carbons. (Note the partial positive charge on **19** and the full negative charge on the same carbon in **21**.) In the classical benzoin condensation, the polarity reversal (termed "umpolung") has been accomplished by the reaction of benzaldehyde with a catalytic amount of cyanide followed by a proton transfer to give an intermediate carbanion **21** that undergoes a nucleophilic addition to another molecule of benzaldehyde. Subsequent decomposition of the tetrahedral intermediate by loss of HCN produces benzoin. This reaction must be performed with considerable care since cyanide is extremely toxic. Fortunately, an alternative, and very safe process for accomplishing the same transformation is now available.

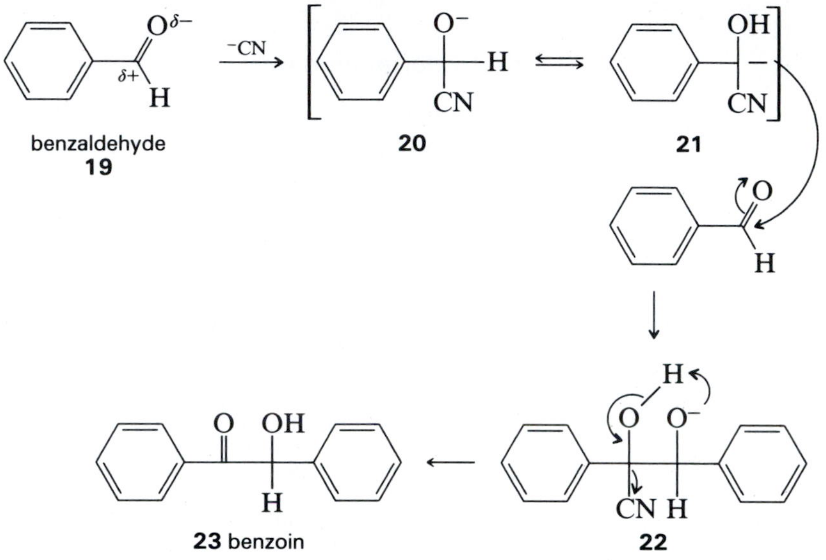

benzaldehyde
19

20

21

23 benzoin

22

Recently, thiamine hydrochloride (vitamin B$_1$) has been shown to be an efficient catalyst for the benzoin condensation. Polarity reversal reactions such as the benzoin condensation are extremely important in many biological processes, and the mechanism for effecting catalysis with polarity reversal is very similar to the mode of action of thiamine as a coenzyme under physiological conditions. Thiamine pyrophosphate (TPP) is a ubiquitous coenzyme in living systems. It is easily formed from the important nutritional factor, thiamine, in humans and other animals. People who do not consume enough thiamine in their diets develop a disease called beriberi (Shinhalese for weak-weak), a disease of the nervous system characterized by partial paralysis of the extremities, emaciation, and anemia. The mechanism of action of TPP has been shown to involve the removal of a relatively acidic proton on the thiazolium ring component of thiamine. This produces a carbanion that behaves similarly to cyanide anion in the classical benzoin condensation.

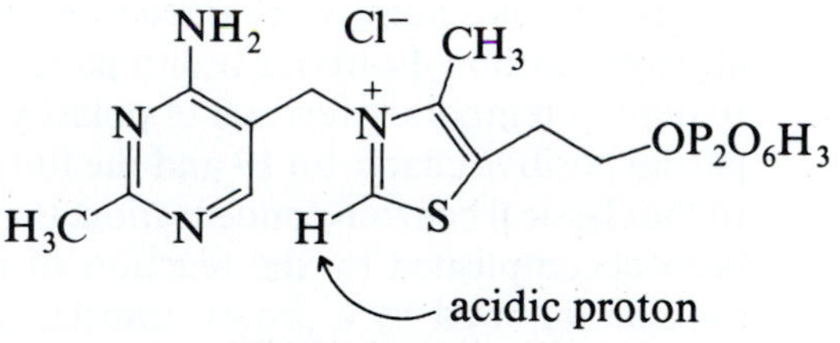

acidic proton

thiamine pyrophosphate

EXPERIMENT 96

Benzoin Condensation of Benzaldehyde Catalyzed by Thiamine

Dissolve 0.35 g (1 mmol)[39] of thiamine hydrochloride (vitamin B_1) in 0.75 mL of water in a 10-mL round-bottomed flask or a small Erlenmeyer flask. Add 3 to 3.5 mL of 95% ethanol and cool the resulting solution with an ice-water bath. Gently swirl the thiamine solution and slowly add 0.75 mL of cold $3M$ sodium hydroxide in water over a 7- to 10-minute period. Add 2.12 g (~2 mL, 20 mmol) of benzaldehyde to the reaction mixture, swirl, and check the pH. If the pH is lower than 8, add more of the $3M$ sodium hydroxide solution until the pH is approximately 8 to 9. The thiamine will not be ionized and the condensation reaction will not proceed if the pH is too low. If the pH is higher than 9, a Cannizzaro reaction (see Experiment 97) will become competitive. Using a beaker of water heated to ~60 to 70°C, heat the reaction mixture for one hour while maintaining the internal reaction temperature at 60 to 63°C. Allow the reaction mixture to cool to room temperature and then cool it further in an ice bath to about 10°C. A white precipitate of product should appear. Some oil, which is probably a mixture of starting aldehyde and reaction byproducts, may also form on the water surface.

Collect the crude product by suction filtration and wash it with 5 mL of ice-cold water. Recrystallize the product from 95% ethanol-water, using a minimum amount of the hot solvent necessary to dissolve the product. Determine the weight, percent yield, and melting point (~137°C) of the benzoin. Record and interpret the IR spectrum (KBr pellet) of benzoin.

In the absence of a reagent such as cyanide or thiamine, benzaldehyde will react with hydroxide to form a tetrahedral intermediate that can provide a source of hydride to reduce another molecule of benzaldehyde to the corresponding alcohol. This is why control of the pH was so critical in the benzoin condensation of benzaldehyde.

benzaldehyde   potassium   benzyl alcohol
                                                      benzoate

---

[39] This reaction also works well when scaled up by a factor of 10 or more. A larger-scale reaction will provide an adequate amount of benzoin for subsequent experiments (see Experiments 98 to 102).

EXPERIMENT 97

Cannizzaro Reaction of Benzaldehyde

Dissolve 3 g[40] of potassium hydroxide in 6 mL of water in a 25-mL round-bottomed flask. Add 2 mL of benzaldehyde. Add a stir bar or boiling chip and attach a reflux condenser. Heat the mixture at reflux for 1 hour. Cool the reaction mixture to room temperature and transfer it to a small separatory funnel. Extract the mixture with two 25 mL volumes of ether. Separate the two phases and acidify the aqueous phase to pH ~ 3 by dropwise addition of concentrated hydrochloric acid. Cool the acidified solution in an ice bath and collect the crystals of benzoic acid by suction filtration. Air dry the benzoic acid and record its mp (~ 122°C, recrystallize from hot water, if necessary), IR spectrum (KBr pellet), and NMR spectrum ($CDCl_3$). While the benzoic acid is air drying, dry the ether extract over anhydrous magnesium sulfate. Gravity filter the ether into a tared, round-bottomed flask and evaporate the ether on a rotary evaporator. Record the weight of the crude benzyl alcohol. Also record its IR spectrum (neat film) and its NMR spectrum ($CDCl_3$) and compare the spectra with those of the acid and starting aldehyde. Especially note the absence of a carbonyl stretching frequency in the IR spectrum of the alcohol relative to the starting aldehyde.

The facile formation of substituted benzoins in the benzoin condensation provides an excellent method for the synthesis of 1,2-diols by reduction of the carbonyl group of the benzoin. Conceptually, reduction of the keto group of benzoin could give a mixture of the *erythro*- and *threo*-diols. However, careful sodium borohydride reduction selectively gives the *erythro*-isomer presumably through a chelated intermediate that encourages reduction from the least-hindered face of the carbonyl group. The structure of the diol can be confirmed by comparison of the melting point of the product with that of either of the authentic *erythro*- (mp 137°C) or *threo*-products (mp 119°C). Alternatively, preparation of the corresponding acetonide allows unambiguous structure determination by NMR spectroscopy since the methyl groups of the acetonide derived from the *threo*-isomer are identical by an axis of rotation whereas the methyl groups in the acetonide derived from the *erythro*-isomer are diastereotopic.

---

[40] This reaction can also be done on a reduced scale (0.2 to 0.5 times) in a conical reaction vial. The extraction can be conveniently performed using a disposable pipette to transfer the layers.

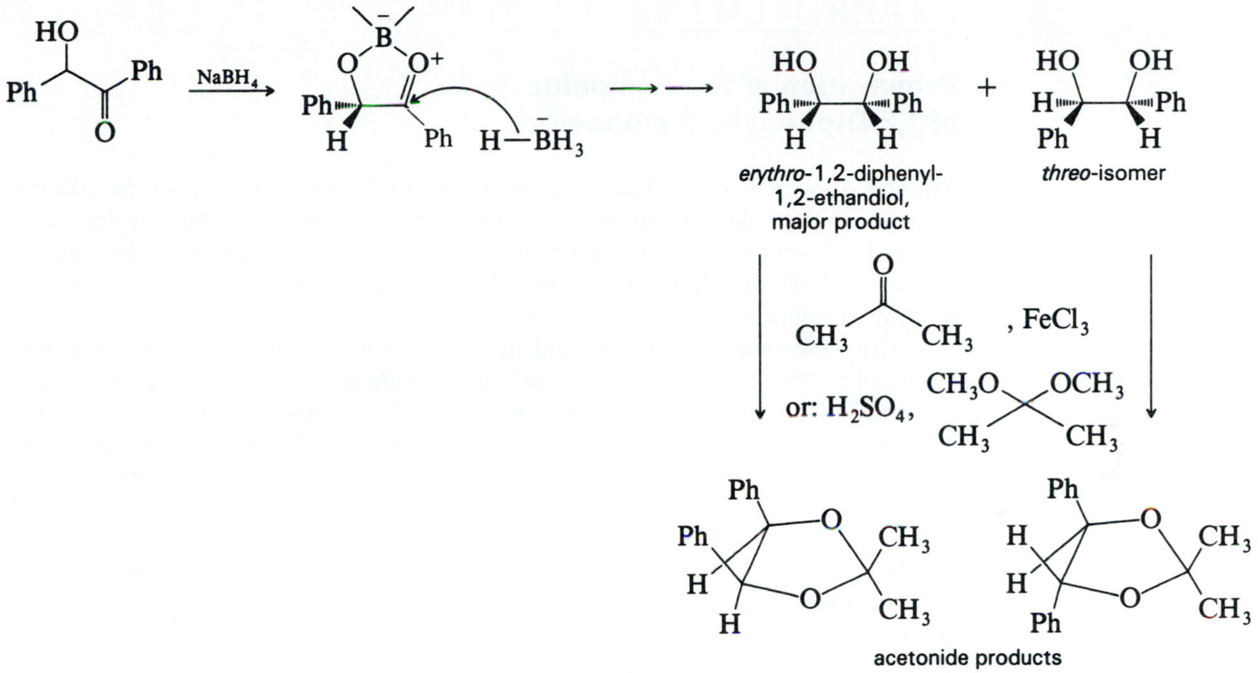

erythro-1,2-diphenyl-
1,2-ethandiol,
major product

threo-isomer

acetonide products

EXPERIMENT 98

Stereoselective Reduction of Benzoin[41]

Place a magnetic stir bar, 2 g (9.42 mmol) of benzoin (from Experiment 96), and 20 mL of absolute ethanol in a 125-mL Erlenmeyer flask. While stirring, carefully add 0.4 g (10.6 mmol) of sodium borohydride portionwise to the mixture. After the addition is complete, stir the mixture for another 15 minutes at room temperature. Cool the flask in an ice bath and decompose the excess sodium borohydride by first adding 30 mL of water followed by the careful and dropwise addition of 1 mL of 6M HCl. The mixture may foam uncontrollably if the acid is added too quickly. Add another 10 mL of water and stir the mixture for 15 minutes. Collect the white precipitate by suction filtration. Wash the product diol with water on the suction funnel and allow the product to air dry. Record the mp (~ 136 to 137°C). If a low or very broad melting point product is obtained, recrystallize the product from acetone-hexanes. Record the IR spectrum (KBr) and compare it to the spectrum of the starting benzoin, noting the absence of the carbonyl stretching band.

---

[41] Adapted from A. T. Rowland, *J. Chem. Ed.*, **60**, 1084 (1983).

EXPERIMENT 99

Preparation of the Acetonide
of 1,2-Diphenyl-1,2-ethanediol

Method A, with FeCl$_3$.   Place 1 g (4.67 mmol) of the diol in a 50- to 100-mL round-bottomed flask fitted with a stir bar, reflux condenser, and drying tube. Temporarily remove the condenser and add 0.3 g of anhydrous ferric chloride followed by 30 mL of anhydrous acetone.   Reattach the reflux condenser and heat the mixture at reflux for 20 to 30 minutes.

Cool the reaction mixture and pour it into a separatory funnel containing 10 mL of a 10% aqueous potassium carbonate solution and 50 mL of water.   Extract the solution with two 20- to 25-mL portions of dichloromethane.   Combine the organic extracts and wash them with 25 mL of water.   Dry the organic layer over anhydrous sodium sulfate and gravity filter it into a tared round-bottomed flask. Evaporate the solvent on the rotary evaporator.   Add 15 mL of boiling pentanes and gravity filter the suspension to remove unreacted diol.   Concentrate the filtrate to 3 to 4 mL and cool it in an ice bath.   If necessary, scratch the inside of the flask to induce crystallization.   Collect the product by suction filtration and wash it with a small amount of ice-cold pentanes.   Determine the mp ($\sim$ 57 to 59°C).   Record the NMR spectrum in CDCl$_3$ and use it to assign the stereochemistry of the product acetonide (*meso*-2,2-dimethyl-4,5-diphenyl-1,3-dixolane) and the diol precursor (*erythro*-1,2-diphenyl-1,2-ethanediol).

Method B, with 2,2-dimethoxypropane (acetone dimethylketal).   Dissolve 1 g (4.67 mmol) of the diol in 15 mL of anhydrous acetone in a 50- to 100-mL round-bottomed flask fitted with a stir bar.   Add 2 mL of acetone dimethyl acetal (2,2-dimethoxypropane).   Immediately stopper the flask and cool it in an ice bath.   Remove the stopper and add 12 drops of concentrated sulfuric acid *(CAUTION: This acid is highly corrosive)*.

Carefully replace the stopper and continue to stir the cooled flask for 10 minutes.   While the acetonide is forming, prepare a solution of 2 g of Na$_2$CO$_3$ in 30 mL of water.   After 10 minutes, transfer the cold acetone solution to a separatory funnel. Rinse the reaction flask with two 25-mL portions of diethyl ether and transfer the rinses to the separatory funnel.   Carefully add the aqueous Na$_2$CO$_3$ solution to the separatory funnel.   Swirl the funnel gently before inserting its stopper.   *(CAUTION: Carbon dioxide evolution may be vigorous and cause loss of material.)*   Insert the stopper and immediately invert the funnel and vent it by opening the stopcock.   With the stopcock open, swirl the funnel again to complete the neutralization of the acid and evolution of carbon dioxide.   Close the stopcock.   Position the funnel upright and remove the stopper.   Drain off the lower aqueous layer and extract the remaining organic layer with two 30 mL-portions of water and then with one portion of saturated sodium chloride solution.   Transfer the organic layer to an Erlenmeyer flask and dry the solution over MgSO$_4$.   Gravity filter the dry solution into a tared round-bottomed flask and evaporate the solvent. Extra care may need to be taken to remove any excess 2,2-dimethoxypropane. Determine the weight of the residual product.

Record the NMR spectrum in CDCl₃ and use it to assign the stereochemistry of the product acetonide (2,2-dimethyl-4,5-diphenyl-1,3-dioxolane) and the diol precursor (1,2-diphenyl-1,2-ethanediol). Examination of the NMR spectrum before recrystallization will also give you an impression of the stereoselectivity of the initial sodium borohydride reduction by possible detection of the other isomeric acetonide. In most cases, essentially one isomer is observed. Recrystallization as described in Method A will give you an isomerically pure product.

In addition to the reduction of substituted benzoins to 1,2-diols, they may also be oxidized to α-diketones, thus further enhancing their value as synthetic intermediates. Benzoin (from Experiment 96) can be oxidized to the α-diketone, benzil, very efficiently with nitric acid or with copper (II) acetate.

benzoin   benzil

EXPERIMENT 100

Oxidation of Benzoin with Nitric Acid

Place 2 g[42] (9.4 mmol) of benzoin and 7.5 mL of concentrated nitric acid *(CAUTION: This acid causes severe burns)* in a 25-mL round-bottomed flask in the hood. Add a stir bar or boiling chip. Attach a condenser fitted with a T-tube and gas trap (Figure 1.14) to remove any vapors generated from the nitric acid. Heat the mixture at reflux on a steam bath or with a sand bath at ~ 100°C for 30 minutes.

Allow the mixture to cool and pour it into 35 to 40 mL of water. Swirl the suspension for a few minutes and then collect the product by suction filtration. Wash the yellow solid with water on the filter funnel. Air dry the product for a few minutes and then press it down on the filter funnel with a spatula to squeeze out excess water. Recrystallize the still moist product from ethanol-water by first dissolving the product in ~ 3 to 5 mL of hot ethanol and then by adding hot water dropwise until the solution just becomes turbid. Clarify the hot solution with a drop or two of hot ethanol and then allow the solution to cool. Collect the product by suction filtration.

---

[42] This reaction can be easily scaled up by a factor of 5 to 10 if the subsequent reactions of benzil are to be performed.

Note the color. Record the yield and mp ($\sim 95\,°$C). Determine the purity by TLC on silica gel using toluene as the eluent ($R_f = \sim 0.6$). Record and interpret the IR spectrum (KBr pellet) of the product. Compare the IR spectrum to that of benzoin.

EXPERIMENT 101

Oxidation of Benzoin with Copper (II) Acetate[43]

Place 2 g (9.4 mmol) of benzoin, 3.75 g (18.8 mmol) of cupric acetate monohydrate, 15 mL of acetic acid, and 5 mL of water in a 50-mL round-bottomed flask fitted with a reflux condenser. Heat the mixture to reflux for 15 minutes. Gravity filter the mixture using a heated funnel to remove the cuprous oxide. Cool the filtrate and collect the product on a Büchner funnel. Wash the product with a small amount of water. Air dry the product and record the yield and mp ($\sim 95\,°$C).

To illustrate further the synthetic utility of the compounds encountered in this series, the next experiment illustrates a base-induced rearrangement of α-diketones. Reaction of benzil with potassium hydroxide initiates a rearrangement to the potassium salt of benzilic acid. Acidification provides the free acid. This reaction is another example of a migration similar to that observed in the pinacol rearrangement (see Experiment 94), except that the migration is induced by an anion instead of a cation.

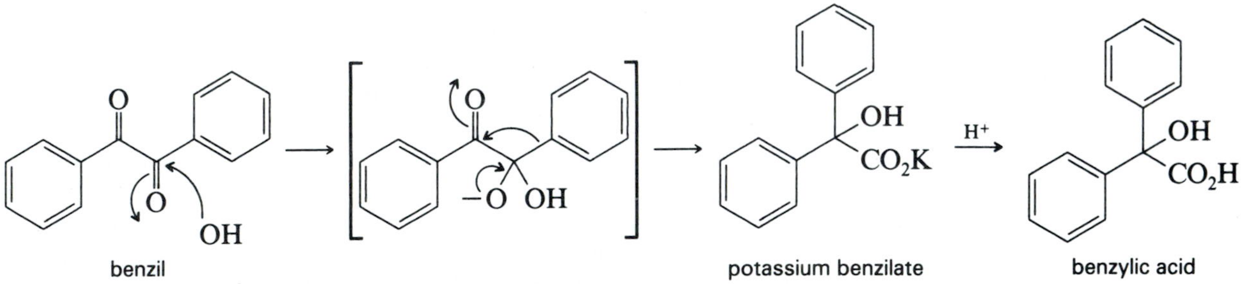

benzil   potassium benzilate   benzylic acid

EXPERIMENT 102

Rearrangement of Benzil to Benzilic Acid

Dissolve 3 g of potassium hydroxide in 5 mL of water in a small Erlenmeyer flask by heating on a hot plate. While the solution is cooling to room temperature, place 2 g of benzil and 7 mL of methanol in a 25-mL round-bottomed flask. Attach a reflux

[43] P. Depreux, G. Bethegnies, and A. Marcincal-Lefebvre, *J. Chem. Ed.*, **65**, 553 (1988).

condenser and heat the mixture until the solid is dissolved. Discontinue heating, cool in an ice bath, and pour the cold potassium hydroxide solution into the methanol solution of benzil. Swirl the solution and note the color change. Add a stir bar or a boiling chip and heat the solution at reflux until the solid goes into solution. This should require no more than 10 minutes. Cool the solution in an ice bath. If no crystals form, add 3 to 4 mL of methanol and cool again. Collect the colorless crystals of potassium benzilate by suction filtration. Air dry the crystals and record the weight. Dissolve the salt in the least amount of hot water and acidify with dropwise addition of concentrated hydrochloric acid until the pH is about 3 to 4. Collect the solid benzilic acid by suction filtration. Recrystallize the product from ethanol water. Record the yield and mp ($\sim 151\,°C$) of the dry, purified benzilic acid. Record (KBr pellet) and interpret the IR spectrum of the benzilic acid. Compare the IR spectrum to that of the starting material.

    α-Diketones are also very useful in the preparation of cyclic compounds of great value. The reaction of urea and benzil in the presence of base also initiates a benzilic acid rearrangement. In this case, however, the initial nucleophile is urea instead of hydroxide. The rearrangement is followed by an intramolecular condensation of the urea to form a heterocyclic ring, 5,5-diphenylhydantoin, eventually. The sodium salt of 5,5-diphenylhydantoin is an anticonvulsant used for the treatment of epilepsy.

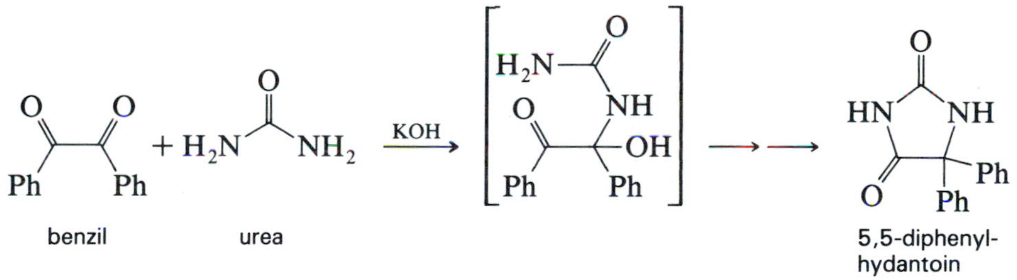

benzil   urea   5,5-diphenyl-hydantoin

EXPERIMENT 103

Preparation of 5,5-Diphenylhydantoin[44]

To a 50-mL round-bottomed flask containing a magnetic stir bar and fitted with a reflux condenser, add 2 g (~ 10 mmol) of benzil and 1.2 g of urea. Add 20 mL of ethanol and 3.3 g (50 mmol) of standard 85% potassium hydroxide in 4 mL of water. Reflux the mixture for 2 to $2\frac{1}{2}$ hours. Cool the mixture and remove some of the

---

[44] Adapted from M. C. Pankaskie and L. Small, *J. Chem. Ed.*, **63**, 650 (1986), and R. C. Hayward, *J. Chem. Ed.*, **60**, 512 (1983).

insoluble solids by gravity filtration. Cool the filtrate further in an ice-water bath and add 6N sulfuric acid slowly to the solution to adjust the pH to ~3. Collect the product by suction filtration. Record the mp (~295 to 300°C) and if necessary recrystallize the hydantoin from 95% ethanol. Record an IR spectrum (KBr pellet) of the hydantoin and compare the carbonyl stretching frequency to that of the starting benzil.

Questions for Experiments 97 to 103

1. Why could cyanide or thiamine not catalyze the condensation of acetone to pinacol?
2. Benzaldehyde can be air oxidized to benzoic acid. What consequence would this oxidation have during an attempted benzoin condensation?
3. Write out the detailed mechanism for the formation of 5,5-diphenylhydantoin from benzil and urea in the presence of potassium hydroxide. What is the initial role of the hydroxide?

In the benzoin condensation the nucleophilic carbon that added to the carbonyl group of the other molecule of benzaldehyde was the carbonyl carbon atom. A large number of condensations are known in which the α-carbon atom of one carbonyl compound adds to another carbonyl group. This class of reactions is known as the *aldol* and related condensations. The aldol condensation refers to a group of reactions that involve the addition of an enol or enolate to the carbonyl group of an aldehyde or ketone to produce a β-hydroxycarbonyl product. This will be illustrated by the preparation of dibenzalacetone.

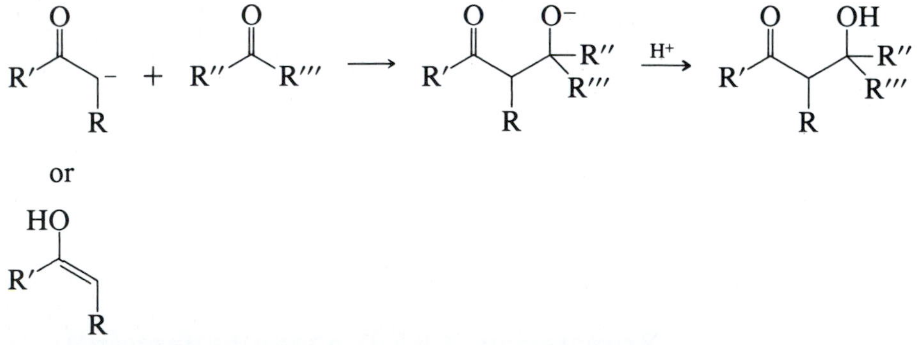

or

Dibenzalacetone can be prepared by the base-catalyzed condensation of acetone with two equivalents of benzaldehyde. This reaction is an example of a double aldol condensation followed by dehydration. Dehydration of aldol products is difficult to avoid when the resulting double bond will be conjugated to an aromatic ring. The side products, formed in small amounts, are those that result from the reaction of acetone with only one equivalent of benzaldehyde and the Cannizzaro reaction of benzaldehyde (see Experiment 97).

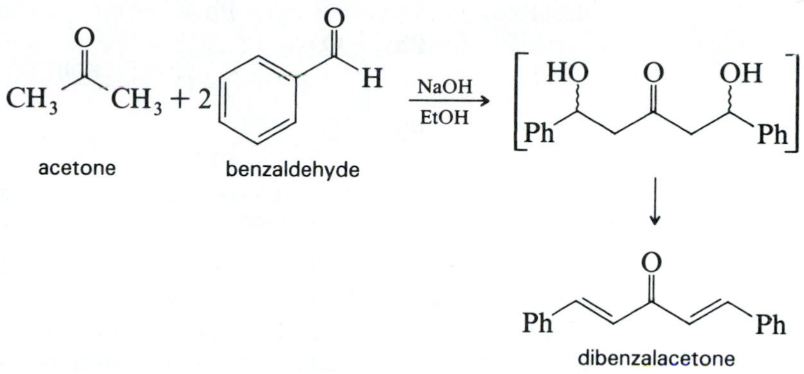

dibenzalacetone

EXPERIMENT 104

Preparation of Dibenzalacetone by an Aldol Condensation

In the hood, prepare a mixture of 1 mL (~ 1 g, 10 mmol) of benzaldehyde, 0.37 mL (0.29 g, 5 mmol) of acetone, and 1 mL of ethanol. Add one half of this mixture to a solution of 1 g of sodium hydroxide in 10 mL of water and 8 mL of ethyl alcohol in a 50-mL Erlenmeyer flask at room temperature. Stir the mixture and after 15 minutes add the rest of the mixture. Rinse the last traces of the carbonyl compound solution into the basic solution with 1 mL of ethanol. Stir the reaction mixture for 30 minutes more and then collect the solid product by suction filtration. Wash the product on the filter funnel with water to remove as much of the aqueous base as possible. Dry the product and recrystallize it from ethanol using the minimum amount of hot solvent to dissolve the product. Record the yield and melting point. Record the IR (KBr pellet) and NMR (CDCl$_3$) spectra of the product. Note especially the position of the carbonyl band in the IR spectrum and the chemical shift of the alkene hydrogens in the NMR. Interpret the positions of these peaks in the spectra.

Another interesting aldol condensation is the condensation of 1,3-diphenyl-2-propanone with benzil, followed by dehydration to produce 2,3,4,5-tetraphenyl-cyclopentadienone, a deep purple compound,[45] which has found extensive use in cycloaddition reactions.

---

[45] E. A. Harrison, *J. Chem. Ed.*, **65**, 828 (1988).

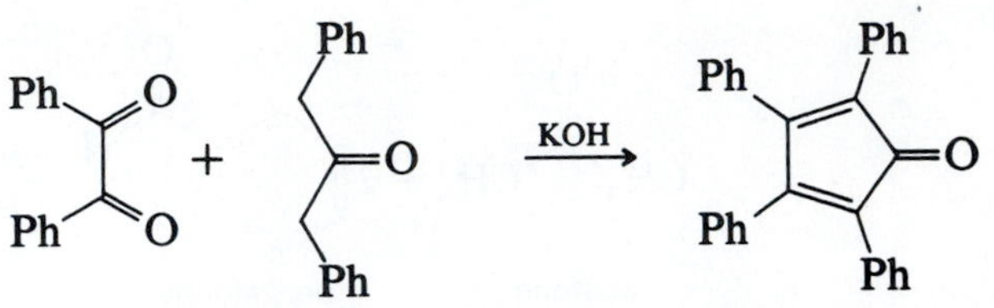

benzil   1,3-diphenyl-2-   2,3,4,5-tetraphenyl-
                propanone   cyclopentadienone

In a variant of the aldol reaction, the enolate component will be derived from an anhydride that will be made *in situ* from the reaction of acetic anhydride with phenylacetic acid. Triethylamine will be used as a base to generate regioselectively a small, but reactive, equilibrium concentration of the enolate. After the initial condensation, an acyl transfer reaction occurs to give the β-acetoxy acid. Elimination of acetic acid results in final production of the α,β-unsaturated acid product. One driving force for the elimination is the formation of a fully conjugated product. This is often referred to as the Perkin reaction.

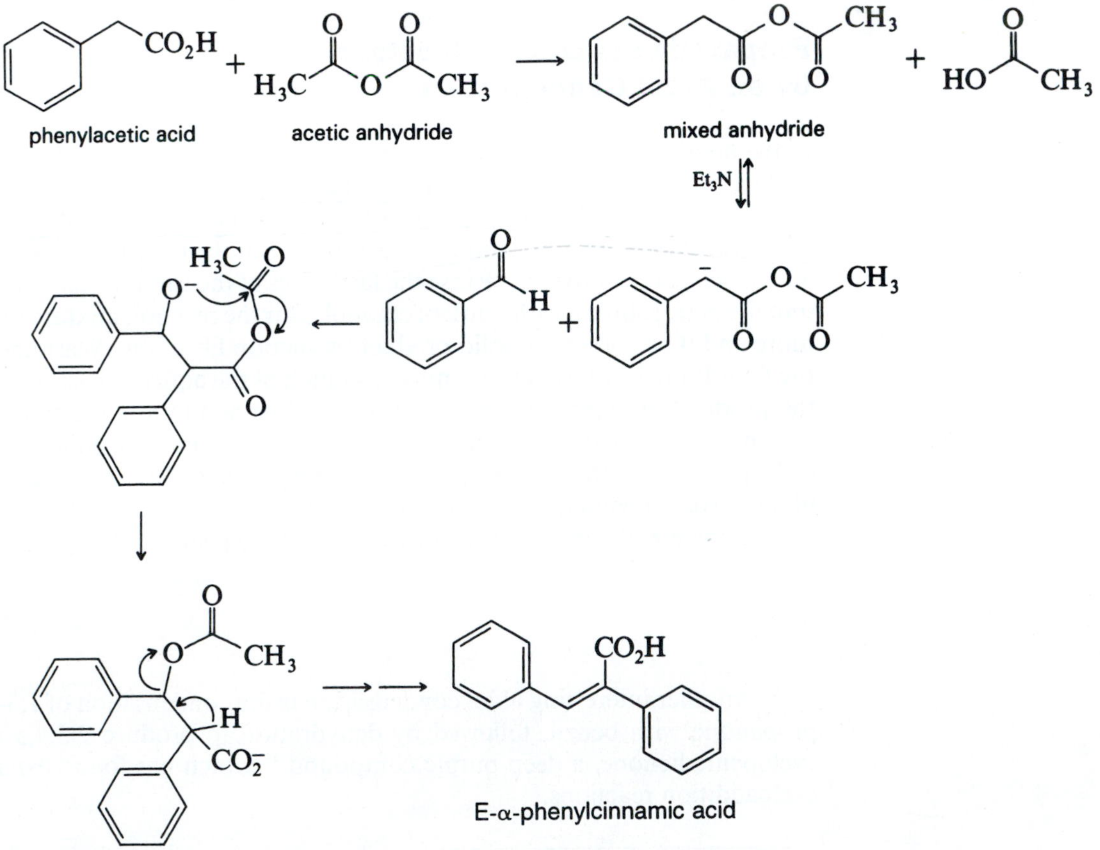

phenylacetic acid   acetic anhydride   mixed anhydride

E-α-phenylcinnamic acid

EXPERIMENT 105

The Perkin Reaction of an Anhydride with Benzaldehyde

Place 0.9 g of phenylacetic acid, 1 mL of benzaldehyde, 0.67 mL of acetic anhydride *(CAUTION: Acetic anhydride is a highly corrosive and dehydrating agent),* and 1 mL of triethylamine in a 25-mL round-bottomed flask in the hood. Attach a reflux condenser and heat the mixture at reflux using a sand bath at ~ 120 to 140°C for 30 to 40 minutes. Cool the flask in an ice bath and add 1.3 mL of concentrated hydrochloric acid. Add about 10 mL of ether and stir or swirl the mixture to dissolve any precipitates that have formed.

Transfer the mixture to a small separatory funnel and add another 10 mL of ether. Wash the organic solution with two 10-mL portions of water. Extract the ether solution with three 7- to 10 mL-portions of 5% sodium hydroxide solution. Combine the aqueous base solutions that contain the salt of the product and carefully acidify the solution with concentrated hydrochloric acid to pH 3. Allow the mixture to stand at room temperature for about 30 minutes and then collect the product by suction filtration. Air dry the crude product and recrystallize the product from ethanol-water. Record the yield, melting point, and IR (KBr pellet) and NMR (CDCl$_3$) spectra.

Questions for Experiment 105

1. Account for the regioselective formation of the enolate shown upon reaction of the mixed anhydride with triethylamine.
2. What is the driving force for the intramolecular acetyl transfer after the initial aldol reaction?

In the following experiment, instead of the nucleophile adding to the carbonyl group, the nucleophile adds to the β-carbon atom of an α,β-unsaturated ketone, which is known as the Michael addition. An aldol condensation of acetone followed by dehydration produces mesityl oxide. 1,4-Addition, or Michael condensation, of diethyl malonate to mesityl oxide (4-methyl-3-penten-2-one) is induced by bases, including alkoxides. Subsequent intramolecular cyclization, hydrolysis, and acid-catalyzed decarboxylation give 5,5-dimethyl-1,3-cyclohexanedione, dimedone.

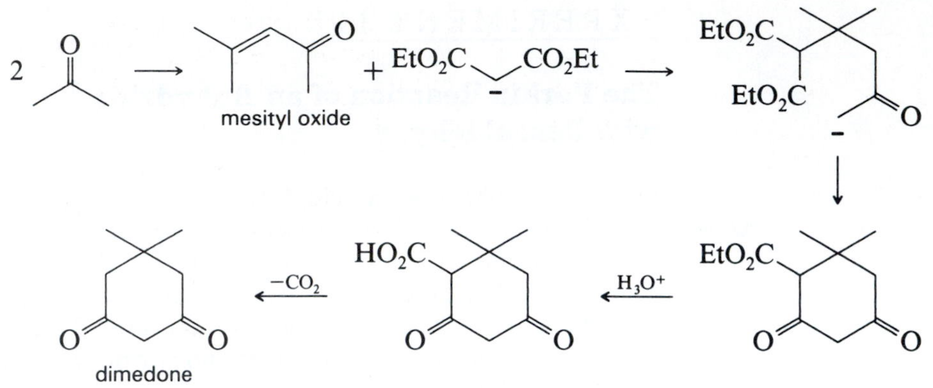

EXPERIMENT 106

Preparation of 5,5-Dimethylcyclohexan-1,3-dione (Dimedone)

Place 5 mL of methanol in a 50-mL round-bottomed flask and add 0.7 g of solid sodium methoxide *(CAUTION: Sodium methoxide is very hydroscopic)*. Add 2 mL of diethyl malonate. Attach a condenser and heat the mixture to reflux. Momentarily discontinue heating and, using a pipette, add 1.4 mL of mesityl oxide dropwise down through the condenser. A vigorous reaction may occur. Once the reaction appears to subside, heat the mixture back to reflux and continue to reflux it for 30 minutes. Again temporarily discontinue heating and add 10 mL of 2N sodium hydroxide. Reflux the mixture for another 90 minutes.

Cool the reaction and evaporate most of the methanol on the rotary evaporator. Heat the mixture again to reflux and slowly add ~ 15 mL of 4N HCl until the mixture is near pH 2 to 3. Cool the mixture in an ice bath and collect the dimedone by suction filtration. Wash the product with a small amount of cold water and recrystallize it from acetone. Record the yield, mp (~ 147 to 148°C), and IR (KBr pellet) and NMR (CDCl$_3$) spectra.

Questions for Experiment 106

1. Write detailed mechanisms for each step of the synthesis of dimedone.
2. Why do β-keto acids decarboxylate so easily, but α-keto acids do not?
3. Why does the anion derived from diethyl malonate add to the β-position (1,4-addition) rather than directly to the carbonyl group of mesityl oxide?

A related experiment[46] involves a phase-transfer catalyzed (PTC) addition of ethyl acetoacetate to crotonaldehyde followed by an intramolecular aldol condensation to produce 5-methyl-6-carbethoxy-2-cyclohexanone. This is an example of the Robinson annulation reaction, which is extremely useful for the formation of cyclohexenone derivatives.

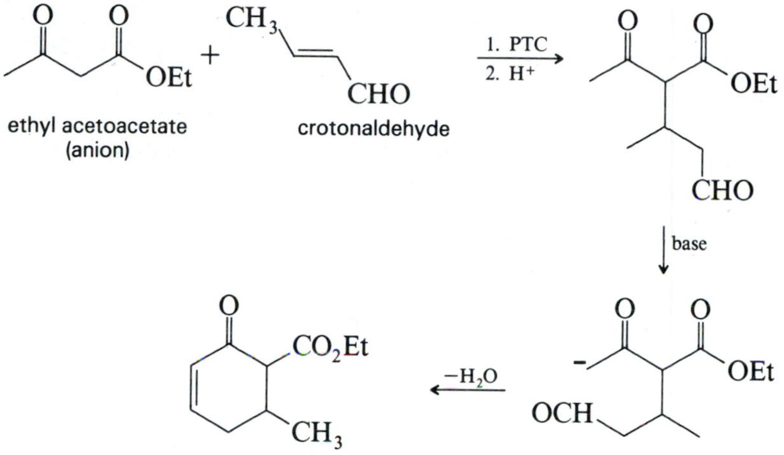

ethyl acetoacetate (anion)   crotonaldehyde

5-methyl-6-carbethoxy-2-cyclohexenone

Another very useful source of nucleophilic carbon for addition to carbonyl groups are the Wittig reagents, or phosphorus ylids. The Wittig reaction involves the reaction of a phosphorus ylid with a ketone or aldehyde to give an alkene. The ylid is prepared by reaction of a tertiary phosphine with an alkyl halide, followed by subsequent treatment with a strong base. The reaction is compatible with a number of other functional groups in the same molecules and thus provides advantages over most of the other methods of alkene formation. Sometimes mixtures of *cis*- and *trans*-alkenes are formed, but many modifications of the basic Wittig reaction have been reported to control the ratio of the stereoisomers formed. The following experiments illustrate the versatility of Wittig and related reactions.

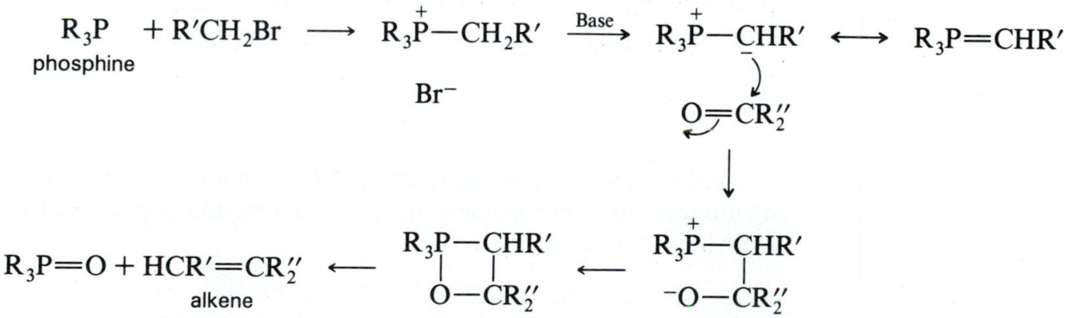

[46] D. S. Soriano, A. M. Lombardi, P. J. Persichini, and D. Nalewajek, *J. Chem. Ed.*, **65**, 637 (1988).

The careful condensation of 9-anthraldehyde with the ylid made from benzyl-triphenylphosphonium chloride provides the *trans* alkene, 9-(2-phenylethenyl)-anthracene.[47]

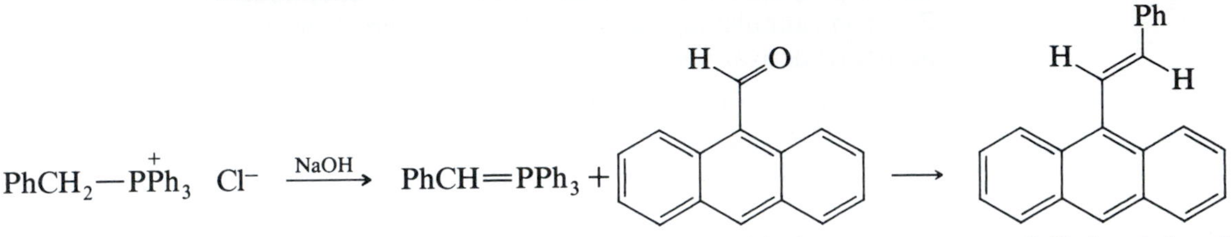

$$PhCH_2\!-\!\overset{+}{P}Ph_3 \quad Cl^- \xrightarrow{NaOH} \quad PhCH\!=\!PPh_3 \; + $$

anthraldehyde

trans-9-(2-phenylethenyl)-anthracene

EXPERIMENT 107

Wittig Condensation

To a 10-mL round-bottomed flask or a small reaction vial, fitted with a magnetic stir bar or stir vane, and in the hood, add 0.97 g (2.5 mmol) of benzyltriphenylphos-phonium chloride (commercially available), 0.57 g (2.5 mmol) of 9-anthraldehyde, and 3 mL of dichloromethane. While vigorously stirring the reaction mixture, slowly add dropwise 1.3 mL of a 50% aqueous sodium hydroxide solution. After the addi-tion is complete, stir the reaction at room temperature for 30 minutes. Remove the organic layer with a disposable pipette (recall that dichloromethane is the lower layer). Save the organic layer and add another 5 mL of dichloromethane to the reaction flask and stir vigorously. Remove this second extract and combine it with the first organic fraction. Repeat the extraction of the aqueous layer with another 2 to 3 mL of dichloromethane. Again combine this extract with the other combined organic extracts. Dry the organic extract over anhydrous calcium chloride and gravity filter it (or transfer with a filter pipette) into a tared round-bottomed flask. Evaporate the organic solvent on the rotary evaporator. Recrystallize the crude product, which also contains triphenylphosphine oxide, from n-propyl alcohol. Record the mp (~ 131 to 133°C) of the yellow crystals. Record the IR (CHCl₃) and NMR spectra (CDCl₃) and use the spectra to assign the stereochemistry about the alkene.

Another suitable Wittig reaction, which avoids the potential problem of form-ing mixtures of alkene isomers, involves the reaction of piperonal with isopropyltri-phenylphosphonium ylid.[48]

---

[47] Adapted from E. F. Silversmith, *J. Chem. Ed.*, **63**, 645 (1986).

[48] R. M. Pike, D. W. Mayo, S. S. Butcher, D. J. Butcher, and R. J. Hinkle, *J. Chem. Ed.*, **63**, 917 (1986).

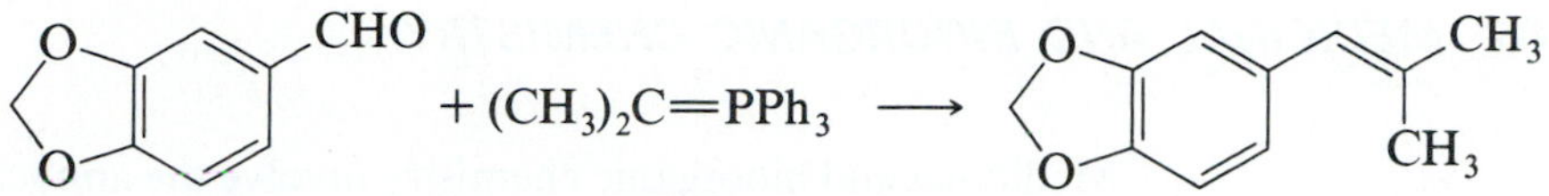

The use of phosphonates in place of phosphine-derived ylids is a useful modification of the Wittig reaction that is often called the Horner or Horner–Emmons modification.[49] This experiment will illustrate this modification by the reaction of cyclohexanone with triethyl phosphonoacetate. The result is a very convenient and practical synthesis of α,β-unsaturated esters.

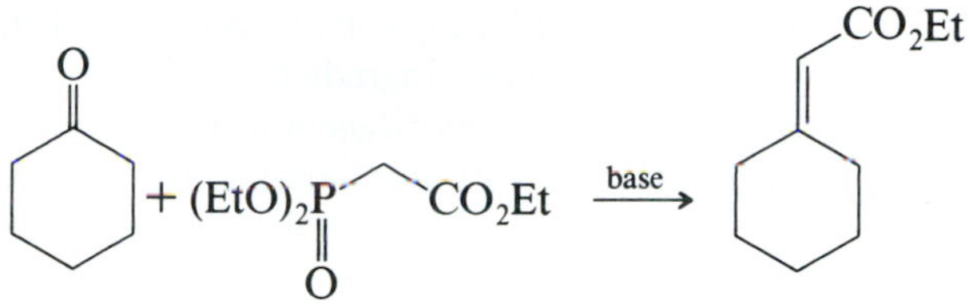

EXPERIMENT 108

Horner–Emmons Modification of the Wittig Reaction

To a dry 50-mL round-bottomed flask fitted with a calcium chloride drying tube and a magnetic stir bar, add 20 mL of 1,2-dimethoxyethane and 0.5 g (10 mmol) of 50% sodium hydride in oil *(CAUTION: Sodium hydride reacts explosively with moisture).* While maintaining the mixture at room temperature, partially remove the drying tube and add dropwise 2.25 g (10 mmol) of triethyl phosphonoacetate. Stir the resulting mixture for 1 hour. Again, partially remove the drying tube and add 1 g (10 mmol) of neat cyclohexanone dropwise. Try to maintain the reaction temperature less than 30°C during the addition. Stir the mixture for an additional 15 minutes. Quench the reaction by slowly adding 20 to 25 mL of cold water. Transfer the mixture to a small separatory funnel and extract with three 15- to 20-mL portions of ether. Combine the ether layers and wash them with 10 mL of saturated sodium chloride solution. Dry the ether solution over anhydrous magnesium sulfate and gravity filter the solution into a tared round-bottomed flask. Evaporate the ether on the rotary evaporator. If possible, distill the product under reduced pressure (see Section 2.3) and record its boiling point (~ 90°C at 10 mm). Record the IR (neat film) and NMR spectra (CDCl$_3$), noting the characteristics of an α,β-unsaturated ester.

---

[49] W. S. Wadsworth, Jr., and W. D. Emmons, *J. Am. Chem. Soc.,* **83,** 1733 (1961).

10.18 MEDICINAL AND BIOORGANIC CHEMISTRY

Medicinal and bioorganic chemistry involve the application of organic chemistry to biochemical problems. This can take the form of synthesis of biochemically active substances, their structure determination, or other studies that utilize organic chemical techniques. In whatever form, organic chemistry is responsible for most of the fundamental knowledge obtained about biochemical systems on the molecular level. This section will illustrate the importance of organic chemistry in biological studies. Experiments 72 and 103 also illustrate the use of organic synthesis in the preparation of biologically important substances.

The principal ingredient in most nonaspirin pain relievers (such as Tylenol) is *p*-acetylaminophenol (acetaminophen). It can be prepared in the laboratory by acetylation of *p*-aminophenol with acetic anhydride. The more nucleophilic amino group attacks one of the carbonyls of the acetic anhydride to form the amide product.

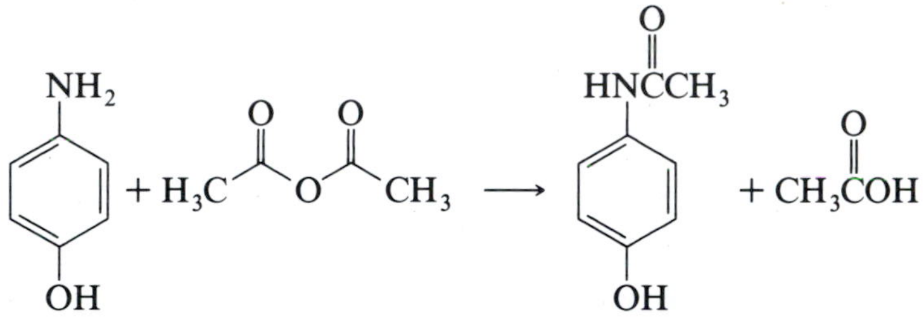

A second ingredient of headache and cold remedies is phenacetin, the ethyl ether prepared from acetaminophen. Sodium ethoxide is used as the base to prepare the phenoxide, since it is readily soluble in ethanol, as are the acetaminophen and ethyl iodide. Thus, all the ingredients form a single phase. The Williamson reaction of the phenoxide with ethyl iodide to produce the ether proceeds in good yield.

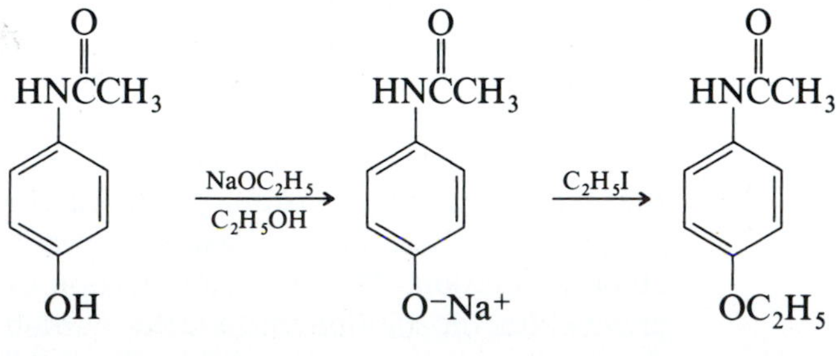

It is common practice in the pharmaceutical industry to synthesize analogs of those compounds that are known to possess desirable medicinal properties. Such analogs differ only slightly from the known compounds and are submitted for testing in hopes that they will possess greater activity or less toxicity than the parent compound. It is usually difficult to predict whether the slight molecular changes will produce a better or worse drug, but in some cases analogs have been logically arrived at which do possess greatly enhanced activity.

Simple analogs of the drugs acetaminophen and phenacetin would be the *ortho* and *meta* isomers. In this experiment you will be given as an unknown either *ortho-*, *meta-*, or *para*-aminophenol.[50] You will acetylate your unknown, thus preparing either acetaminophen or one of its analogs. Then you will perform the Williamson reaction on your product, leading to phenacetin or one of its analogs. By comparing the melting ranges of your purified products with the *approximate* known melting points of the possible analogs you might obtain (listed below), you should be able to identify which compounds you have synthesized.

X	$-OH$	$-OC_2H_5$
ortho	207°	77°
meta	143°	94°
para	166°	133°

EXPERIMENT 109

Acetylation of an Aminophenol

In the hood, suspend 1.1 g of your aminophenol in 3 mL of water in a 10- to 25-mL round-bottomed flask containing a magnetic stir bar and add 1.2 mL of acetic anhydride. Stir the flask vigorously and warm on the steam bath or in a 100°C sand bath. The solid should dissolve. After 10 minutes ice cool the mixture and suction filter the product, using 1 mL of cold water to wash the solid. Dry it as much as possible by drawing air through the crystals for a few minutes. Weigh the entire crude product and then separate 1 g for use in the Williamson reaction. During the 45-minute reflux period of the Williamson step, recrystallize your remaining crude product from a minimal amount of hot water, using a Craig tube and decolorizing with charcoal, if necessary. Isolate your purified material and allow it to dry thoroughly. Determine the fraction of crude material you recrystallized. How much pure, dry product would you have obtained had you purified your entire sample? What percent yield would this have been? Record this *calculated percent yield* of the acetylation product, along with the melting range of the dry product.

---

[50] Instructor's note: This experiment can be performed without the use of different aminophenols as unknowns. Acetaminophen and phenacetin can be prepared using the described procedures starting with *p*-aminophenol.

EXPERIMENT 110

Williamson Ether Synthesis with an Acetamidophenol

Note: This procedure employs 1 g of acetylated aminophenol; if you have less, scale down the other reagents proportionately.

Obtain ~ 160 mg of sodium by picking up small pieces with tweezers, wiping off the mineral oil with a dry towel, and placing them in a *dry,* tared, small reaction vial or a 25-mL round-bottomed flask and weigh.

C A U T I O N : Sodium reacts violently with water and should not be allowed to contact water or unprotected skin. If the sodium is supplied in large pieces covered with oil, cut into small pieces under the oil and transfer several pieces to a beaker containing hexanes to remove the oil. Small pieces may then be rapidly transferred to a smaller flask containing hexanes to be weighed while being protected from atmospheric moisture. Finally, transfer the weighed pieces to the *dry* reaction flask.

Attach a reflux condenser, start the cooling water, and add 4 mL of absolute ethanol through the condenser; then attach a $CaCl_2$ drying tube atop the condenser. A vigorous reaction of the sodium with ethanol will occur. If all the sodium does not dissolve after the reaction subsides, warm the flask on a steam bath. Cool the solution to room temperature and add 1 g of your acetamidophenol. Then add 0.8 mL of ethyl iodide *(CAUTION: Ethyl iodide is a powerful alkylating agent)* slowly through the condenser. Reflux for 45 minutes and then slowly add 10 mL of water through the top of the condenser to the refluxing solution. Allow the mixture to cool, which should lead to the formation of crystals. Cool the mixture in an ice bath and collect the crystals by suction filtration using a little cold water to wash them. Dissolve your product in a minimal amount of warm ethanol. If necessary, add about 0.1 g of decolorizing charcoal and filter. Keeping the solution hot, slowly add hot water until the solution *just begins* to grow cloudy. Allow the solution to cool, with scratching to induce crystallization. Ice cool and suction filter the mixture. Allow the crystals to dry and determine their weight and melting range. Record the IR and NMR spectra of the product.

Questions for Experiments 109 and 110

1. In the acetylation step, why does the amino group acetylate rather than the phenol?
2. In the Williamson reaction, why does the phenol group alkylate rather than the hydroxyl group of ethanol? Why does the phenol group alkylate rather than the amide nitrogen?
3. Why is the reaction of NaOEt with an iodoaniline derivative not used to make phenacetin as shown in the following equation?

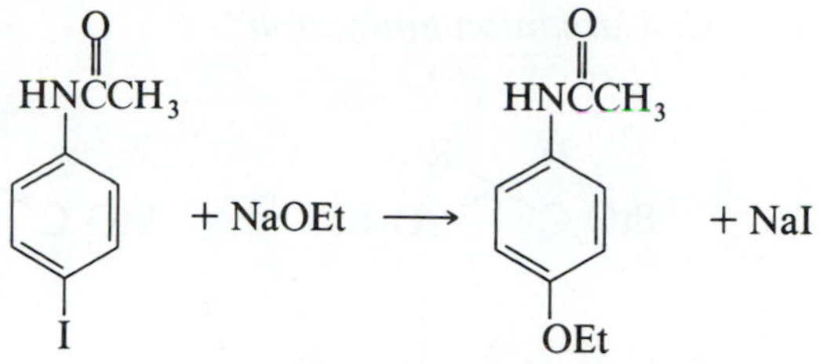

4. Identify the amide carbonyl absorption in the IR spectrum and account for its position.
5. Note the positions of the acetyl methyl and the peaks corresponding to the ethyl group and account for their chemical shift. Account for the multiplicity observed for the resonances due to the ethyl group and the aromatic protons.

———————————————

Amino acids are the constituents or building blocks of proteins. Because amino acids are essential for life, their biosynthesis and chemical synthesis have been studied intensely. The schemes below depict some of the typical chemical syntheses of amino acids. All these procedures are capable of synthesizing only racemic amino acids, whereas nature uses the L-form almost exclusively in protein formation. Methods of resolving D,L-amino acids have been developed and a considerable amount of recent synthetic effort has gone into the design of asymmetric chemical syntheses of amino acids.

A. Aminolysis of α-halo acids

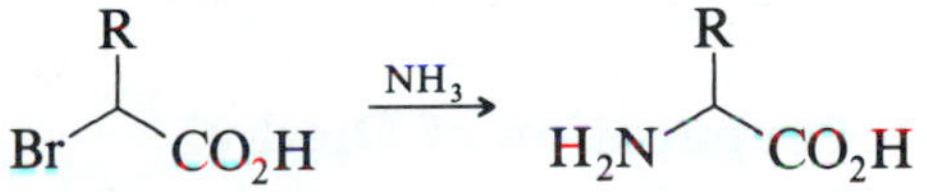

B. Gabriel phthalimide synthesis

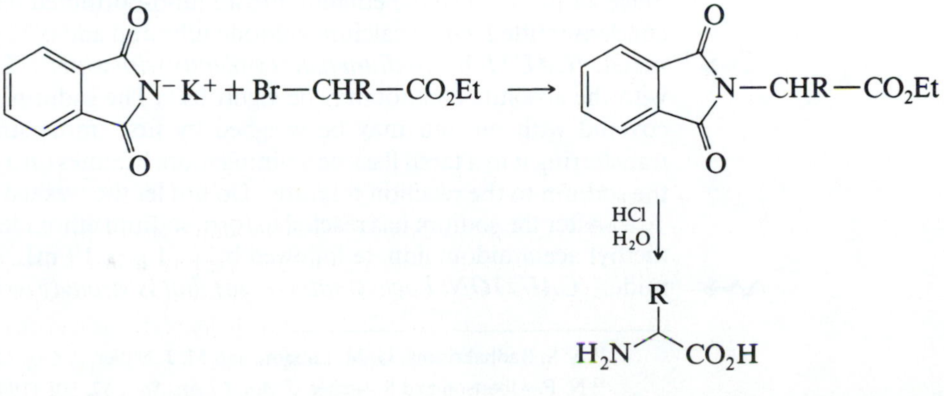

C. Carbanion amination[51]

$$\text{EtO}_2\text{C} \overset{\text{H} \quad \text{R}}{\diagdown\diagup} \text{CO}_2\text{Et} \xrightarrow{\text{base}} \text{EtO}_2\text{C} \overset{\text{R}}{\overset{|}{-}} \text{CO}_2\text{Et} + \text{H}_2\text{N}-\text{O} \diagdown \text{NO}_2$$

$$\text{H}_2\text{N} \overset{\text{R}}{\diagdown\diagup} \text{CO}_2\text{H} \xleftarrow[\text{H}_2\text{O}]{\text{HCl}} \text{EtO}_2\text{C} \overset{\text{H}_2\text{N} \quad \text{R}}{\diagdown\diagup} \text{CO}_2\text{Et} + {}^-\text{O} \diagdown \text{NO}_2$$

D. Carbanion alkylation: The Sörensen amino acid synthesis involves the alkylation of diethyl acetamidomalonate.

$$\text{EtO}_2\text{C} \overset{\text{H} \quad \text{NH}-\text{COCH}_3}{\diagdown\diagup} \text{CO}_2\text{Et} \xrightarrow[\text{2. R}-\text{X}]{\text{1. NaOEt}} \text{EtO}_2\text{C} \overset{\text{R} \quad \text{NH}-\text{COCH}_3}{\diagdown\diagup} \text{CO}_2\text{Et} \xrightarrow[\text{H}_2\text{O}]{\text{HCl}} \text{H}_2\text{N} \overset{\text{R}}{\diagdown\diagup} \text{CO}_2\text{H}$$

The actual procedure that will be used in this experiment will be a slight modification of Sorensen's method similar to that described by Albertson and Archer in 1945.[52]

Aminoacid Synthesis: Preparation of Phenylalanine

EXPERIMENT 111

Preparation of Diethyl Benzylacetamidomalonate

Place 25 mL of absolute ethanol into a round-bottomed flask in the hood. Attach a condenser fitted with a calcium chloride tube and add 0.25 g (~ 10 mmol) of sodium metal. *(CAUTION: Sodium reacts violently with water.)* The reaction of the sodium with the absolute ethanol may be vigorous. (The sodium metal should be supplied covered with oil, but may be weighed by first immersing it in hexanes, rapidly transferring it to a tared flask containing more hexanes on a balance, and then adding the sodium to the reaction mixture. Do not let the washed sodium stand in the air.)

After the sodium has reacted to form sodium ethoxide, add 2.17 g (10 mmol) of diethyl acetamidomalonate followed by 1.71 g (1.19 mL, 10 mmol) of benzyl bromide. *(CAUTION: Lachrymator—gas that is strongly irritant to the eyes (e.g., tear*

---

[51] A. S. Radhakrishna, G. M. Loudon, and M. J. Miller, *J. Org. Chem.*, **44**, 4836 (1980).

[52] N. F. Albertson and S. Archer, *J. Am. Chem. Soc.*, **67**, 308 (1945).

gas.) All alkyl halides should be handled in the hood with extreme care. Rinse out any glassware containing the bromide into waste bottles containing excess aqueous base or base in ethanol.

 Heat the reaction mixture to reflux for 2 hours. Then add about 25 mL of water and extract the solution with three 25-mL portions of ether or ethyl acetate. *(CAUTION: Some benzyl bromide may remain unreacted.)* Wash the combined extracts with 25 mL of water followed by 25 mL of saturated sodium chloride solution. Dry the organic layer over anhydrous magnesium sulfate. Gravity filter the solution into a tared round-bottomed flask and evaporate the solvent on the rotary evaporator. The crude product should be obtained in high yield (~90%) either as a thick oil or a solid with a mp of about 100 to 104°C. If necessary, the compound can be recrystallized from water to give a pure sample (mp 104 to 106°C), but the crude material should be adequate for the next reaction. Save a small amount of material for IR (neat film or KBr pellet) and NMR ($CDCl_3$) analysis. Record and interpret the spectra.

EXPERIMENT 112

Preparation of Phenylalanine

Add the crude product from the alkylation reaction to 10 to 20 mL of 6N hydrochloric acid and reflux for 2 to 3 hours. The resulting hydrochloride salt may be converted to the free amino acid by either of the following procedures. However, if done carefully, method (b) generally gives better results.

(a) Cool the reaction mixture and adjust the pH to the isoelectric point (pH = 6) with concentrated NH_4OH. Further cool the solution in an ice bath and collect the crystals by filtration. Record the melting point (~257°C).

(b) Evaporate the aqueous acid hydrolysis solution to dryness on the rotary evaporator. Dissolve the residue in 0.5 mL of water. Add 10 mmol of triethylamine (use your hood!). Swirl for a few minutes to give a white paste of the triethylamine hydrochloride and the neutral (zwitterionic) amino acid. Add 5 to 10 mL more of acetone and swirl the suspension for a few minutes more. Filter off the solid mixture (triethylamine hydrochloride and the phenylalanine) using a small Büchner funnel. Wash the solid on the funnel with three portions of chloroform. Interestingly, the chloroform dissolves the triethylamine hydrochloride but not the phenylalanine. Wash the remaining solid (neutral phenylalanine) with 10 mL of ether and air dry the product. Record the yield and melting point of the phenylalanine. If possible, obtain IR (KBr pellet) and NMR spectra (D_2O or DMSO-d_6) and compare these spectra with those obtained from the previous alkylation step.

If possible, analyze your phenylalanine by paper chromatography. For comparison, obtain some authentic phenylalanine (R = CH_2Ph) and glycine (R = H) from your instructor.

Questions for Experiments 111 and 112

1. Why are the amino acids prepared by the procedures described above racemic?

2. Write the mechanisms for all the steps in the Sorenesen synthesis of phenylalanine.

3. What is the purpose of the triethylamine in method (b) of the isolation of phenylalanine? Why should excess be avoided?

4. Why will the phenylalanine not precipitate at pH values above or below pH 6 in method (a) of the isolation?

5. Amides are generally more difficult to hydrolyze than esters. Thus, partial hydrolysis of diethyl benzylacetamidomalonate followed by decarboxylation might yield *N*-acetyl-phenylalanine. What effect would this incomplete reaction have on the ability to isolate pure phenylalanine using the procedures described?

6. Paper chromatography of the phenylalanine product may show that it is contaminated with glycine (R = H). What is the source of the glycine?

Peptide Synthesis: Preparation of Benzoylglycylphenylalanine Methyl Ester

Large peptides, or proteins, are the main constituents of the dry weight of animal cells and are primarily responsible for the structure and function of living matter. One objective of protein chemists, therefore, is to explain the properties of these macromolecules in terms of their actual structure. This involves the determination and examination of their parts or building blocks, the amino acids, the arrangement of the parts, and how such an arrangement can be functional.

A protein may contain varying numbers (from two, a dipeptide, to thousands) of any of the 20 common amino acids. The ordered combination of amino acids results in the formation of amide or peptide bonds in a specific sequence. It is this unique amino acid sequence that imparts the required informational character to the protein and determines not only its higher order structure, but also its function. The chemical synthesis of peptides or proteins, therefore, also requires the ability to form peptide bonds (connect amino acids) in the proper order.

This experiment will involve the synthesis of a protected dipeptide *N*-benzoyl-glycylphenylalanine methyl ester (also called hippurylphenylalanine methyl ester). The general strategy used for dipeptide synthesis requires that one amino acid be protected at its amino end and the other at the carboxyl end to avoid formation of polymeric mixtures. In this experiment, the amino group of glycine, the simplest amino acid, will be protected with a benzoyl group. This protected amino acid, called hippuric acid, is a natural product. In fact, it is a metabolism product in which benzoic acid is coupled to glycine in the body to assist the removal of benzoic acid. (Read the label of your favorite beverage. It may include sodium benzoate as a preservative.) Experiments have shown that after ingestion of sodium benzoate, the human body converts the benzoic acid essentially quantitatively to hippuric acid so that it can be excreted in the urine. Hippuric acid is also commercially available and serves as a convenient N-protected amino acid. Since phenylalanine will be the carboxyl ("C") terminal residue of the dipeptide, its carboxylic acid may be protected from interfering reactions by prior conversion to the methyl ester.

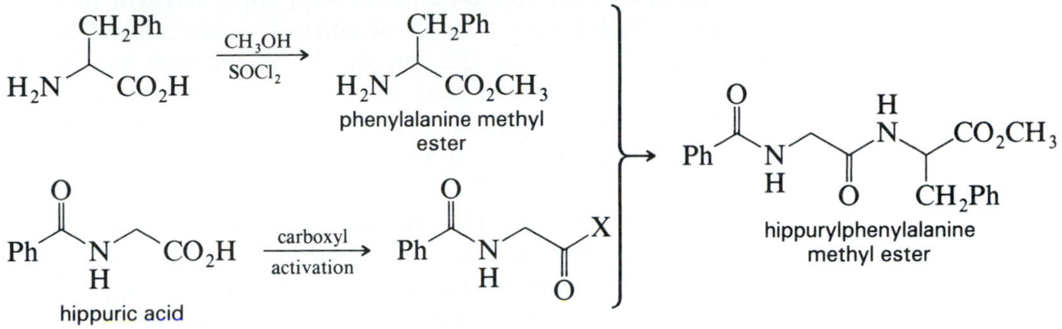

EXPERIMENT 113

Preparation of Phenylalanine Methyl Ester

In the hood, carefully add 0.5 mL of thionyl chloride (*CAUTION: Thionyl chloride is highly toxic and corrosive*) to 20 mL of methanol in a 50-mL round-bottomed flask fitted with a reflux condenser and cooled in an ice-salt bath to about −10°C (internal temperature). Add 0.83 g (5 mmol) of phenylalanine from the previous experiment or from commercial sources. Remove the ice-salt bath and heat the mixture at reflux for 2 hours. (Alternatively, the reaction mixture can be stored at room temperature until the next laboratory period.) Concentrate the mixture on the rotary evaporator to about 2 to 5 mL and then precipitate the product by slow addition of anhydrous, peroxide-free ether (see Section 1.15.8 for the peroxide test). If necessary, recrystallize the phenylalanine methyl ester hydrochloride from methanol/ether again and record its melting point (∼ 160°C). Record the IR spectrum (KBr pellet) and compare it to that of the initial phenylalanine.

EXPERIMENT 114

Preparation of a Peptide by a Mixed Anhydride Method

Peptide bonds can be formed by a variety of carboxyl activation procedures.[53] Two of the most commonly used are mixed anhydrides and carbodiimides (i.e., dicyclohexylcarbodiimide). In this experiment, isobutyl chloroformate will be used to form

---

[53] Y. S. Klausner and M. Bodansky, *Synthesis,* 453 (1972).

a mixed anhydride of hippuric acid ($X = OCO_2iBu$). Subsequent reaction with the amino group of the phenylalanine methyl ester will produce the peptide.

Equip a 25-mL round-bottomed flask with a magnetic stirring bar and a Claisen adapter. Attach a drying tube to one arm of the adapter and a septum to the top of the other. Dissolve 412 mg (2.3 mmol) of hippuric acid and 0.36 mL (263 mg, 2.6 mmol) of triethylamine in 5 mL of a 1 : 1 mixture of anhydrous dimethylformamide (DMF) and tetrahydrofuran (THF). Other solvents such as acetonitrile are also acceptable if they are anhydrous. Cool the solution in an ice-salt bath to $\sim -10°C$. Add 0.3 mL (314 mg, 2.3 mmol) of neat isobutyl chloroformate with a syringe by injecting directly through the septum. Stir the mixture for about 30 minutes. A precipitate may form and make stirring difficult. Meanwhile, dissolve 500 mg (~ 2.3 mmol) of phenylalanine methyl ester hydrochloride in 5 mL more of the same solvent and add this solution rapidly to the original reaction mixture followed by a dropwise addition of 0.36 mL (2.6 mmol) of triethylamine. Warm the mixture to room temperature and stir it for $1\frac{1}{2}$ hours. Dilute the mixture with 15 mL of water and extract it with three 10-mL portions of ether. Wash the combined ether extracts with two 5-mL portions of 1N hydrochloric acid, two 5-mL portions of 10% aqueous sodium bicarbonate, and then 5 mL of saturated aqueous sodium chloride. Dry the organic layer over anhydrous magnesium sulfate. Gravity filter the solution into a clean, tared round-bottomed flask and evaporate the solvent.

Analyze the residue, which is mainly the peptide product, by TLC on silica gel. If necessary and if time allows, purify the peptide ester by flash chromatography on silica gel. Record and interpret the NMR ($CDCl_3$) and IR (neat or KBr pellet if the product solidifies) spectra of the product.

Questions for Experiments 113 and 114

1. For the esterification of phenylalanine, thionyl chloride was added to methanol before the phenylalanine. This reaction does not involve the formation of an acid chloride. How does the reaction work? Write a mechanism for the esterification.

2. Write a mechanism for the carboxyl activation of hippuric acid with isobutyl chloroformate or with DCC. Show how the active form is converted to the peptide when treated with phenylalanine methyl ester.

3. Why was triethylamine added to the reaction during the formation of the peptide?

Related References

1. For alternative methods for the preparation of amino acid esters see C. L. Borders, Jr., D. M. Blech, and K. D. McElvany, *J. Chem. Ed.,* **61,** 814 (1984).

2. For a related synthesis of acetylpropylphenylalanine methyl ester see P. E. Young and A. Campbell, *J. Chem. Ed.,* **59,** 701 (1982).

3. For a convenient synthesis of the peptide sweetener aspartame (aspartylphenylalanine methyl ester) see G. Lindeberg, *J. Chem. Ed.,* **64,** 1062 (1987).

4. Analysis of aspartame and its hydrolysis products is also very instructive. See A. R. Conklin, *J. Chem. Ed.,* **64,** 1065 (1987).

Synthesis of a β-Lactam

β-Lactam is the common name for 2-azetidinone, a four-membered ring amide. β-Lactam rings are the functional components of the most important and widely used antibiotics in the world. Two of the most common forms of β-lactams are the penicillins and cephalosporins. While nontoxic to most humans, these molecules inhibit bacterial cell-wall synthesis and thereby interfere with infectious bacterial growth. These compounds have saved millions of lives over the last few generations. Most clinically used β-lactam antibiotics are obtained by fermentation or semisynthesis (chemical modification of the natural products). However, new forms of these antibiotics are needed to overcome the resistance developed by several types of bacteria to the common antibiotics. Versatile methods for the asymmetric synthesis of the core β-lactam ring have recently been developed.[54,55] This series of experiments will illustrate one method of preparing precursors to several new types of antibiotics, including monobactams and oxamazins.[56] The process involves amino

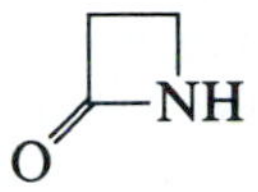

2-azetidinone (β-lactam) 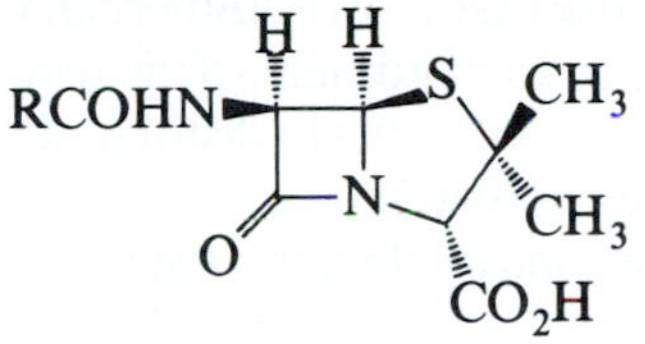 · a penicillin 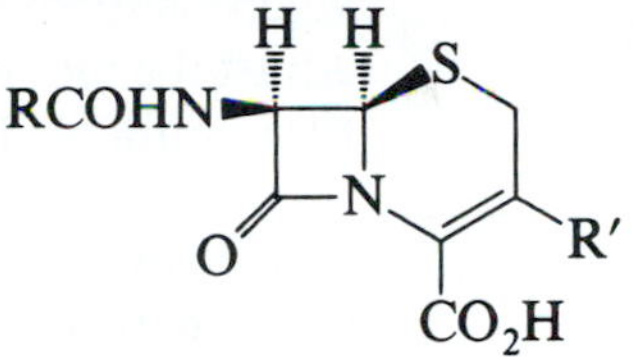 · a cephalosporin

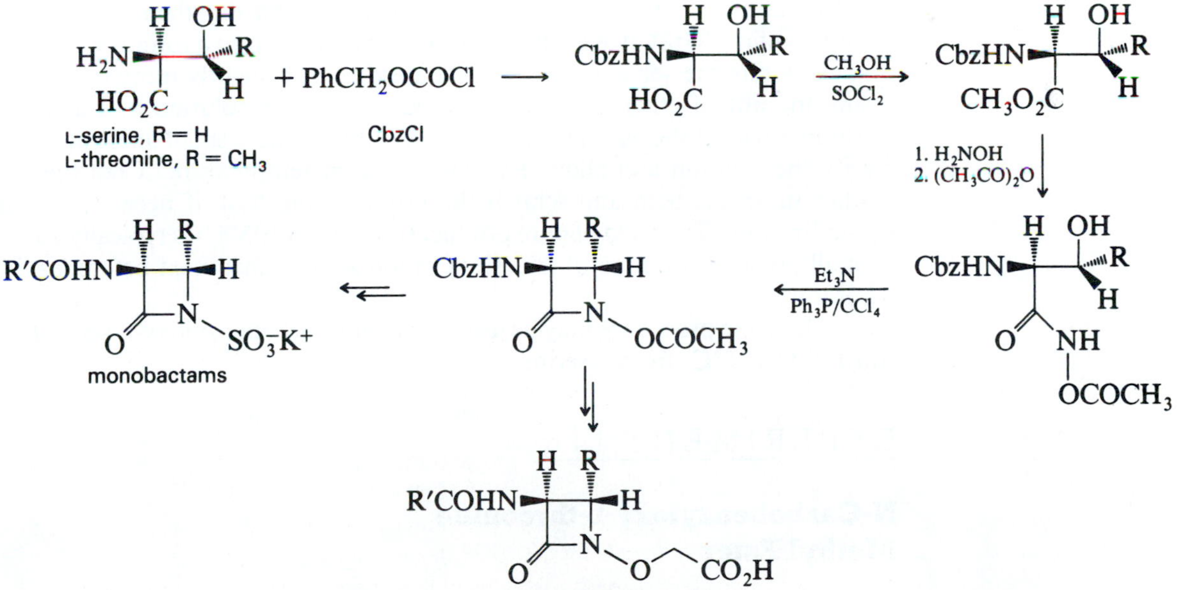

L-serine, R = H
L-threonine, R = CH₃

CbzCl

monobactams

an oxamazin

[54] M. J. Miller, A. Biswas, and M. A. Krook, *Tetrahedron*, **39**, 2571 (1983).
[55] M. J. Miller, *Accts. Chem. Res.*, **19**, 49 (1986).
[56] S. R. Woulfe and M. J. Miller, *J. Med. Chem.*, **28**, 1447 (1985).

protection and carboxyl esterification of an amino acid precursor (serine or threonine) followed by simple conversion to a hydroxamic acid and cyclization to the β-lactam.

EXPERIMENT 115

Preparation of Carbobenzyloxy-L-threonine

To a suspension of 2.1 g (25 mmol) of potassium carbonate in 15 mL of water in a 25- to 50-mL round-bottomed flask in the hood, add 1.2 g (10 mmol) of L-threonine and stir the mixture vigorously. Carefully add 2.0 g (1.67 mL, ~ 11 mmol) of carbobenzyloxy chloride (CbzCl) in about five portions over 30 minutes while continuing the vigorous stirring. (Note that most commercial CbzCl is only about 95% pure. If so, adjust the amount added accordingly. This reagent should be used as supplied from the commercial source. Attempts to distill it often lead to vigorous decomposition.)

Stir the mixture another hour at room temperature. Extract the mixture twice with 5- to 10-mL portions of ether to remove any excess CbzCl and benzyl alcohol (formed from hydrolysis and decarboxylation of the CbzCl). Cool the aqueous portion in an ice bath and carefully acidify it to pH 2 to 3 by dropwise addition of 6N HCl. An oil may separate. Extract the mixture with two 15- to 25-mL portions of ethyl acetate. Combine the ethyl acetate extracts and dry them over anhydrous sodium sulfate. Gravity filter the ethyl acetate mixture into a round-bottomed flask and evaporate the solvent. Dissolve the residue with slightly more than the minimum amount of ether in the same flask. Warm the solution and add hexanes portionwise until the warm solution is just turbid. Add a small amount of ether to clarify the solution and allow it to cool to room temperature. Cool the mixture further in an ice bath and scratch the inside of the flask if necessary to induce crystallization. The yield of pure product (mp 103 to 105°C) is typically 75 to 85%. If at all possible continue to the esterification as described next before leaving the laboratory.

The same procedure may be used to prepare carbobenzyloxy-L-serine (R = H, mp 119 to 121°C) from L-serine.

EXPERIMENT 116

N-Carbobenzyloxy-L-threonine Methyl Ester

Add N-carbobenzyloxy-L-threonine (1.25 g, 5 mmol) to 15 mL of anhydrous methanol in a 50-mL round-bottomed flask. Cool the suspension to 0°C in an ice bath and very carefully add 0.4 mL (0.65 g, 5.4 mmol) of neat thionyl chloride dropwise.

(CAUTION: The rapid addition of thionyl chloride will cause a rapid, exothermic reaction with methanol and result in splattering.) After the addition is complete, remove the ice bath and store the lightly stoppered flask at room temperature until the next laboratory period.

Evaporate the solvent and dissolve the residue in 25 mL of ethyl acetate. Transfer the solution to a small separatory funnel and extract the solution with 10 mL of water, two 10-mL portions of 5% sodium bicarbonate, and 10 mL of brine (saturated sodium chloride). Dry the ethyl acetate solution over anhydrous magnesium sulfate and gravity filter it into a 50-mL round-bottomed flask. Evaporate the solvent to leave the methyl ester as a thick oil or crude solid and then recrystallize the product from ether-hexanes. Determine the yield (typically 80 to 90%) and the melting point (88 to 90°C). Record the NMR (CDCl$_3$) and infrared (KBr) spectra and compare these to the spectra of the starting material, noting the change from the acid to the ester.

The same procedure can be used to prepare the corresponding serine methyl ester. This compound is usually obtained as a thick oil.

EXPERIMENT 117

Preparation of O-Acetyl-N-carbobenzyloxy-L-threonine Hydroxamate

Add Cbz-L-theonine methyl ester (1 g, 3.75 mmol) to 15 mL of anhydrous methanol in a 50-mL round-bottomed flask cooled to 0°C in an ice bath. In a 50-mL Erlenmeyer flask in the hood, prepare methanolic hydroxylamine by adding 15 mmol of KOH dissolved in 5 to 10 mL of warm methanol to a solution of 7.5 mmol of hydroxylamine hydrochloride in another 5 to 10 mL of warm methanol. (If necessary, heat the separate mixtures in methanol to boiling to assist dissolving the KOH and hydroxylamine hydrochloride. Then cool as close to room temperature without recrystallizing the KOH or hydroxylamine hydrochloride and add the KOH solution to the hydroxylamine solution *in the hood*. A precipitate of KCl will form.)

Pour the methanolic hydroxylamine mixture with the suspended KCl directly into the flask containing the cold threonine methyl ester. Remove the ice bath. After 30 to 45 minutes add 7.5 mmol (~0.71 mL) of acetic anhydride. After 5 minutes *(no longer)*, pour the reaction mixture into a separatory funnel containing enough of equal volumes of 5% sodium bicarbonate and ethyl acetate to form two layers. Separate the layers and extract the organic layer with another 15 mL of 5% sodium bicarbonate. Combine the aqueous layers and acidify them to pH 4 by the dropwise addition of 6N HCl. Extract the aqueous solution with two 20 mL-portions of ethyl acetate. Combine the ethyl acetate layers and wash them with 10 mL of brine. Dry the ethyl acetate with anhydrous magnesium sulfate. Gravity filter the mixture into a round-bottomed flask and evaporate the solvent. Recrystallize the residue in the same flask from ethyl acetate-hexanes. Record the yield (60 to 70%) and melting point (111 to 113°C). Record the NMR (CDCl$_3$) and IR (KBr) spectra, noting the absence of the methyl group from the starting material and the presence of the acetyl methyl group (δ 2.1) in the NMR and new carbonyl stretches in the IR (1800 and 1660 cm^{-1}). This compound should not be stored more than several days since it is susceptible to hydrolysis of the acetate and somewhat prone to undergo Lossen rearrangement (see page 499).

The corresponding serine hydroxamate can also be prepared (mp 120 to 121°C).

EXPERIMENT 118

Preparation of N-Acetoxy-[4(S)-methyl-3-(S)-benzyloxyformamido]-2-azetidinone

Dissolve 0.5 g (1.6 mmol of the O-acetyl hydroxamate in 15 mL of anhydrous acetonitrile in a *dry* 50-mL round-bottomed flask containing a $\frac{1}{2}$-in. stir bar in the hood containing 1 mL of carbon tetrachloride *(CAUTION: Carbon tetrachloride is toxic)*. Add 0.23 mL (~1.6 mmol) of triethylamine and 420 mg (1.6 mmol) of triphenylphosphine simultaneously. Attach a calcium chloride drying tube. Stir the solution for 1 to 2 hours at room temperature. Evaporate the solvent and dissolve the residue in 25 mL of ethyl acetate. Transfer the solution to a small separatory funnel and extract the solution with 10 mL of water and 10 mL of brine. Dry the solution over anhydrous magnesium sulfate and gravity filter the solution into a 50-mL round-bottomed flask. Evaporate the solvent. Separate the β-lactam product from triphenylphosphine oxide by flash chromatography (see Section 2.6) using silica gel and ethyl acetate-hexanes (1 : 2) as the eluent. The product is usually obtained as a clear oil.

The serine-derived β-lactam can be prepared in the same manner (mp 130 to 131°C).

Questions for Experiments 115 to 118

1. Write a mechanism for the reaction of CbzCl with L-threonine. Why is little or no product obtained that contains a Cbz group on the hydroxyl group of the threonine?

2. Write the mechanism for the conversion of CbzCl to benzyl alcohol upon reaction with water.

3. The conversion of Cbz-L-threonine to the corresponding methyl ester with SOCl$_2$ in methanol does not proceed by way of the acid chloride. What is the role of the thionyl chloride?

4. Write the mechanism for the potential Lossen rearrangement of the O-acetyl hydroxamate derivative of L-threonine. If the rearrangement occurred in water, what would be the structure of the rearrangement product?

5. Account for the high carbonyl stretching frequency ($1780 \ cm^{-1}$) in the IR spectrum of the β-lactam product.

6. Why are β-lactams more reactive than normal amides?

Carbohydrate Chemistry: Acetylation of Glucose

Carbohydrates serve both as structural components and as a source of energy in biological systems. The carbohydrates contain highly oxygenated (hydrated) carbon frameworks. While it is often difficult to distinguish all the individual functional groups in carbohydrates chemically, different classes of the same functional group can often be selectively modified. Often reactions of carbohydrates require careful selection of the reaction conditions. The total dehydration of sugar can be catalyzed by concentrated sulfuric acid to give water and carbon.

All five of the hydroxyl groups of D-glucose react with acetic anhydride to give a pentaacetate. However, depending on whether acid or base is used as the catalyst, two isomeric pentaacetates can be obtained. Furthermore, the isomer formed with base catalysis can be isomerized with acids to the isomer formed with acid catalysis. These results can be explained by recalling that, in solution, D-glucose is almost entirely in the pyranose form as a pair of rapidly equilibrating isomers, the α- and β-anomers. This equilibrium is established by either acids or bases. Since the hydroxyl group at the anomeric carbon of the β-anomer is less sterically hindered than that of the α-anomer, β-D-glucose is converted to its pentaacetate more rapidly by either acid or base catalysis. As the β-anomer is converted to its pentaacetate, more α-anomer is converted to the β-anomer to reestablish the equilibrium. If the reaction is run under conditions where the β-pentaacetate cannot be converted to the α-isomer, all the glucose, even though it started as a mixture of α- and β-anomers, will eventually be converted to the β-pentaacetate. Indeed, this is what happens under base catalysis since there is no mechanistic path by which the base can isomerize the β-pentaacetate to the α-isomer.

D-glucose, α-anomer   D-glucose, β-anomer

However, in the presence of acid, the β-pentaacetate can be protonated and establish an equilibrium through a cationic intermediate that can react with acetate or acetic acid to form either the α- or β-pentaacetate. Thus, the two esters can be equilibrated in the presence of acids. The composition of the product is determined by the thermodynamic stabilities of the two isomers. In this case, the α-isomer is more stable since it is essentially the only product obtained when the esterification is performed with acid catalysis or when the β-isomer is heated with acid.

EXPERIMENT 119

Preparation of α-D-Glucose Pentaacetate[57]

In the hood, place 12.5 mL of a zinc chloride–acetic anhydride solution in a 25-mL round-bottomed flask.[58] Attach a reflux condenser and heat the mixture gently on a steam bath. Add 2.5 g of powdered D-glucose in small portions to the flask by removing the condenser for each addition. If the glucose is added too rapidly, the reaction may become too vigorous.

After all the glucose has been added, heat the mixture on a steam bath for 1 hour. Pour the cooled reaction mixture into 125 mL of ice water. Cool the mixture and stir it occasionally for 45 minutes or until the oil solidifies. This may require considerable patience since the solidification often requires 30 to 45 minutes. Collect the product by suction filtration and recrystallize it from 20 to 25 mL of methanol-water ($\sim 1:2$). Determine the yield and melting point of the product. If possible, record the optical rotation of the product in methanol (see Section 3.6).

EXPERIMENT 120

Preparation of β-D-Glucose Pentaacetate

While the product in Experiment 119 is solidifying, mix 2.5 g of D-glucose and 2 g of sodium acetate in a 50-mL flask with a reflux condenser. Add 12.5 mL of acetic anhydride and heat the mixture on a steam bath for $1\frac{1}{2}$ hours. Swirl the mixture occasionally to maintain a clear solution. Slowly pour the mixture into 125 mL of ice water with stirring. Suction filter, air dry, and recrystallize the product from methanol-water ($\sim 1:2$) as described for the other isomer. Record the yield and melting point of the product. If possible, record the optical rotation of the product in methanol (see Section 3.6).

---

[57] Adapted from A. Ault, *Techniques and Experiments for Organic Chemistry,* 4th ed. (Boston: Allyn and Bacon, 1983), p. 394.

 [58] A large stock solution is prepared by adding 42 g of anhydrous zinc chloride to 1 L of acetic anhydride in a 50-mL round-bottomed flask. (*CAUTION: This mixture is corrosive and a powerful dehydrating agent.*)

EXPERIMENT 121

Isomerization of β- to α-Pentaacetate

Place about one-half of the β-D-glucose pentaacetate in a 25-mL round-bottomed flask fitted with a reflux condenser. Add 5 mL of the acetic anhydride–zinc chloride solution and heat the mixture on a steam bath. Isolate the product by the same methods used above and determine the melting point. If possible, confirm the isomerization by recording the optical rotation of the product.

Additional carbohydrate experiment. An interesting laboratory exercise for the determination of carbohydrate structures has been reported.[59] The procedures include reducing test, monosaccharide composition, thin-layer chromatographic (TLC) linkage analysis from acid and enzymatic (α-glucosidase, β-glucosidase, and invertase) hydrolysis, and determination of optical rotation.

Questions for Experiments 119 to 121

1. Provide a mechanism for the acid- or base-catalyzed equilibration of the two anomeric forms of D-glucose.
2. Explain (a mechanism will help) why only acid can equilibrate the two pentaacetates.
3. Why must anhydrous reagents be used in this experiment?

Alternative Bioorganic Laboratory Experiments

Several other experiments described earlier in this chapter can certainly be classified as medicinal or bioorganic chemistry. For example, the synthesis of sulfanilamide and tetraphenylporphyrin are biologically relevant.

References

The following list of references also includes very interesting bioorganic experiments that can be effectively carried out in undergraduate laboratories.

1. Letcher, R. M., "Structure Elucidation of a Natural Product," *J. Chem. Ed.,* **60,** 79 (1983). This experiment describes the isolation of limonene from citrus peel and provides several interesting reactions that help elucidate the structure of limonene.
2. Schwarz, M., and J. A. Klun, "Synthesis and Evaluation of the Sex Pheromone of the Bagworm Moth," *J. Chem. Ed.,* **63,** 1014 (1986).
3. Cormier, R. A., and J. N. Hoba, "Laboratory Synthesis of Insect Pheromones," *J. Chem. Ed.,* **61,** 927 (1984).

---

[59] B. J. White and J. F. Robyt, *J. Chem. Ed.,* **65,** 164 (1988).

4. Bartlett, P. A., et al., "Synthesis of Frontalin, the Aggregation Pheromone of the Southern Pine Beetle," *J. Chem. Ed.,* **61,** 816 (1984).

5. Nitta, Y., E. Izbicka-Dimitrejevic, and D. W. Bolen, "Interaction of α-Chymotrypsin with a Strained Cyclic Ether," *J. Chem. Ed.,* **61,** 929 (1984).

10.19   POLYMERS

A large number of materials of biological and economic importance are polymers. Polymers are usually very high molecular weight macromolecules that consist of repeating units of much smaller molecular weight. Proteins (see Experiments 110 to 113) are extremely important biologically occurring polymers of amino acids that make up a major potion of our hair, skin, tissue, and cellular constituents. Polysaccharides are other biologically important polymers made from smaller carbohydrate (sugar) units (see Experiments 119 to 121). A third class of biologically important polymers consists of the polynucleotides such as deoxyribonucleic acid (DNA) and ribonucleic acid (RNA). These materials contain all the genetic information of a living cell. DNA and RNA consist of linear chains of smaller molecules called nucleotides.

Synthetic polymers are of vast economic importance in manufacturing processes. Materials such as plastics, synthetic fibers, resins, and adhesives consist of polymers with average molecular weights in excess of 10,000.

This experiment gives examples of addition-type and condensation-type polymers and copolymers, as well as providing exposure to the techniques of bulk, solution, and interfacial polymerization. Polystyrene (bulk), Nylon 6,10, and styrene-maleic anhydride copolymer are prepared.

Students should work in pairs. One partner should prepare polystyrene and the other the styrene-maleic anhydride copolymer; both should pull the nylon.

EXPERIMENT 122

Polymerization of Styrene

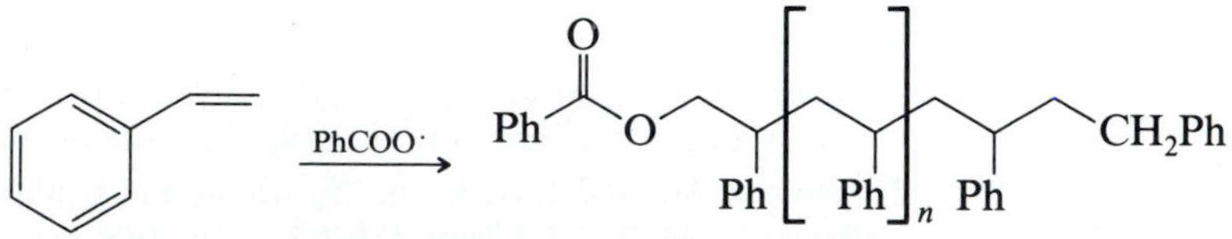

This experiment uses dibenzoyl peroxide to initiate polymerization of styrene. The following comments should be read and clearly understood before beginning the experiment.

C A U T I O N : Dibenzoyl peroxide (sometimes called benzoyl peroxide) is a solid that is flammable when dry and may be exploded by shock or friction. It should never be left in contact with cloth or wood. Wash any area where it has spilled. Benzoyl peroxide is a strong oxidizing agent. Heat, shock, or contact with other materials may cause fire or explosive decomposition.

THE FOLLOWING SAFETY PRECAUTIONS SHOULD BE EXERCISED IN HANDLING BENZOYL PEROXIDE

1. Keep away from heat, sparks, and open flames.
2. *Do not* add to hot material.
3. *Do not* grind or subject the benzoyl peroxide to any type of frictional heat.
4. Prevent contamination with readily oxidizable materials.
5. Avoid contact with skin, eyes, and clothing.
6. Store in paper containers and keep the containers closed to prevent drying out.

Into a 10- × 70-mm test tube, place about 2 mL of styrene together with ~ 20 mg dibenzoyl peroxide (enough to fit on the end of a small metal spatula; if too little catalyst is used the polystyrene will not harden, and if too much catalyst is used bubbles of carbon dioxide will be trapped in the polymer). Stopper the tube with a cotton plug and suspend it by means of a piece of wire into a steam bath. Maintain it at about 100°C for at least 75 minutes.

Remove the test tube from the steam bath and cool it in ice. The solid plug of polystyrene should separate from the glass. If the plug is not sufficiently loose to be tapped out of the test tube, carefully break open the tube on the inner rim of a ceramic crock. Being careful not to burn the material, test the melting properties of the polystyrene by heating a small chip on the end of a metal spatula with a very small burner flame. Grind a few small chunks of the material in a mortar and test its solubility in toluene, acetone, methyl ethyl ketone, and methanol. Based on the results of the solubility tests, comment on the polar or nonpolar nature of the polymer.

EXPERIMENT 123

Preparation of Polyhexamethylene Sebacamide (Nylon 6,10)

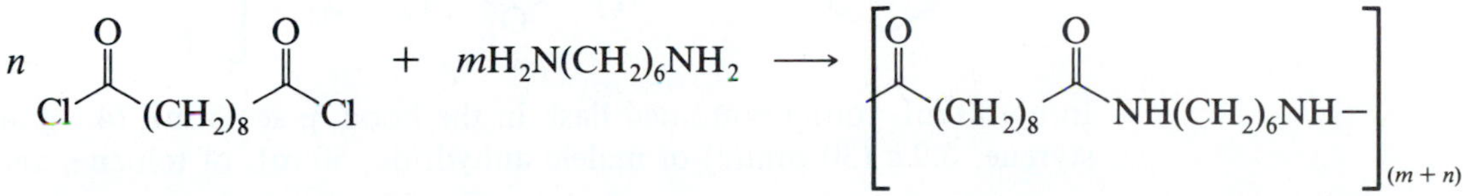

(CAUTION: Disposable gloves should be worn as both reactants are skin irritants.)
Dissolve 1.5 mL (1.38 g, 7.0 mmol) of sebacoyl chloride in 50 mL of tetrachloro-

ethylene[60] in a 100-mL beaker. Next, dissolve 2.2 g (18.9 mmol) of 1,6-hexanediamine (hexamethylenediamine) in 25 mL of water and carefully layer this solution on top of the sebacoyl chloride-tetrachlorethylene solution. This can best be accomplished by slowly pouring the solution along a stirring rod or through a funnel, the lower end of which is located just above the surface of the tetrachloroethylene. A film of nylon polymer will form immediately at the interface between the two liquid layers.

With a bent metal spatula, hook the polymer film at the center and raise it from the beaker as a continuously forming rope that can then be collected in a second beaker. After the nylon-forming process has been completed, stir the mixture to ensure complete reaction between the diacid chloride and the diamine, and *pour the residue into the bottle provided for waste. Do not pour the residue into the sink!*

Wash the polymer several times with 1:1 water-acetone (wear your gloves to avoid contact with any of the polymer-forming reagents and solvents), then wash it with water, and finally press it between several layers of paper towels. Allow the nylon to air dry, or dry in an oven at about 60°C.

Take a small piece of the dry nylon on a spatula tip or in a metal spoon and carefully heat it over the burner (*CAUTION: Keep away from flammable materials and solvents in the laboratory*) until it melts. With a second spatula tip or a boiling chip, see if you can draw a fiber out of the melt. Test the solubility of the dry polymer in acetone, toluene, methyl ethyl ketone, and methanol.

10.19.1 Copolymerization of Styrene with Maleic Anhydride

EXPERIMENT 124

Preparation of Styrene-Maleic Anhydride Random Copolymer

In a 100-mL round-bottomed flask in the hood, place 5 mL (4.6 g, 44 mmol) of styrene, 2.9 g (30 mmol) of maleic anhydride, 50 mL of toluene, and 200 mg of

---

[60] Trichloroethylene, or 1,1,1-trichloroethane, may also be used to dissolve sebacoyl chloride. If carbon tetrachloride is used, 65 mL is recommended. (*CAUTION: Halogenated hydrocarbons are toxic. Always handle them in a well ventilated hood.*)

dibenzoyl peroxide. Swirl the flask until the materials are dissolved. Add a boiling chip and attach a reflux condenser. Heat the reaction mixture on a steam bath for 15 minutes. The toluene may not boil, but the reaction should proceed to give a white flocculent copolymer in the flask. Cool the flask in an ice bath. Add 20 mL of methanol and stir the resulting slurry. Filter the slurry by suction filtration. Wash the product with a small amount of methanol and allow it to air dry. Test the melting property of the material on a spatula tip and compare it with that of the polystyrene. Test the solubility in toluene, acetone, methyl ethyl ketone, and methanol.

Questions for Experiments 122 to 124

1. Write a mechanism for the benzoyl peroxide polymerization of styrene and label the steps as initiation, propagation, and termination.
2. Polymerization of caprolactam produces Nylon 6. Indicate the repeating unit, and state how Nylon 6 differs in sequence from Nylon 6,10.
3. Why is a molar excess of 1,6-hexanediamine required in making Nylon 6,10?
4. In what way would you expect the polymer chains of a textile fiber, such as Nylon 6,6, to differ from those of a rubber, such as poly-(*cis*-isoprene)?
5. Assuming exactly a 1 : 1 copolymer of maleic anhydride and styrene, calculate the yield of the copolymer based on starting maleic anhydride.

Additional Polymer-Related Experiments

1. For the synthesis of a plasticizer, dioctyl phthalate, and evaluation of its effects on the physical properties of polystyrenes, see A. Caspar, J. Gillois, G. Guillerm, M. Savignac, and L. Vo-Quang, *J. Chem. Ed.,* **63,** 811 (1986).
2. For uses of a vinylpyridine polymer as a useful base in syntheses, see D. Getman, D. Hagery, H. Wilson, and W. F. Wood, *J. Chem. Ed.,* **61,** 551 (1984).

Appendix

Vapor Pressure-Temperature Nomograph

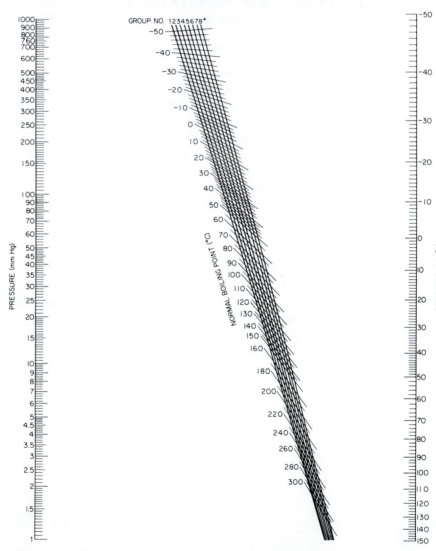

From *Industrial and Engineering Chemistry,* Vol. 38, p. 320, 1946.

---

* See Table A.1.

Table A.1. Groups of Compounds Represented in Nomographs

Group 1	Group 3	Group 5
Anthracene	Acetaldehyde	Ammonia
Anthraquinone	Acetone	Benzyl alcohol
Butylethylene	Amines	Methylamine
Carbon disulfide	Chloroanilines	Phenol
Phenanthrene	Cyanogen chloride	Propionic acid
Sulfur monochloride	Esters	
Trichloroethylene	Ethylene oxide	
	Formic acid	Group 6
Group 2	Hydrogen cyanide	Acetic anhydride
Benzaldehyde	Mercuric chloride	Isobutyric acid
Benzonitrile	Methyl benzoate	Water
Benzophenone	Methyl ether	
Camphor	Methyl ethyl ether	
Carbon suboxide	Naphthols	Group 7
Carbon sulfoselenide	Nitrobenzene	Benzoic acid
Chlorohydrocarbons	Nitromethane	Butyric acid
Dibenzyl ketone	Tetranitromethane	Ethylene glycol
Dimethylsilicane		Heptanoic acid
Ethers	Group 4	Isocaproic acid
Halogenated hydrocarbons	Acetic acid	Methyl alcohol
Hydrocarbons	Acetophenone	Valeric acid
Hydrogen fluoride	Cresols	
Methyl ethyl ketone	Cyanogen	
Methyl salicylate	Dimethylamine	Group 8
Nitrotoluenes	Dimethyl oxalate	n-Amyl alcohol
Nitrotoluidines	Ethylamine	Ethyl alcohol
Phosgene	Glycol diacetate	Isoamyl alcohol
Phthalic anhydride	Methyl formate	Isobutyl alcohol
Quinoline	Nitrosyl chloride	Mercurous chloride
Sulfides	Sulfur dioxide	n-Propyl alcohol

Vapor Pressure-Temperature Nomograph

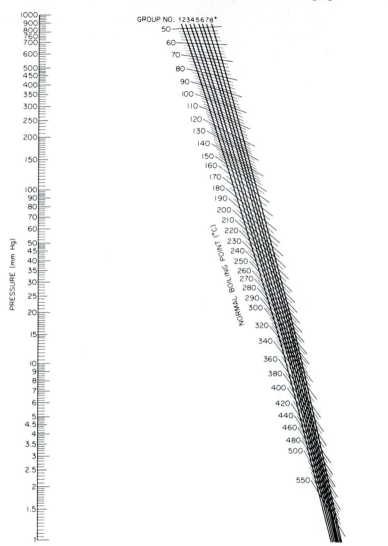

From *Industrial and Engineering Chemistry,* Vol. 38, p. 320, 1946.

* See Table A.1.

Index

LIST OF REAGENTS

Organic Reagents

Acetic anhydride
Alizarin
Allyl alcohol
Aniline
p-Anisidine hydrochloride
Benzenesulfonyl chloride
Benzoyl chloride
Benzyl amine
δ-Benzylthiuronium chloride
p-Bromophenacyl chloride
t-Butyl alcohol
Chloroacetic acid
Diethylene glycol
2,2-Dimethoxypropane
Dimethylsulfoxide
2,4-Dinitrobenzenesulfonyl chloride
3,5-Dinitrobenzoyl chloride
2,4-Dinitrochlorobenzene
2,4-Dinitrophenylhydrazine
Dioxane
Ethylene glycol
Girard's Reagent T
Hydrazine
Hydroxylamine hydrochloride
Maleic anhydride
Mercaptoacetic acid
Methyl iodide
1-Naphthol
2-Naphthol
1-Naphthylisocyanate
2-Naphthylisothiocyanate
Ninhydrin
p-Nitrobenzoyl chloride
p-Nitrobenzyl chloride
p-Nitrophenylhydrazine
3-Nitrophthalic anhydride
Oxalic acid
Phenacyl chloride
Phenylhydrazine
σ-Phenylenediame

Phenyl isocyanate
Phenyl isothiocyanate
Phthalic anhydride
Picric acid
Pyridine
Semicarbazide hydrochloride
Silver 3,5-dinitrobenzoate
Sodium acetate
Sodium potassium tartrate
Tetranitromethane
Thiourea
Toluene
p-Tolyl isocyanate
p-Toluenesulfonic acid
p-Toluenesulfonyl chloride
2,4,7-Trinitrofluorenone
Xanthydrol

Acids and Bases

Acetic acid
Ammonium hydroxide
Hydriodic acid
Hydrochloric acid
Nitric acid
85% Phosphoric acid
Sulfuric acid

Solvents

Acetone
Carbon disulfide
Carbon tetrachloride
Chloroform
95% Ethanol
Ethanol (absolute)
Methanol
Methylene chloride
Pentanes
Petroleum ether (hexanes)
Tetrahydrofuran
Toluene